NANOSCIENCE AND TECHNOLOGY

Series Editors:
P. Avouris B. Bhushan D. Bimberg K. von Klitzing H. Sakaki R. Wiesendanger

The series NanoScience and Technology is focused on the fascinating nano-world, mesoscopic physics, analysis with atomic resolution, nano and quantum-effect devices, nanomechanics and atomic-scale processes. All the basic aspects and technology-oriented developments in this emerging discipline are covered by comprehensive and timely books. The series constitutes a survey of the relevant special topics, which are presented by leading experts in the field. These books will appeal to researchers, engineers, and advanced students.

Bharat Bhushan
Harald Fuchs

Applied Scanning Probe Methods VII

Biomimetics and Industrial Applications

With 223 Figures and 18 Tables
Including 9 Color Figures

 Springer

Editors:

Professor Bharat Bhushan
Nanotribology Laboratory for Information
Storage and MEMS/NEMS (NLIM)
W 390 Scott Laboratory, 201 W. 19th Avenue
The Ohio State University, Columbus
Ohio 43210-1142, USA
e-mail: Bhushan.2@osu.edu

Professor Dr. Harald Fuchs
Center for Nanotechnology (CeNTech)
and Institute of Physics
University of Münster
Gievenbecker Weg 11, 48149 Münster, Germany
e-mail: fuchsh@uni-muenster.de

Series Editors:

Professor Dr. Phaedon Avouris
IBM Research Division
Nanometer Scale Science & Technology
Thomas J. Watson Research Center, P.O. Box 218
Yorktown Heights, NY 10598, USA

Professor Bharat Bhushan
Nanotribology Laboratory for Information
Storage and MEMS/NEMS (NLIM)
W 390 Scott Laboratory, 201 W. 19th Avenue
The Ohio State University, Columbus
Ohio 43210-1142, USA

Professor Dr. Dieter Bimberg
TU Berlin, Fakutät Mathematik,
Naturwissenschaften,
Institut für Festkörperphysik
Hardenbergstr. 36, 10623 Berlin, Germany

Professor Dr., Dres. h. c. Klaus von Klitzing
Max-Planck-Institut für Festkörperforschung
Heisenbergstrasse 1, 70569 Stuttgart, Germany

Professor Hiroyuki Sakaki
University of Tokyo
Institute of Industrial Science,
4-6-1 Komaba, Meguro-ku, Tokyo 153-8505, Japan

Professor Dr. Roland Wiesendanger
Institut für Angewandte Physik
Universität Hamburg
Jungiusstrasse 11, 20355 Hamburg, Germany

DOI 10.1007/11785705
ISSN 1434-4904
ISBN-10 3-540-37320-9 Springer Berlin Heidelberg New York
ISBN-13 978-3-540-37320-9 Springer Berlin Heidelberg New York

Library of Congress Control Number: 2006932716

Springer is a part of Springer Science+Business Media

springer.com

© Springer-Verlag Berlin Heidelberg 2007

Typesetting and production: LE-TₑX Jelonek, Schmidt & Vöckler GbR, Leipzig
Cover: WMX Design, Heidelberg
Printed on acid-free paper   2/3100/YL - 5 4 3 2 1 0

Preface

The scanning probe microscopy field has been rapidly expanding. It is a demanding task to collect a timely overview of this field with an emphasis on technical developments and industrial applications. It became evident while editing Vols. I–IV that a large number of technical and applicational aspects are present and rapidly developing worldwide. Considering the success of Vols. I–IV and the fact that further colleagues from leading laboratories were ready to contribute their latest achievements, we decided to expand the series with articles touching fields not covered in the previous volumes. The response and support of our colleagues were excellent, making it possible to edit another three volumes of the series. In contrast to topical conference proceedings, the applied scanning probe methods intend to give an overview of recent developments as a compendium for both practical applications and recent basic research results, and novel technical developments with respect to instrumentation and probes.

The present volumes cover three main areas: novel probes and techniques (Vol. V), charactarization (Vol. VI), and biomimetics and industrial applications (Vol. VII).

Volume V includes an overview of probe and sensor technologies including integrated cantilever concepts, electrostatic microscanners, low-noise methods and improved dynamic force microscopy techniques, high-resonance dynamic force microscopy and the torsional resonance method, modelling of tip cantilever systems, scanning probe methods, approaches for elasticity and adhesion measurements on the nanometer scale as well as optical applications of scanning probe techniques based on nearfield Raman spectroscopy and imaging.

Volume VI is dedicated to the application and characterization of surfaces including STM on monolayers, chemical analysis of single molecules, STM studies on molecular systems at the solid–liquid interface, single-molecule studies on cells and membranes with AFM, investigation of DNA structure and interactions, direct detection of ligand protein interaction by AFM, dynamic force microscopy as applied to organic/biological materials in various environments with high resolution, noncontact force microscopy, tip-enhanced spectroscopy for investigation of molecular vibrational excitations, and investigation of individual carbon nanotube polymer interfaces.

Volume VII is dedicated to the area of biomimetics and industrial applications. It includes studies on the lotus effect, the adhesion phenomena as occurs in gecko feet, nanoelectromechanical systems (NEMS) in experiment and modelling, application of STM in catalysis, nanostructuring and nanoimaging of biomolecules for biosensors, application of scanning electrochemical microscopy, nanomechanical investigation of pressure sensitive adhesives, and development of MOEMS devices.

As in the previous volumes a distinction between basic research fields and industrial scanning probe techniques cannot be made, which is in fact a unique factor in nanotechnology in general. It also shows that these fields are extremely active and that the novel methods and techniques developed in nanoprobe basic research are rapidly being transferred to applications and industrial development.

We are very grateful to our colleagues who provided in a timely manner their manuscripts presenting state-of-the-art research and technology in their respective fields. This will help keep research and development scientists both in academia and industry well informed about the latest achievements in scanning probe methods. Finally, we would like to cordially thank Dr. Marion Hertel, senior editor chemistry, and Mrs. Beate Siek of Springer for their continuous support and advice without which these volumes could have never made it to market on time.

July, 2006

Prof. Bharat Bhushan, USA
Prof. Harald Fuchs, Germany
Prof. Satoshi Kawata, Japan

Contents – Volume VII

21   Lotus Effect: Roughness-Induced Superhydrophobicity
Michael Nosonovsky, Bharat Bhushan   1

21.1   Introduction   1

21.2   Contact Angle Analysis   4
21.2.1   Homogeneous Solid–Liquid Interface   5
21.2.2   Composite Solid–Liquid–Air Interface   8
21.2.3   Stability of the Composite Interface   11

21.3   Calculation of the Contact Angle for Selected Rough Surfaces
and Surface Optimization   19
21.3.1   Two-Dimensional Periodic Profiles   20
21.3.2   Three-Dimensional Surfaces   23
21.3.3   Surface Optimization for Maximum Contact Angle   29

21.4   Meniscus Force   31
21.4.1   Sphere in Contact with a Smooth Surface   31
21.4.2   Multiple-Asperity Contact   33

21.5   Experimental Data   34

21.6   Closure   37

References   38

22   Gecko Feet: Natural Attachment Systems for Smart Adhesion
Bharat Bhushan, Robert A. Sayer   41

22.1   Introduction   41

22.2   Tokay Gecko   42
22.2.1   Construction of Tokay Gecko   42
22.2.2   Other Attachment Systems   44
22.2.3   Adaptation to Surface Roughness   45
22.2.4   Peeling   47
22.2.5   Self-Cleaning   48

22.3   Attachment Mechanisms   51
22.3.1   Unsupported Adhesive Mechanisms   52

22.3.2   Supported Adhesive Mechanisms   54

22.4   Experimental Adhesion Test Techniques and Data   56
22.4.1   Adhesion Under Ambient Conditions   56
22.4.2   Effects of Temperature   58
22.4.3   Effects of Humidity   58
22.4.4   Effects of Hydrophobicity   60

22.5   Design of Biomimetic Fibrillar Structures   60
22.5.1   Verification of Adhesion Enhancement
          of Fabricated Surfaces Using Fibrillar Structures   60
22.5.2   Contact Mechanics of Fibrillar Structures   62
22.5.3   Fabrication of Biomimetric Gecko Skin   65

22.6   Closure   69

References   73

23   Novel AFM Nanoprobes
          Horacio D. Espinosa, Nicolaie Moldovan, K.-H. Kim   77

23.1   Introduction and Historic Developments   77

23.2   DPN and Fountain Pen Nanolithography   81
23.2.1   NFP Chip Design – 1D and 2D Arrays   84
23.2.2   Microfabrication of the NFP   94
23.2.3   Independent Lead Zirconate Titanate Actuation   99
23.2.4   Applications   102
23.2.5   Perspectives of NFP   108

23.3   Ultrananocrystalline-Diamond Probes   109
23.3.1   Chip Design   111
23.3.2   Molding and Other Fabrication Techniques   112
23.3.3   Performance Assessment and Wear Tests   115
23.3.4   Applications   118
23.3.5   Perspectives for Diamond Probes   128

References   129

24   Nanoelectromechanical Systems – Experiments and Modeling
          Horacio D. Espinosa, Changhong Ke   135

24.1   Introduction   135

24.2   Nanoelectromechanical Systems   136
24.2.1   Carbon Nanotubes   136
24.2.2   Fabrication Methods   137
24.2.3   Inducing and Detecting Motion   140
24.2.4   Functional NEMS Devices   146

24.2.5   Future Challenges   163

24.3   Modeling of NEMS   165
24.3.1   Multiscale Modeling   166
24.3.2   Continuum Mechanics Modeling   176

References   190

**25   Application of Atom-resolved Scanning Tunneling Microscopy
in Catalysis Research**
Jeppe Vang Lauritsen, Ronny T. Vang, Flemming Besenbacher   . . .   197

25.1   Introduction   197

25.2   Scanning Tunneling Microscopy   199

25.3   STM Studies of a Hydrotreating Model Catalyst   200

25.4   Selective Blocking of Active Sites on Ni(111)   207

25.5   High-Pressure STM: Bridging the Pressure Gap in Catalysis   214

25.6   Summary and Outlook   220

References   221

26   Nanostructuration and Nanoimaging of Biomolecules for Biosensors
*Claude Martelet, Nicole Jaffrezic-Renault, Yanxia Hou,
Abdelhamid Errachid, François Bessueille*   225

26.1   Introduction and Definition of Biosensors   225
26.1.1   Definition   225
26.1.2   Biosensor Components   225
26.1.3   Immobilization of the Bioreceptor   226

26.2   Langmuir–Blodgett and Self-Assembled Monolayers
          as Immobilization Techniques   227
26.2.1   Langmuir–Blodgett Technique   227
26.2.2   Self-Assembled Monolayers   236
26.2.3   Characterization of SAMs and LB Films   248

26.3   Prospects and Conclusion   253

References   255

27   Applications of Scanning Electrochemical Microscopy (SECM)
Gunther Wittstock, Malte Burchardt, Sascha E. Pust   259

27.1   Introduction   260
27.1.1   Overview   260

27.1.2   Relation to Other Methods   261
27.1.3   Instrument and Basic Concepts   262

27.2   Application in Biotechnology and Cellular Biology   266
27.2.1   Investigation of Immobilized Enzymes   266
27.2.2   Investigation of Metabolism of Tissues and Adherent Cells   277
27.2.3   Investigation of Mass Transport Through Biological Tissue   284

27.3   Application to Technologically Important Electrodes   288
27.3.1   Investigation of Passive Layers and Local Corrosion Phenomena .   288
27.3.2   Investigation of Electrocatalytically Important Electrodes   290

27.4   Conclusion and Outlook: New Instrumental Developments
          and Implication for Future Applications   293

References   294

**28   Nanomechanical Characterization of Structural
          and Pressure-Sensitive Adhesives**
          Martin Munz, Heinz Sturm   301

28.1   Introduction   303

28.2   A Brief Introduction to Scanning Force Microscopy (SFM)   305
28.2.1   Various SFM Operation Modes   305
28.2.2   Contact Mechanics   308
28.2.3   Extracting Information from Thermomechanical Noise   310

28.3   Fundamental Issues of Nanomechanical Studies in the Vicinity
          of an Interface   311
28.3.1   Identification of the Interface   312
28.3.2   Implications of the Interface for Indentation Measurements   314

28.4   Property Variations Within Amine-Cured Epoxies   320
28.4.1   A Brief Introduction to Epoxy Mechanical Properties   320
28.4.2   Epoxy Interphases   323

28.5   Pressure-Sensitive Adhesives (PSAs)   329
28.5.1   A Brief Introduction to PSAs   329
28.5.2   Heterogeneities of an Elastomer–Tackifier PSA as Studied
          by Means of M-LFM   331
28.5.3   The Particle Coalescence Behavior of an Acrylic PSA as Studied
          by Means of Intermittent Contact Mode   337
28.5.4   Evidence for the Fibrillation Ability of an Acrylic PSA
          from the Analysis of the Noise PSD   340

28.6   Conclusions   342

References   343

29   **Development of MOEMS Devices and Their Reliability Issues**
            Bharat Bhushan, Huiwen Liu   349

29.1   Introduction to Microoptoelectromechanical Systems   349

29.2   Typical MOEMS Devices: Structure and Mechanisms   351
29.2.1   Digital Micromirror Device and Other Micromirror Devices   351
29.2.2   MEMS Optical Switch   353
29.2.3   MEMS-Based Interferometric Modulator Devices   355
29.2.4   Grating Light Valve Technique   356
29.2.5   Continuous Membrane Deformable Mirrors   357

29.3   Reliability Issues of MOEMS   358
29.3.1   Stiction-Induced Failure of DMD   358
29.3.2   Thermomechanical Issues with Micromirrors   360
29.3.3   Friction- and Wear-Related Failure   361
29.3.4   Contamination-Related Failure   361

29.4   Summary   363

References   364

Subject Index   367

Contents – Volume V

1   Integrated Cantilevers and Atomic Force Microscopes
Sadik Hafizovic, Kay-Uwe Kirstein, Andreas Hierlemann   1

1.1   Overview   1

1.2   Active Cantilevers   2
1.2.1   Integrated Force Sensor   4
1.2.2   Integrated Actuation   8

1.3   System Integration   10
1.3.1   Analog Signal Processing and Conditioning   10
1.3.2   Digital Signal Processing   13

1.4   Single-Chip CMOS AFM   16
1.4.1   Measurements   19

1.5   Parallel Scanning   19

1.6   Outlook   21

References   21

2   Electrostatic Microscanner
Yasuhisa Ando   23

2.1   Introduction   23

2.2   Displacement Conversion Mechanism   24
2.2.1   Basic Conception   24
2.2.2   Combination with Comb Actuator   25
2.2.3   Various Types of Displacement Conversion Mechanism   27

2.3   Design, Fabrication Technique, and Performance   29
2.3.1   Main Structure of 3D Microstage   29
2.3.2   Amplification Mechanism of Scanning Area   31
2.3.3   Fabrication Using ICP-RIE   34
2.3.4   Evaluation of Motion of 3D Microstage   37

2.4   Applications to AFM   39
2.4.1   Operation by Using Commercial Controller   39

2.4.2   Evaluation of Microscanner Using Grating Image   41
2.4.3   SPM Operation Using Microscanner   45

References   49

3   Low-Noise Methods for Optical Measurements of Cantilever Deflections
Tilman E. Schäffer   51

3.1   Introduction   51

3.2   The Optical Beam Deflection Method   52
3.2.1   Gaussian Optics   52
3.2.2   Detection Sensitivity   54

3.3   Optical Detection Noise   55
3.3.1   Noise Sources   55
3.3.2   Shot Noise   55

3.4   The Array Detector   56

3.5   Dynamic Range and Linearity   59
3.5.1   The Two-Segment Detector   59
3.5.2   The Array Detector   61

3.6   Detection of Higher-Order Cantilever Vibration Modes   62
3.6.1   Normal Vibration Modes   63
3.6.2   Optimization of the Detection Sensitivity   64

3.7   Calculation of Thermal Vibration Noise   66
3.7.1   Focused Optical Spot of Infinitesimal Size   66
3.7.2   Focused Optical Spot of Finite Size   67

3.8   Thermal Spring Constant Calibration   69

References   70

4 Q-controlled Dynamic Force Microscopy in Air and Liquids
Hendrik Hölscher, Daniel Ebeling, Udo D. Schwarz   75

4.1   Introduction   75

4.2   Theory of Q-controlled Dynamic Force Microscopy   76
4.2.1   Equation of Motion of a Dynamic Force Microscope
        with Q-control   76
4.2.2   Active Modification of the Q-factor   78
4.2.3   Including Tip–Sample Interactions   80
4.2.4   Prevention of Instabilities by Q-control in Air   82
4.2.5   Reduction of Tip–Sample Indentation and Force
        by Q-control in Liquids   86

4.3   Experimental Applications of Q-control   89

4.3.1   Examples for Q-control Applications in Ambient Conditions . . .   90
4.4   Summary.   94

References   95

5   High-Frequency Dynamic Force Microscopy
         Hideki Kawakatsu   99

5.1   Introduction   99

5.2   Instrumental   99
5.2.1   Cantilever   99
5.2.2   Detection.   102
5.2.3   Excitation   105
5.2.4   Circuitry   106

5.3   Experimental.   107
5.3.1   Low-Amplitude Operation   107
5.3.2   Manipulation.   108
5.3.3   Atomic-Resolution Lateral Force Microscopy   108
5.3.4   Other Techniques for High Frequency Motion Detection   108

5.4   Summary and Outlook   109

References   110

6   Torsional Resonance Microscopy and Its Applications
         Chanmin Su, Lin Huang, Craig B. Prater, Bharat Bhushan   113

6.1   Introduction to Torsional Resonance Microscopy   113

6.2   TRmode System Configuration   115

6.3   Torsional Modes of Oscillation   119

6.4   Imaging and Measurements with TRmode.   123
6.4.1   TRmode in Weakly-Coupled Interaction Region   123
6.4.2   TRmode Imaging and Measurement in Contact Mode   127

6.5   Applications of TRmode Imaging   129
6.5.1   High-Resolution Imaging Application   129
6.5.2   Electric Measurements Under Controlled Proximity by TRmode .   132
6.5.3   In-Plane Anisotropy.   138

6.6   Torsional Tapping Harmonics for Mechanical Property
         Characterization.   140
6.6.1   Detecting Cantilever Harmonics Through Torsional Detection . . .   142
6.6.2   Reconstruction of Time-Resolved Forces   142
6.6.3   Force-Versus-Distance Curves   143
6.6.4   Mechanical Property Measurements and Compositional Mapping .   144

6.7   Conclusion   145

References   146

7   Modeling of Tip-Cantilever Dynamics in Atomic Force Microscopy
           Yaxin Song, Bharat Bhushan   149

7.1   Introduction   155
7.1.1   Various AFM Modes and Measurement Techniques   155
7.1.2   Models for AFM Cantilevers   161
7.1.3   Outline   163

7.2   Modeling of AFM Tip-Cantilever Systems in AFM   163
7.2.1   Tip–Sample Interaction   164
7.2.2   Point-Mass Model   166
7.2.3   The 1D Beam Model   168
7.2.4   Pure Torsional Analysis of TRmode   171
7.2.5   Coupled Torsional-Bending Analysis   177

7.3   Finite Element Modeling of Tip-Cantilever Systems   187
7.3.1   Finite Element Beam Model of Tip-Cantilever Systems   188
7.3.2   Modeling of TappingMode   192
7.3.3   Modeling of Torsional Resonance Mode   196
7.3.4   Modeling of Lateral Excitation Mode   199

7.4   Atomic-Scale Topographic and Friction Force Imaging in FFM . .   200
7.4.1   FFM Images of Graphite Surface   202
7.4.2   Interatomic Forces Between Tip and Surface   204
7.4.3   Modeling of FFM Profiling Process   205
7.4.4   Simulations on Graphite Surface   208

7.5   Quantitative Evaluation of the Sample's Mechanical Properties . .   213

7.6   Closure   216

A   Appendices   217
A.1   Stiffness and Mass Matrices of 3D Beam Element   217
A.2   Mass Matrix of the Tip   218
A.3   Additional Stiffness and Mass Matrices
           Under Linear Tip–Sample Interaction   219

References   220

**8   Combined Scanning Probe Techniques
           for In-Situ Electrochemical Imaging at a Nanoscale**
           Justyna Wiedemair, Boris Mizaikoff, Christine Kranz   225

8.1   Overview   227

8.2   Combined Techniques   228

8.2.1	Integration of Electrochemical Functionality	230
8.2.2	Combined Techniques Based on Force Interaction	231
8.2.3	Combined Techniques Based on Tunneling Current	232
8.2.4	Combined Techniques Based on Optical Near-Field Interaction	233
8.2.5	Theory	234
8.2.6	Combined Probe Fabrication	234
8.3	Applications	243
8.3.1	Model Systems	244
8.3.2	Imaging Enzyme Activity	246
8.3.3	AFM Tip-Integrated Biosensors	249
8.3.4	Combined SPM for Imaging of Living Cells	253
8.3.5	Measurement of Local pH Changes	255
8.3.6	Corrosion Studies	257
8.4	Outlook: Further Aspects of Multifunctional Scanning Probes	259
References		261

9 New AFM Developments to Study Elasticity and Adhesion at the Nanoscale
Robert Szoszkiewicz, Elisa Riedo

	Robert Szoszkiewicz, Elisa Riedo	269
9.1	Introduction	270
9.2	Contact Mechanics Theories and Their Limitations	271
9.3	Modulated Nanoindentation	273
9.3.1	Force-Indentation Curves	273
9.3.2	Elastic Moduli	276
9.4	Ultrasonic Methods at Local Scales	278
9.4.1	Brief Description of Ultrasonic Methods	278
9.4.2	Applications of Ultrasonic Techniques in Elasticity Mapping	281
9.4.3	UFM Measurements of Adhesion Hysteresis and Their Relations to Friction at the Tip-Sample Contact	282
References		284

10 Near-Field Raman Spectroscopy and Imaging
Pietro Giuseppe Gucciardi, Sebastiano Trusso, Cirino Vasi, Salvatore Patanè, Maria Allegrini

	Salvatore Patanè, Maria Allegrini	287
10.1	Introduction	287
10.2	Raman Spectroscopy	289
10.2.1	Classical Description of the Raman Effect	289
10.2.2	Quantum Description of the Raman Effect	291
10.2.3	Coherent Anti-Stokes Raman Scattering	295

10.2.4   Experimental Techniques in Raman Spectroscopy   296

10.3   Near-Field Raman Spectroscopy   299
10.3.1   Theoretical Principles of the Near-Field Optical Microscopy   300
10.3.2   Setups for Near-Field Raman Spectroscopy   302

10.4   Applications of Near-Field Raman Spectroscopy   306
10.4.1   Structural Mapping   307
10.4.2   Chemical Mapping   312
10.4.3   Probing Single Molecules by Surface-Enhanced
          and Tip-Enhanced Near-Field Raman Spectroscopy   314
10.4.4   Near-Field Raman Spectroscopy and Imaging
          of Carbon Nanotubes   321
10.4.5   Coherent Anti-Stokes Near-Field Raman Imaging   324

10.5   Conclusions   326

References   326

Subject Index   331

Contents – Volume VI

**11   Scanning Tunneling Microscopy of Physisorbed Monolayers:
From Self-Assembly to Molecular Devices**
Thomas Müller   1

11.1   Introduction   1

11.2   Source of Image Contrast: Geometric and Electronic Factors . . .   2

11.3   Two-Dimensional Self-Assembly:
Chemisorbed and Physisorbed Systems   4

11.4   Self-Assembly on Graphite   6
11.4.1   Alkane Functionalization and Driving Forces for Self-Assembly .   6
11.4.2   Expression of Chirality   11

11.5   Beyond Self-Assembly   14
11.5.1   Postassembly Modification   14
11.5.2   Templates for Bottom-Up Assembly   21

11.6   Toward Molecular Devices   23
11.6.1   Ring Systems and Electronic Structure   23
11.6.2   Model Systems for Molecular Electronics   25

11.7   Summary and Conclusions   28

References   28

**12   Tunneling Electron Spectroscopy Towards Chemical Analysis
of Single Molecules**
Tadahiro Komeda   31

12.1   Introduction   31

12.2   Vibrational Excitation Through Tunneling Electron Injection . . .   32
12.2.1   Characteristic Features of the Scanning Tunneling Microscope
as an Electron Source   32
12.2.2   Electron-Induced Vibrational Excitation Mechanism   33

12.3   IET Process of Vibrational Excitation   36
12.3.1   Basic Mechanism of Vibrational Excitation in the IET Process . .   37

12.3.2   IETS with the Setup of STM   39
12.3.3   Instrumentation of IETS with the Use of STM   40
12.3.4   Examples of STM-IETS Measurements   41
12.3.5   Theoretical Treatment of STM-IETS Results   44
12.3.6   IETS Mapping   48

12.4   Manipulation of Single Molecule Through Vibrational Excitation .   49
12.4.1   Desorption via Vibrational Excitation   49
12.4.2   Vibration-Induced Hopping   51
12.4.3   Vibration-Induced Chemical Reaction   54

12.5   Action Spectroscopy   55
12.5.1   Rotation of *cis*-2-Butene Molecules   56
12.5.2   Complimentary Information of Action Spectroscopy and IETS . .   57

12.6   Conclusions   60

References   61

13   STM Studies on Molecular Assembly at Solid/Liquid Interfaces
          Ryo Yamada, Kohei Uosaki   65

13.1   Introduction   65

13.2   STM Operations in Liquids   66
13.2.1   Instruments   66
13.2.2   Preparation of Substrates   67

13.3   Surface Structures of Substrates   68
13.3.1   Introduction   68
13.3.2   Structures of Au(111)   68
13.3.3   Structures of Au(100)   68

13.4   SA of Organic Molecules   69
13.4.1   Introduction   69
13.4.2   Assembly of Chemisorbed Molecules: Alkanethiols   70
13.4.3   Assembly of Physisorbed Molecules: *n*-Alkanes   80

13.5   SA of Inorganic Complexes   84
13.5.1   Introduction   84
13.5.2   Assembly of Metal Complexes   85
13.5.3   Assembly of Metal Oxide Clusters: Polyoxometalates   92

13.6   Conclusions   96

References   96

14 **Single-Molecule Studies on Cells and Membranes
Using the Atomic Force Microscope**
*Ferry Kienberger, Lilia A. Chtcheglova, Andreas Ebner,
Theeraporn Puntheeranurak, Hermann J. Gruber,
Peter Hinterdorfer* 101

14.1 Abstract 101

14.2 Introduction 102

14.3 Principles of Atomic Force Microscopy 103

14.4 Imaging of Membrane–Protein Complexes 104
14.4.1 Membranes of Photosynthetic Bacteria and Bacterial S-Layers . . 104
14.4.2 Nuclear Pore Complexes 106
14.4.3 Cell Membranes with Attached Viral Particles 106

14.5 Single-Molecule Recognition on Cells and Membranes 110
14.5.1 Principles of Recognition Force Measurements 110
14.5.2 Force-Spectroscopy Measurements on Living Cells. 113

14.6 Unfolding and Refolding of Single-Membrane Proteins 117

14.7 Simultaneous Topography and Recognition
Imaging on Cells (TREC). 119

14.8 Concluding Remarks 122

References 123

15 **Atomic Force Microscopy of DNA Structure and Interactions**
Neil H. Thomson. 127

15.1 Introduction: The Single-Molecule, Bottom-Up Approach 127

15.2 DNA Structure and Function 129

15.3 The Atomic Force Microscope 131

15.4 Binding of DNA to Support Surfaces 137
15.4.1 Properties of Support Surfaces for Biological AFM 137
15.4.2 DNA Binding to Surfaces 138
15.4.3 DNA Transport to Surfaces 142

15.5 AFM of DNA Systems 143
15.5.1 Static Imaging versus Dynamic Studies 143
15.5.2 The Race for Reproducible Imaging of Static DNA 144
15.5.3 Applications of Tapping-Mode AFM to DNA Systems 146

15.6 Outlook 157

References 159

16   **Direct Detection of Ligand–Protein Interaction Using AFM**
           Małgorzata Lekka, Piotr Laidler, Andrzej J. Kulik   165

16.1   Cell Structures and Functions   166
16.1.1   Membranes and their Components: Lipids and Proteins   166
16.1.2   Glycoproteins   167
16.1.3   Immunoglobulins   169
16.1.4   Adhesion Molecules   170
16.1.5   Plant Lectins   173

16.2   Forces Acting Between Molecules   175
16.2.1   Repulsive Forces   177
16.2.2   Attractive Forces   179

16.3   Force Spectroscopy   181
16.3.1   Atomic Force Microscope   182
16.3.2   Force Curves Calibration   187
16.3.3   Determination of the Unbinding Force   188
16.3.4   Data Analysis   189

16.4   Detection of the Specific Interactions on Cell Surface   193
16.4.1   Isolated Proteins   194
16.4.2   Receptors in Plasma Membrane of Living Cells   196

16.5   Summary   201

References   202

17   **Dynamic Force Microscopy for Molecular-Scale Investigations
           of Organic Materials in Various Environments**
           Hirofumi Yamada, Kei Kobayashi   205

17.1   Brief Overview   205

17.2   Principles and Instrumentation of Frequency Modulation Detection
           Mode Dynamic Force Microscopy   206
17.2.1   Transfer Function of the Cantilever as a Force Sensor   206
17.2.2   Detection Methods of Resonance Frequency Shift
           of the Cantilever   208
17.2.3   Instrumentation of the Frequency Modulation Detection Mode . .   210
17.2.4   Frequency Modulation Detector   212
17.2.5   Phase-Locked-Loop Frequency Modulation Detector   212
17.2.6   Relationship Between Frequency Shift and Interaction Force . . .   214
17.2.7   Inversion of Measured Frequency Shift to Interaction Force   216

17.3   Noise in Frequency Modulation Atomic Force Microscopy   217
17.3.1   Thermal Noise Drive   217
17.3.2   Minimum Detectable Force in Static Mode   218

17.3.3   Minimum Detectable Force
Using the Amplitude Modulation Detection Method 218
17.3.4   Minimum Detectable Force
Using the Frequency Modulation Detection Method 219
17.3.5   Effect of Displacement-Sensing Noise
on Minimum Detectable Force 220
17.3.6   Comparison of Minimum Detectable Force
for Static Mode and Dynamic Modes 223

17.4   High-Resolution Imaging
of Organic Molecules in Various Environments 225
17.4.1   Alkanethiol Self-Assembled Monolayers 225
17.4.2   Submolecular-Scale Contrast in Copper Phthalocyanines 228
17.4.3   Atomic Force Microscopy Imaging in Liquids 230

17.5   Investigations of Molecular Properties 233
17.5.1   Surface Potential Measurements 233
17.5.2   Energy Dissipation Measurements 241

17.6   Summary and Outlook 243

References 244

18   Noncontact Atomic Force Microscopy
Yasuhiro Sugawara 247

18.1   Introduction 247

18.2   NC-AFM System the Using FM Detection Method 247

18.3   Identification of Subsurface Atom Species 249

18.4   Tip-Induced Structural Change on a Si(001) Surface at 5 K 251

18.5   Influence of Surface Stress on Phase Change
in the Si(001) Step at 5 K 252

18.6   Origin of Anomalous Dissipation Contrast
on a Si(001) Surface at 5 K 253

18.7   Summary 254

References 255

**19   Tip-Enhanced Spectroscopy for Nano Investigation
of Molecular Vibrations**
Norihiko Hayazawa, Yuika Saito 257

19.1   Introduction 257

19.2   TERS (Reflection and Transmission Modes) 258

19.2.1   Experimental Configuration of TERS 258
19.2.2   Transmission Mode 259
19.2.3   Reflection Mode 260

19.3   How to Fabricate the Tips? 261
19.3.1   Vacuum Evaporation and Sputtering Technique 261
19.3.2   Electroless Plating 261
19.3.3   Etching of Metal Wires Followed by Focused Ion Beam Milling . . 262
19.3.4   Other Methods 263

19.4   Tip-Enhanced Raman Imaging 263
19.4.1   Selective Detection of Different Organic Molecules 264
19.4.2   Observation of Single-Walled Carbon Nanotubes 265

19.5   Polarization-Controlled TERS 268
19.5.1   Polarization Measurement by Using a High NA Objective Lens . . 268
19.5.2   Metallized Tips and Polarizations 269
19.5.3   Example of p- and s-Polarization Measurements in TERS 271

19.6   Reflection Mode for Opaque Samples 272
19.6.1   TERS Spectra of Strained Silicon 272
19.6.2   Nanoscale Characterization of Strained Silicon 274

19.7   For Higher Spatial Resolution 275
19.7.1   Tip-Pressurized Effect 275
19.7.2   Nonlinear Effect 278

19.8   Conclusion 282

References 283

**20   Investigating Individual Carbon Nanotube/Polymer Interfaces
           with Scanning Probe Microscopy**
           Asa H. Barber, H. Daniel Wagner, Sidney R. Cohen 287

20.1   Mechanical Properties of Carbon-Nanotube Composites 288
20.1.1   Introduction 288
20.1.2   Mechanical Properties of Carbon Nanotubes 288
20.1.3   Carbon-Nanotube Composites 290

20.2   Interfacial Adhesion Testing 292
20.2.1   Historical Background 292
20.2.2   Shear-Lag Theory 293
20.2.3   Kelly–Tyson Approach 294
20.2.4   Single-Fiber Tests 294

20.3   Single Nanotube Experiments 296
20.3.1   Rationale and Motivation 296
20.3.2   Drag-out Testing (Ex Situ Technique) 297
20.3.3   Pull-out Testing (In Situ) 298

20.3.4   Comparison of In Situ and Ex Situ Experiments   304
20.3.5   Wetting Experiments   306

20.4   Implication of Results and Comparison with Theory   311
20.4.1   Interfaces in Engineering Composites   311
20.4.2   Simulation of Carbon-Nanotube/Polymer
           Interfacial Adhesion Mechanisms   312
20.4.3   Discussion of Potential Bonding Mechanisms at the Interface . . .   314

20.5   Complementary Techniques   314
20.5.1   Raman Spectroscopy   314
20.5.2   Scanning Electron Microscopy   316
20.5.3   Overall Conclusions   320

References   320

Subject Index   325

Contents – Volume I

Part I   Scanning Probe Microscopy

1   Dynamic Force Microscopy
André Schirmeisen, Boris Anczykowski, Harald Fuchs   3

2   Interfacial Force Microscopy: Selected Applications
Jack E. Houston   41

3   Atomic Force Microscopy with Lateral Modulation
Volker Scherer, Michael Reinstadtler, Walter Arnold   75

4   Sensor Technology for Scanning Probe Microscopy
Egbert Oesterschulze, Rainer Kassing   117

5   Tip Characterization for Dimensional Nanometrology
John S. Villarrubia   147

Part II   Characterization

6   Micro/Nanotribology Studies Using Scanning Probe Microscopy
Bharat Bhushan   171

7   Visualization of Polymer Structures with Atomic Force Microscopy
Sergei Magonov   207

8   Displacement and Strain Field Measurements from SPM Images
Jürgen Keller, Dietmar Vogel, Andreas Schubert, Bernd Michel . .   253

9   AFM Characterization of Semiconductor Line Edge Roughness
Ndubuisi G. Orji, Martha I. Sanchez, Jay Raja,
Theodore V. Vorburger   277

10   Mechanical Properties of Self-Assembled Organic Monolayers:
Experimental Techniques and Modeling Approaches
Redhouane Henda   303

11 **Micro-Nano Scale Thermal Imaging Using Scanning Probe Microscopy**
Li Shi, Arun Majumdar 327

12 **The Science of Beauty on a Small Scale. Nanotechnologies Applied to Cosmetic Science**
Gustavo Luengo, Frédéric Leroy 363

Part III Industrial Applications

13 **SPM Manipulation and Modifications and Their Storage Applications**
Sumio Hosaka 389

14 **Super Density Optical Data Storage by Near-Field Optics**
Jun Tominaga 429

15 **Capacitance Storage Using a Ferroelectric Medium and a Scanning Capacitance Microscope (SCM)**
Ryoichi Yamamoto 439

16 **Room-Temperature Single-Electron Devices formed by AFM Nano-Oxidation Process**
Kazuhiko Matsumoto 459

Subject Index 469

Contents – Volume II

1   Higher Harmonics in Dynamic Atomic Force Microscopy
Robert W. Stark, Martin Stark   1

2   Atomic Force Acoustic Microscopy
Ute Rabe   37

3   Scanning Ion Conductance Microscopy
Tilman E. Schäffer, Boris Anczykowski, Harald Fuchs   91

4   Spin-Polarized Scanning Tunneling Microscopy
Wulf Wulfhekel, Uta Schlickum, Jurgen Kirschner   121

5   Dynamic Force Microscopy and Spectroscopy
Ferry Kienberger, Hermann Gruber, Peter Hinterdorfer   143

**6   Sensor Technology for Scanning Probe Microscopy
and New Applications**
*Egbert Oesterschulze, Leon Abelmann, Arnout van den Bos,
Rainer Kassing, Nicole Lawrence, Gunther Wittstock,
Christiane Ziegler*   165

7   Quantitative Nanomechanical Measurements in Biology
Małgorzata Lekka, Andrzej J. Kulik   205

**8   Scanning Microdeformation Microscopy:
Subsurface Imaging and Measurement of Elastic Constants
at Mesoscopic Scale**
Pascal Vairac, Bernard Cretin   241

**9   Electrostatic Force and Force Gradient Microscopy:
Principles, Points of Interest and Application to Characterisation
of Semiconductor Materials and Devices**
Paul Girard, Alexander Nikolaevitch Titkov   283

**10   Polarization-Modulation Techniques in Near-Field Optical Microscopy
for Imaging of Polarization Anisotropy in Photonic Nanostructures**
*Pietro Giuseppe Gucciardi, Ruggero Micheletto, Yoichi Kawakami,
Maria Allegrini*   321

11　　**Focused Ion Beam as a Scanning Probe: Methods and Applications**
Vittoria Raffa, Piero Castrataro, Arianna Menciassi, Paolo Dario .　361

Subject Index 413

Contents – Volume III

12 **Atomic Force Microscopy in Nanomedicine**
Dessy Nikova, Tobias Lange, Hans Oberleithner,
Hermann Schillers, Andreas Ebner, Peter Hinterdorfer   1

13 **Scanning Probe Microscopy:**
From Living Cells to the Subatomic Range
Ille C. Gebeshuber, Manfred Drack, Friedrich Aumayr,
Hannspeter Winter, Friedrich Franek 27

14 **Surface Characterization and Adhesion and Friction Properties**
of Hydrophobic Leaf Surfaces and Nanopatterned Polymers
for Superhydrophobic Surfaces
Zachary Burton, Bharat Bhushan 55

15 **Probing Macromolecular Dynamics and the Influence**
of Finite Size Effects
Scott Sills, René M. Overney 83

16 **Investigation of Organic Supramolecules by Scanning Probe Microscopy**
in Ultra-High Vacuum
Laurent Nony, Enrico Gnecco, Ernst Meyer 131

17 **One- and Two-Dimensional Systems: Scanning Tunneling Microscopy**
and Spectroscopy of Organic and Inorganic Structures
Luca Gavioli, Massimo Sancrotti 183

18 **Scanning Probe Microscopy Applied to Ferroelectric Materials**
Oleg Tikhomirov, Massimiliano Labardi, Maria Allegrini 217

19 **Morphological and Tribological Characterization of Rough Surfaces**
by Atomic Force Microscopy
Renato Buzio, Ugo Valbusa 261

20 **AFM Applications for Contact and Wear Simulation**
Nikolai K. Myshkin, Mark I. Petrokovets, Alexander V. Kovalev . . 299

21 **AFM Applications for Analysis of Fullerene-Like Nanoparticles**
Lev Rapoport, Armen Verdyan 327

22		**Scanning Probe Methods in the Magnetic Tape Industry**
James K. Knudsen 343

Subject Index 371

Contents – Volume IV

23 **Scanning Probe Lithography for Chemical,
Biological and Engineering Applications**
Joseph M. Kinsella, Albena Ivanisevic   1

24 **Nanotribological Characterization of Human Hair and Skin
Using Atomic Force Microscopy (AFM)**
Bharat Bhushan, Carmen LaTorre 35

25 **Nanofabrication with Self-Assembled Monolayers
by Scanning Probe Lithography**
Jayne C. Garno, James D. Batteas 105

26 **Fabrication of Nanometer-Scale Structures
by Local Oxidation Nanolithography**
Marta Tello, Fernando García, Ricardo García 137

27 **Template Effects of Molecular Assemblies Studied
by Scanning Tunneling Microscopy (STM)**
Chen Wang, Chunli Bai 159

28 **Microfabricated Cantilever Array Sensors for (Bio-)Chemical Detection**
Hans Peter Lang, Martin Hegner, Christoph Gerber 183

29 **Nano-Thermomechanics: Fundamentals and Application
in Data Storage Devices**
B. Gotsmann, U. Dürig 215

30 **Applications of Heated Atomic Force Microscope Cantilevers**
Brent A. Nelson, William P. King 251

Subject Index 277

List of Contributors – Volume VII

Flemming Besenbacher
Interdisciplinary Nanoscience Center (iNANO), University of Aarhus
DK-8000 Aarhus C, Denmark
e-mail: fbe@inano.dk

François Bessueille
LSA, Université Lyon I, 43 Boulevard du 11 Novembre 1918
69622 Villeurbanne Cedex, France
e-mail: francois.bessueille@univ-lyon1.fr

Bharat Bhushan
Nanotribology Laboratory for Information Storage and MEMS/NEMS (NLIM)
W 390 Scott Laboratory, 201 W. 19th Avenue, Ohio State University
Columbus, Ohio 43210-1142, USA
E-mail: bhushan.2@osu.edu

Malte Burchardt
Faculty of Mathematics and Sciences
Department for Pure and Applied Chemistry
and Institute of Chemistry and Biology of the Marine Environment (ICBM)
Carl von Ossietzky University of Oldenburg
D-26111 Oldenburg, Germany
e-mail: malteburchardt@gmx.de

Abdelhamid Errachid]
Laboratory of NanoBioEngineering, Barcelona Science Park
Edifici Modular, C/Josep Samitier 1–5, 08028-Barcelona, Spain
e-mail: aerrachid@pcb.ub.es

Horacio D. Espinosa
Department of Mechanical Engineering, Northwestern University
2145 Sheridan Rd., Evanston, IL 60208-3111, USA
e-mail: espinosa@northwestern.edu

Yanxia Hou
Ecole Centrale de Lyon, STMS/CEGELY
36 Avenue Guy de Collongue, F-69131 Ecully Cedex, France
e-mail: yanxiahou24@yahoo.com

Nicole Jaffrezic-Renault
Ecole Centrale de Lyon, STMS/CEGELY
36 Avenue Guy de Collongue, F-69131 Ecully Cedex, France
e-mail: Nicole.Jaffrezic@ec-lyon.fr

Changhong Ke
Department of Mechanical Engineering, Northwestern University
2145 Sheridan Rd., Evanston, IL 60208-3111, USA
e-mail: c-ke@northwestern.edu

Keun-Ho Kim
Department of Mechanical Engineering, Northwestern University
2145 Sheridan Rd., Evanston, IL 60208-3111, USA
e-mail: kkim@nualumni.edu

Jeppe Vang Lauritsen
Interdisciplinary Nanoscience Center (iNANO)
Department of Physics and Astronomy
University of Aarhus, DK-8000 Aarhus C, Denmark
e-mail: jvang@phys.au.dk

Huiwen Liu
Nanotribology Laboratory for Information Storage and MEMS/NEMS (NLIM)
The Ohio State University
650 Ackerman Road, Suite 255, Columbus, Ohio 43202, USA
e-mail: Huiwen.Liu@seagate.com

Claude Martelet
Ecole Centrale de Lyon, STMS/CEGELY
36 Avenue Guy de Collongue, F-69131 Ecully Cedex, France
France
e-mail: Claude.Martelet@ec-lyon.fr

Nicolaie Moldovan
Department of Mechanical Engineering, Northwestern University
2145 Sheridan Rd., Evanston, IL 60208-3111, USA
e-mail: n-moldovan@northwestern.edu

Martin Munz
National Physcial Laboratory (NPL), Quality of Life Division
Hampton road, Teddington, Middlesex TW11 0LW, UK
e-mail: martin.munz@npl.co.uk

Michael Nosonovsky
Nanomechanical Properties Group
Materials Science and Engineering Laboratory
National Institute of Standards and Technology
100 Bureau Dr., Mail Stop 8520, Gaithersburg, MD 20899-8520, USA
e-mail: michael.nosonovsky@nist.gov

Sascha E. Pust
Faculty of Mathematics and Sciences
Department for Pure and Applied Chemistry
and Institute of Chemistry and Biology of the Marine Environment (ICBM)
Carl von Ossietzky University of Oldenburg
D-26111 Oldenburg, Germany
E-mail: sascha.pust@uni-oldenburg.de

Robert A. Sayer
Nanotribology Lab for Information Storage and MEMS/NEMS (NLIM)
The Ohio State University
650 Ackerman Road, Suite 255, Columbus, Ohio 43202, USA
e-mail: Sayer.11@osu.edu

Heinz Sturm
Federal Institute for Materials Research (BAM), VI.25
Unter den Eichen 87, D-12205 Berlin, Germany
e-mail: heinz.sturm@bam.de

Ronnie T. Vang
Interdisciplinary Nanoscience Center (iNANO)
Department of Physics and Astronomy
University of Aarhus, DK-8000 Aarhus C, Denmark
e-mail: rtv@inano.dk

Gunther Wittstock
Faculty of Mathematics and Sciences
Department for Pure and Applied Chemistry
and Institute of Chemistry and Biology of the Marine Environment (ICBM)
Carl von Ossietzky University of Oldenburg, D-26111 Oldenburg, Germany
e-mail: gunther.wittstock@uni-oldenburg.de

List of Contributors – Volume V

Maria Allegrini
Dipartimento di Fisica "Enrico Fermi", Università di Pisa
Largo Bruno Pontecorvo, 3, 56127 Pisa, Italy
e-mail: maria.allegrini@df.unipi.it

Yasuhisa Ando
Tribology Group, National Institute of Advanced Industrial Science and Technology
1-2 Namiki, Tsukuba, Ibaraki, 305-8564, Japan
e-mail: yas.ando@aist.go.jp

Bharat Bhushan
Nanotribology Laboratory for Information Storage and MEMS/NEMS (NLIM)
W 390 Scott Laboratory, 201 W. 19th Avenue, Ohio State University
Columbus, Ohio 43210-1142, USA
e-mail: bhushan.2@osu.edu

Daniel Ebeling
Center for NanoTechnology (CeNTech), Heisenbergstr. 11, 48149 Münster
e-mail: Daniel.Ebeling@uni-muenster.de

Pietro Guiseppe Gucciardi
CNR-Istituto per i Processi Chimico-Fisici, Sezione di Messina
Via La Farina 237, I-98123 Messina, Italy
e-mail: gucciardi@its.me.cnr.it

Sadik Hafizovic
ETH-Zurich, PEL, Hoenggerberg HPT H6, 8093 Zurich, Switzerland
e-mail: hafizovi@phys.ethz.ch

Andreas Hierlemann
ETH-Zurich, PEL, Hoenggerberg HPT H6, 8093 Zurich, Switzerland
e-mail: hierlema@phys.ethz.ch

Hendrik Hölscher
Center for NanoTechnology (CeNTech), Heisenbergstr. 11, 48149 Münster
e-mail: Hendrik.Hoelscher@uni-muenster.de

Lin Huang
Veeco Instruments, 112 Robin Hill Road, Santa Barbara, CA 93117, USA
e-mail: lhuang@veeco.com

Hideki Kawakatsu
Institute of Industrial Science, University of Tokyo
Komaba 4-6-1, Meguro-Ku, Tokyo 153-8505, Japan
e-mail: kawakatu@iis.u-tokyo.ac.jp

Kay-Uwe Kirstein
ETH-Zurich, PEL, Hoenggerberg HPT H6, 8093 Zurich, Switzerland
e-mail: kirstein@phys.ethz.ch

Christine Kranz
School of Chemistry and Biochemistry, Georgia Institute of Technology
311 Ferst Dr., Atlanta GA 30332-0400, USA
e-mail: Christine.Kranz@chemistry.gatech.edu

Boris Mizaikoff
School of Chemistry and Biochemistry, Georgia Institute of Technology
311 Ferst Dr., Atlanta GA 30332-0400, USA
e-mail: Boris.Mizaikoff@chemistry.gatech.edu

Salvatore Patanè
Dipartimento di Fisica della Materia e Tecnologie Fisiche Avanzate
Università di Messina, Salita Sperone 31, I-98166 Messina, Italy
e-mail: patanes@unime.it

Craig Prater
Veeco Instruments, 112 Robin Hill Road, Santa Barbara, CA 93117, USA
e-mail: cprater@veeco.com

Elisa Riedo
Georgia Institute of Technology, School of Physics
837 State Street, Atlanta, GA 30332-0430, USA
e-mail: elisa.riedo@physics.gatech.edu

Tilman E. Schäffer
Institute of Physics and Center for Nanotechnology, University of Münster
Heisenbergstr. 11, 48149 Münster, Germany
e-mail: tilman.schaeffer@uni-muenster.de

Udo D. Schwarz
Department of Mechanical Engineering, Yale University
P.O. Box 208284, New Haven, CT 06520-8284, USA
e-mail: Udo.Schwarz@yale.edu

Yaxin Song
Nanotribology Lab for Information Storage and MEMS/NEMS (NLIM)
The Ohio State University
650 Ackerman Road, Suite 255, Columbus, Ohio 43202, USA
e-mail: Song.220@osu.edu

Chanmin Su
Veeco Instruments, 112 Robin Hill Road, Santa Barbara, CA 93117, USA
e-mail: csu@veeco.com

Robert Szoszkiewicz
Georgia Institute of Technology, School of Physics
837 State Street, Atlanta, GA 30332-0430, USA
e-mail: robert.szoszkiewicz@physics.gatech.edu

Sebastiano Trusso
CNR-Istituto per i Processi Chimico-Fisici, Sezione di Messina
Via La Farina 237, I-98123 Messina, Italy
e-mail: trusso@its.me.cnr.it

Cirino Vasi
CNR-Istituto per i Processi Chimico-Fisici, Sezione di Messina
Via La Farina 237, I-98123 Messina, Italy
e-mail: vasi@its.me.cnr.it

Justyna Wiedemair
School of Chemistry and Biochemistry, Georgia Institute of Technology
311 Ferst Dr., Atlanta GA 30332-0400, USA
e-mail: Justyna.Wiedemair@chemistry.gatech.edu

List of Contributors – Volume VI

Asa H. Barber
Queen Mary, University of London, Department of Materials
Mile End Road, London E1 4NS, UK
e-mail: a.h.barber@qmul.ac.uk

Lilia A. Chtcheglova
Institute for Biophysics, Johannes Kepler University of Linz
Altenbergerstr. 69, A-4040 Linz, Austria
e-mail: lilia.chtcheglova@jku.at

Sidney R. Cohen
Chemical Research Support, Weizmann Institute of Science
Rehovot 76100, Israel
e-mail: Sidney.cohen@weizmann.ac.il

Andreas Ebner
Institute for Biophysics, Johannes Kepler University of Linz
Altenbergerstr. 69, A-4040 Linz, Austria
e-mail: andreas.ebner@jku.at

Hermann J. Gruber
Institute for Biophysics, Johannes Kepler University of Linz
Altenbergerstr. 69, A-4040 Linz, Austria
e-mail: hermann.gruber@jku.at

Norihiko Hayazawa
Nanophotonics Laboratory
RIKEN (The Institute of Physical and Chemical Research)
2-1 Hirosawa, Wako, Saitama, 351-0198, Japan
e-mail: hayazawa@riken.jp

Peter Hinterdorfer
Institute for Biophysics, Johannes Kepler University of Linz
Altenbergerstr. 69, A-4040 Linz, Austria
e-mail: peter.hinterdorfer@jku.at

Ferry Kienberger
Institute for Biophysics, Johannes Kepler University of Linz
Altenbergerstr. 69, A-4040 Linz, Austria
e-mail: ferry.kienberger@jku.at

Kei Kobayashi
International Innovation Center, Kyoto University, Katsura, Nishikyo
Kyoto 615-8520, Japan
e-mail: keicoba@iic.kyoto-u.ac.jp

Tadahiro Komeda
Institute of Multidisciplinary Research for Advanced Materials (IMRAM)
Tohoku University
2-1-1, Katahira, Aoba, Sendai, 980-0877 Japan
e-mail: komeda@tagen.tohoku.ac.jp

Andrzej J. Kulik
Ecole Polytechnique Fédérale de Lausanne, EPFL – IPMC – NN
1015 Lausanne, Switzerland
e-mail: Andrzej.Kulik@epfl.ch

Piotr Laidler
Institute of Medical Biochemistry, Collegium Medicum Jagiellonian University
Kopernika 7, 31-034 Kraków, Poland
e-mail: mblaidle@cyf-kr.edu.pl

Małgorzata Lekka
The Henryk Niewodniczański Institute of Nuclear Physics
Polish Academy of Sciences, Radzikowskiego 152, 31–342 Kraków, Poland
e-mail: Malgorzata.Lekka@ifj.edu.pl

Thomas Mueller
Veeco Instruments, 112 Robin Hill Road, Santa Barbara, CA 93117, USA
e-mail: tmueller@veeco.com

Theeraporn Puntheeranurak
Institute for Biophysics, Johannes Kepler University of Linz
Altenbergerstr. 69, A-4040 Linz, Austria
e-mail: theeraporn.puntheeranurak@jku.at

Yuika Saito
Nanophotonics Laboratory
RIKEN (The Institute of Physical and Chemical Research)
2-1 Hirosawa, Wako, Saitama, 351-0198, Japan

Yasuhiro Sugawara
Department of Applied Physics, Graduate School of Engineering,
Osaka University, Yamada-oka 2-1, Suita, Osaka 565-0871, Japan
e-mail: sugawara@ap.eng.osaka-u.ac.jp

Neil H. Thomson
Molecular and Nanoscale Physics Group, University of Leeds
EC Stoner Building, Woodhouse Lane, Leeds, LS2 9JT, UK
e-mail: n.h.thomson@leeds.ac.uk

Kohei Uosaki
Division of Chemistry, Graduate School of Science, Hokkaido University
N10 W8, Sapporo, Hokkaido, 060-0810, Japan
e-mail: uosaki@pcl.sci.hokudai.ac.jp

H. Daniel Wagner
Dept. Materials and Interfaces, Weizmann Institute of Science
Rehovot 76100, Israel
e-mail: daniel.wagner@weizmann.ac.il

Hirofumi Yamada
Department of Electronic Science & Engineering, Kyoto University
Katsura, Nishikyo, Kyoto 615-8510, Japan
e-mail: h-yamada@kuee.kyoto-u.ac.jp

Ryo Yamada
Division of Material Physics, Graduate School of Engineering Science
Osaka University, Machikaneyama-1-3, Toyonaka, Osaka, 060-0810, Japan
e-mail:yamada@molectronics.jp

21 Lotus Effect: Roughness-Induced Superhydrophobicity

Michael Nosonovsky · Bharat Bhushan

21.1
Introduction

The rapid advances in nanotechnology since the 1980s, including such applications as microelectromechanical/nanoelectromechanical systems (MEMS/NEMS), have stimulated development of new materials and the design of surfaces, which should meet new sets of requirements [3, 6, 11]. In MEMS/NEMS, the surface-to-volume ratio grows with miniaturization and surface phenomena dominate. One of the crucial surface properties for materials in microscale/nanoscale applications is water-repellency, nonwetting or hydrophobicity. It is also usually desirable to reduce wetting in liquid-flow applications and some conventional applications, such as glass windows, in order for liquid to flow away along their surfaces. Wetting may lead to formation of menisci at the interface between solid bodies during sliding contact, which increases adhesion/friction. As a result of this, the wet friction force is greater than the dry friction force, which is usually undesirable [4, 5, 7]. On the other hand, high adhesion is desirable in some applications, such as adhesive tapes and adhesion of cells to biomaterial surfaces; therefore, enhanced wetting by changing roughness would be desirable in these applications [35, 36]. There is also an analogy between the motion of a droplet on a rough solid surface and the motion of gas bubbles in a liquid near rough walls of a vessel or channel, which allows us to apply the results of droplet motion analysis to study the mobility of nanobubbles [37].

Wetting is characterized by the contact angle, which is defined as the measurable angle which a liquid makes with a solid. The contact angle depends on several factors, such as roughness and the manner of surface preparation, and its cleanliness [1, 25]. If the liquid wets the surface (referred to as wetting liquid or hydrophilic surface), the value of the static contact angle is $0 \leq \theta \leq 90°$, whereas if the liquid does not wet the surface (referred to as nonwetting liquid or hydrophobic surface), the value of the contact angle is $90° < \theta \leq 180°$. The term hydrophobic/hydrophilic, which was originally applied only to water (the Greek word *hydro* means "water"), is often used to describe the contact of a solid surface with any liquid. Surfaces with a contact angle between 150° and 180° are called superhydrophobic. Superhydrophobic surfaces also have very low water contact angle hysteresis. The contact angle hysteresis is the difference between the advancing and the receding contact angles. If additional liquid is added to a sessile drop, the contact line advances and each time motion ceases, the drop exhibits an advancing contact angle. Alternatively, if liquid is removed from the drop, the contact angle decreases to a receding value before the contact retreats. For a droplet moving along the solid surface, the contact angle at the front of the droplet (advancing contact angle) is greater than that at the back of the droplet (receding contact angle), owing to roughness and surface heterogeneity, resulting in the contact angle hysteresis. For liquid-flow applications, in addition to high contact angle another wetting property of interest is a very low water roll-off angle, which denotes the angle to which a surface must be tilted for roll-off of water drops (i.e., very low water contact angle hysteresis) [22,28].

One of the ways to increase the hydrophobic or hydrophilic properties of the surface is to increase surface roughness; so roughness-induced hydrophobicity has become a subject of extensive investigation. Wenzel (1936) found that the contact angle of a liquid with a rough surface is different from that with a smooth surface. Cassie and Baxter (1944) showed that air (or gas) pockets may be trapped in the cavities of a rough surface, resulting in a composite solid–liquid-air interface, as opposed to a homogeneous solid–liquid interface. Shuttleworth and Bailey (1948) studied spreading of a liquid over a rough solid surface and found that the contact angle at the absolute minimum of the surface energy corresponds to the values predicted by [51] or [15]. Johnson and Dettre (1964) showed that the homogeneous and composite interfaces correspond to the two metastable equilibrium states of a droplet. Bico et al. (2002), Marmur (2003, 2004), Lafuma and Quéré (2003), Patankar (2003, 2004a), He et al. (2003), and other authors investigated recently metastability of artificial superhydrophobic surfaces and showed that whether the interface is homogeneous or composite may depend on the history of the system (in particular, whether the liquid was applied from the top or from the bottom). Extrand (2002) pointed out that whether the interface is homogeneous or composite depends on droplet size, owing to gravity. It was suggested also that the so-called two-tiered (or double) roughness, created by superposition of two roughness patterns at different length scales [24,41,47], and fractal roughness [45] may lead to superhydrophobicity. Herminghaus (2000) showed that certain self-affine profiles may result in superhydrophobic surfaces even for wetting liquids; in the case the local equilibrium condition for the triple line (line of contact between solid, liquid, and air) is satisfied. Nosonovsky and Bhushan (2005, 2006a) pointed out that such configurations, although formally possible, are likely to be unstable. Nosonovsky and Bhushan (2006a,b) proposed a stochastic model for wetting of rough surfaces with a certain probability associated with every equilibrium state. According to their model, the overall contact angle with a two-dimensional rough profile is calculated by assuming that the overall configuration of a droplet is a result of superposition of numerous metastable states. The probability-based concept is consistent with the experimental data [29], which suggest that transition between the composite and homogeneous interfaces is gradual, rather than instant.

It has been demonstrated experimentally, that roughness changes the contact angle in accordance with the Wenzel model. Yost et al. (1994) found that roughness enhances wetting of a copper surface with Sn–Pb eutectic solder, which has a contact angle of 15–20° for a smooth surface. Shibushi et al. (1996) measured the contact angle of various liquids (mixtures of water and 1,4-dioxane) on alkylketen dimmer (AKD) substrate (contact angle not larger than 109° for a smooth surface). They found that for wetting liquids the contact angle decreases with increasing roughness, whereas for nonwetting liquids it increases. Semal et al. (1999) investigated the effect of surface roughness on contact angle hysteresis by studying a sessile droplet of squalane spreading dynamically on multilayer substrates (behenic acid on glass) and found that microroughness increases slow the rate of droplet spreading. Erbil et al. (2003) measured the contact angle of polypropylene (contact angle of 104° for a smooth surface) and found that the contact angle increases with increasing roughness. Burton and Bhushan (2005) measured the variation of the contact angle with the roughness of patterned surfaces and found that in the case of hydrophilic surfaces, it decreases with increasing roughness, and for hydrophobic surfaces, it increases with increasing roughness.

In the last decade, material scientists paid attention to natural surfaces, which are extremely hydrophobic. Among them are leaves of water-repellent plants such as *Nelumbo nucifera* (lotus) and *Colocasia esculenta*, which have high contact angles with water [2, 34, 50]. First, the surface of the leaves is usually covered with a range of different waxes made from a mixture of large hydrocarbon molecules, measuring around 1 nm in diameter, that have a strong phobia of being wet. Second, the surface is very rough owing to so-called papillose epidermal cells, which form asperities or papillae (Fig. 21.1). In addition to the microscale roughness of the leaf due to the papillae, the surface of the papillae is also rough with submicron-sized asperities [50]. Thus, they have hierarchical micro- and nano-sized structures. The water droplets on these surfaces readily sit on the apex of nanostructures because air bubbles fill in the valleys of the structure under the droplet; therefore, these leaves exhibit considerable superhydrophobicity. The water droplets on the leaves remove any contaminant particles from there surfaces when they roll off, leading to self-cleaning ability, also referred to as the lotus effect. Other examples of biological surfaces include duck feathers and butterfly wings. Their corrugated surfaces provide air pockets that prevent water from completely touching the surface. Study and simulation of biological objects with desired properties is referred to as "biomimetics", which comes from the Greek word *biomimesis* meaning to mimic life. As far as realization of strongly water repellent artificial surfaces is concerned, they can be constructed by chemically treating surfaces with low-surface-energy substances, such as poly(tetrafluoroethylene), silicon, or wax, or by fabricating extremely rough hydrophobic surfaces directly [23, 28, 33, 45]. Sun et al. (2005) studied an artificial poly(dimethylsiloxane) (PDMS) replica of a lotus leaf surface and found a water contact angle of 160° for the rough surface, whereas for the smooth PDMS surface it is about 105°.

As stated earlier, when two solids come in contact in the presence of a wetting liquid, a meniscus is often formed [4, 5, 7]. A meniscus results in a normal meniscus force, which, in turn, results in an increase in tangential friction force. The magnitude of the meniscus force depends on the number of asperity contacts and asperity radii, which depend on roughness, and on the surface tension of the liquid and the contact angle. The contact angle, as stated before, depends on surface roughness, and thus roughness affects the wet friction force; this effect was neglected earlier [35].

In this chapter, a model to provide a relationship between local roughness and contact angle as well as contact angle hysteresis is presented and experimental data are discussed. Both homogeneous and composite interfaces are studied. The model is used to analyze different possible roughness distributions and to calculate

Fig. 21.1. (a) Scanning electron microscope (SEM) image of papillae on water-repellent plant leaves of *Colocasia esculenta* and *Nelumbo nucifera* (lotus). (b) Distribution of the papillae on the *C. esculenta* leaf surface which forms a pattern, close to hexagonal [35]

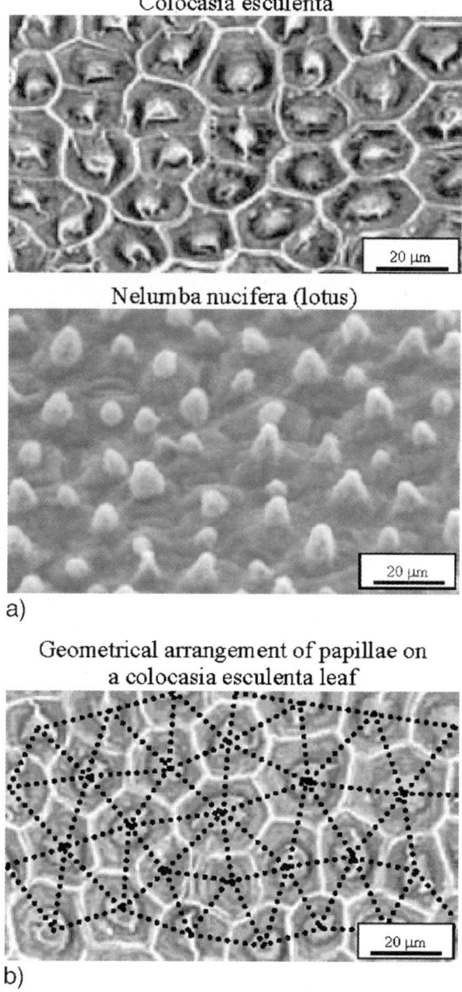

the effect of local roughness distribution on the contact angle, which allows us to make recommendations for optimized biomimetic superhydrophobic surfaces. The analysis is also compared with experimental data on natural and artificial water-repellent surfaces. Furthermore, the effect of roughness on the meniscus force and on the friction force during liquid-mediated contact is considered and discussed.

21.2
Contact Angle Analysis

As pointed out in the preceding section, a liquid in contact with a solid surface may form both homogeneous (solid–liquid) and composite (solid–liquid–air) interfaces. Both cases are studied in this section.

21.2.1
Homogeneous Solid–Liquid Interface

In this section, the dependence of the contact angle on the surface tension is considered for a liquid in contact with a smooth and a rough solid surface, forming a homogeneous interface. The effects of the contact area and of sharp edges on the contact angle are discussed.

21.2.1.1
Smooth Surface

It is well known that the surface atoms or molecules of liquids or solids have energy above that of similar atoms and molecules in the interior, which results in surface tension or free surface energy being an important surface property. This property is characterized quantitatively by the surface tension or free surface energy γ, which is equal to the work that is required to create a unit area of the surface at constant volume and temperature. The units of γ are joules per square meter or newtons per meter and it can be interpreted either as energy per unit surface area or as tension force per unit length of a line at the surface. When a solid is in contact with a liquid, the molecular attraction will reduce the energy of the system below that for the two separated surfaces. This may be expressed by the Dupré equation:

$$W_{SL} = \gamma_{SA} + \gamma_{LA} - \gamma_{SL} , \tag{21.1}$$

where W_{SL} is the work of adhesion per unit area between two surfaces, γ_{SA} and γ_{SL} are the surface energies (surface tensions) of the solid against air and liquid, and γ_{LA} is the surface energy (surface tension) of liquid against air [25].

If a droplet of liquid is placed on a solid surface, the liquid and solid surfaces come together under equilibrium at a characteristic angle called the static contact angle θ_0 (Fig. 21.2). The contact angle can be determined from the condition of the total energy of the system being minimized [1, 25]. The total energy E_{tot} is given by

$$E_{tot} = \gamma_{LA}(A_{LA} + A_{SL}) - W_{SL}A_{SL} , \tag{21.2}$$

where A_{LA} and A_{SL} are the contact areas of the liquid with the solid and air, respectively. It is assumed that the droplet is small enough so that the gravitational potential energy can be neglected. At the equilibrium $dE_{tot} = 0$, which yields

$$\gamma_{LA}(dA_{LA} + dA_{SL}) - W_{SL}dA_{SL} = 0 . \tag{21.3}$$

For a droplet of constant volume, it is easy to show using geometrical considerations that

$$dA_{LA}/dA_{SL} = \cos\theta_0 . \tag{21.4}$$

Combining (21.1), (21.3), and (21.4), we obtain the so-called Young equation for the contact angle:

$$\cos\theta_0 = \frac{\gamma_{SA} - \gamma_{SL}}{\gamma_{LA}} . \tag{21.5}$$

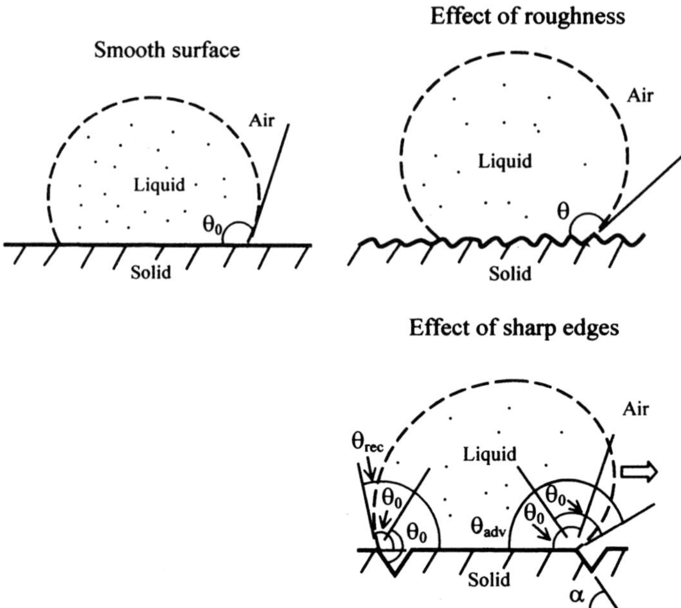

Fig. 21.2. Droplet of liquid in contact with a solid surface – smooth surface, contact angle θ_0; rough surface, contact angle θ – and a surface with sharp edges. For a droplet moving from *left* to *right* on a sharp edge (shown by an *arrow*), the contact angle at a sharp edge may be any value between the contact angle with the horizontal plane and with the inclined plane. This effect results in a difference between advancing ($\theta_{adv} = \theta_0 + \alpha$) and receding ($\theta_{rec} = \theta_0 - \alpha$) contact angles [35]

Equation (21.5) provides with the value of the static contact angle for given surface tensions. Note that although the term "air" is used throughout this chapter, the analysis does not change in the case of another gas, such as a liquid vapor.

For microscale/nanoscale droplets, the value of the contact angle may be different from that for macroscale droplets. The effect of line tension is believed to be responsible for that, since the energy of the molecules in the vicinity of the triple line is different from that in the bulk or at the surface; however, this effect is significant only at the nanometer range [42]. Checco et al. (2003) showed that the atomic force microscopy (AFM) data on the scale dependence of the contact angle for nanosized and microsized alkane droplets may rather be explained by the effect of surface heterogeneity, assuming that tiny droplets first appear at the most wettable spots.

21.2.1.2
Rough Surface

The Young equation (21.5) is valid only in the case of a flat solid surface. If the surface is rough, the roughness affects the contact angle in two ways: owing to the increased contact area A_{SL} and owing to the effect of sharp edges. We consider the effect of the contact area first.

Let us consider a rough solid surface, which consists of asperities and valleys, with a typical size of roughness details smaller than the size of the droplet (Fig. 21.1b). For a droplet of constant volume in contact with a rough surface without air pockets, the cosine of the contact angle is given as the ratio of the differentials of the liquid–air contact area and the area under the droplet, based on the geometrical considerations, which is given as [51]

$$\cos\theta = dA_{LA}/dA_F = \frac{A_{SL}}{A_F} dA_{LA}/dA_{SL} = R_f \cos\theta_0 \,, \tag{21.6}$$

where θ is the contact angle for the rough surface and A_F is the flat solid–liquid contact area or a projection of the solid–liquid area A_{SL} on the horizontal plane. R_f is a roughness factor defined as

$$R_f = \frac{A_{SL}}{A_F} \,. \tag{21.7}$$

Equation (21.6) shows that if the liquid wets a flat surface ($\cos\theta_0 > 0$), it will wet also the rough surface with a contact angle of $\theta < \theta_0$, since $A_{SL}/A_F > 1$. Furthermore, for nonwetting liquids ($\cos\theta_0 < 0$), the contact angle with a rough surface will be greater than that with the flat surface, $\theta > \theta_0$. The dependence of the contact angle on the roughness factor is presented in Fig. 21.3 for different values of θ_0 (Nosonovsky and Bhushan 2005).

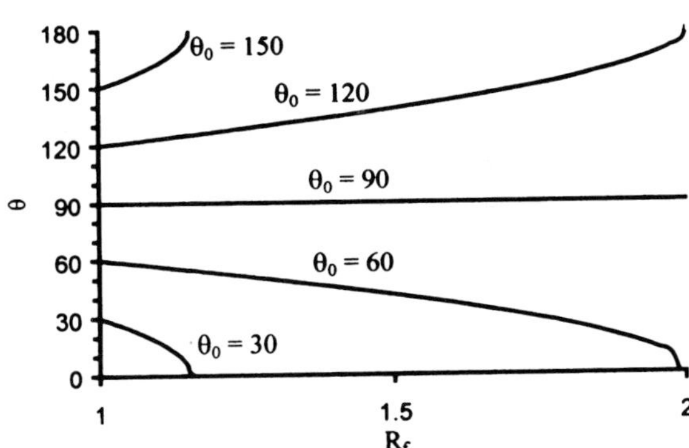

Fig. 21.3. Contact angle for a rough surface (θ) as a function of the roughness factor (R_f) for various contact angles for a smooth surface (θ_0) [35]

21.2.1.3
Contact Angle Hysteresis

There are several reasons which may lead to the contact angle hysteresis. These include surface roughness and heterogeneity.

For a rough surface, a sharp edge can pin the line of contact of the solid, liquid, and air (also known as the triple line) at a position far from stable equilibrium, i.e., at contact angles different from θ_0 [19]. This effect is illustrated in the bottom sketch of Fig. 21.2, which shows a droplet propagating along a solid surface with grooves. At the edge point, the contact angle is not defined and can have any value between the values corresponding to the contact with the horizontal and inclined surfaces. For a droplet moving from left to right, the triple line will be pinned at the edge point until it is able to proceed to the inclined plane. As observed from Fig. 21.2, the change of the surface slope (α) at the edge is the reason for the pinning. Because of the pinning, the value of the contact angle at the front of the droplet (dynamic maximum advancing contact angle or $\theta_{\text{adv}} = \theta_0 + \alpha$) is greater than θ_0, whereas the value of the contact angle at the back of the droplet (dynamic minimum receding contact angle or $\theta_{\text{rec}} = \theta_0 - \alpha$) is smaller than θ_0. This phenomenon is known as the contact angle hysteresis [19,25,26]. A hysteresis domain of the dynamic contact angle is thus defined by the difference $\theta_{\text{adv}} - \theta_{\text{rec}}$. The liquid can travel easily along the surface if the contact angle hysteresis is small. It is noted that the static contact angle lies within the hysteresis domain; therefore, increasing the static contact angle up to the values of a superhydrophobic surface (approaching $180°$) will result also in reduction of the contact angle hysteresis. The contact angle hysteresis can also exist even if the surface slope changes smoothly, without sharp edges. For a chemically heterogeneous surface, the change of local contact angle due to heterogeneity has a similar effect as the change of the surface slope.

21.2.2
Composite Solid–Liquid–Air Interface

In this section, the dependence of the contact angle on the surface tension is considered for a liquid in contact with a smooth and a rough solid surface, forming a composite solid–liquid–air interface with air pockets trapped at the valleys between the asperities.

21.2.2.1
Contact Angle

It is noted that (21.6) is valid only for moderate values of R_{f}, when $-1 \leq R_{\text{f}} \cos \theta_0 \leq 1$. For high roughness, a wetting liquid will be completely absorbed by the rough surface cavities. As will be shown in this section, a nonwetting liquid cannot penetrate into surface cavities with large slopes, resulting in the formation of air pockets, leading to a composite solid–liquid–air interface, shown for a sawtooth and a smooth profile in Fig. 21.4a. The solid–liquid contact zones are located at the peaks of the asperities, whereas the air pockets and solid–air contact zones are in the valleys. The solid–liquid contact area will not further increase with increasing roughness [19].

Formation of the composite interface is shown in Fig. 21.4a for a sawtooth profile with slope α and for a smooth profile. In order to determine whether the interface is solid–liquid or composite, the change of net energy dE_{tot} should be considered, which corresponds to the displacement ds of the liquid–air surface along the inclined groove wall. For the solid–liquid interface, $dE_{\text{tot}} < 0$; therefore

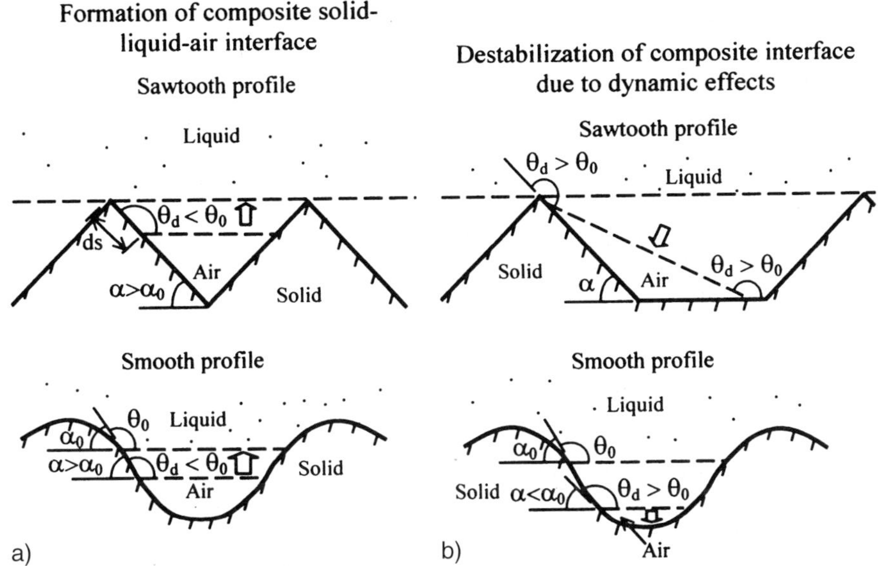

Fig. 21.4. (a) Formation of a composite solid–liquid–air interface for sawtooth and smooth profiles and (b) destabilization of the composite interface for the sawtooth and smooth profiles owing to dynamic effects. Dynamic contact angle $\theta_d > \theta_0$ corresponds to an advancing liquid–air interface, whereas $\theta_d < \theta_0$ corresponds to a receding interface [35]

it is more energetically profitable for the liquid to advance and fill the groove. For the composite interface, however, $dE_{\text{tot}} > 0$; therefore it is more energetically profitable for the liquid to recede and leave the groove. The change of energy depends on slope α, and the critical value of the slope, α_0, can be found by setting $dE_{\text{tot}} = 0$:

$$dE_{\text{tot}} = dA_{\text{SL}}\gamma_{\text{SL}} + dA_{\text{LA}}\gamma_{\text{LA}} = -2\,ds\cos\alpha_0\,ds + 2\,ds(\gamma_{\text{SL}} - \gamma_{\text{SA}})$$
$$= -2\gamma_{\text{LA}}\cos\alpha_0\,ds + 2(\gamma_{\text{SL}} - \gamma_{\text{SA}})\,ds = 0 \ . \tag{21.8}$$

Substituting (21.5) into (21.8), we obtain $\cos\alpha_0 = -\cos\theta_0$ or

$$\alpha_0 = 180° - \theta_0 \ . \tag{21.9}$$

For small slopes ($\alpha < \alpha_0$), $dE_{\text{tot}} < 0$ and the interface is solid–liquid, whereas for large slopes ($\alpha > \alpha_0$), $dE_{\text{tot}} > 0$ and the interface is composite, as shown in Fig. 21.4a. For a profile of arbitrary smooth shape, a composite interface is possible if the slope exceeds the critical value α_0 at some point. In this case, the liquid would recede and leave space in the groove, as shown in Fig. 21.4a for a smooth profile.

Cassie and Baxter (1944) extended the Wenzel equation, which was originally developed for the homogeneous solid–liquid interface, for the composite interface. For this case, there are two sets of interfaces: a liquid–air interface with the ambient and a flat composite interface under the droplet involving solid–liquid, liquid–air, and solid–air interfaces. For fractional flat geometrical areas of the solid–liquid and liquid–air interfaces under the droplet, f_{SL} and f_{LA}, the flat area of the composite interface is

$$A_C = f_{SL}A_C + f_{LA}A_C = R_f A_{SL} + f_{LA}A_C \,. \tag{21.10}$$

In order to calculate the contact angle in a manner similar to the derivation of (21.6), the differential area of the liquid–air interface under the droplet, $f_{LA}\,dA_C$, should be subtracted from the differential of the total liquid–air area dA_{LA}, which yields

$$\cos\theta = \frac{dA_{LA} - f_{LA}\,dA_C}{dA_C} = \frac{dA_{SL}}{dA_F}\frac{dA_F}{dA_C}\frac{dA_{LA}}{dA_{SL}} - f_{LA} = R_f f_{SL}\cos\theta_0 - f_{LA}\,. \tag{21.11}$$

According to (21.11), in the limit of high R_f, f_{SL} approaches zero, whereas f_{LA} approaches unity, and hence θ approaches 180°. However, the Cassie–Baxter model does not provide any particular form of the dependence of f_{SL} and f_{LA} on R_f and does not explain under which conditions the composite interface forms.

Shuttleworth and Bailey (1948) studied spreading of a liquid over a rough solid surface and found that the contact angle at the absolute minimum of the surface energy corresponds to the values given by (21.2) (for the homogeneous interface) or (21.5) (for the composite interface). According to their analysis, spreading of a liquid continues, until simultaneously (21.1) (Young's equation) is satisfied locally at the triple line and the minimal surface condition is satisfied over the entire liquid–air interface. The minimal surface condition states that the sum of the inverse principal radii of curvature, r_1 and r_2, is constant at any point, and thus governs the shape of the liquid–air interface:

$$\frac{1}{r_1} + \frac{1}{r_2} = \text{const.} \tag{21.12}$$

Johnson and Detre (1964) showed that the homogeneous and composite interfaces correspond to the two states of a droplet. Even though it may be geometrically possible for the system to become composite, it may be energetically profitable for the liquid to penetrate into valleys between asperities and to form the homogeneous interface. Swain and Lipowski (1998) extended the Wenzel and Cassie–Baxter model for the case of a chemically heterogeneous surface. Marmur (2003) formulated geometrical conditions for a surface under which the energy of the system has a local minimum and the composite interface may exist. Patankar (2004a) pointed out that whether the homogeneous or the composite interface exists depends on the system's history, i.e., on whether the droplet was formed at the surface or deposited. However, the aforementioned analyses do not provide us with the answer as to which of the two possible configurations, homogeneous or composite, will actually form.

In order to investigate whether the composite interface can exist, its stability should be analyzed. It was assumed in the preceding paragraphs that the composite interface is stable, and the derivation was based on the assumption that the advancing liquid–air interface remains a horizontal plane. This assumption is hard to justify in the case of large distances between the asperities or in the case when the slope changes slowly. In these cases, the liquid–air interface can easily be destabilized owing to an imperfectness of the profile shape or owing to dynamic effects, such as surface waves. This may result in formation of the homogeneous solid–liquid interface. On the basis of these considerations, Fig. 21.4b shows schematically such destabilization of the composite interface, for the case of a large distance between the asperities and for the case of a smooth profile. The liquid advances, if the liquid–air interface reaches a position at which its local angle with the solid surface is greater than θ_0, and eventually the liquid fills the valley. Therefore, (21.11) should not be used to predict the contact angle, since it ignores the possibility of destabilization of the composite interface. It is suggested that a stochastic approach should be developed with account for possibility of composite interface destabilization and different probabilities of different states of equilibrium [36, 37].

It is noted that the composite interface may normally exist only for hydrophobic materials, because in the case of a hydrophilic material the liquid will advance and fill the valleys between the asperities. It has been reported, however, that the so-called metastable equilibrium states with a composite interface may be experimentally observed in some cases even for hydrophilic materials [29], resulting in high contact angles and effectively making a hydrophilic surface hydrophobic. However, these states are highly unstable and can hardly be practically utilized. Although their nature is not completely understood, they exist apparently owing to pinning of the triple line as a consequence of nanoroughness and surface heterogeneity.

21.2.2.2
Effect of the Composite Interface on the Contact Angle Hysteresis

The transition to the composite solid–liquid–air interface, which was considered earlier, results in a dramatic reduction of the solid–liquid area of contact and, thus, adhesion of the droplet to the solid surface. Therefore, transition to the composite interface can decrease the contact angle hysteresis, so that a droplet would move easily. Reduction of hysteresis due to transition to a composite interface was observed experimentally [26, 29].

21.2.3
Stability of the Composite Interface

As pointed out in the preceding sections, although a composite interface may be possible geometrically, it may be unstable, so the stability of a homogeneous interface should be analyzed. The stability of the homogeneous interface is not considered here, because such an interface is usually stable. While transition from the composite interface to the homogeneous is a well-known phenomenon, the opposite transition has never been observed experimentally. This is due to very high activation energy required to induce the composite state. Stability implies that in addition to satisfying the equilibrium condition for the net energy

$$dE = 0 \, , \tag{21.13}$$

the stable configuration should satisfy the minimum net energy condition

$$d^2E > 0 \, . \tag{21.14}$$

The interface may be destabilized owing to small perturbations caused by various external influences and effects, for example, by capillary or gravitational waves. Furthermore, the configuration may have many stable equilibrium conditions and be transformed from one stable position to another owing to the external effects, with a certain probability of finding the system at a given state. These phenomena are considered in the following section.

21.2.3.1
Destabilization Due to Capillary and Gravitational Waves

A wave may form at the liquid–air interface owing to gravitational or capillary forces

$$z = A \cos(kx - \omega t) \, , \tag{21.15}$$

where z is the vertical displacement, and k and ω are the wavenumber and frequency, which are related to each other by

$$\omega^2 = gk + \frac{\gamma_{LA}}{\rho} k^3 \, , \tag{21.16}$$

where g is the gravity constant, ρ is the liquid density and γ_{LA} is the liquid–air interface energy [30]. For most microscale/nanoscale applications, the effect of gravity is small and the frequency is given by

$$\omega = \sqrt{\frac{\gamma_{LA} k^3}{\rho}} \, . \tag{21.17}$$

The capillary waves may lead to composite interface destabilization, as will be shown later [36].

It is assumed that the interface energy γ_{LA} is a constant for given materials and that it is size-independent. Generally speaking, this is not true for a very small thickness of liquid comparable with the range of intermolecular forces. However, in the present work we are assuming that the relevant size of the surface roughness, as well as the thickness of the liquid layer, is greater than the range of the intermolecular forces and therefore that γ_{LA} is constant.

Consider a sawtooth profile (Fig. 21.5a) with teeth height $a \tan \alpha/2$ and distance between teeth d. The teeth represent asperities of a rough surface. It is assumed that the model of the sawtooth profile can capture important features of more complicated rough surfaces. The horizontal liquid–air interface is located at a distance z from the valley and has small waves of amplitude A and wavenumber k. The total change of the energy of the system, from the energy of the homogeneous solid–liquid interface, is given by the sum of surface changes throughout the inclined and horizontal portions of the surface and corresponding liquid–air parts, plus the energy of the waves. The surface energy changes at inclined and horizontal portions of the surfaces and at corresponding liquid–air parts are given by corresponding surfaces' lengths times the corresponding interface energies. The length of the inclined portion of the interface is $z/\sin \alpha$ and the length of the corresponding section of the wavy surface is $(z/\tan\alpha)(kL_0/2\pi)$, where L_0 is the length of the liquid–air interface per wave period, given by the integral

$$L_0 = \int_0^{2\pi/k} \sqrt{1 + (Ak)^2 \sin^2(xk)}\,dx = 4\sqrt{1 + (Ak)^2}\, E\left(\frac{Ak}{\sqrt{1 + (Ak)^2}}\right),$$

$$(21.18)$$

where $E(x)$ is the elliptical integral of the second kind [36]. The length of the horizontal portion of the interface is d, and the length of the corresponding section of the wavy surface is $dkL_0/(2\pi)$. The energy change is given by

$$
\begin{aligned}
U(z) &= -\frac{2z}{\sin\alpha}(\gamma_{SL} - \gamma_{SA}) + \frac{2z}{\tan\alpha}\frac{kL_0}{2\pi}\gamma_{LA} \\
&\quad - d\left[(\gamma_{SL} - \gamma_{SA}) - \gamma_{LA}\frac{kL_0}{2\pi}\right]H(z) + E_0 \\
&= \frac{2z\gamma_{LA}}{\sin\alpha}\left(\cos\theta_0 + \cos\alpha\frac{kL_0}{2\pi}\right) + d\gamma_{LA}\left(\cos\theta_0 + \frac{kL_0}{2\pi}\right)H(z) + E_0
\end{aligned}
$$

$$(21.19)$$

and E_0 is the energy of the waves, γ_{SL} and γ_{SA} are interface energies for the solid–liquid and solid–air interfaces, respectively, and $H(z)$ is the step function, such that $H(z) = 0$ for $z \leq 0$ and $H(z) = 1$ for $z > 0$. Young's equation for the contact angle is used in the derivation of (21.19) [1, 25]:

$$\cos\theta_0 = \frac{\gamma_{SA} - \gamma_{SL}}{\gamma_{LA}}.$$

$$(21.20)$$

It is assumed in (21.19), that $z > A$.

In the limiting case of a flat liquid–air interface ($A = 0$), the surface energy is given by [36]

$$
\begin{aligned}
U(z) &= -\frac{2z}{\sin\alpha}(\gamma_{SL} - \gamma_{SA}) + \frac{2z}{\tan\alpha}\gamma_{LA} - d\left[(\gamma_{SL} - \gamma_{SA}) - \gamma_{LA}\right]H(z) \\
&= \frac{2z\gamma_{LA}}{\sin\alpha}(\cos\theta_0 + \cos\alpha) + d\gamma_{LA}(\cos\theta_0 + 1)H(z)
\end{aligned}
$$

$$(21.21)$$

for $A < z$.

For $z > 0$, the energy may increase or decrease with increasing z, depending on the sign of $(\cos\theta_0 + \cos\alpha)$, since both γ_{LA} and $\sin\alpha$ are positive. In particular, if $180° - \alpha > \theta_0$, the energy increases with z, whereas otherwise it decreases. The stable position corresponds to the minimum value of the energy, which is $z = a/(2\tan\alpha)$ (liquid staying at the tops of the asperities) for $180° - \alpha < \theta_0$ and $z = 0$ for $180° - \alpha > \theta_0$ (homogeneous solid–liquid interface).

For the wavy liquid–air interface, based on (21.19), the energy may increases with increasing z, if $\cos\theta_0 + \cos\alpha kL_0/(2\pi) > 0$, for $z > A$. However, for

Fig. 21.5. Sawtooth profile (**a**) with a smooth liquid–air interface, (**b**) with a wavy liquid–air interface and (**c**) with a stochastic distribution of air pockets [36]

Composite solid-liquid-air interface

Smooth liquid-air interface

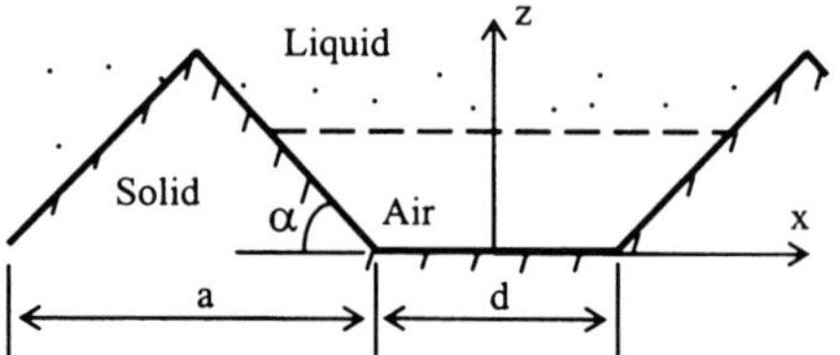

Wavy liquid-air interface

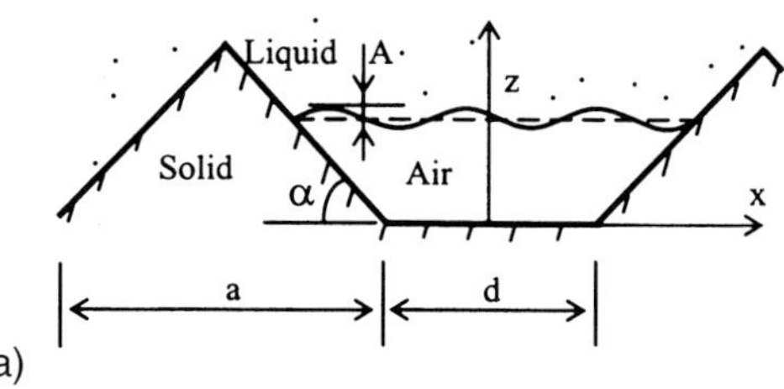

a)

Stochastic distribution of air pockets

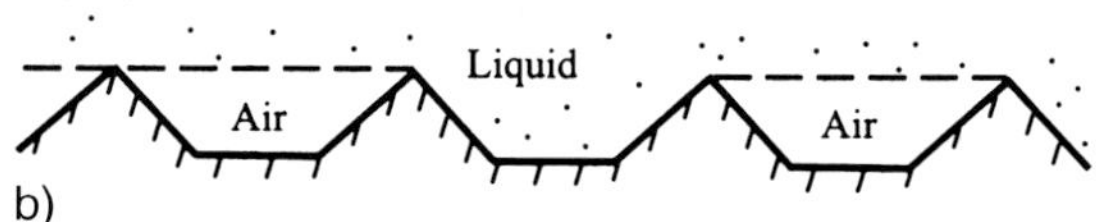

b)

$z < A$, the waves touch the horizontal part of the interface, and only the fraction $[\pi - \arccos(z/A)]/\pi$ of the interface is liquid–air. In this case the energy change is given by

$$U(z) = \frac{z\gamma_{\mathrm{LA}}}{\sin\alpha}\left[\cos\theta_0 + \frac{\pi - \arccos(z/A)}{\pi}\left(\frac{kL_{\mathrm{LA}}}{2\pi}\right)\cos\alpha\right]$$

$$+ \, d\gamma_{\mathrm{LA}}\left(\cos\theta_0 + \frac{kL_{\mathrm{LA}}}{2\pi}\right)\frac{\pi - \arccos(z/A)}{\pi}H(z) + E_0 \qquad (21.22)$$

for $A > z$, where L_{LA} is the length of the liquid–air part of the interface (the wave not touching the solid horizontal part of the interface) given by [36]

$$L_{\mathrm{LA}} = \int_{-\arccos(z/A)/k}^{\arccos(z/A)/k} \sqrt{1 + (Ak)^2 \sin^2(xk)}\,\mathrm{d}x\ . \qquad (21.23)$$

The energy change U as a function of liquid–air interface position z is presented in Fig. 21.6 for the smooth interface (21.21) and for the wavy liquid–air interface

Fig. 21.6. Energy change as a function of interface position for smooth and wavy liquid–air interfaces, $d\gamma_{LA}(\cos\theta_0 + 1) = 0.9$, $\gamma_{LA}(\cos\theta_0 + \cos\alpha)/\sin\alpha = -0.2$, $\gamma_{LA}[\cos\theta_0 + (kL_0/2\pi)\cos\alpha]/\sin\alpha = -0.15$, $E_0 = 0.015$, $A = 0.1$ [36]

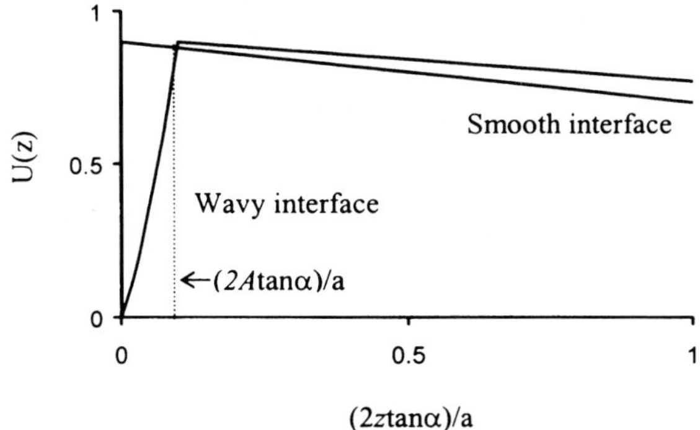

((21.19), (21.22)). It is noted that in the case when the waves are introduced, $U(z)$ has a local minimum at $z = 0$, which corresponds to the homogeneous solid–liquid interface, and in the case of $\cos\theta_0 + \cos\alpha kL_0/(2\pi) < 0$ it has another minimum at $z = a/(2\tan\alpha)$, which corresponds to the composite solid–liquid–air interface (liquid staying at the tops of the asperities). The interface position z is normalized in such a manner that the first equilibrium position ($z = 0$) corresponds to zero and the second equilibrium position [$z = a/(2\tan\alpha)$] corresponds to the value of unity in Fig. 21.6. In this case, the system has two equilibriums and may be, with a certain probability, in either one or the other position. The interface consists of many asperities and valley, some of the valleys have a homogeneous interface, whereas others have a composite interface (Fig. 21.5b). It is assumed that the probability p for the interface to be composite depends on geometrical parameters of the interface and the values of the energy which correspond to the equilibrium states [36].

21.2.3.2
Stochastic Model

In this section, the mechanism of destabilization of the composite interface due to liquid–air interface waves will be considered, and a statistical model for the interface destabilization will be discussed [36]. It was shown in the previous section that the interface may have two stable states. The first stable state corresponds to the homogeneous interface with energy level

$$U(0) = 0 \, . \tag{21.24}$$

The second state corresponds to the composite interface ($z = a/2\tan\alpha$) with energy level obtained from (21.19)

$$U(a/2\tan\alpha) = \frac{a\gamma_{LA}}{\tan\alpha\sin\alpha}\left(\cos\theta_0 + \cos\alpha\frac{kL_0}{2\pi}\right)$$

$$+ \, d\gamma_{LA}\left(\cos\theta_0 + \frac{kL_0}{2\pi}\right) + E_0 \, . \tag{21.25}$$

A certain probability p may be associated with each of the two stable states of energy. Assuming that the waves with the energy E_0 have a similar effect on the system as the thermal fluctuation of an ideal gas with energy kT, the Maxwell–Boltzmann statistical distribution may be applied [21]. On the basis of the Maxwell–Boltzmann distribution, the probability is exponentially dependent upon the energy level:

$$p = B \exp\left(-\frac{U}{E_0}\right) , \tag{21.26}$$

where B is a normalization constant [36]. Substituting (21.25) into (21.26), we obtain the probability of the composite interface:

$$p(\phi) = C \exp(-\phi/\phi_0) , \tag{21.27}$$

where

$$\phi = d/a , \tag{21.28}$$

$$\phi_0 = \frac{E_0}{a\gamma_{\mathrm{LA}} \left(\cos\theta_0 + \frac{kL_0}{2\pi}\right)} , \tag{21.29}$$

and

$$C = B \exp\left[-\frac{a\gamma_{\mathrm{LA}}}{E_0 \tan\alpha \sin\alpha} \left(\cos\theta_0 + \cos\alpha \frac{kL_0}{2\pi}\right) - 1\right] . \tag{21.30}$$

21.2.3.3
Analysis of Rough Profiles

In this section, a patterned rough surface will be analyzed. Consider a periodic sawtooth profile with a distance between asperities d and width a, as shown in Fig. 21.5. Let us assume that the probability of the interface remaining composite, p, decreases exponentially with distance between asperities according to (21.27) [36]. The roughness factor for the homogeneous interface based on (21.7), is given by

$$R_{\mathrm{f}} = \frac{d + a/\cos\alpha}{d + a} = \frac{\phi + 1/\cos\alpha}{\phi + 1} . \tag{21.31}$$

The total fraction of valleys which are covered with liquid (homogeneous interface) is given by $1 - p$, whereas the fraction of the valleys which have air pockets (composite interface) is given by p, obtained from (21.27). On the basis of this, the fractional areas are given by

$$\begin{aligned}
f_{\mathrm{SL}} &= \frac{(1 - p)(d + a/\cos\alpha)}{(1 - p)(d + a/\cos\alpha) + p(d + a)} \\
&= \frac{(1 - p)(\phi + 1/\cos\alpha)}{(1 - p)(\phi + 1/\cos\alpha) + p(\phi + 1)}
\end{aligned} \tag{21.32}$$

and

$$f_{LA} = \frac{p(d+a)}{(1-p)(d+a/\cos\alpha) + p(d+a)}$$

$$= \frac{p(\phi+1)}{(1-p)(\phi+1/\cos\alpha) + p(\phi+1)} . \tag{21.33}$$

Substituting (21.31)–(21.33) into (21.11), we obtain the expression for the contact angle:

$$\cos\theta = \frac{(1-p)(\phi+1/\cos\alpha)^2 \cos\theta_0 - p(\phi+1)^2}{(1-p)(\phi+1/\cos\alpha)(\phi+1) + p(\phi+1)^2} . \tag{21.34}$$

The results for the contact angle as a function of ϕ are presented in Fig. 21.7a. It is observed that higher roughness (lower ϕ) corresponds to higher contact angles [36].

Comparison of the models, based on (21.6) (homogeneous interface), (21.11) (solid–liquid–air composite interface) and (21.32) (stochastic interface) is presented

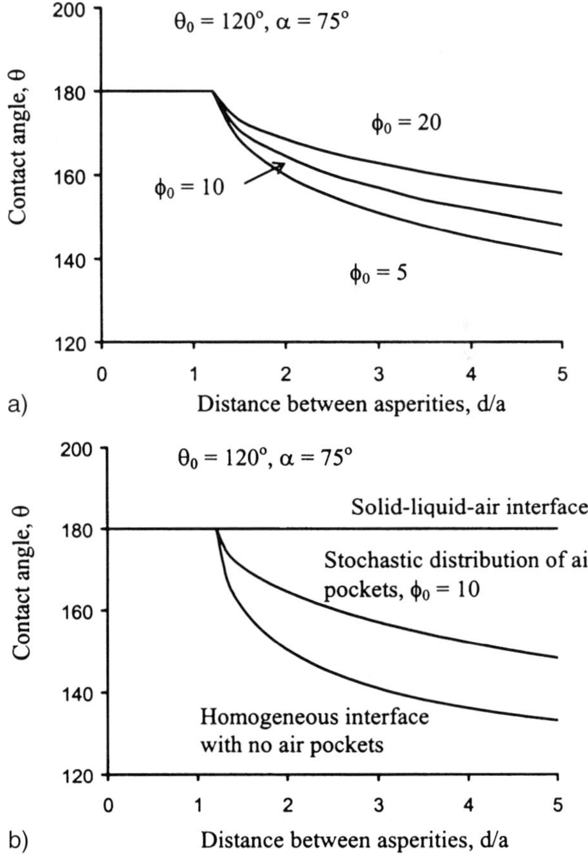

Fig. 21.7. (a) Contact angle as a function of distance between asperities for the stochastic model and (b) comparison of the interface with no air pockets, a composite liquid–air interface and a stochastic distribution of air pockets [36]

in Fig. 21.7b. It is observed that for high roughness (small ϕ), $|R_\mathrm{f} \cos \theta_0| > 1$ and all three models predict $\theta = 180°$. However, for greater distance between the asperities (higher ϕ), the composite interface model, which does not account for the possibility of destabilization, still predicts $\theta = 180°$, if $\theta_0 + \alpha > 180°$, whereas the homogeneous interface model predicts a rapid decrease of θ down to the value of θ_0, owing to a decreasing roughness factor. The stochastic model yields the values of the contact angle close to the composite interface model for short distances between the asperities (small ϕ); however, with increasing ϕ, the probability of destabilization of the composite interface grows, and eventually the values of the contact angle approach those predicted by the homogeneous interface model.

21.2.3.4
Effect of Gravity

In the preceding analysis we ignored the effect of gravity by assuming that the gravity force is small compared with the surface tension forces. However, for big droplets, this assumption may not be correct. If the weight of a droplet exceeds the vertical component of the total surface tension force at the triple line, a droplet suspended at the tops of the asperities will collapse [22]. Thus, a maximum critical size of the droplet R_max exits above which the droplet cannot remain suspended on the tops of the asperities. Let us investigate how this maximum size depends on the period of the asperities l. Consider a rough surface with roughness period l and amplitude $\hat{z}$, which corresponds to a maximum droplet size R_max. The weight of the droplet is proportional to its volume and to the third power of R_max:

$$W \propto R_\mathrm{max} . \tag{21.35}$$

For the maximum value of the droplet radius, the weight is equal to the total vertical component of the surface tension and is proportional to the surface tension times the cosine of the contact angle times the total perimeter of the triple lines:

$$W \propto \gamma_\mathrm{SL} t N \cos \theta , \tag{21.36}$$

where N is the number of asperities under the droplet [37].

Consider another rough surface with period αl and amplitude $\alpha \hat{z}$ which has the same roughness factor and the same contact angle. The length of the triple line for each asperity will be αt. The number of asperities under the droplet is proportional to the second power of R_max, divided by the second power of α:

$$N \propto R_\mathrm{max}^2 / \alpha^2 . \tag{21.37}$$

Combining (21.35)–(21.37), we obtain

$$R_\mathrm{max} \propto 1/\alpha . \tag{21.38}$$

This result suggests that with increasing size of asperities, the ability of a rough surface to form a composite interface decreases and larger droplets collapse. Therefore, smaller asperities make a composite interface more likely, owing to gravity. Increasing the droplet size has the same effect as increasing the period of roughness [37].

21.2.3.5
Effect of Dual Roughness

It is known that natural superhydrophobic surfaces have hierarchical roughness at two scale ranges: microroughness and nanoroughness (Fig. 21.8), although the reason for this is still obscure [8, 14, 24, 29, 47]. It was suggested by Nosonovsky and Bhushan (2006c) that nanoroughness is required to enhance the stability of the composite interface. A superhydrophobic surface should be able to form a composite interface in order to have a high contact angle and low contact angle hysteresis. However, the composite interface is often fragile and can be irreversibly transformed into the homogeneous interface, thus damaging superhydrophobicity. The mechanisms which lead to destabilization of the composite interface are scale-dependent. To effectively resist these scale-dependent mechanisms, a multiscale (hierarchical) roughness is required. High asperities resist the capillary waves, while nanobumps prevent nanodroplets from filling the valleys between asperities and pin the triple line in the case of a hydrophilic spot. Such multiscale roughness was found in natural and successful artificial superhydrophobic surfaces [8, 14, 20, 24, 34, 37, 45, 47].

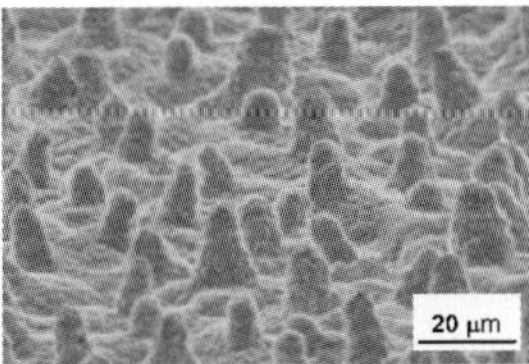
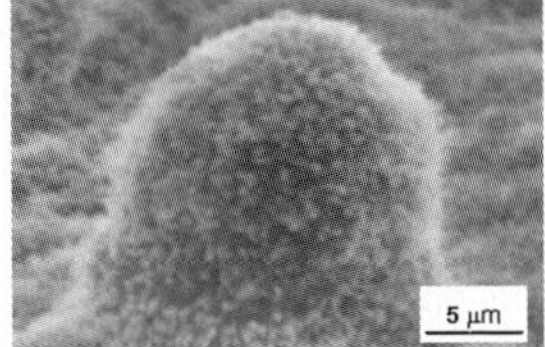

Fig. 21.8. Scanning electron micrographs of *N. nucifera* (lotus) leaf surface show hierarchical roughness with nanobumps (average peak-to-valley height 780 nm, midwidth 40 nm, peak radius 150 nm) on the top of microbumps (papillae, average peak-to-valley height 13 μm, midwidth 10 μm, peak radius 3 μm). Nanobumps data obtained with atomic force microscopy, scan size 2×2 μm; microbumps measured using an optical profiler [8]

21.3
Calculation of the Contact Angle for Selected Rough Surfaces and Surface Optimization

The contact angle of a liquid with a number of rough surfaces is calculated in this section. The model, presented in the preceding sections, combines the effect of surface area, possibility of formation of a composite interface, and the effect of sharp edges. Several selected rough surfaces are considered, shown in Fig. 21.5. First, two-dimensional surface profiles are analyzed, followed by more complex three-dimensional surfaces. On the basis of the analysis, roughness optimization for contact angle is conducted.

21.3.1
Two-Dimensional Periodic Profiles

21.3.1.1
Sawtooth Periodic Profile

Let us consider a surface with a sawtooth profile with a tooth angle (or the absolute value of the slope) of α (Fig. 21.9). Using (21.7), we calculate the roughness factor as

$$R_{\mathrm{f}} = \frac{A_{\mathrm{SL}}}{A_{\mathrm{F}}} = (\cos\alpha)^{-1} \, . \tag{21.39}$$

Using (21.6), we obtain the contact angle as

$$\cos\theta = \frac{\cos\theta_0}{\cos\alpha} \, . \tag{21.40}$$

An increase of α above the critical level α_0, given by (21.9), will result in a transition from a complete solid–liquid contact to a composite solid–liquid–air interface, and (21.40) cannot be used any further. Substituting the value of slope $\alpha = \alpha_0$ found from (21.9) in (21.40), we can obtain the value of θ, which corresponds to α_0, and this gives that the critical value α_0 corresponds to the contact angle $\theta = 180°$. This means that by increasing the tooth angle toward the critical value, a surface with the contact angle approaching $180°$ can be produced, for a given θ_0. However, sharp edges, which may lead to pinning of the triple line, make the sawtooth profile undesirable. In addition to this, the sawtooth profile provides roughness only in the direction perpendicular to the grooves, so the anisotropy of the profile will result in anisotropy of the contact angle [17].

21.3.1.2
General Periodic Profile

For a general form of the surface $z(x, y)$, the solid–liquid area of contact is equal to

$$A_{\mathrm{SL}} = A_{\mathrm{F}} \int\!\!\int_{A_{\mathrm{F}}} \sqrt{1 + \left(\frac{\partial z}{\partial x}\right)^2 + \left(\frac{\partial z}{\partial y}\right)^2} \, \mathrm{d}x\,\mathrm{d}y \, . \tag{21.41}$$

A periodic two-dimensional surface profile with period λ can be presented as a Fourier series:

$$z(x) = \sum_{n=1}^{\infty} A_n \sin\frac{2\pi n x}{\lambda} + B_n \cos\frac{2\pi n x}{\lambda} \, . \tag{21.42}$$

The derivatives of $z(x)$ are given as

$$\frac{\mathrm{d}z}{\mathrm{d}x} = \frac{2\pi}{\lambda} \sum_{n=1}^{\infty} A_n n \cos\frac{2\pi n x}{\lambda} - B_n n \sin\frac{2\pi n x}{\lambda} \, ,$$

$$\frac{\mathrm{d}z}{\mathrm{d}y} = 0 \, . \tag{21.43}$$

Sawtooth periodic profile

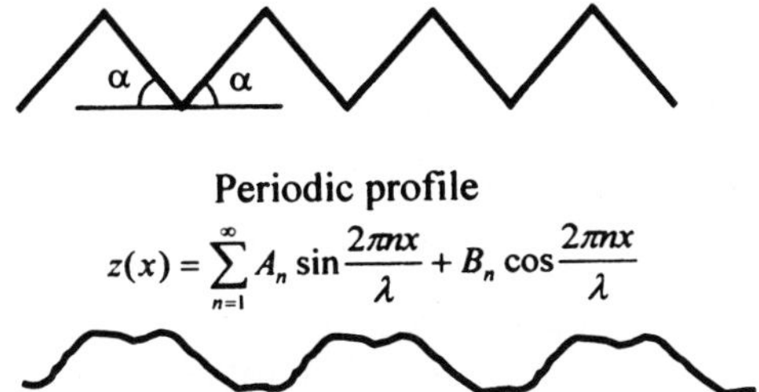

Periodic profile

$$z(x) = \sum_{n=1}^{\infty} A_n \sin\frac{2\pi nx}{\lambda} + B_n \cos\frac{2\pi nx}{\lambda}$$

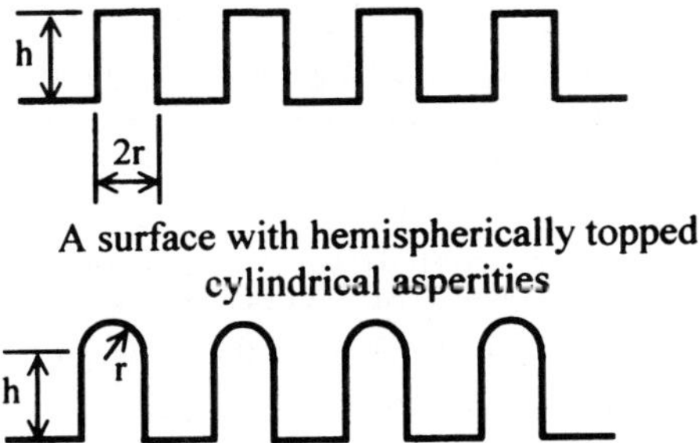

A surface with rectangular asperities

A surface with hemispherically topped cylindrical asperities

A surface with conical or pyramidal asperities

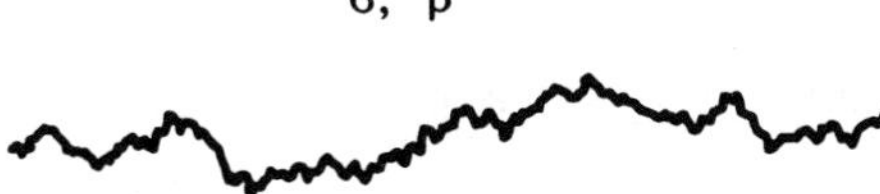

Random Gaussian surface

$$\sigma, \ \beta^{\bullet}$$

Fig. 21.9. Various rough surfaces [35]

Substitution of (21.43) and (21.41) into (21.7) provides us with an expression for the roughness factor of a periodic profile:

$$R_{\mathrm{f}} = 1 + \frac{1}{\lambda} \int_0^\lambda \sqrt{1 + \frac{4\pi^2}{\lambda^2} \left(\sum_{n=1}^\infty A_n n \cos \frac{2\pi n x}{\lambda} - B_n n \sin \frac{2\pi n x}{\lambda} \right)^2} \; \mathrm{d}x \; . \tag{21.44}$$

It is possible to determine whether the composite interface is possible, by considering the slope of the profile. In order for the composite interface to form, the absolute value of the slope must exceed the critical angle α_0, given by (21.9), at any point:

$$\left| \frac{2\pi}{\lambda} \sum_{n=1}^\infty A_n n \cos \frac{2\pi n x}{\lambda} - B_n n \sin \frac{2\pi n x}{\lambda} \right| > \tan(\alpha_0) = \tan(-\theta_0) \; . \tag{21.45}$$

As an example, let us consider a sinusoidal profile

$$z(x) = A_1 \sin \frac{2\pi x}{\lambda} \; . \tag{21.46}$$

By substituting (21.46) into (21.44) and integrating, we can obtain a closed-form solution:

$$
\begin{aligned}
R_{\mathrm{f}} &= \frac{1}{\lambda} \int_0^\lambda \sqrt{1 + (2\pi A_1/\lambda)^2 \cos^2(2\pi x/\lambda)} \; \mathrm{d}x \\
&= \frac{1}{2\pi} \int_0^{2\pi} \sqrt{1 + (2\pi A_1/\lambda)^2 \cos^2 x} \; \mathrm{d}x \\
&= \frac{2}{\pi} \sqrt{1 + (2\pi A_1/\lambda)^2} \int_0^{\pi/2} \sqrt{1 - \frac{(2\pi A_1/\lambda)^2}{1 + (2\pi A_1/\lambda)^2} \sin^2 x} \; \mathrm{d}x \\
&= \frac{2}{\pi} \sqrt{1 + (2\pi A_1/\lambda)^2} E\left(\frac{(2\pi A_1/\lambda)}{\sqrt{1 + (2\pi A_1/\lambda)^2}} \right) ,
\end{aligned}
\tag{21.47}
$$

where $E(x)$ is the so-called elliptical integral of the second kind, the values of which are tabulated in handbooks:

$$E(k) = \int_0^{\pi/2} \sqrt{1 - k^2 \sin^2 x} \, \mathrm{d}x \; . \tag{21.48}$$

The maximum absolute value of the slope of the sinusoidal profile (21.46) is achieved at $x = 0$ and is equal to $2\pi A_1/\lambda$. With an increase of A_1/λ, the slope increases, and a composite interface may be formed (Fig. 21.4a). For the composite interface to form, the slope at some points should exceed the critical value α_0, given by (21.9). By using (21.45) and setting $x = 0$ and using (21.45), we find the condition for the existence of the composite interfaceas

$$\frac{2\pi A_1}{\lambda} > \tan(-\theta_0) \ . \tag{21.49}$$

The contact angle can be calculated by substituting R_f from (21.47) into (21.6). The dependence of the contact angle on the amplitude of the sinusoidal profile is presented in Fig. 21.10a. It is observed that lower values of θ correspond to lower values of θ_0 at the transition to the composite interface, unlike in the case of the sawtooth surface, which has critical values corresponding to $\theta_0 = 180°$. For $\theta_0 = 100°$ the critical value of R_f is 5.67 ($\theta = 131°$), for $\theta_0 = 120°$ the critical value of R_f is 1.73 ($\theta = 140°$), and for $\theta_0 = 150°$ the critical value of R_f is 0.58 ($\theta = 159°$). Further increase of A_1/λ may lead to a corresponding increase of R_f and θ, on the basis of (21.38) [54]. However, as discussed earlier, the composite interface can be destabilized and (21.38) cannot be applied; therefore, the sinusoidal interface is not recommended for producing superhydrophobic surfaces. In addition to this, the sinusoidal profile provides roughness only in the direction perpendicular to the grooves, which would result in contact angle anisotropy.

21.3.2
Three-Dimensional Surfaces

The analysis of profiles provides us with critical values of the roughness parameters in the case when the contact line is parallel to the grooves. Three-dimensional surfaces, which constitute a more general case, with various typical shapes of asperities are considered in this section.

21.3.2.1
Array of Asperities of Identical Shape and Size

Let us consider a rough surface with rectangular asperities, which have a square foundation with side $2r$ and height h (Fig. 21.9). For each asperity, the area of the surface is given by

$$A_{\mathrm{asp}} = 8rh + 4r^2 \ , \tag{21.50}$$

whereas the flat projection area is $4r^2$. Under the assumption that asperities are randomly distributed throughout the surface with a density of η asperities per unit area, the total contact surface area is given by

$$A_{\mathrm{SL}} = A_F + A_F\eta(8rh + 4r^2) - A_F 4\eta r^2 = A_F(1 + 8\eta r^2) \ . \tag{21.51}$$

The roughness factor is found using (21.7) and (21.51):

$$R_f = 1 + 8\eta rh = 1 + 2p^2 h/r \ , \tag{21.52}$$

where p is a packing parameter for asperities with a square foundation, $p = 2r\sqrt{\eta}$. The packing parameter is equal to the fraction of the surface area which is covered by asperities.

Fig. 21.10. (a) Contact angle for a rough surface (θ) as a function of surface parameters for the surface with a sinusoidal profile, rectangular (*dotted line*), hemispherically topped cylindrical (*solid line*), and conical/pyramidal asperities. (**b**) Dependence of the roughness factor (R_f) and the contact angle for a rough surface (θ) on roughness parameters for a Gaussian surface [35]

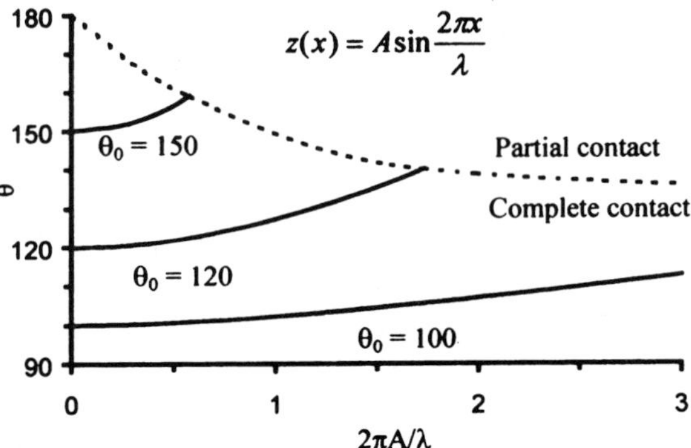

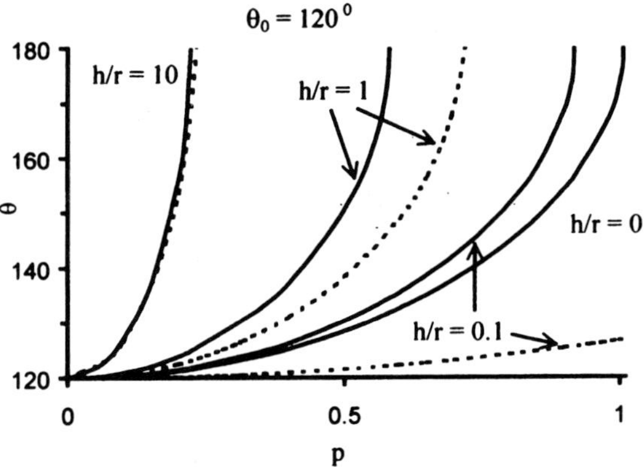

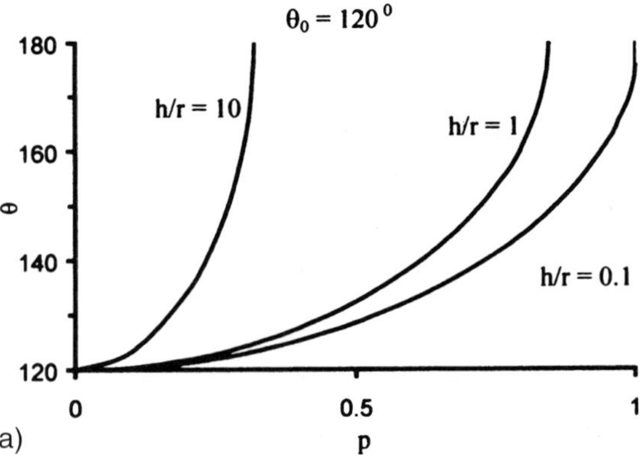

Fig. 21.10. (continued)

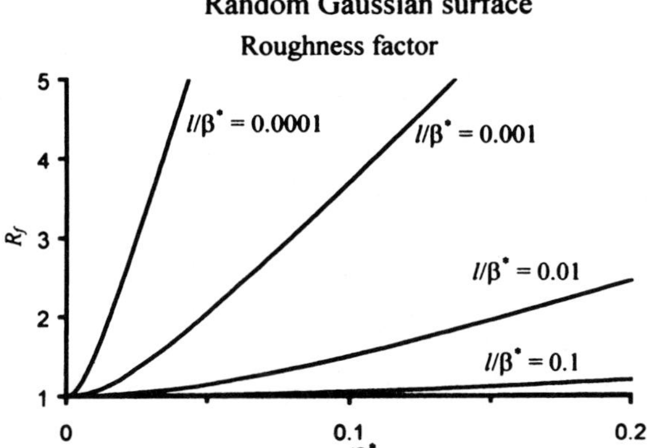

In a similar manner R_f can be calculated for asperities with a cylindrical foundation of height h and hemispherical top of radius r (Fig. 21.9). For each asperity, the area of the surface is given by

$$A_{asp} = 2\pi r^2(1 + h/r) \,, \tag{21.53}$$

whereas the flat projection area is given by πr^2. Under the assumption that asperities are randomly distributed throughout the surface with a density of η asperities per unit area, the total contact surface area is given by

$$A_{SL} = A_F + A_F\eta 2\pi r^2(1 + h/r) - A_F\eta\pi r^2 = A_F\left[1 + \eta\pi r^2(1 + 2h/r)\right] \,. \tag{21.54}$$

The roughness factor is found using (21.7) and (21.54):

$$R_f = 1 + \eta\pi r^2(1 + 2h/r) = 1 + p^2(1 + 2h/r) \,, \tag{21.55}$$

where the packing parameter for asperities with a circular foundation is $p = r\sqrt{\pi\eta}$.

For conical asperities of height h, radius r, and side length $L = \sqrt{h^2 + r^2}$ (Fig. 21.5), it can be obtained in a similar manner:

$$A_{\mathrm{asp}} = \pi r^2 (1 + L/r) \, , \tag{21.56}$$

$$\begin{aligned} A_{\mathrm{SL}} &= A_{\mathrm{F}} + A_{\mathrm{F}} \eta \pi r^2 (1 + L/r) - A_{\mathrm{F}} \eta \pi r^2 \\ &= A_{\mathrm{F}}(1 + \eta \pi r L) = A_{\mathrm{F}} \left(1 + \eta \pi r^2 \sqrt{1 + (h/r)^2} \right) \, , \end{aligned} \tag{21.57}$$

and

$$R_{\mathrm{f}} = 1 + \eta \pi r L = 1 + \eta \pi r^2 \sqrt{1 + (h/r)^2} = 1 + p^2 \sqrt{1 + (h/r)^2} \, , \tag{21.58}$$

where the packing parameter for asperities with a circular foundation is $p = r\sqrt{\pi \eta}$.

For pyramidal asperities with a square foundation of width $2a$ and height h, the corresponding quantities are given as

$$A_{\mathrm{asp}} = 4r^2 \left[1 + \sqrt{1 + (h/r)^2} \right] \, , \tag{21.59}$$

$$\begin{aligned} A_{\mathrm{SL}} &= A_{\mathrm{F}} + 4 A_{\mathrm{F}} \eta r^2 \left[1 + \sqrt{1 + (h/r)^2} \right] - 4 A_{\mathrm{F}} \eta r^2 \\ &= A_{\mathrm{F}} \left(1 + 4 \eta r^2 \sqrt{1 + (h/r)^2} \right) \, , \end{aligned} \tag{21.60}$$

and

$$R_{\mathrm{f}} = 1 + 4 \eta r^2 \sqrt{1 + (h/r)^2} = 1 + p^2 \sqrt{1 + (h/r)^2} \, , \tag{21.61}$$

where the packing parameter for asperities with a square foundation is $p = 2r\sqrt{\eta}$.

The dependence of the contact angle on the normalized radius of the asperities (taken as p) for $\theta_0 = 120°$ and for different ratios of h/r is presented in Fig. 21.10a, determined on the basis of (21.6), (21.52), (21.55), (21.59), and (21.61), for the rectangular, hemispherically topped, conical and pyramidal asperities. It is observed that with an increase of p, the value of the contact angle increases and reaches $180°$. For higher aspect ratios, the increase of θ is faster.

In order to determine the critical values of roughness parameters which correspond to the transition to the composite interface, it should be analyzed whether the local slope can exceed the critical value α_0 and whether the composite interface is likely to remain stable. It is difficult to conduct such an analysis owing to its complexity; however, an estimate can be made using the fact that with increasing average absolute value of the slope of the surface both the local slope increases and the destabilization of the composite interface becomes less likely, since the surface is less smooth. On the basis of this, we assume here that, in a similar manner as for the two-dimensional profiles, the absolute value of the surface slope is responsible for the transition to the composite liquid–solid–air interface, and consider an average absolute value of the slope. For rectangular, hemispherically topped, conical, and pyramidal asperities, the mean absolute value of the slope, m, is equal to the density of the asperities and the flat projection area times the average absolute value of the slope (equal to twice the aspect ratio):

$$m = \eta \pi r^2 (h/r) = \eta \pi h r \, . \tag{21.62}$$

The critical value can be found using a similar approach as in derivation of (21.45)

$$m_0 = \eta \pi h r = \tan(180 - \theta_0) = \tan(-\theta_0) \,. \tag{21.63}$$

From (21.52), (21.55), (21.58), (21.61), and (21.63), it may be shown that, for the selected value of $\theta_0 = 120°$, for rectangular, hemispherically topped, conical, and pyramidal asperities, the contact angle may approach $180°$ before the critical value of the roughness is reached for the values of h/r shown in Fig. 21.10a [35].

The equations developed here are used to calculate the contact angle for a lotus leaf and this is compared with measured data. The lotus leaf has almost hemispherically topped asperities (papillae) which are covered with wax ([34, 50], Fig. 21.1). The static contact angle for a water droplet against a paraffin wax surface was reported by [18] as $104°$ and by [27] as $103°$. On the basis of the data reported by [50], the number of asperities (papillae) can be estimated for the lotus leaf as 3400 per square millimeter ($\eta = 0.0034\,\mu\mathrm{m}^{-2}$), the average radius of the hemispherically topped asperities r is $10\,\mu\mathrm{m}$, and aspect ratio h/r is approximately 1. From (21.6) and (21.27), these values of the parameters correspond to the roughness factor $R_\mathrm{f} \sim 4$ and the contact angle $\theta = 165°$ (using $\theta_0 = 104°$ for wax). During the measurements, conducted in our laboratory, the value of the static contact angle for deionized water on a lotus leaf was found to be $156 \pm 2°$. Neinhuis and Barthlott (1997) reported a contact angle value of $162°$ for a water droplet on a lotus leaf.

21.3.2.2
Random Rough Surface

A nominally flat random rough surface can be considered as a superposition of a flat plane and a two-dimensional random process which is characterized by a height distribution and an autocorrelation function. Many engineering and natural rough surfaces can be characterized by a Gaussian height distribution and an exponential autocorrelation function [4, 5]. In this case, a rough surface is described by only two parameters: the standard deviation of asperity heights, σ, and the correlation length, β^*. The correlation length is a spatial parameter and it can be viewed as a measure of randomness. β^* is responsible for the horizontal scale of the surface, whereas σ is responsible for the vertical scale of the surface. Measured roughness is dependent on the short- and long-wavelength limit of measurement [4, 5].

The absolute value of slope of a Gaussian surface also has a Gaussian distribution with the mean

$$m = \frac{\sigma}{l}\sqrt{\frac{1 - \left[\exp(-\beta^*/l)\right]^2}{\pi}} \,, \tag{21.64}$$

where l is the sampling interval or short-wavelength limit, which is a distance between data points during a measurement [52]. For a surface, the sampling interval is given by a low-wavelength limit of the Gaussian roughness, and is comparable with the atomic dimensions [35].

An element of the area of a surface with slopes of $\partial z/\partial x$ and $\partial z/\partial y$ in x- and y-directions is given by

$$dA = \sqrt{1 + (\partial z/\partial x)^2 + (\partial z/\partial y)^2}\, dx\, dy \ . \tag{21.65}$$

The distribution of $\sqrt{1 + (\partial z/\partial x)^2 + (\partial z/\partial y)^2}$ is not Gaussian in general, but in most applications the slope is small and the mean value of slope m can be taken to calculate the mean value of $\sqrt{1 + (\partial z/\partial x)^2 + (\partial z/\partial y)^2}$. It can also be assumed that the slopes in the x- and y-directions are the same. Using (21.63) and substituting m given by (21.64) and integrating, we can calculate [35] the roughness factor as

$$R_f = 1 + \frac{1}{A_{SL}} \int\!\!\int_{A_f} \sqrt{1 + (\partial z/\partial x)^2 + (\partial z/\partial y)^2}\, dx\, dy$$

$$= 1 + \sqrt{1 + 2m^2} = \sqrt{1 + 2\frac{\sigma^2}{l^2}\frac{1 - \exp(-l/\beta^*)^2}{\pi}} \ . \tag{21.66}$$

For small l/β^*

$$R_f \approx \sqrt{1 + 2\left(\frac{\sigma}{\beta^*}\right)^2 \frac{2}{\pi(l/\beta^*)}} \ . \tag{21.67}$$

Furthermore, for small values of $\sigma^2/(l\beta^*)$

$$R_f \approx 1 + \frac{2}{\pi}\left(\frac{\sigma}{\beta^*}\right)^2\left(\frac{\beta^*}{l}\right) \ . \tag{21.68}$$

In order to estimate the critical value of the roughness parameter, we assume, as in the previous subsection, that the average absolute value of the surface slope is responsible for the transition to the composite solid–liquid–air interface. The absolute value of slope is given by (21.64), so, in a similar manner as in the derivation of (21.63), the critical value of the Gaussian surface roughness parameter is

$$m_0 = \left(\frac{\sigma}{l}\sqrt{\frac{1 - \left[\exp(-\beta^*/l)\right]^2}{\pi}}\right)_0 = \tan(-\theta_0) \ . \tag{21.69}$$

The dependence of the roughness factor on σ/β^* is presented in Fig. 21.10b on the basis of (21.66). The dependence of the contact angle on σ/β^* using the roughness factor is presented in Fig. 21.10b. It is observed that both the roughness factor and the contact angle increase with increasing σ/β^*. From (21.69), it may be shown that for the selected value of $\theta_0 = 120°$, the contact angle may approach $180°$ prior to the critical values of the roughness parameter being reached for the values of σ/β^* and l/β^* shown. It is noted that for most natural and engineering Gaussian surfaces, the ratio σ/β^* is much less than 0.1 and the average value of the slope is small ($m \ll 1$). Therefore, although the roughness is below the critical value, it is difficult to achieve high contact angles with Gaussian random surfaces with a realistic value of σ/β^* [35].

21.3.3
Surface Optimization for Maximum Contact Angle

Among the several types of the surfaces considered in the preceding subsections, the highest contact angles are achieved with the sawtooth profile and rectangular, hemispherically topped, conical, and pyramidal asperities. As stated earlier, the sawtooth profile is undesirable owing to its sharp edges, which may pin the triple line, and because the grooves may reinforce wetting. Therefore, the rectangular, hemispherically topped, conical, and pyramidal asperities should be considered as the most appropriate for producing the highest contact angles. In order to prevent the contact angle hysteresis, it is desirable to avoid asperities with sharp edges, which may cause pinning of the triple line. Therefore, hemispherically topped asperities are the most appropriate. A case will also be made later for pyramidal asperities.

Two-tiered roughness involving two wavelengths has been considered by some authors (e.g., Herminghaus 2000) to decrease wetting, however, it is more likely to involve sharp edges, which are undesirable, and lead to unstable composite solid–liquid–air interface.

On the basis of (21.55), (21.58), and (21.61) and the results shown in Fig. 21.10, the maximum contact angle can be achieved by increasing the aspect ratio h/r and the packing parameter p. The maximum aspect ratio may be achieved by increasing asperity height. The maximum packing parameter may be achieved by packing the asperities as tightly as possible. The square of the packing parameter p^2 is equal to the ratio of the foundation area of the asperities to the total surface area; therefore, a higher value of p corresponds to a higher packing density. For asperities with a circular foundation, the square pattern of the distribution of the asperities results in packing of $1/(2r)$ rows per unit area with $1/(2r)$ asperities per unit length in the row. A higher density of the packing of the asperities can be achieved by a hexagonal distribution of asperities (Fig. 21.11). This distribution pattern results in packing of $1/(\sqrt{3}r)$ rows of asperities per unit length with $1/(2r)$ asperities per unit length in the row, or $\eta = 1/(2\sqrt{3}r^2)$, which yields

$$p = r\sqrt{\pi\eta} = \sqrt{\frac{\pi}{2\sqrt{3}}} \approx 0.952 \,. \tag{21.70}$$

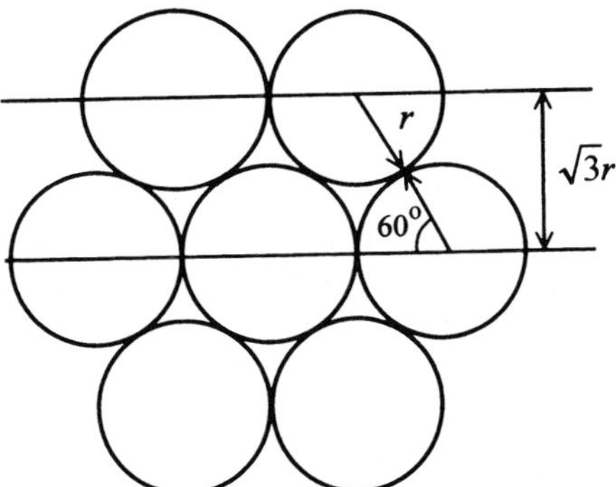

Fig. 21.11. Hexagonal pattern of packing of circular asperities for highest packing density [35]

Therefore, the recommendation for surface optimization is to take hexagonally packed hemispherically topped asperities with a high aspect ratio (needlelike). It is noted that certain leaves tend to have the distribution of the papillae close to hexagonal ([35]; Fig. 21.1).

An alternative shape which provides packing density $p = 1$ is given by pyramidal asperities with a square foundation. In order to avoid pinning due to sharp edges, the tops may be rounded with the hemispheres. Rectangular asperities do not provide space for liquid to penetrate; therefore, in the case of asperities with a square foundation, the pyramidal shape should be used. It should be noted that valleys with rounded edges have the same effect on the contact angle as asperities do [35].

Optimized surfaces

Hemispherically topped cylindrical asperities

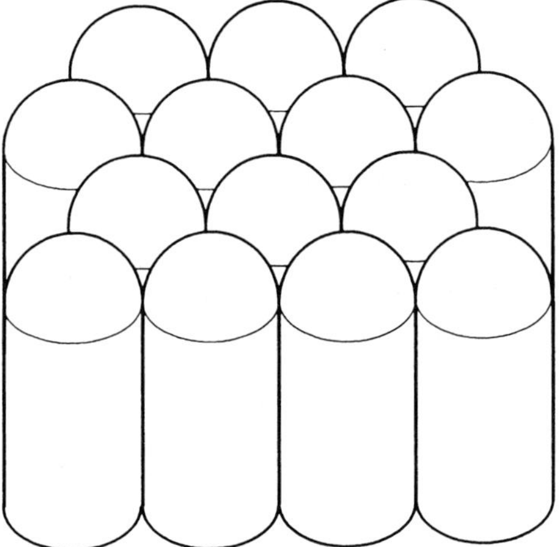

Hemispherically topped pyramidal asperities

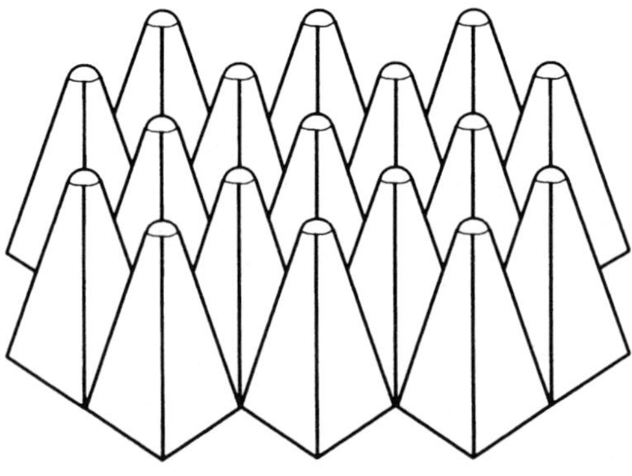

Fig. 21.12. Optimized spaced roughness distribution – hemispherically topped cylindrical asperities and pyramidal asperities with a square foundation and rounded tops. A square base gives higher packing density but introduces undesirable sharp edges [35]

The foundation radius of individual asperities, r (for circular foundation), or foundation side length, $2r$ (for square foundation), should be small compared with typical droplets. The upper limit of the droplet size may be estimated on the basis of the requirement that the gravity effect is small compared with that of the surface tension. The gravitational energy of the droplet is given by its density ρ multiplied by the volume, the gravitational constant $g = 9.81 \text{ m/s}^2$, and the radius r,

$$W_g = \frac{4}{3}\pi r^3 \rho g r \ , \tag{21.71}$$

whereas the energy due to the surface tension can be estimated by the droplet surface area multiplied by the surface tension:

$$W_g = 4\pi r^2 \gamma_{LA} \ . \tag{21.72}$$

From $W_g \ll W_s$, we find the maximum droplet radius which can exist is

$$r_{max} \ll \sqrt{\frac{3\gamma_{LA}}{\rho g}} \ . \tag{21.73}$$

Typical quantities for water of $\rho = 1000 \text{ kg/m}^3$ and $\gamma_{LA} = 72 \text{ mJ/m}^2$ result in $r_{max} \ll 4.7$ mm. Although the small droplets will tend to unite into bigger ones, the minimum droplet radius is limited only by the molecular scale, so it is desirable to have r as small as possible.

To summarize, the highest possible contact angle and the lowest contact angle hysteresis, which is desirable in applications, may be achieved by using hemispherically topped asperities with a hexagonal packing pattern or by using pyramidal asperities with a rounded top. These recommendations can be used for producing superhydrophobic surfaces ([35]; Fig. 21.12).

For wetting liquids, roughness results in a decreased contact angle, in accordance with (21.6). Therefore, in order to create a superhydrophobic surface using the effect of roughness, a hydrophobic film is required. Hydrophobic coating is a well-known method of increasing the water-repellency of a material [43].

21.4
Meniscus Force

When two solids come in contact, a meniscus can form owing to condensation of liquid or because the liquid film may be present initially, if the liquid is wetting. For nonwetting liquids, a meniscus will not be formed. The meniscus causes an increase of the friction force. In this section the effect of surface roughness on the meniscus force will be considered for the case of a sphere in contact with a flat surface (single asperity contact) and for the case of multiple-asperity contact.

21.4.1
Sphere in Contact with a Smooth Surface

Consider a sphere with radius R in contact with a flat surface with a meniscus (Fig. 21.13a). The shape and size of the meniscus, as well as the total energy of the

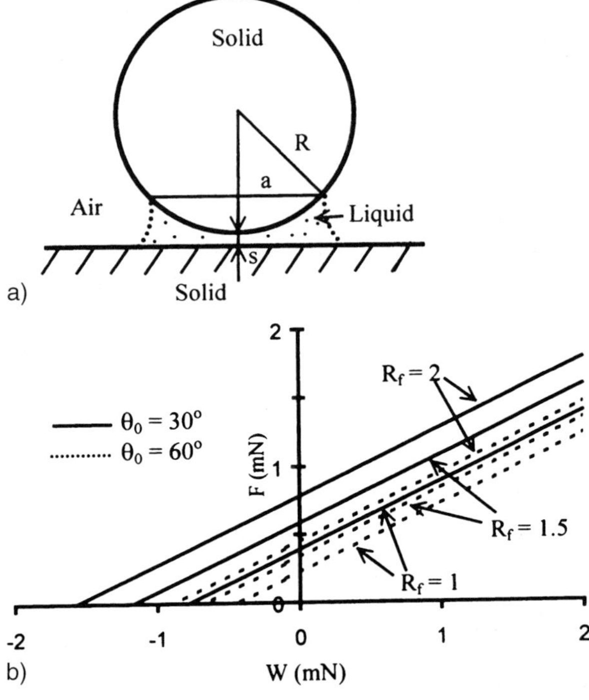

Fig. 21.13. (a) Meniscus formation during wet contact of a flat surface with a sphere. (b) Dependence of the friction force (F) on the normal load (W) for single-asperity contact, $\gamma_{LA} = 0.073\,\mathrm{J/m^2}$ (water), $R = 1$ mm, $\mu = 0.5$, for different values of $R_f = R_{f1} = R_{f2}$, $\theta_0 = \theta_1 = \theta_2$ [35]

system, depend on the separation distance s between the flat surface and the center of the sphere. The normal meniscus force F_m which acts upon the sphere and the flat surface can be calculated as the derivative of the energy E_{tot} by s:

$$F_m = \frac{dE_{\text{tot}}}{ds} \,. \tag{21.74}$$

There are two solid–liquid interfaces, with the sphere and with the flat surface. The areas of these interfaces are approximately equal to πa^2, where a is the meniscus radius. Under the assumption that the ratio a/R is small and $A_{LA} \ll A_{SL}$, the total energy E_{tot} is equal to the sum of the surface energies at the solid–liquid interface. On the basis of these assumptions, E_{tot} can be calculated, using (21.2), as

$$E_{\text{tot}} = \pi a^2 (\gamma_{SL1} - \gamma_{SA1} + \gamma_{SL2} - \gamma_{SA2}) = \pi a^2 \gamma_{LA}(\cos\theta_1 + \cos\theta_2) \,, \tag{21.75}$$

where the indices 1 and 2 correspond to the sphere and the flat surface, and a is defined in Fig. 21.13 [35]. The volume of liquid V is a function of s and a and is given as the sum of the cylindrical volume with height $s + a^2/(2R)$ and cross-sectional area πa^2 minus a volume of the spherical segment of height $a^2/(2R)$:

$$V = \pi a^2 s + \frac{\pi a^4}{4R} \,. \tag{21.76}$$

The volume of the liquid remains constant during the contact, so (21.76) may be viewed as a quadratic equation for a^2, which is solved as

$$a^2 = -2Rs \pm 2R\sqrt{s^2 + V/(\pi R)} \,. \tag{21.77}$$

The derivative of a^2 by s, $\mathrm{d}(a^2)/\mathrm{d}s$ for the sphere touching the plane ($s = 0$) is given as

$$\frac{\mathrm{d}a^2}{\mathrm{d}s} = -2R \; . \tag{21.78}$$

By using the derivative of (21.75) and the expression in (21.78) in (21.74), we get

$$F_\mathrm{m} = 2\pi R \gamma_\mathrm{LA}(\cos\theta_1 + \cos\theta_2) \; . \tag{21.79}$$

Equation (21.79) (also known as the equation for the Laplace pressure) provides us with the value of the normal force due to the meniscus. If the sphere and the surface are rough, with roughness factors of R_f1 and R_f2, respectively, the roughness must be taken into account:

$$F_\mathrm{m} = 2\pi R \gamma_\mathrm{LA1}(R_\mathrm{f1}\cos\theta_1 + R_\mathrm{f2}\cos\theta_2) \; . \tag{21.80}$$

In the presence of a meniscus, the friction force is given by [4,5]

$$F = \mu\,(W + F_\mathrm{m}) \; . \tag{21.81}$$

The coefficient of friction in the presence of the meniscus force, μ_wet, is calculated using only the applied normal load, as normally measured in experiments:

$$\mu_\mathrm{wet} = \mu\left(1 + \frac{F_\mathrm{m}}{W}\right) \; . \tag{21.82}$$

Equation (21.55) shows that μ_wet is greater than μ, because F_m is not taken into account for calculation of the normal load in the wet contact.

The effect of a meniscus on friction force for different surface roughness is presented in Fig. 21.13b. It is observed that a roughness factor of $R_\mathrm{f} = 2$ may result in a significant change of the friction force due to the meniscus. In applications, it is usually desirable to decrease the meniscus forces; therefore, a smooth surface is preferable in the case of a single asperity contact.

21.4.2
Multiple-Asperity Contact

In the case of multiple-asperity contact, a statistical approach is used to model the contact. For a random surface with a certain σ and β^*, the average peak radius R_p and the number of contacts N depend on roughness owing to the so-called scale effect [9]. Bhushan and Nosonovsky (2004) showed that the average peak radius $\overline{R_\mathrm{p}}$ is related to σ and β^*, whereas the number of contacts, for moderate loads, is proportional to the load, divided by σ and β^*:

$$\overline{R_\mathrm{p}} \propto \frac{(\beta^*)^2}{\sigma} \; , \tag{21.83}$$

and

$$N \propto \frac{W}{\sigma\beta^*} \; . \tag{21.84}$$

For asperities of equal peak radius R_p, the meniscus force is given by [4,5,49]

$$F_m = 2\pi R_p \gamma_{LA} (\cos\theta_1 + \cos\theta_2) N \propto \frac{\beta^* W}{\sigma^2} (\cos\theta_1 + \cos\theta_2) \; . \tag{21.85}$$

The size of menisci is comparable with the size of individual contacts [35].

We consider a rough surface which consists of the short-wavelength roughness superimposed over the long-wavelength roughness with a typical size of the roughness smaller than the meniscus size (Fig. 21.14a). The nanoscale roughness of the two bodies is characterized by the roughness factors R_{f1} and R_{f2}. We further assume that asperities have an average peak radius $\overline{R_p}$. Substituting (21.6) into (21.85), we obtain

$$F_m = 2\pi \overline{R_p} \gamma_{LA} (\cos\theta_1 + \cos\theta_2) N \propto \frac{\beta^* W}{\sigma^2} (R_{fn1} \cos\theta_1 + R_{fn2} \cos\theta_2) \; . \tag{21.86}$$

The meniscus force as a function of σ^2/β^*, which is a measure of roughness, is presented in Fig. 21.14b [35]. It is observed that with increasing roughness σ^2/β^*, the meniscus force decreases. A high nanoscale roughness factor may slightly increase the meniscus force. Since it is usually desirable to reduce the meniscus force, a rough surface with high σ^2/β^* is preferable in the case of multiple-asperity contact [35].

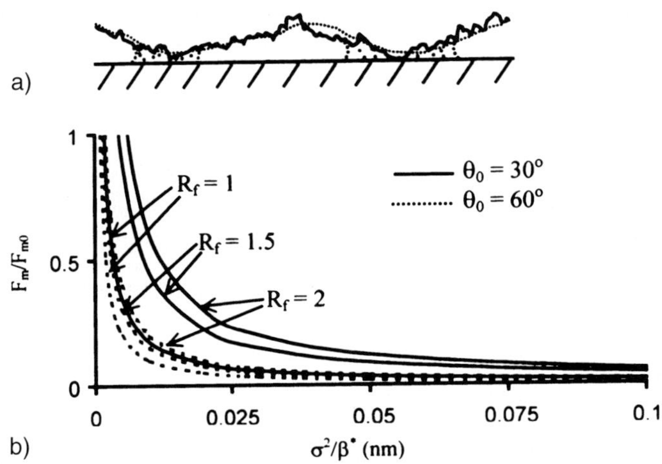

Fig. 21.14. (**a**) Menisci formation during wet contact of a smooth surface with a rough surface with a short-wavelength roughness superimposed over long-wavelength roughness (*dotted line*). (**b**) Dependence of the meniscus force (normalized by F_{m0}, meniscus force value at $\sigma^2/\beta^* = 0.001$ nm) on roughness σ^2/β^*, for different values of $R_f = R_{f1} = R_{f2}$, $\theta_0 = \theta_1 = \theta_2$ [35]

21.5
Experimental Data

Experimental data on natural, artificial and biomimetic rough surfaces which demonstrate the dependence of the contact angle on roughness are discussed in this section.

Burton and Bhushan (2006) measured the roughness of several water-repellent plants, including *N. nucifera* (lotus) and *C. esculenta* using scanning electron microscopy (SEM), optical microscopy, and AFM. Figure 21.15 shows optical profiler images of papillae on the surface of a lotus leaf. It is observed that the papilla surface itself is rough. Asperities on a fresh lotus leaf are too large to be measured by AFM; therefore, measurements have to be made on a dried leaf which does not have tall asperities. Furthermore, the top and the bottom of the leaf have to be measured separately. Figure 21.16 shows AFM scans of a dried lotus leaf. The two-dimensional profiles are taken from the top scan and the bottom scan for each scan size and are spliced together to get the total profile of the leaf [14]. The contact angle with water for lotus was found to be approximately 156°.

Burton and Bhushan (2005) measured the contact angle with artificial patterned poly(methyl methacrylate) (PMMA) surfaces with bumps with a low aspect ratio (LAR; 1:1 height-to-diameter ratio) and a high aspect ratio (HAR; 3:1 height-to-diameter ratio). PMMA was chosen because it is a polymer often used in BioMEMS/BioNEMS devices. The diameter of the asperities near the top is approximately 100 nm and the pitch of the asperities (the distance between each asperity) is approximately 500 nm. Figure 21.17 shows SEM images of the two types of patterned structures, LAR and HAR, on a PMMA surface. According to the model presented earlier, by introducing roughness to a flat surface, the hydrophobicity will either increase or decrease depending on the initial contact angle on a flat surface. The material chosen was initially hydrophilic, so to obtain a sample that is hydrophobic, a self-assembled monolayer (SAM) was deposited on the sample surfaces. The samples chosen for the SAM deposition were the flat film and the HAR for each polymer. The SAM perfluorodecyltriethoxysilane (PFDTES) was deposited on the polymer surface using a vapor phase deposition technique. PFDTES was chosen because of the hydrophobic nature of the surface. It should be noted that the bumps should be as close as possible to provide a large surface area. In order to benefit from an increase in the contact angle and a decrease in the contact area with an increase in surface roughness, the pitch of the bumps should be smaller than the water droplet and the size of the contacting body.

Figure 21.18 shows the static contact angle for various samples. These values correlate well with the model describing roughness with hydrophobicity. The contact

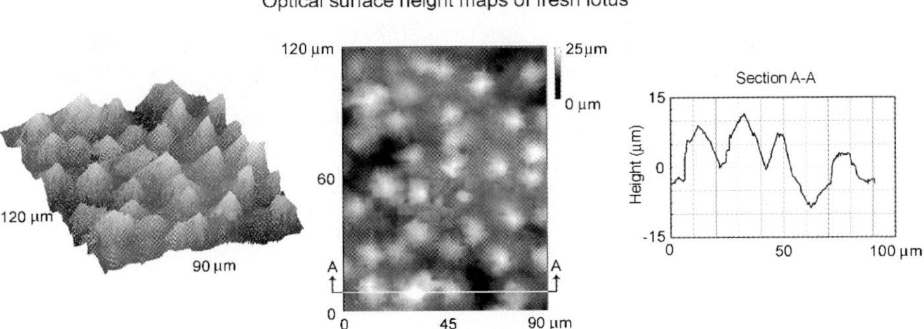

Fig. 21.15. Surface height map and a 2D profile of a lotus leaf using an optical profiler [14]

AFM surface height maps of dried lotus

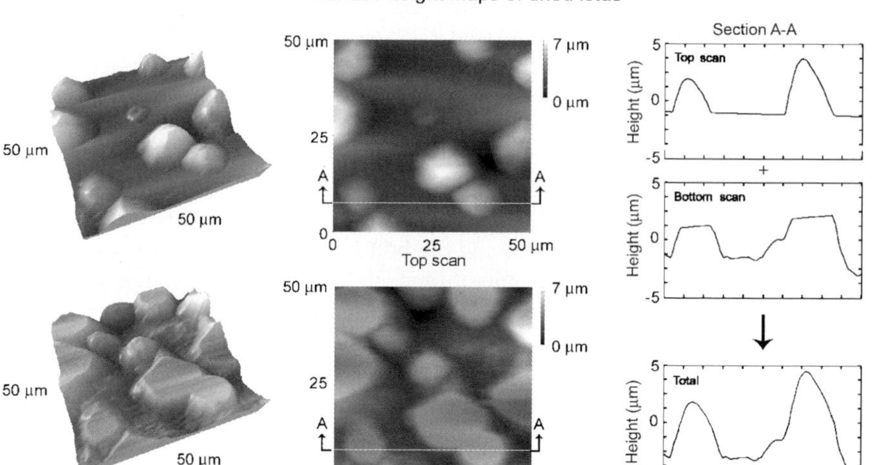

Fig. 21.16. Surface height map and a 2D profile of a lotus leaf using an atomic force microscope (*AFM*) [14]

PMMA low aspect ratio (LAR)

PMMA high aspect ratio (HAR)

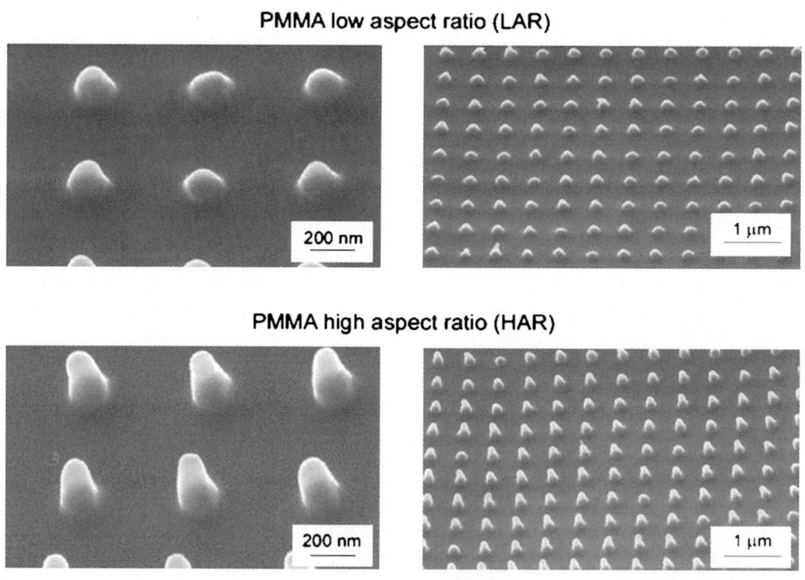

Fig. 21.17. SEM micrograph of the patterned poly(methyl methacrylate) (*PMMA*) surface. Both low aspect ratio (*LAR*) and high aspect ratio (*HAR*) are shown from two magnifications to see the asperity shape and the asperity pattern on the surface [13]

Fig. 21.18. The contact angle for different materials and for different roughness [13]

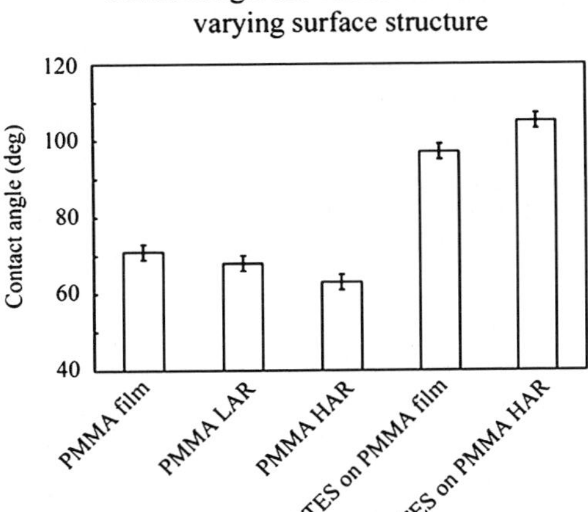

angle decreased with increased roughness for the hydrophilic PMMA film. For a hydrophobic surface, the model predicts an increase of contact angle with roughness, which is what happens when PMMA HAR is coated with PFDTES.

Sun et al. (2005) produced an artificial lotus leaf from PDMS; by nanocasting, making positive and negative replicas of an actual leaf. Flat PDMS has a contact angle with water of about $105°$, comparable with that of wax ($104°$ according to [18] and $103°$ according to [27]). They found a contact angle of $160°$ for the positive leaf replica and only $110°$ for the negative replica. The value for the positive replica is comparable with that for the lotus leaf. Since both positive and negative replicas have the same roughness factor but reversed asperities and valleys, this result suggests that air pockets at the valleys (composite interface) play a main role in increasing the contact angle for lotus leaf.

These experimental data are consistent with the model presented in the preceding sections.

21.6
Closure

Roughness affects the contact angle between a liquid and a solid surface in several ways, such as the increased area of the solid–liquid interface and the effect of sharp edges. For nonwetting liquids, the contact angle with a rough surface is greater than that with a flat surface and may approach $180°$ in the case of superhydrophobic surfaces. This effect is known for water-repellent leaves of plants, such as lotus, which have high roughness and high contact angles, and can be utilized for producing superhydrophobic surfaces. The effect of sharp edges is in pinning of the triple line and it may result in a difference between advancing and receding contact angles (contact angle hysteresis), which impedes fluid flow. This is not desirable in most applications, which require easy fluid flow.

For very rough surfaces, a composite solid–liquid–air interface can form, which affects the contact angle as well as the contact angle hysteresis. The composite interface can be destabilized owing to external perturbations, such as capillary waves. For a system with multiple states of equilibrium, a stochastic model has been developed, which assumes that a certain probability, depending on the net energy of the system, is associated with every state.

On the basis of an analysis of various modeled roughness distributions, a surface with roughness-induced superhydrophobic properties should satisfy the following requirements. First, asperities must have high aspect ratio to provide a large surface area. Second, sharp edges should be avoided, to prevent pinning of the triple line. Third, asperities should be tightly packed in to minimize the distance between them and to avoid destabilization of the composite interface. Fourth, asperities should be small compared with a typical droplet size. And fifth, in the case of a *hydrophilic* surface, a hydrophobic film must be applied in order to have initial $\theta > 90°$. These recommendations can be utilized for producing superhydrophobic surfaces. Remarkably, all these conditions are satisfied by biological water-repellent surfaces, such as leaves: they have tightly packed hemispherically topped papillae with high (on the order of unity) aspect ratios and a wax coating. Verification of the model was conducted using experimental data for the contact angle of a water droplet on a lotus leaf surface and artificial surfaces.

In addition, for wetting liquids, the effect of roughness on meniscus force was considered and it was found that roughness with typical asperity size greater than of the meniscus may decrease the meniscus force, whereas the nanoscale roughness, with typical asperity size smaller than that of the meniscus, may slightly increase the meniscus force, which is usually undesirable.

Experimental measurements on nanopatterned surfaces and replicas are consistent with the model.

References

1. Adamson AV (1990) Physical Chemistry of Surfaces, Wiley, New York
2. Barthlott W, Neinhuis C (1997) Purity of the Sacred Lotus, or Escape from Contamination in Biological Surfaces, Planta 202:1–8
3. Bhushan B (1998) Tribology Issues and Opportunities in MEMS, Kluwer, Dordrecht
4. Bhushan B (1999) Principles and Applications of Tribology, Wiley, New York
5. Bhushan B (2002) Introduction to Tribology, Wiley, New York
6. Bhushan B (2004) Springer Handbook of Nanotechnology, Springer, Berlin, Heidelberg, New York
7. Bhushan B (2005) Nanotribology and Nanomechanics – An Introduction, Springer, Berlin, Heidelberg, New York
8. Bhushan B, Jung YC (2006) Micro- and Nanoscale Characterization of Hydrophobic and Hydrophilic Leaf Surfaces. Nanotechnology 17:2758–2772
9. Bhushan B, Nosonovsky M (2003) Scale Effects in Friction Using Strain Gradient Plasticity and Dislocation-Assisted Sliding (Microslip), Acta Mater 51:4331–4345
10. Bhushan B, Nosonovsky M (2004) Scale Effects in Dry and Wet Friction, Wear, and Interface Temperature, Nanotechnology 15:749–761
11. Bhushan B, Israelachvili JN, Landman U (1995) Nanotribology: Friction, Wear and Lubrication at the Atomic Scale. Nature 374:607–616

12. Bico J, Thiele U, Quéré D (2002) Wetting of Textured Surfaces. Colloids Surf A 206:41–46
13. Burton Z, Bhushan B (2005) Hydrophobicity, Adhesion, and Friction Properties of Nanopatterned Polymers and Scale Dependence for Micro- and Nanoelectromechanical Systems. Nanoletters 5:1607–1613
14. Burton Z, Bhushan B (2006) Surface Characterization and Adhesion and Friction Properties of Hydrophobic Leaf Surfaces. Ultramicroscopy 106:709–719
15. Cassie A, Baxter S (1944) Wettability of Porous Surfaces. Trans Faraday Soc 40:546–551
16. Checco A, Guenoun P, Daillant J (2003) Nonlinear Dependence of the Contact Angle of Nanodroplets on Contact Line Curvatures. Phys Rev Lett 91:186101
17. Chen Y, He B, Lee J, Patankar NA (2005) Anisotropy in the Wetting of Rough Surfaces. J Colloid Interface Sci 281:458–465
18. Craig RG, Berry GC, Peyton FA (1960) Wetting of Poly-(Methyl Methacrylate) and Polystyrene by Water and Saliva. J Phys Chem 64:541–543
19. Eustathopoulos N, Nicholas MG, Drevet B (1999) Wettability at High Temparatures, Pergamon, Amsterdam
20. Erbil HY, Demirel AL, Avci Y (2003) Transformation of a Simple Plastic into a Superhydrophobic Surface. Science 299:1377–1380
21. Eyring H (1964) Statistical Mechanics and Dynamics, Wiley, New York
22. Extrand CW (2002) Model for Contact Angle and Hysteresis on Rough and Ultraphobic Surfaces. Langmuir 18:7991–7999
23. He B, Patankar NA, Lee J (2003) Multiple Equllibrium Droplet Shapes and Design Criterion for Rough Hydrophobic Surfaces. Langmuir 19:4999–5003
24. Herminghaus S (2000) Roughness-Induced Non-Wetting. Europhys Lett 52:165–170
25. Israelachvili JN (1992) Intermolecular and Surface Forces, 2nd edition, Academic, London
26. Johnson RE, Dettre RH (1964) Contact Angle Hysteresis, In: Fowkes FM (ed) Contact Angle, Wettability, and Adhesion. Advances in Chemistry Series 43. American Chemical Society Washington DC, p 112–135
27. Kamusewitz H, Possart W, Paul D (1999) The Relation Between Young's Equilibrium Contact Angle and the Hysteresis on Rough Paraffin Wax Surfaces. Colloids Surf A 156:271–279
28. Kijlstra J, Reihs K, Klami A (2002) Roughness and Topology of Ultra-Hydrophobic surfaces. Colloids Surf A 206:521–529
29. Lafuma A, Quéré D (2003) Superhydrophobic States. *Nature Materials* 2:457–460
30. Landau L, Lifshitz E (1959) Fluid Mechanics, Pergamon, London
31. Marmur A (2003) Wetting on Hydrophobic Rough Surfaces: to Be Heterogeneous or Not to Be? Langmuir 19:8343–8348
32. Marmur A (2004) The Lotus Effect: Superhydrophobicity and Metastability, Langmuir 20:3517–3519
33. Miwa M, Nakajima A, Fujishima A, Hashimoto K, Watanabe T (2000) Effects of the Surface Roughness on Sliding Angles of Water Droplets on Superhydrophobic Surfaces. Langmuir 16:5754–5760
34. Neinhuis C, Barthlott W (1997) Characterization and Distribution of Water-Repellent, Self-Cleaning Plant Surfaces. Ann Bot 79:667–677
35. Nosonovsky M, Bhushan B (2005) Roughness Optimization for Biomimetic Superhydrophobic Surfaces. Microsyst Technol 11:535–549
36. Nosonovsky M, Bhushan B (2006a) Stochastic Model for Metastable Wetting of Roughness-Induced Superhydrophobic Surfaces. Microsyst Technol 12:231–237
37. Nosonovsky M, Bhushan B (2006b) Wetting of Rough Three-Dimensional Superhydrophobic Surfaces. Microsyst Technol 12:273–281
38. Nosonovsky M, Bhushan B (2006c) Hierarchical Roughness Makes Superhydrophobic States Stable. Microelectronic Engineering (in press)

39. Patankar NA (2003) On the Modeling of Hydrophobic Contact Angles on Rough Surfaces. Langmuir 19:1249–1253
40. Patankar NA (2004a) Transition Between Superhydrophobic States on Rough Surfaces. Langmuir 20:7097–7102
41. Patankar NA (2004b) Mimicking the Lotus Effect: Influence of Double Roughness Structures and Slender Pillars. Langmuir 20:8209–8213
42. Quéré D (2004) Surface Wetting: Model Droplets. Nate Mater 3:79–80
43. Satas D (ed) (1991) Coating Technology Handbook, Dekker, New York
44. Semal S, Blake TD, Geskin V, de Ruijter ML, Castelein G, De Coninck J (1999) Influence of Surface Roughness on Wetting Dynamics. Langmuir 15:8765–8770
45. Shibuichi S, Onda T, Satoh N, Tsujii K (1996) Super-Water-Repellent Surfaces Resulting from Fractal Structure. J Phys Chem 100:(1951)2–(1951)7
46. Shuttleworth R, Bailey GLJ (1948) The Spreading of a Liquid over a Rough Solid. Discuss Faraday Soc 3:16–22
47. Sun M, Luo C, Xu L, Ji H, Ouyang Q, Yu D, Chen Y (2005) Artificial Lotus Leaf by Nanocasting. Langmuir 21:8978–8981
48. Swain PS, Lipowsky R (1998) Contact Angles on Heterogeneous Surfaces: a New Look at Cassie's and Wenzel's Laws. Langmuir 14:6772–6780
49. Tian X, Bhushan B (1996) The Micro-Meniscus Effect of a Thin Liquid Film on the Static Friction of Rough Surface Contact. J Phys D: Appl Phys 29:163–178
50. Wagner P, Furstner R, Barthlott W, Neinhuis C (2003) Quantitative Assessment to the Structural Basis of Water Repellency in Natural and Technical Surfaces. J Exp Bot 54:1295–1303
51. Wenzel RN (1936) Resistance of Solid Surfaces to Wetting by Water. Ind Eng Chem 28:988–994
52. Whitehouse DJ, Archard JF (1970) The Properties of Random Surfaces of Significance in Their Contact. Proc R Soc Lond Ser A 316:97–121
53. Yost FG, Michael JR, Eisenmann ET (1995) Extensive Wetting due to Roughness. Acta Metall Mater 45:299–305
54. Zhou XB, De Hosson JTM (1995) Influence of Surface Roughness on the Wetting Angle. Acta Metall Mater 45:299–305

22 Gecko Feet: Natural Attachment Systems for Smart Adhesion

Bharat Bhushan · Robert A. Sayer

22.1
Introduction

Almost 2500 years ago, the ability of the gecko to "run up and down a tree in any way, even with the head downwards" was observed by Aristotle [2]. This phenomenon is not limited to geckos, but occurs in several animals and insects as well. This universal attachment ability will be referred to as reversible adhesion or smart adhesion [15]. Many insects (i.e., flies and beetles) and spiders have been the subject of investigation. Geckos, however, have been the most widely studied owing to the fact that they exhibit the most versatile and effective adhesive known in nature. As a result, the vast majority of this chapter will be concerned with gecko feet.

Although there are over 1000 species of geckos [30,48], the Tokay gecko (*Gekko gecko*) has been the main focus of scientific research [34,41]. The Tokay gecko is the second largest gecko species, attaining lengths of approximately 0.3–0.4 and 0.2–0.3 m for males and females, respectively. They have a distinctive blue or gray body with orange or red spots and can weigh up to 300 g [76]. These geckos have been the most widely investigated species of gecko owing to the availability and size of these creatures.

Even though the adhesive ability of geckos has been known since the time of Aristotle, little was understood about this phenomenon until the late nineteenth century when microscopic hairs covering the toes of the gecko were first noted. The development of electron microscopy in the 1950s enabled scientists to view a complex hierarchical morphology that covers the skin on the gecko's toes. Over the past century and a half, scientific studies have been conducted to determine the factors that allow the gecko to adhere and detach from surfaces at will, including surface structure [3,5,41,59–61,63,80], the mechanisms of adhesion [6–8,19,26, 34,39,59,64,67,73,78], and adhesive strength [3,6,34,38,39,41].

There is great interest among the scientific community to further study the characteristics of gecko feet in the hope that this information can be applied to the production of microsurfaces/nanosurfaces capable of recreating the adhesive forces generated by these lizards. Common man-made adhesives such as tape or glue involve the use of wet adhesives that permanently attach two surfaces. However, replication of the characteristics of gecko feet would enable the development of a superadhesive polymer tape capable of clean, dry adhesion [25,52,53,68,70,71,81]. These reusable adhesives have potential for use in everyday objects such as tape, fasteners, and toys [23] and in advanced technology such as microelectric and space applications [52,81]. Replication of the dynamic climbing and peeling ability of geckos could find use in the treads of wall-climbing robots [51,71].

22.2
Tokay Gecko

22.2.1
Construction of Tokay Gecko

The explanation for the adhesive properties of gecko feet can be found in the surface structure of the skin on the toes of the gecko. The skin comprises a complex fibrillar structure of lamellae (scansors), setae, branches, and spatulae [59]. As shown in Figs. 22.1 and 22.2 [4, 6, 24], the gecko consists of an intricate hierarchy of structures beginning with lamellae, soft ridges that are 1–2 mm in length [59] that are located on the attachment pads (toes) that compress easily so that contact can be made with rough bumpy surfaces. Tiny curved hairs known as setae extend from the lamellae. These setae are typically 30–130 μm in length and 5–10 μm in diameter [34, 59, 60, 80]. At the end of each seta, 100–1000 spatulae (called because of its shape) [34, 59] with a diameter of 0.1–0.2 μm [59] branch out and form the points of contact with the surface. The tips of the spatulae are approximately 0.2–0.3 μm in width [59], 0.5 μm in length, and 0.01 μm in thickness [57].

Several studies have been conducted to determine the number and size of the setae and spatulae of the gecko. Scanning electron microscopy has been employed to visually determine the values listed in Table 22.1. The setal density was originally reported to be 5000 setae per square millimeter by [59]. This value has been used in various scientific studies [6]. On the basis of pictures obtained with a scanning electron microscope (SEM), a more accurate value of about 14,000 setae per square millimeter has been proposed by [63] and verified

Table 22.1. Surface characteristics of Tokay gecko feet

	Size	Density	Adhesive force
Seta	30–130[a–d]/5–10[a–d] length/diameter (μm)	$\sim$ 14,000[f,h] setae/mm^2	194 μN[h]
Branch	20–30[a]/1–2[a] length/diameter (μm)	–	–
Spatula	2–5[a]/0.1–0.2[a,e] length/diameter (μm)	100–1000[a,b] spatulae per seta	–
Tip of spatula	$\sim$ 0.5[a,e]/0.2–0.3[a,d]/$\sim$ 0.01[e] length/width/thickness (μm)	–	11 nN[i]

Young's modulus of surface material for keratin is 1–20 GPa [9, 61]
[a] [59]
[b] [34]
[c] [60]
[d] [80]
[e] [57]
[f] [63]
[g] [5]
[h] [6]
[i] [38]

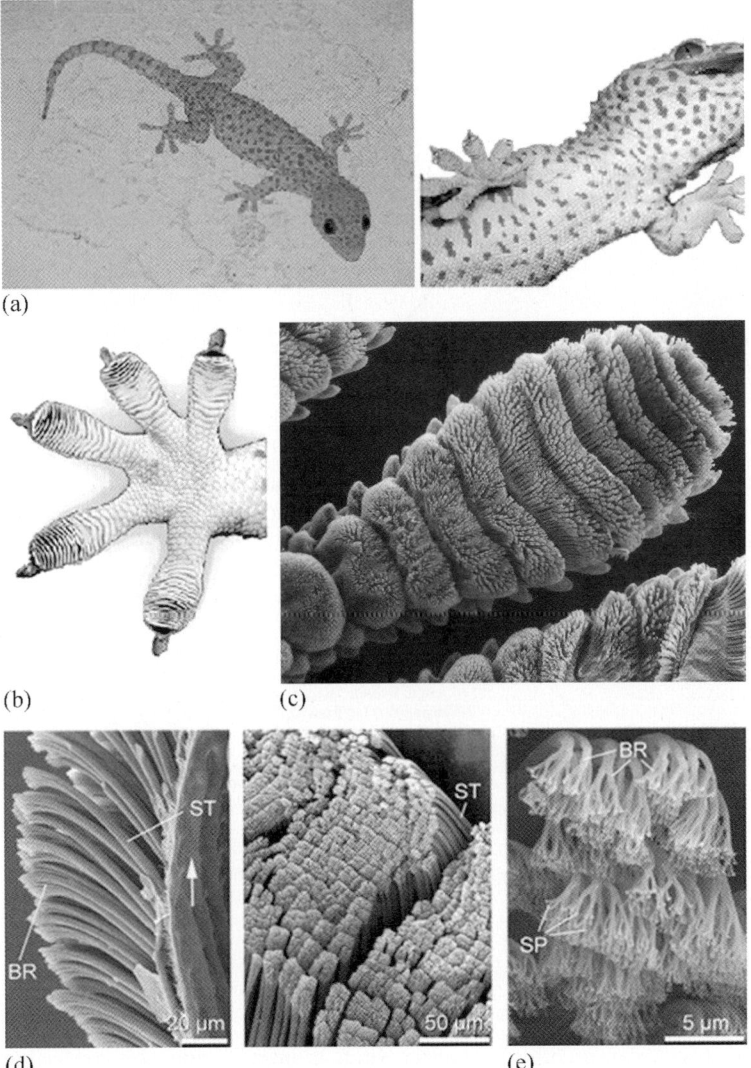

Fig. 22.1. (**a**) A Tokay gecko [6]. The hierarchical structures of a gecko foot; (**b**) a gecko foot [6] and (**c**) a gecko toe [4]. Each toe contains hundreds of thousands of setae and each seta contains hundreds of spatulae. Scanning electron microscope (SEM) micrographs of (**d**) the setae [24] and (**e**) the spatulae [24]. ST seta, SP spatula, BR branch

by [5]. The attachment pads on two feet of the Tokay gecko have an area of about $220\,mm^2$, which can produce a clinging ability of about $20\,N$, the vertical force required to pull a lizard down a nearly vertical ($85°$) surface [41]. In isolated setae a 2.5-μN preload yielded adhesion of 20 to $40\,\mu N$ and thus the adhesion coefficient, which represents the strength of adhesion as a function of preload as 8 to 16 [7].

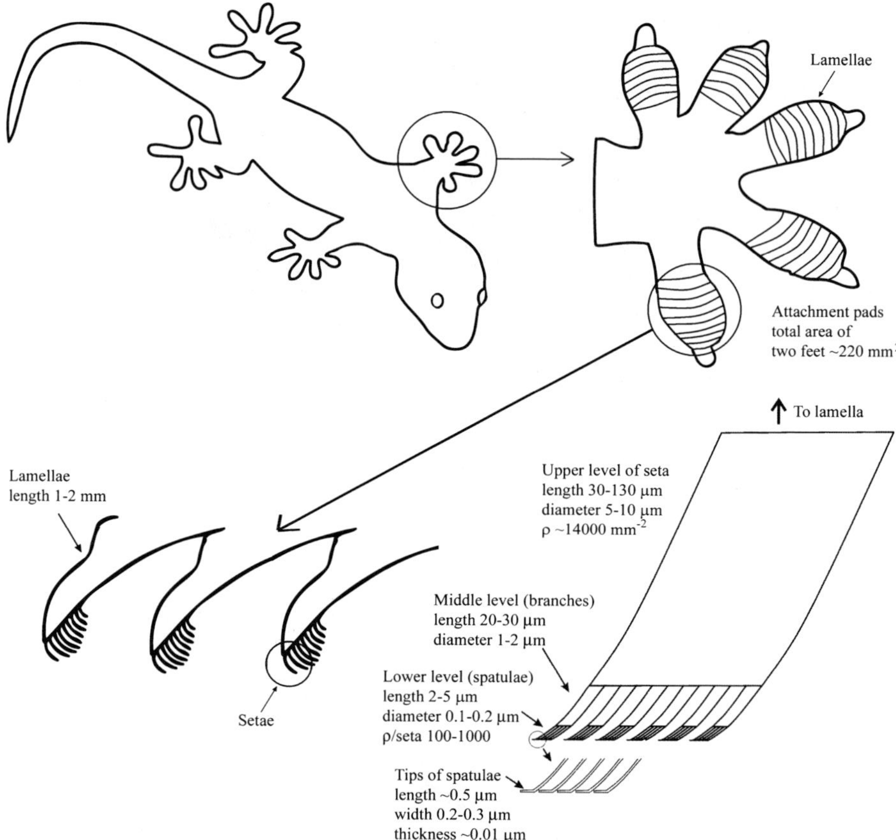

Fig. 22.2. A Tokay gecko including the overall body, one foot, a cross-sectional view of the lamellae, and an individual seta

22.2.2
Other Attachment Systems

Attachment systems in other creatures such as insects and spiders have similar structures to that of gecko skin. The microstructures utilized by beetles, flies, spiders, and geckos can be seen in Fig. 22.3a. As the size (mass) of the creature increases, the radius of the terminal attachment elements decreases. This allows a greater number of setae to be packed into an area, hence increasing the real area of contact and the adhesive strength. It was determined by [3] that the density of the terminal attachment elements, ρ_A, per square meter strongly increases with increasing body mass, m, in kilograms. In fact, a master curve can be fit between all the different species (Fig. 22.3b):

$$\log \rho_A = 13.8 + 0.669 \log m \ . \tag{22.1}$$

The correlation coefficient, r, of the master curve is equal to 0.919. Flies and beetles have the largest attachment pads and the lowest density of terminal attachment

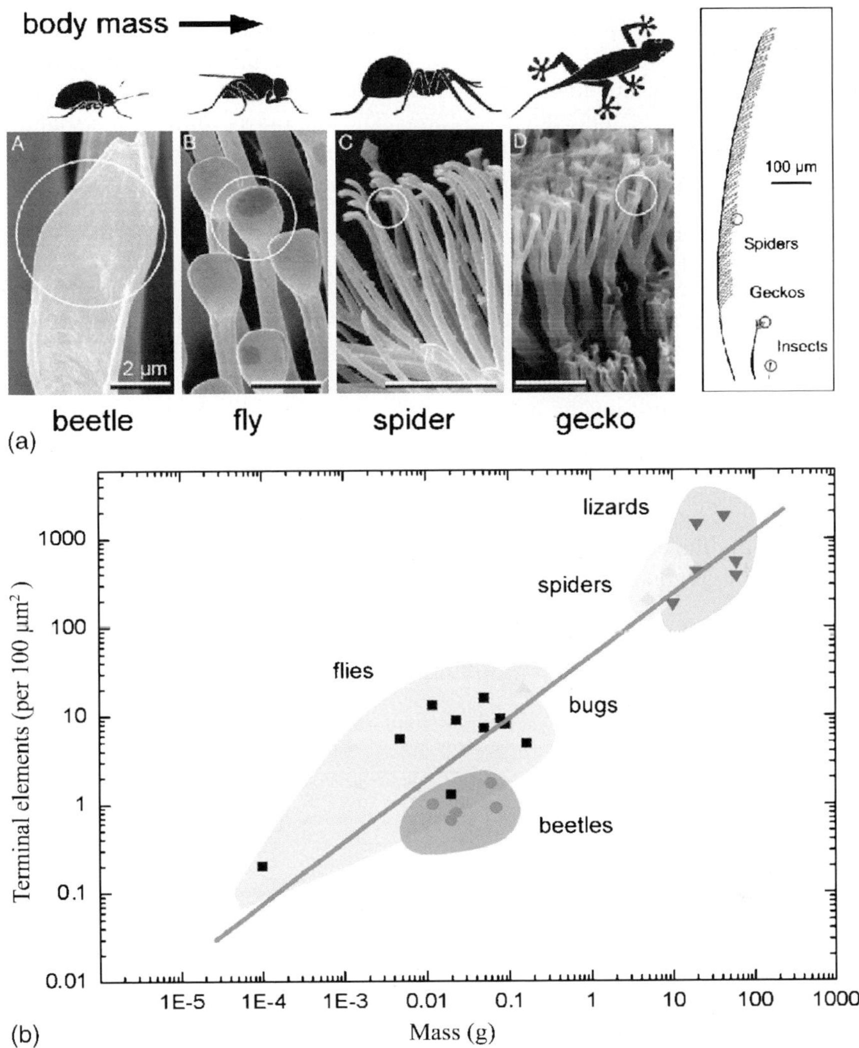

Fig. 22.3. (**a**) Terminal elements of the hairy attachment pads of a beetle, fly, spider, and gecko and (**b**) the dependence of terminal element density on body mass [3]

elements. Spiders have highly refined attachment elements that cover the leg of the spider. Lizards have both the highest body mass and the greatest density of terminal elements (spatulae).

22.2.3
Adaptation to Surface Roughness

Typical rough, rigid surfaces are only able to make intimate contact with a mating surface equal to a very small portion of the perceived apparent area of contact. In fact, the real area of contact, A_r, is typically 2–6 orders of magnitude less than the apparent area of contact, A_a [12, 13]. Autumn et al. [7] proposed that divided contacts serve as a means for increasing adhesion. Arzt et al. [3] used a thermodynamical surface energy approach to calculate adhesive force. The authors assumed that a spatula is a hemisphere with radius R. For calculation of the adhesive force of a single contact, F_a, Johnson–Kendall–Roberts (JKR) theory was used [47]:

$$F_a = -(3/2)\,\pi\gamma R\;,\tag{22.2}$$

where γ is surface energy per unit area. Equation (22.2) shows that adhesive force of a single contact is proportional to a linear dimension of the contact. For a constant area divided into a large number of contacts or setae, n, the radius of a divided contact, R_1, is given by $R_1 = R/\sqrt{n}$; therefore, the adhesive force of (22.2) can be modified for multiple contacts such that

$$F_a' = -(3/2)\,\pi\gamma\left(R/\sqrt{n}\right)n = \sqrt{n}\,F_a\;,\tag{22.3}$$

where F_a' is the total adhesive force from the divided contacts. Thus, the total adhesive force is simply the adhesive force of a single contact multiplied by the square root of the number of contacts. However, this model only considers contact with a flat surface.

On natural rough surfaces the compliance and adaptability of setae are the primary sources of high adhesion. Intuitively, the hierarchical structure of gecko

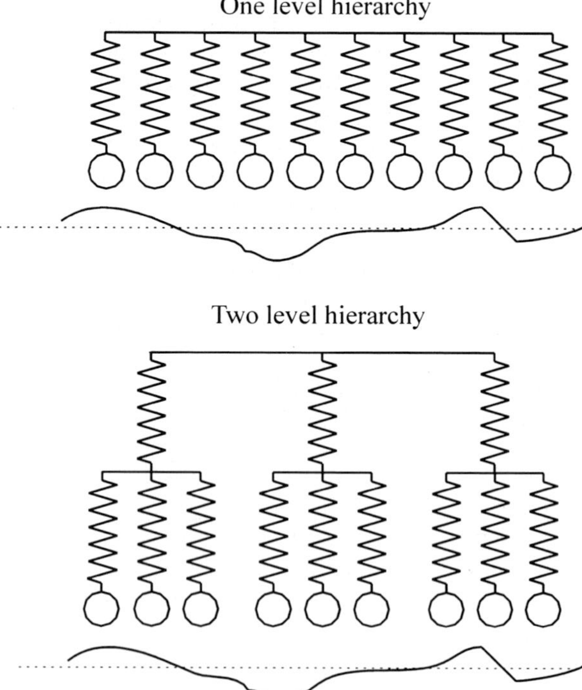

Fig. 22.4. One- and two-level spring models for the simulation of a seta of a Tokay gecko in contact and a random rough surface [15]

setae allows for greater contact with a natural rough surface than a nonbranched attachment system. Two-dimensional profiles of surfaces that a gecko may encounter were obtained using a stylus profiler. These profiles along with the surface-selection methods and surface parameters for scan lengths of 80, 400, and 2000 μm are presented in the "Appendix". Bhushan et al. [15] used the spring model of Fig. 22.4 to simulate the contact between a gecko seta and random rough surfaces similar to those found in the "Appendix". The results of this model suggest that as levels of hierarchy are added to a surface, the adaptation range to the roughness of that surface increases. The lamellae can adapt to the waviness of the surface, while the setae and spatulae allow for the adaptation to microroughness and nanoroughness, respectively. Through the use of the hierarchy of the structures of its skin, a gecko is able to bring a much larger percentage of its skin in contact with the mating surface.

Material properties also play an important role in adhesion. A soft material is able to achieve greater contact with a mating surface than a rigid material (Sect. 22.5.2). Gecko skin is composed of β-keratin, which has a Young's modulus in the range 1–20 GPa [9, 61]. Gecko setae have a Young's modulus much lower than that of the bulk material. Autumn [4] has experimentally determined that setae have an effective modulus of about 100 kPa. By combining optimal surface structure and material properties, Mother Nature has created an evolutionary superadhesive.

22.2.4
Peeling

Although geckos are capable of producing large adhesive forces, they retain the ability to remove their feet from an attachment surface at will by peeling action. Orientation of spatulae facilitates peeling. Autumn et al. [6] were the first to experimentally show that adhesive force of gecko setae is dependent on the three-dimensional orientation as well as the preload applied during attachment (Sect. 22.4.1.1). Owing to this fact, geckos have developed a complex foot motion during walking. First the toes are carefully uncurled during attachment. The maximum adhesion occurs at an attachment angle of 30° – the angle between a seta and the mating surface. The gecko is then able to peel its foot from surfaces one row of setae at a time by changing the angle at which its setae contact a surface. At an attachment angle greater than 30° the gecko will detach from the surface.

Shah and Sitti [65] determined the theoretical preload required for adhesion as well as the adhesive force generated for setal orientations of 30°, 40°, 50°, and 60° in order for a solid material (elastic modulus, E, Poisson's ratio, v) to make contact with the rough surface described by

$$f(x) = h \sin^2\left(\frac{\pi x}{\lambda}\right), \tag{22.4}$$

where h is the amplitude and λ is the wavelength of the roughness profile. For a solid adhesive block to achieve intimate contact with the rough surface neglecting surface forces, it is necessary to apply a compressive stress, S_c [45]

$$S_c = \frac{\pi E h}{2\left(1 - v^2\right)\lambda}. \tag{22.5}$$

Equation (22.5) can be modified to account for fibers oriented at an angle θ. The preload required for contact is summarized in Fig. 22.5a. As the orientation angle decreases, so does the required preload. Similarly, adhesive strength is influenced by fiber orientation. As seen in Fig. 22.5b, the greatest adhesive force occurs at $\theta = 30°$.

Gao et al. [24] created a finite-element model of a single gecko seta in contact with a surface. A tensile force was applied to the seta at various angles, θ, as shown in Fig. 22.5c. For forces applied at an angle less than 30°, the dominant failure mode was sliding. In contrast, the dominant failure mode for forces applied at angles greater than 30° was detachment. This verifies the results of [6] that detachment occurs at attachment angles greater than 30°.

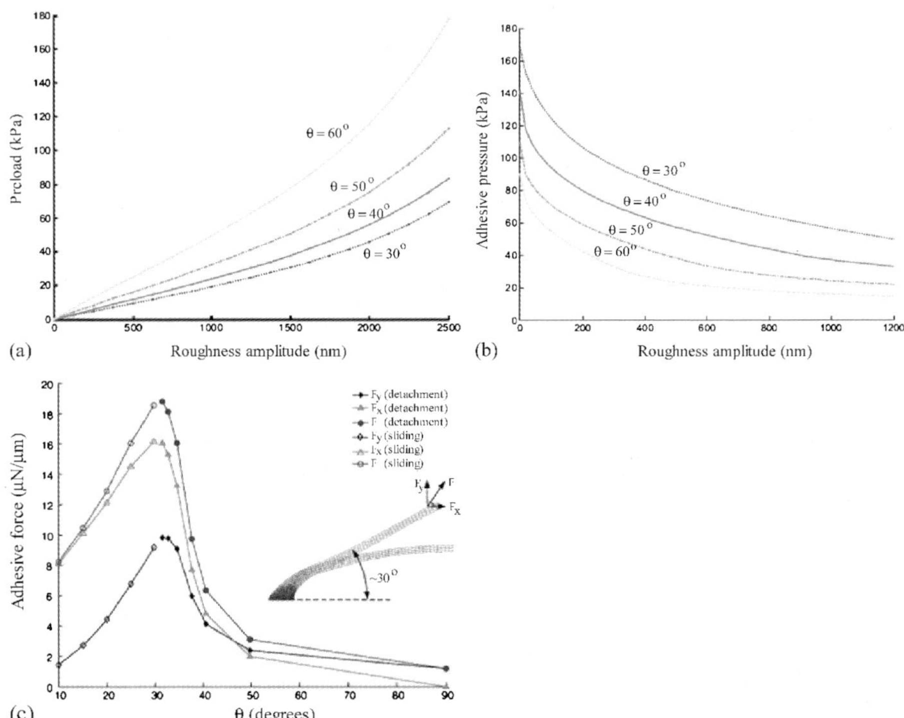

Fig. 22.5. Contact mechanics results for the effect of fiber orientation on (**a**) preload and (**b**) adhesive force for roughness amplitudes ranging from 0 to 2500 nm [65]. (**c**) Finite-element analysis of the adhesive force of a single seta as a function of pull direction [24]

22.2.5
Self-Cleaning

Natural contaminants (dirt and dust) as well as man-made pollutants are unavoidable and have the potential to interfere with the clinging ability of geckos. Particles found in the air consist of particulates that are typically less than 10 μm in diameter while those found on the ground can often be larger [35, 44]. Intuitively, it seems that the great adhesive strength of gecko feet would cause dust and other particles to become trapped in the spatulae and that they would have no way of being removed without some sort of manual cleaning action on the part of the gecko. However, geckos are not known to groom their feet like beetles [74] nor do they secrete sticky fluids to remove adhering particles like ants [22] and tree frogs [31], yet they retain adhesive properties. One potential source of cleaning is during the time when the lizards undergo molting, or the shedding of the superficial layer of epidermal cells. However, this process only occurs approximately once per month [77]. If molting were the sole source of cleaning, the gecko would rapidly lose its adhesive properties as it is exposed to contaminants in nature [32]. Hansen and Autumn [32] tested the hypothesis that gecko setae become cleaner with repeated use – a phenomenon known as self-cleaning.

The cleaning ability of gecko feet was first tested experimentally. The test procedures employed by [32] will be summarized. Silica–alumina ceramic microspheres of 2.5-μm radius were applied to clean setal arrays. Figure 22.6a shows the setal arrays immediately after dirtying and after five simulated steps. It can be noticed that a significant fraction of the particles has been removed after five steps as compared with the original dirtied arrays. The maximum shear stress that these "dirty" arrays could withstand was measured using a piezoelectric force sensor. After each step that the gecko took, the shear stress was once again measured. As seen in Fig. 22.6b, after only four steps, the gecko foot is clean enough to withstand its own body weight.

In order to understand this cleaning process, substrate–particle interactions must be examined. The interaction energy between a dust particle and a wall and spatulae can be modeled as shown in Fig. 22.7. The interaction between a spherical dust particle and the wall, W_{pw}, can be expressed as [43]

$$W_{\mathrm{pw}} = \frac{-A_{\mathrm{pw}} R_{\mathrm{p}}}{6 D_{\mathrm{pw}}} , \qquad (22.6)$$

where p and w refer to the particle and wall, respectively. A is the Hamaker constant, R_{p} is the radius of the particle, and D_{pw} is the separation distance between the particle and the wall. Similarly, the interaction energy between a spherical dust particle and a spatula, s, assuming that the spatula tip is spherical, is [43]

$$W_{\mathrm{ps}} = \frac{-A_{\mathrm{ps}} R_{\mathrm{p}} R_{\mathrm{s}}}{6 D_{\mathrm{ps}} (R_{\mathrm{p}} + R_{\mathrm{s}})} . \qquad (22.7)$$

The ratio of the two interaction energies, N, can be expressed as

$$N = \frac{W_{\mathrm{pw}}}{W_{\mathrm{ps}}} = \left(1 + \frac{R_{\mathrm{p}}}{R_{\mathrm{s}}}\right) \frac{A_{\mathrm{pw}} D_{\mathrm{ps}}}{A_{\mathrm{ps}} D_{\mathrm{pw}}} . \qquad (22.8)$$

When the energy required to separate a particle from the wall is greater than that required to separate it from a spatula, self-cleaning will occur. For example, if

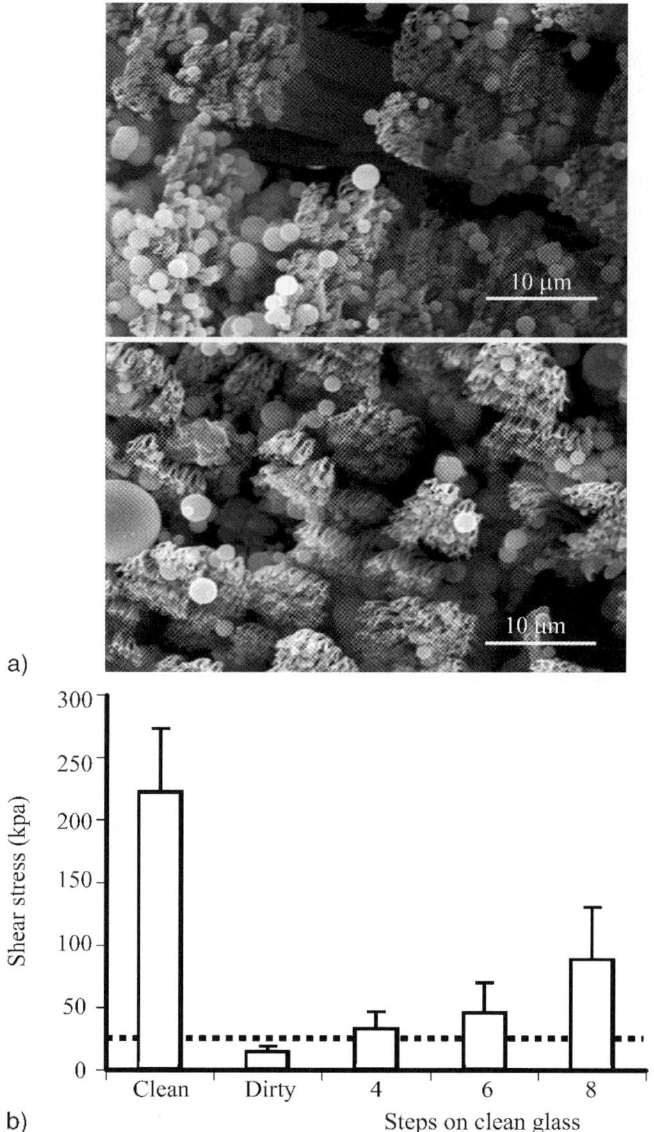

Fig. 22.6. (a) SEM images of spatulae after dirtying with microspheres (*top*) and after five simulated steps (*bottom*). (b) Mean shear stress exerted by a gecko on a surface after dirtying. The *dotted line* represents sufficient recovery to support body weight by a single toe [32]

$R_p = 2.5\,\mu\text{m}$ and $R_s = 0.1\,\mu\text{m}$ [59, 80], self-cleaning will occur as long as no more than 26 spatulae are attached to the dust particle at one time assuming similar Hamaker constants and gap distances. The maximum number of spatulae, as well as the percentage of available spatulae, in contact with a particle for self-cleaning to occur is tabulated in Table 22.2. It can be seen that very small particles (less than

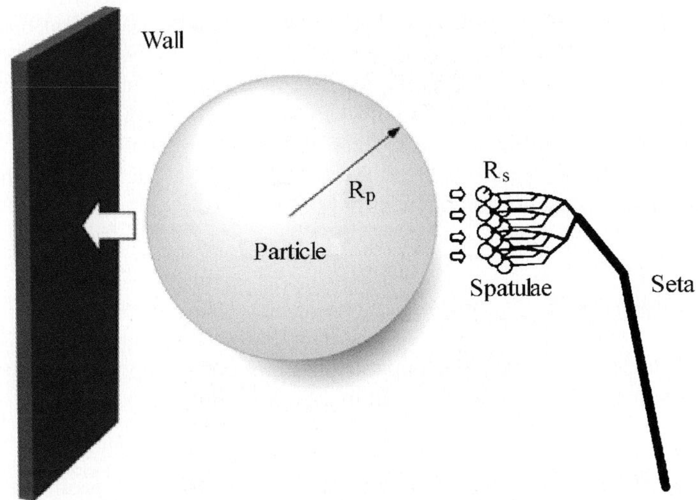

Fig. 22.7. Model of interactions between gecko spatulae of radius R_s, a spherical dirt particle of radius R_p, and a planar wall that enables self-cleaning [32]

Table 22.2. Maximum number of spatulae that can be in contact with a contaminant particle in order for self-cleaning to occur

Radius of particle (μm)	Maximum no. of spatulae in contact with particle	Area of sphere (μm^2)	Spatulae available	Percentage needed to adhere
0.1	2	0.03	0.25	804
0.5	6	0.79	6.2	96
1	11	3	25	44
2.5	26	19	156	17
5	51	79	622	8
10	101	314	2488	4
20	201	1257	9952	2

Spatula radius, R_s, is 0.1 μm

0.5-μm diameter) do not come into contact with enough spatulae to adhere. Owing to the curvature of larger particles relative to the planar field of the spatulae, very few spatulae are able to come into contact with the particle. As a result, Hansen and Autumn [32] concluded that self-cleaning should occur for all spherical spatulae interacting with all spherical particles.

22.3
Attachment Mechanisms

When asperities of two solid surfaces are brought into contact with each other, chemical and/or physical attractions occur. The force developed that holds the two surfaces together is known as adhesion. In a broad sense, adhesion is considered to be either physical or chemical in nature [10–13, 16, 37, 43, 83]. Chemical interactions such as electrostatic attraction charges [64] as well as intermolecular forces [34] including van der Waals and capillary forces have all been proposed as potential adhesion mechanisms in gecko feet. Others have hypothesized that geckos adhere to surfaces through the secretion of sticky fluids [67, 78], suction [67], increased frictional force [36], and microinterlocking [19].

22.3.1
Unsupported Adhesive Mechanisms

Several of the aforementioned mechanisms have not been supported in testing. The rejected mechanisms of adhesion are summarized in Table 22.3.

Table 22.3. Proposed mechanisms of adhesion utilized by gecko feet and experimental evidence against and in favor of the proposed theories

Unsupported adhesive mechanisms			
Mechanism	Proposed by	Experimental evidence against	Disproven by
Secretion of sticky fluids	N/A	Geckos lack glands on their toes that produce sticky fluids capable of adhesion	[67, 78]
Suction	[67]	The adhesive force of a gecko is not affected in high-vacuum experiments	[19]
Electrostatic attraction	[64]	Geckos are able to adhere to surfaces in ionized air (which would eliminate electrostatic attraction)	[19]
Increased frictional force	[36]	Observations that a gecko can adhere upside down, even though frictional force only acts parallel to a surface	Numerous observers
Micro-interlocking	[19]	Measurements of large adhesive forces of a gecko seta on molecularly smooth SiO_2	[6]
Supported adhesive mechanisms			
Mechanism	Proposed by	Experimental evidence for	Supported by
Van der Waals forces (primary)	[34]	Overall and setal adhesion matches the theoretical adhesion values predicted by van der Waals forces	[6]
Capillary forces (secondary)	[34]	Adhesive force of a single gecko spatula was affected by relative humidity present in the air	[39]

22.3.1.1
Secretion of Sticky Fluids

Although several insects and frogs rely on sticky fluids to adhere to surfaces, geckos lack glands on their toes capable of producing these fluids [67, 78]. As a result, this hypothesis has been ruled out.

22.3.1.2
Suction

Simmermacher [67] proceeded to propose that geckos make use of miniature suction cups as an adhesive mechanism. Suction cups operate under the principle of microcapillary evacuation. When a suction cup comes into contact with a surface, air is forced out of the contact area, creating a pressure differential. The adhesive force generated is simply the pressure differential multiplied by the apparent area of contact [10].

Suction cups lose their adhesive strength when used under high-vacuum conditions since a pressure differential can no longer be developed. This mechanism of adhesion can be easily investigated by comparing adhesive force under a vacuum to adhesive force under atmospheric conditions. Experiments carried out in a vacuum by [19] did not show a difference between the adhesive force under these conditions compared with ambient conditions, thus rejecting suction as an adhesive mechanism.

22.3.1.3
Electrostatic Attraction

Electrostatic attraction occurs when two dissimilar heteropolar surfaces come in contact. Electrostatic forces are produced by one or more valence electrons transferring completely from one atom to another. When the separation between two surfaces is approximately equal to the atomic spacing (0.3 nm), the bond generated is quite strong and resembles that within the bulk material. If an insulator (e.g., gecko setae) is brought into contact with a metal, there is a large separation of charge at the interface that produces an electrostatic attraction [18, 20, 46, 72, 79]. Rubbing action during activities such as walking and running would increase the fraction of charged surface area. This is often referred to as the "triboelectric" effect [33, 66].

Schmidt [64] proposed electrostatic attraction as the mechanism of adhesion used by the gecko attachment system. Dellit [19] conducted experiments to determine the electrostatic contribution to gecko adhesion. This testing utilized X-ray bombardment to create ionized air and hence eliminate electrostatic attraction. It was determined that geckos were still able to adhere to surfaces in these conditions and, therefore, electrostatic charges could not be the sole cause of attraction.

22.3.1.4
Increased Frictional Force

It has also been postulated that the adhesive strength of gecko attachment pads arises from a large frictional force due to a large real area of contact [36]. The hierarchical structure of lamellae–setae–branches–spatulae enables a gecko to create a real area of contact with a mating surface that is orders of magnitude greater than that with a nondivided surface. Since the coefficients of static and kinetic friction are dependent on contact area [12, 13], a large real area of contact would cause a large coefficient of friction. Under this theory, a gecko would be able to climb vertical walls if the frictional force exceeded the weight of the lizard. Although large frictional forces could enable geckos to walk up vertical surfaces, they would not account for a gecko's ability to cling to surfaces upside down. Frictional force only acts in the direction parallel to the contact surface, yet to hang upside down an adhesive force is required perpendicular to the surface. As a result, frictional force has been discounted as a potential mechanism.

22.3.1.5
Microinterlocking

Dellit [19] proposed that the curved shape of setae act as microhooks that catch on rough surfaces. This process known as microinterlocking would allow geckos to attach to rough surfaces. Autumn et al. [6] demonstrated the ability of a gecko to generate large adhesive forces when in contact with a molecularly smooth SiO_2 microelectromechanical system (MEMS) semiconductor. Since surface roughness is necessary for microinterlocking to occur, it has been ruled out as a mechanism of adhesion.

22.3.2
Supported Adhesive Mechanisms

Two mechanisms, van der Waals forces and capillary forces, remain as the potential sources of gecko adhesion. These attachment mechanisms are described in detail in the following sections and are summarized in Table 22.3.

22.3.2.1
Van der Waals Forces

Van der Waals bonds are secondary bonds that are weak in comparison with other physical bonds such as covalent, hydrogen, ionic, and metallic bonds. Unlike other physical bonds, van der Waals forces are always present regardless of separation and are effective from very large separations (approximately 50 nm) down to atomic separation (approximately 0.3 nm). The van der Waals force per unit area between two parallel surfaces, f_{vdW}, is given by [29, 42, 43]

$$f_{vdW} = \frac{A}{6\pi D^3} \quad \text{for } D < 30\,\text{nm} , \tag{22.9}$$

where A is the Hamaker constant and D is the separation between surfaces.

Hiller [34] showed experimentally that the surface energy of a substrate is responsable for Gecko adhesion. One potential adhesive mechanism would then be van der Waals forces [6, 73]. Assuming van der Waals forces to be the dominant adhesive mechanism utilized by geckos, we can calculate the adhesive force of a gecko.

Typical values of the Hamaker constant range from 4×10^{-20} to 4×10^{-19} J [43]. In the calculation, the Hamaker constant is assumed to be 10^{-19} J, the surface area of a spatula is taken to be 2×10^{-14} m^2 [5, 59, 80], and the separation between the spatula and the contact surface is estimated to be 0.6 nm. This equation yields the force of a single spatula to be about 0.5 μN. By applying the surface characteristics of Table 22.1, we find the maximum adhesive force of a gecko to be 150–1500 N for varying spatula density of 100–1000 spatulae per seta. If an average value of 550 spatulae per seta is used, the adhesive force of a single seta is approximately 270 μN, which is in agreement with the experimental value obtained by Autumn et al. [6], which will be discussed in Sect. 22.4.1.1.

Another approach to calculate adhesive force is to assume that spatulae are cylinders that terminate in hemispherical tips. By using (22.2) and assuming that the radius of each spatula is about 100 nm and that the surface energy is expected to be 50 mJ/m^2 [3], we can predict the adhesive force of a single spatula to be 0.02 μN. This result is an order of magnitude lower than the first approach calculated for the higher value of A. For a lower value of 10^{-20} J for the Hamaker constant, the adhesive force of a single spatula is comparable to that obtained using the surface-energy approach.

Several experimental results favor van der Waals forces as the dominant adhesive mechanism, including temperature testing [8] and adhesive-force measurements of a gecko seta with both hydrophilic and hydrophobic surfaces [6]. These data will be presented in Sects. 22.4.2–22.4.4.

22.3.2.2
Capillary Forces

It has been hypothesized that capillary forces that arise from liquid-mediated contact could be a contributing or even the dominant adhesive mechanism utilized by gecko spatulae [3, 73]. Experimental adhesion measurements (Sects. 22.4.3 and 22.4.4) conducted on surfaces with different hydrophobicities and at various humidities [39] support this hypothesis as a contributing mechanism. During contact, any liquid that wets or has a small contact angle on surfaces will condense from vapor in the form of an annular-shaped capillary condensate. Owing to the natural humidity present in the air, water vapor will condense to liquid on the surface of bulk materials. During contact this will cause the formation of adhesive bridges (menisci) owing to the proximity of the two surfaces and the affinity of the surfaces for condensing liquid [21, 58, 82].

Capillary forces can be divided into two components: a meniscus force from surface tension and a rate-dependent viscous force [11–13]. The total adhesive force is simply the sum of the two components. The meniscus contribution to adhesion between a spherical surface and a flat plate, F_{M}, is given by [50]

$$F_{\mathrm{M}} = 2\pi R \gamma_{\mathrm{l}} \left(\cos\theta_1 + \cos\theta_2 \right) , \tag{22.10}$$

where R is the radius of the sphere, γ_{l} is the surface tension of the liquid, and θ_1 and θ_2 are the contact angles of the sphere and the plate, respectively. It should be noted that meniscus force is independent of film thickness. Consequently, even a film as thin as a single monolayer can significantly influence the attraction between two surfaces [10–12, 43].

The viscous component of liquid-mediated adhesion is given by [50]

$$F_v = \frac{\beta \eta_l}{t_s} \,, \tag{22.11}$$

where β is a proportionality constant, η_l is the dynamic viscosity of the liquid, and t_s is the time to separate the two surfaces. t_s is inversely related to the velocity of the interface during detachment. Furthermore, the fluid quantity has a weak dependence on viscous force.

22.4
Experimental Adhesion Test Techniques and Data

Experimental measurements of the adhesive force of a single gecko seta [6] and a single gecko spatula [38, 39] have been made. The effect of the environment, including temperature [8, 49] and humidity [39], has been studied. Some of the data have been used to understand the adhesive mechanism utilized by the gecko attachment system – van der Waals or capillary forces. The majority of experimental results point towards van der Waals forces as the dominant mechanism of adhesion [6, 8]. Recent research suggests that capillary forces can be a contributing adhesive factor [39].

22.4.1
Adhesion Under Ambient Conditions

Two feet of a Tokay gecko are capable of producing about 20 N of adhesive force with a pad area of about 220 mm^2 [41]. Under the assumption that there are 14,000 setae per square millimeter, the adhesive force from a single hair should be approximately 7 µN. It is likely that the magnitude is actually greater than this value because it is unlikely that all setae are in contact with the mating surface [6]. Setal orientation greatly influences adhesive strength. This dependency was first noted by Autumn et al. [6]. It was determined that the greatest adhesion occurs at 30°. In order to determine the adhesive mechanism(s) utilized by gecko feet, it is important to know the adhesive force of a single seta; hence, the adhesive force of gecko foot hair has been the focus of several investigations [6, 38].

22.4.1.1
Adhesive Force of a Single Seta

Autumn et al. [6] used both a MEMS force sensor and a wire as a force gauge to determine the adhesive force of a single seta. The MEMS force sensor is a dual-axis atomic force microscope (AFM) cantilever with independent piezoresistive sensors which allows simultaneous detection of vertical and lateral forces [17].

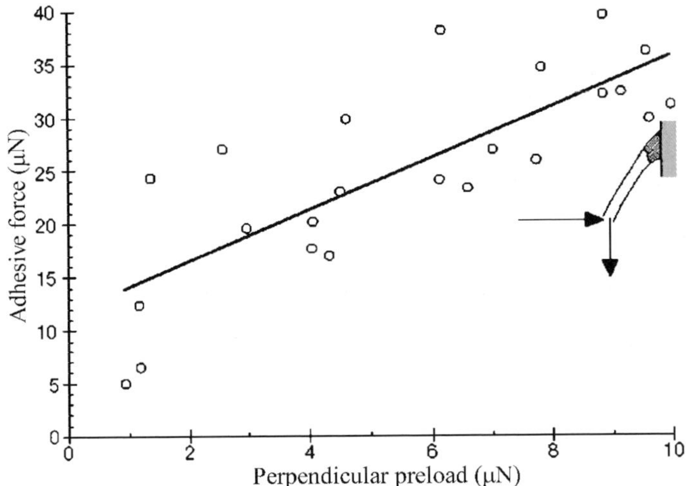

Fig. 22.8. Adhesive force of a single gecko seta as a function of applied preload. The seta was first pushed perpendicularly against the surface and then pulled parallel to the surface [6]

The wire force gage consisted of an aluminum bonding wire which displaced under a perpendicular pull. Autumn et al. [6] discovered that setal force actually depends on the three-dimensional orientation of the seta as well as on the preloading force applied during initial contact. Setae that were preloaded vertically to the surface exhibited only one tenth of the adhesive force ($0.6 \pm 0.7\,\mu$N) compared with setae that were pushed vertically and then pulled horizontally to the surface ($13.6 \pm 2.6\,\mu$N). The dependence of adhesive force of a single gecko spatula on the perpendicular preload is illustrated in Fig. 22.8. The adhesive force increases linearly with the preload, as expected [10–12]. The maximum adhesive force of a single gecko foot hair occurred when the seta was first subjected to a normal preload and then slid $5\,\mu$m along the contacting surface. Under these conditions, the adhesive force measured $194\pm25\,\mu$N (approximately 10 atm adhesive pressure).

22.4.1.2
Adhesive Force of a Single Spatula

Huber et al. [38] used atomic force microscopy to determine the adhesive force of individual gecko spatulae. A seta with four spatulae was glued to an AFM tip. The seta was then brought into contact with a surface and a compressive preload of 90 nN was applied. The force required to pull the seta off of the surface was then measured. As seen in Fig. 22.9, there are two distinct peaks on the graph – one at 10 nN and the other at 20 nN. The first peak corresponds to one of the four spatulae adhering to the contact surface, while the peak at 20 nN corresponds to two of the four spatulae adhering to the contact surface. The average adhesive force of a single spatula was found to be 10.8 ± 1 nN. The measured value is in agreement with the measured adhesive strength of an entire gecko (on the order of 10^9 spatulae on a gecko).

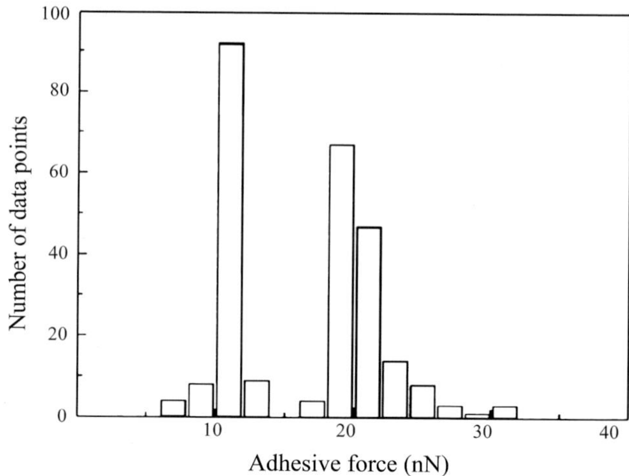

Fig. 22.9. Adhesive force of a single gecko spatula. The peak at 10 nN corresponds to the adhesive force of one spatula and the peak at 20 nN corresponds to the adhesive force of two spatulae [39]

22.4.2
Effects of Temperature

Environmental factors are known to affect several aspects of vertebrate function, including speed of locomotion, digestion rate, and muscle contraction, and as a result several studies have been completed to investigate environmental impact on these functions. Relationships between the environment and other properties such as adhesion have been far less studied [8]. Only two known studies exist that examine the effect of temperature on the clinging force of the gecko [8, 49]. Losos [49] examined adhesive ability of large live geckos at temperatures up to 17 °C. Bergmann and Irschick [8] expanded upon this research for body temperatures ranging from 15 to 35 °C. The geckos were incubated until their body temperature reached the desired level. The clinging ability of these animals was then determined by measuring the maximum force exerted by the geckos as they were pulled off a custom-built force plate. The clinging force of a gecko for the experimental test range is plotted in Fig. 22.10. It was determined that variation in temperature is not statistically significant for the adhesive force of a gecko. From these results, it was concluded that the temperature independence of adhesion supports the hypothesis of clinging as a passive mechanism (i.e., van der Waals forces). Both studies only measured overall clinging ability on the macroscale. There have not been any investigations into the effects of temperature on the clinging ability of a single seta on the microscale and therefore testing in this area is extremely important.

22.4.3
Effects of Humidity

Huber et al. [39] employed similar methods to Huber et al. [38] (discussed in Sect. 22.4.1.2) in order to determine the adhesive force of a single spatula at varying humidity. Measurements were made using an AFM placed in an air-tight chamber.

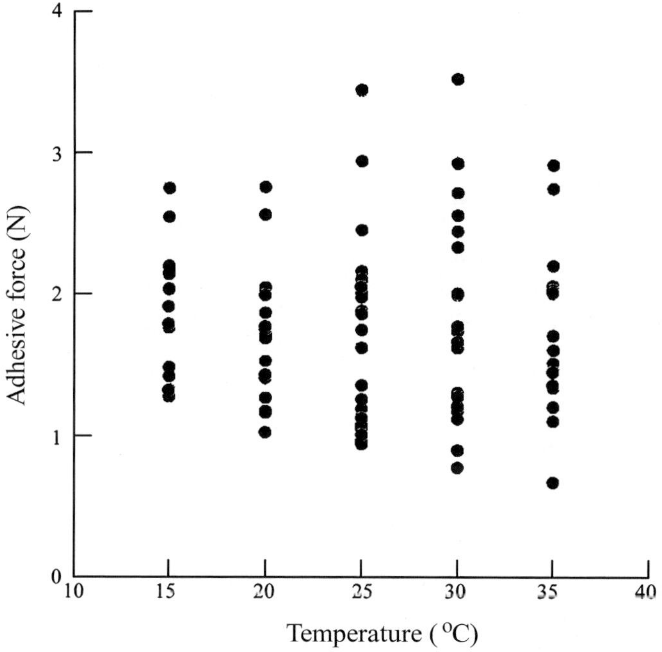

Fig. 22.10. Adhesive force of a gecko as a function of temperature [8]

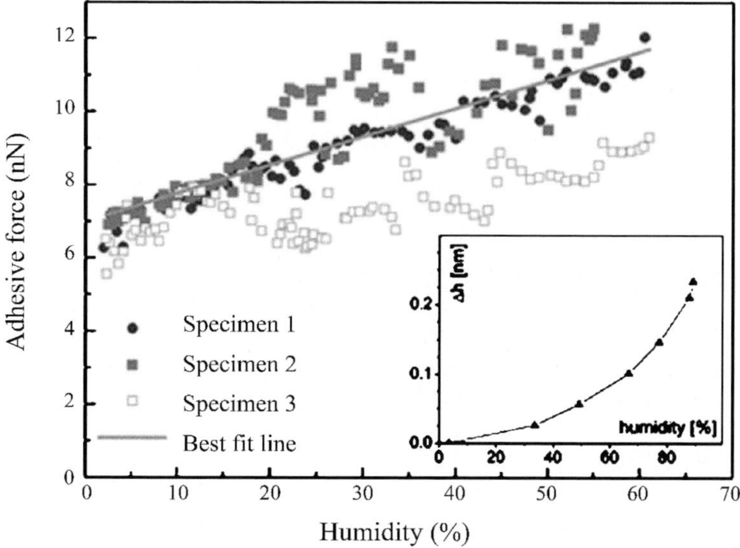

Fig. 22.11. Humidity effects on spatular pull-off force. *Inset*: The increase in water film thickness on a Si wafer with increasing humidity [39]

The humidity was adjusted by varying the flow rate of dry nitrogen into the chamber. The air was continuously monitored with a commercially available hygrometer. All tests were conducted at ambient temperature.

As seen in Fig. 22.11, even at low humidity, the adhesive force is large. An increase in humidity further increases the overall adhesive force of a gecko spatula. The pull-off force roughly doubled as the humidity was increased from 1.5 to 60%. This humidity effect can be explained in two possible ways: (1) by standard capillarity or (2) by a change of the effective short-range interaction due to absorbed monolayers of water – in other words, the water molecules increase the number of van der Waals bonds that are made. On the basis of these data, van der Waals forces are the primary adhesive mechanism and capillary forces are a secondary adhesive mechanism.

22.4.4
Effects of Hydrophobicity

To further test the hypothesis that capillary forces play a role in gecko adhesion, the spatular pull-off force was determined for contact with both hydrophilic and hydrophobic surfaces. As seen in Fig. 22.12a, the capillary adhesion theory predicts that a gecko spatula will generate a greater adhesive force when in contact with a hydrophilic surface compared with a hydrophobic surface, while the van der Waals adhesion theory predicts that the adhesive force between a gecko spatula and a surface will be the same regardless of the hydrophobicity of the surface [7]. Figure 22.12b shows the adhesive pressure of a whole gecko and the adhesive force of a single seta on hydrophilic and hydrophobic surfaces. The data show that the adhesive values are the same on both surfaces. This supports the van der Waals prediction of Fig. 22.12a. Huber et al. [39] found that the hydrophobicity of the attachment surface had an effect on the adhesive force of a single gecko spatula as shown in Fig. 22.12c. These results show that adhesive force has a finite value for superhydrophobic surface and increases as the surface becomes hydrophilic. It is concluded that van der Waals forces are the primary mechanism and capillary forces further increase the adhesive force generated.

22.5
Design of Biomimetic Fibrillar Structures

22.5.1
Verification of Adhesion Enhancement
of Fabricated Surfaces Using Fibrillar Structures

In order to create a material capable of dry adhesion, one would want to mimic the hierarchical structures found on the attachment pads of insects and lizards. Peressadko and Gorb [55] investigated whether adhesion enhancement was experienced through a division of contact area or fibrillar structure. The adhesive strength of a patterned surface and a smooth surface (roughness amplitude, $R_a = 0.5$ nm) of poly(vinyl siloxane) (PVS) was tested on both a smooth and a curved glass surface.

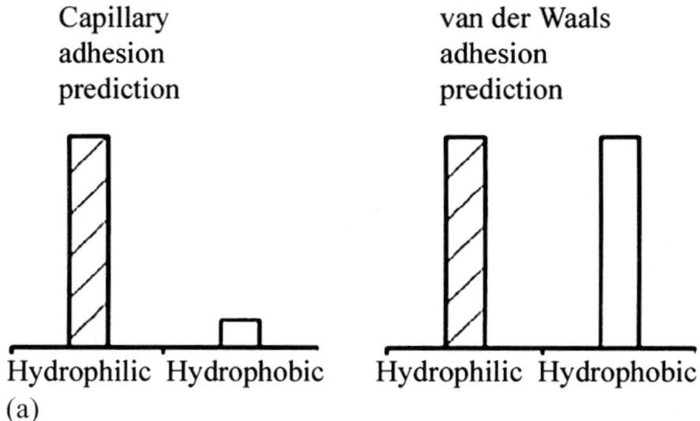

Results supporting van der Waals forces

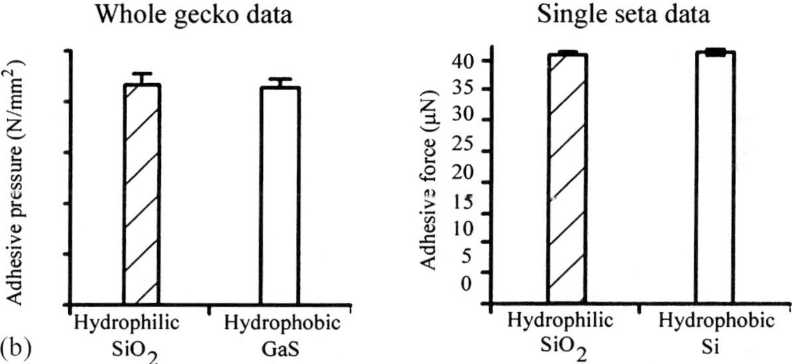

Results supporting capillary forces

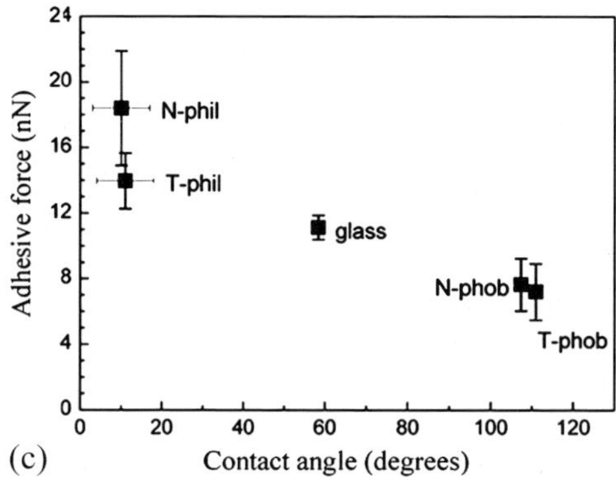

Fig. 22.12. (**a**) Capillary and van der Waals adhesion predictions for the relative magnitude of the adhesive force of gecko setae with hydrophilic and hydrophobic surfaces [7]. (**b**) Results of adhesion testing for a whole gecko and a single seta with hydrophilic and hydrophobic surfaces [7]. (**c**) Results of adhesive force testing with surfaces with different contact angles [39]

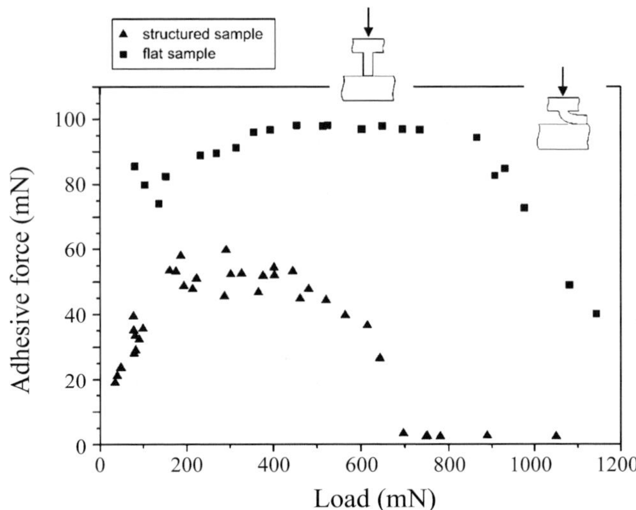

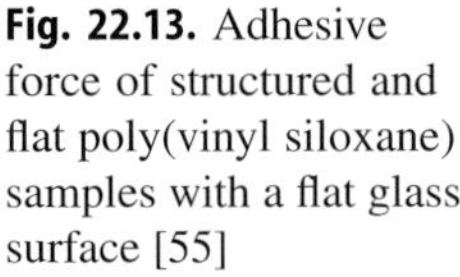

Fig. 22.13. Adhesive force of structured and flat poly(vinyl siloxane) samples with a flat glass surface [55]

Both PVS surfaces were molded. The patterned surface consisted of 72 columns (height 400 μm, cross section 250 μm × 125 μm). The samples were loaded perpendicular to the glass surface. During unloading, the adhesive force was measured. As seen in Fig. 22.13, the adhesive strength of the structured sample was several times greater than that of the flat sample. The adhesive strength of the fibrillar sample decreases at a load beyond 800 mN. This decline in adhesion is due to column buckling. Although the testing only dealt with surfaces made of PVS, one can assume that similar adhesion enhancement would result in structured samples of any material.

22.5.2
Contact Mechanics of Fibrillar Structures

In order for a fibrillar microstructure to act as a good adhesive, it is necessary that the materials be compliant. This allows the fibrillar interface to make contact at as many points as possible. The mechanics of adhesion between a fibrillar structure and a rough surface have been a topic of investigation for many researchers [24, 27, 28, 45, 56, 70]. In order to better understand the mechanics of adhesion, the approach of [45] will be described. The fibrillar surface is modeled in two dimensions, as shown in Fig. 22.14a, and is described by the length, L, and width, $2a$, of the fibrils, and by the area fraction of the interface covered by fibril ends, f. Fibrils composed of two different materials are investigated, a soft material and a hard material. Both of these materials are assumed to be linear-elastic and have properties corresponding to those given in Table 22.4.

If one wants to achieve the intimate contact between an elastic solid and a wavy surface given by (22.4), a compressive stress, S_c, must be applied to the surface. Intimate contact is achieved when [40]

$$\frac{\pi E h}{2(1 - \nu^2) S_c \lambda} < 1 \, , \tag{22.12}$$

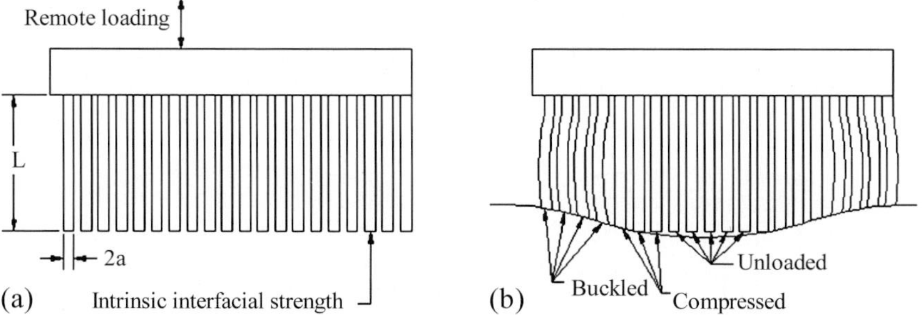

Fig. 22.14. Geometry of (**a**) a model fibrillar structure and (**b**) a fibrillar mat loaded in compression against a rough surface

Table 22.4. Material properties for a soft, good adhesive and a stiff, weak adhesive [45]

Parameter	Soft, good adhesive	Stiff, weak adhesive
Young's modulus, E (Pa)	10^6	10^9
Interfacial fracture energy, Γ_0 (J/m^2)	100	1
Interfacial strength, σ^* (Pa)	10^6	10^3
Applied stress, σ (Pa)	10^4	10^4

where E is Young's modulus and v is Poisson's ratio. By substituting the values of Table 22.4 into this equation, we can see that soft and hard materials can tolerate surface roughness aspect ratios, h/λ, of approximately 5×10^{-3} and 5×10^{-6}, respectively.

If the fibrillar surface is to make contact with a rough surface, the fibrillar mat will be loaded as seen in Fig. 22.14b. The buckling stress, S_b, is given by [75]

$$\frac{S_b}{E} = \frac{1}{3} f \pi^2 \left(\frac{a}{L} \right)^2 . \tag{22.13}$$

If f is taken to be 0.75, the width-to-length ratio, a/L, must be less than or equal to 0.064 for the soft material and 0.002 for the hard material in order for uniform contact to occur.

When long, slender beams (such as setae or nanobumps) are in close proximity, the potential for two adjacent members to adhere laterally to each other arises as depicted in Fig. 22.15. Hui et al. [40] determined a relation to check for this phenomenon:

$$\frac{L}{2a} \left(\frac{2\gamma_s}{3Ea} \right)^{1/4} < \left(\frac{w}{a} \right)^{1/2} , \tag{22.14}$$

where w is the gap between fibrils and γ_s is the surface energy (0.05 J/m^2). Under the assumption that $w/a = 1$, the width-to-length ratio must be greater than or equal to 0.25 and 0.045 for the soft and hard materials, respectively.

It is evident when comparing the results of (22.13) and (22.14) that it is not possible for all of the fibrils to be in contact with the sinusoidal surface of (22.4). As a result, there is a trade-off between the aspect ratio of the fibrils and their adaptability to a rough surface. If the aspect ratio of the fibrils is too large, they can adhere to each other or even collapse under their own weight, as shown in Fig. 22.16a. If the aspect ratio is too small (Fig. 22.16b), the structures will lack the necessary compliance to conform to a rough surface.

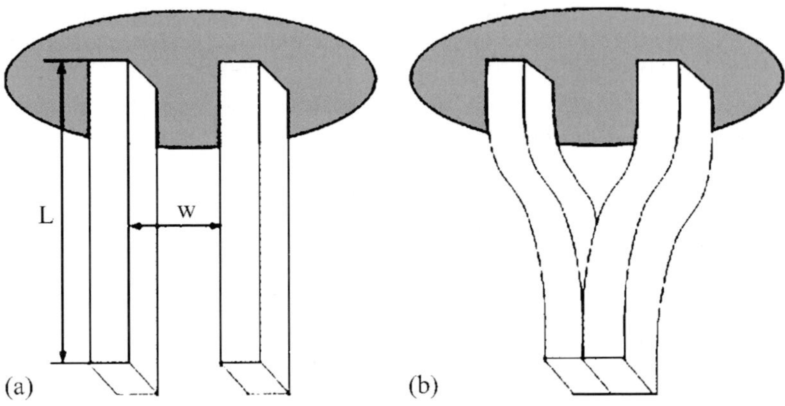

Fig. 22.15. Model of two adjacent fibers adhering laterally to each other [24]

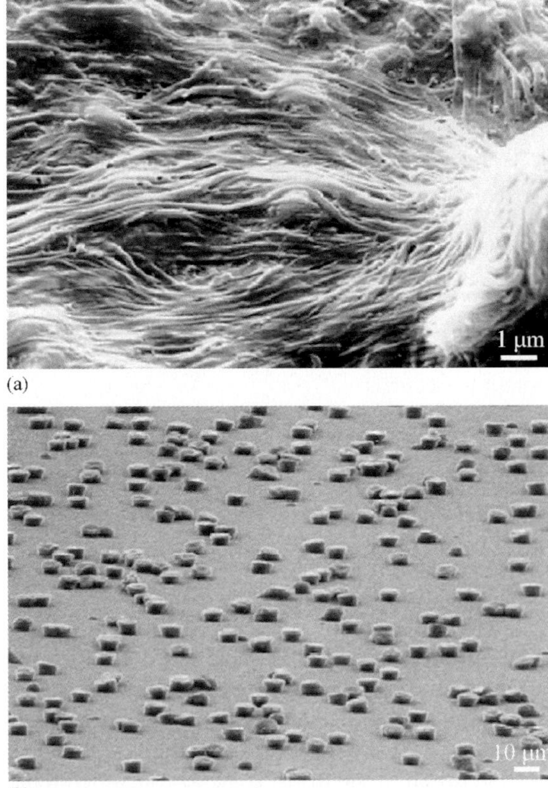

Fig. 22.16. SEM micrographs of (a) high aspect ratio polymer fibrils that have collapsed under their own weight and (b) low aspect ratio polymer fibrils that are incapable of adapting to rough surfaces [69]

22.5.3
Fabrication of Biomimetric Gecko Skin

On the basis of studies found in the literature, the dominant adhesive mechanism utilized by geckos and other spider-attachment systems appears to be van der Waals forces. The complex divisions of the gecko skin (lamellae–setae–branches–spatulae) enable a large real area of contact between the gecko skin and mating surface. Hence, a hierarchical fibrillar microstructure/nanostructure is desirable for dry, superadhesive tapes [45]. The development of a nanocomposite capable of replicating this adhesive force developed in nature is limited by current fabrication techniques.

On the microscale/nanoscale, typical machining methods (e.g., forging, drilling, grinding, lapping) are not possible. In order to create nanobumps, other manufacturing techniques are required and have been the subject of numerous studies [25,52,53,68,70,71,81].

22.5.3.1
Single-Level Hierarchical Structures

Previously, AFM tips were used to create a set of dimples on a wax surface in a process like that depicted in Fig. 22.17. These dimples served as a mold for creating polymer nanopyramids [70]. The adhesive force to an individual pyramid was measured using another AFM cantilever. The force was found to be about 200 μN. Although each pyramid of the material is capable of producing similar forces to that of a gecko seta, it failed to replicate adhesion on a large scale. This was due to the lack of flexibility in the pyramids. In order to ensure that the largest possible area of contact occurs between the tape and mating surface, a soft, compliant fibrillar structure is desired [45]. As shown in previous calculations, the van der Waals adhesive force for two parallel surfaces is inversely proportional to the cube of the distance between two surfaces.

Geim et al. [25] created arrays of nanohairs using electron-beam lithography and dry etching in oxygen plasma (Fig. 22.18a). The original arrays were created on a rigid silicon wafer. This design was only capable of creating 0.01 N of adhesive force for a 1-cm^2 patch. The nanohairs were then transferred from the silicon wafer to a soft bonding substrate. A 1-cm^2 sample was able to create 3 N of adhesive force under the new arrangement. This is approximately one third the adhesive strength of a gecko. Bunching (as described earlier) was determined to greatly reduce the durability and adhesive strength of the polymer tape. The bunching can be clearly seen in Fig. 22.18b.

Multiwalled carbon nanotube (MWCNT) hairs have been used to create superadhesive tapes [81]. The first step in the creation of this surface involves the growth of 50–100 μm MWCNTs on quartz or silicon substrates through chemical vapor deposition. Patterns are then created using a combination of photolithography and a wet and/or dry etching. SEM images of the nanotube surfaces can be seen in Fig. 22.19. On a small scale (nanometer level), the MWCNT surface was able to achieve adhesive forces 200 times greater than those of gecko foot hairs. However, the MWCNT surface could not replicate large scale gecko adhesion due to a lack of compliance.

Directed self-assembly could be used to produce regularly spaced fibers [62,68]. In this technique, a thin liquid polymer film is coated on a flat conductive substrate.

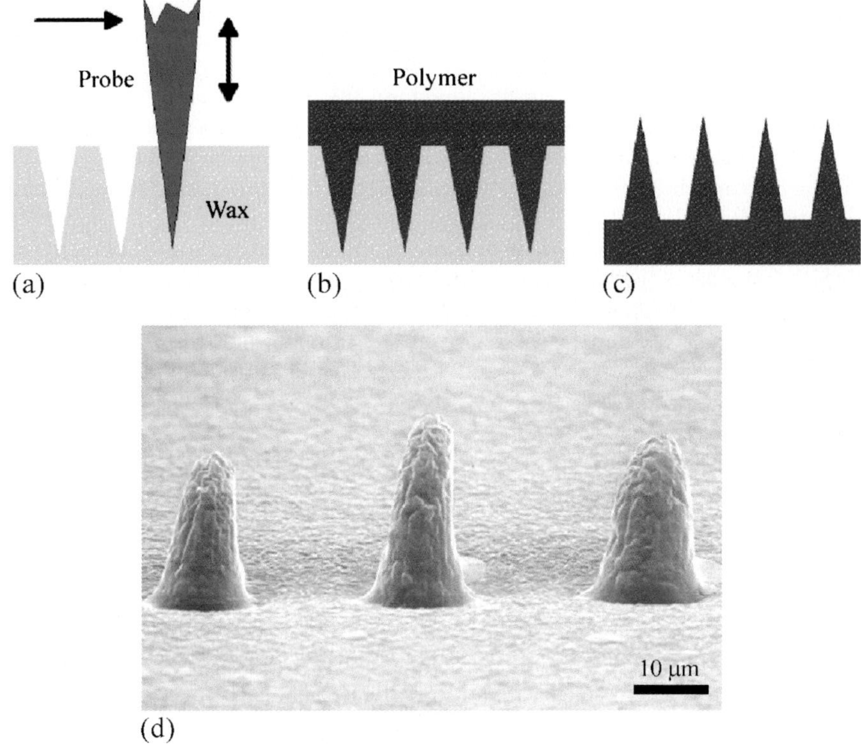

Fig. 22.17. (**a**) Indenting a flat wax surface using a sharp probe (nanotip indenting), (**b**) molding with a polymer, and (**c**) separating the polymer from the wax surface by peeling. (**d**) SEM image of three pillars created by nanotip indentation [68]

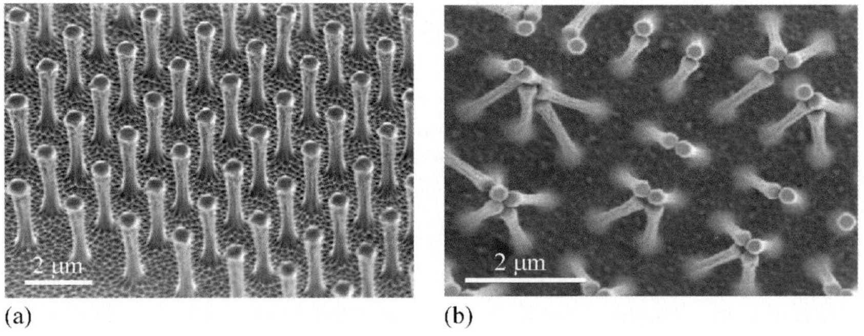

Fig. 22.18. SEM images of (**a**) an array of polyimide nanohairs and (**b**) bunching of the nanohairs, which leads to a reduction in adhesive force [25]

As demonstrated in Fig. 22.20, a closely spaced metal plate is used to apply a direct current electric field on the polymer film. Owing to instabilities, pillars will begin to grow. Self-assembly is desirable because the components spontaneously assemble, typically by bouncing around in a solution or gas phase until a stable structure of

Fig. 22.19. SEM images of multiwalled carbon nanotube structures: *left* grown on silicon by vapor deposition and *right* transferred into a poly(methyl methacrylate) matrix and then exposed on the surface after solvent etching [81]

Fig. 22.20. Directed self-assembly-based method of producing high aspect ratio microhairs/nanohairs [68]

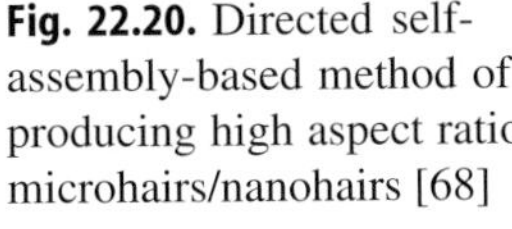
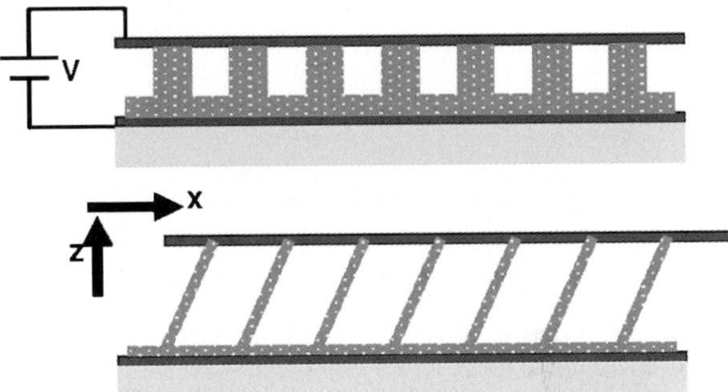

minimum energy is reached. This method is crucial in biomolecular nanotechnology, and has the potential to be used in precise devices [1]. These surface coatings have been demonstrated to be both durable and capable of creating superhydrophobic conditions and have been used to form clusters on the nanoscale [54].

22.5.3.2
Multilevel Hierarchical Structures

The aforementioned fabricated surfaces only have one level of hierarchy. Although these surfaces are capable of producing high adhesion on the microscale/nanoscale, all have failed in producing large-scale adhesion owing to a lack of compliance and bunching. In order to overcome these problems, Northen and Turner [52, 53] created a multilevel compliant system by employing a microelectromechanical-based approach. They created a layer of nanorods which they deemed "organorods"

(Fig. 22.21a). These organorods are comparable in size to that of gecko spatulae (50–200 nm in diameter and 2-µm tall). They sit atop a silicon dioxide chip (approximately 2-µm thick and 100–150 µm across a side), which was created using photolithography (Fig. 22.21b). Each chip is supported on top of a pillar (1 µm in diameter and 50 µm tall) that attaches to a silicon wafer (Fig. 22.21c). The multilevel structures have been created across a 100-mm wafer (Fig. 22.21d).

Adhesion testing was performed using a nanorod surface on a solid substrate and on the multilevel structures. As seen in Fig. 22.22, adhesive pressure of the multilevel structures was several times higher than that of the surfaces with only one level of hierarchy. The durability of the multilevel structure was also much greater than that of the single-level structure. The adhesion of the multilevel structure did not change between iterations one and five. During the same number of iterations, the adhesive pressure of the single-level structure decreased to zero.

Sitti [68] proposed a nanomolding technique for creating structures with two levels of hierarchy. In this method, two different molds are created – one with pores on the order of microns in diameter and a second with pores of nanometer-scale diameter. As seen in Fig. 22.23, the two molds would be bonded to each other and then filled with a liquid polymer. According to [68], the method would enable the manufacturing of a high volume of synthetic gecko foot hairs at low cost.

The literature clearly indicates that in order to create a dry superadhesive, a fibrillar surface construction is necessary to maximize the van der Waals forces by decreasing the distance between the two surfaces. It is also desirable to have a su-

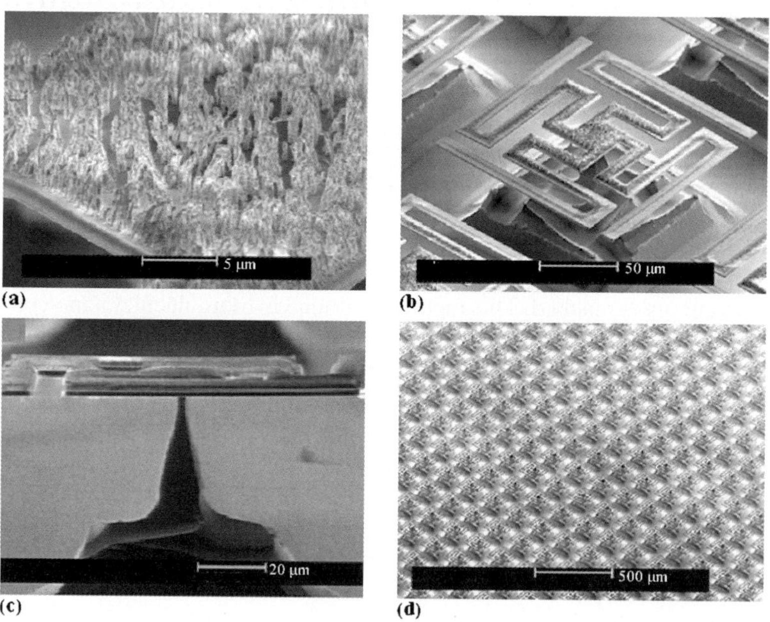

Fig. 22.21. Multilevel fabricated adhesive structure composed of (**a**) organorods, (**b**) silicon dioxide chips, and (**c**) support pillars. (**d**) This structure was repeated multiple times over a silicon wafer [52, 53]

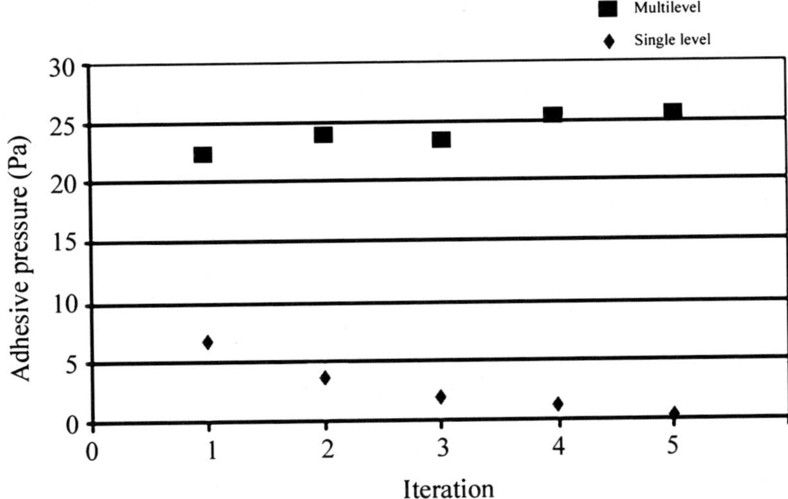

Fig. 22.22. Adhesion test results of a multilevel hierarchical structure (*top*) and a single-level hierarchical structure (*bottom*) repeated for five iterations [53]

Fig. 22.23. Proposed process of creating multilevel synthetic gecko foot hairs using nanomolding. Micron- and nanometer-sized pore membranes are bonded together (*top*) and filled with liquid polymer (*middle*). The membranes are then etched away leaving the polymer surface (*bottom*) [68]

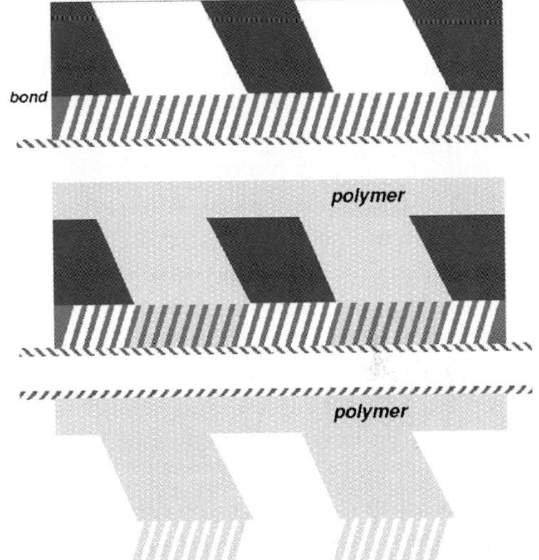

perhydrophobic surface in order to utilize self-cleaning. The material must be soft enough to conform to rough surfaces yet hard enough to avoid bunching, which will decrease the adhesive force.

22.6
Closure

The adhesive properties of geckos and other creatures, such as flies, beetles, and spiders, are due to the hierarchical structures present on each creature's attachment pads. Geckos have developed the most intricate adhesive structures of any of the aforementioned creatures. The attachment system consists of ridges called lamellae that are covered in microscale setae that branch off into nanoscale spatulae. Each structure plays an important role in adapting to surface roughness, bringing the spatulae in close proximity with the mating surface. These structures as well as material properties allow the gecko to obtain a much larger real area of contact between its feet and a mating surface than is possible with a nonfibrillar material. Two feet of a Tokay gecko have about 220 mm^2 of attachment pad area on which the gecko is able to generate approximately 20 N of adhesive force. Although capable of generating high adhesive forces, a gecko is able to detach from a surface at will – an ability known as reversible adhesion or smart adhesion. Detachment is achieved by a peeling motion of the gecko's feet from a surface.

Van der Waals forces are widely accepted in the literature as the dominant adhesive mechanism utilized by hierarchical attachment systems. Capillary forces created by humidity naturally present in the air can further increase the adhesive force generated by the spatulae. Experimental results have supported the adhesive theories of intermolecular forces (van der Waals) as a primary adhesive mechanism and capillary forces as a secondary mechanism, and have been used to rule out several other mechanisms of adhesion, including the secretion of sticky fluids, suction, and increased frictional forces. Atomic force microscopy has been employed by several investigators to determine the adhesive strength of gecko foot hairs. The measured values of the maximum adhesive force of a single seta (194 µN) and of a single spatula (11 nN) are comparable to the van der Waals prediction of 270 µN and 11 nN for a seta and a spatula, respectively. The adhesive force generated by a seta increases with preload and reaches a maximum when both perpendicular and parallel preloads are applied. Although gecko feet are strong adhesives, they remain free of contaminant particles through self-cleaning. Spatular size along with material properties enable geckos to easily expel any dust particles that come into contact with their feet.

There is great interest among the scientific community to create surfaces that replicate the adhesive strength of gecko feet. These surfaces would be capable of reusable dry adhesion and would have uses in a wide range of applications from everyday objects such as tape, fasteners, and toys to microelectric and space applications and even wall-climbing robots. In the design of fibrillar structures, it is necessary to ensure that the fibrils are compliant enough to easily deform to the mating surface's roughness profile, yet rigid enough to not collapse under their own weight. Spacing of the individual fibrils is also important. If the spacing is too small, adjacent fibrils can attract each other through intermolecular forces, which will lead to bunching.

Nanoindentation, electron-beam lithography, and growing of carbon nanotube arrays are all methods that have been used to create fibrillar structures. The limitations of current machining methods on the microscale/nanoscale have resulted in the majority of fabricated surfaces consisting of only one level of hierarchy. Although typically capable of producing a large adhesive force with an individual fibril, all of these surfaces have failed to generate large adhesive forces on the macroscale. Bunching, lack of compliance, and lack of durability are all problems that have arisen with the aforementioned structures. Recently, a multilayered compliant system was created using a microelectromechanical-based approach in combination with nanorods. This method as well as other proposed methods of multilevel nanomolding and directed self-assembly show great promise in the creation of adhesive structures with multiple levels of hierarchy, much like those of gecko feet.

Appendix

Several natural (sycamore tree bark and siltstone) and artificial (dry wall, wood laminate, steel, aluminum, and glass) surfaces were chosen to determine the microscale surface parameters of typical rough surfaces that a gecko might encounter. An Alpha-step 200 stylus profiler (Tencor Instruments, Mountain View, CA, USA) was used to obtain surface profiles of three different length scales: 80 μm, which is

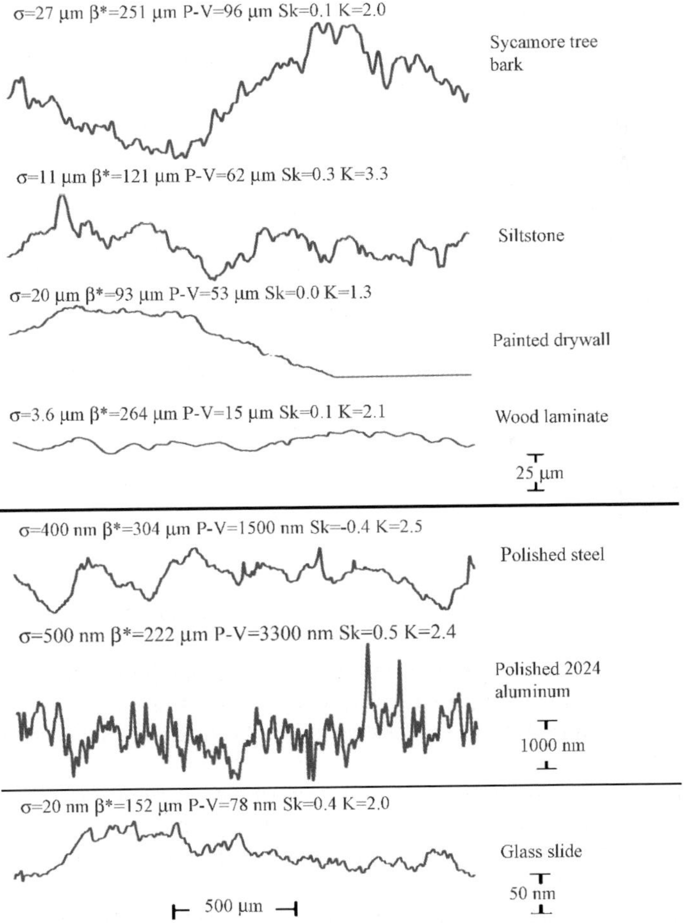

Fig. 22.24. (a) Surface height profiles of various random rough surfaces of interest at a 2000-μm scan length and (b) a comparison of the profiles of two surfaces at 80, 400, and 2000-μm scan lengths [15]

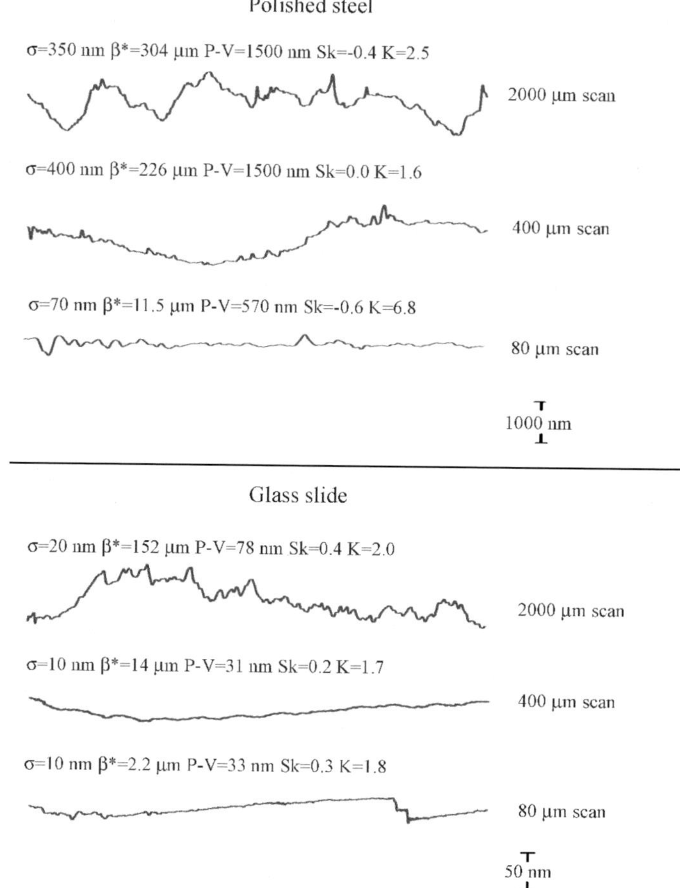

Fig. 22.24. (continued)

approximately the size of a single gecko seta; 2000 μm, which is close to the size of a gecko lamella; and an intermediate scan length of 400 μm. The radius of the stylus tip was 1.5–2.5 μm and the applied normal load was 30 μN (3 mg). The surface profiles were then analyzed using a specialized computer program to determine the root-mean-square amplitude, σ, correlation length, β^*, peak-to-valley distance, P–V, skewness, Sk, and kurtosis, K.

Samples of surface profiles and their corresponding parameters at a scan length of 2000 μm can be seen in Fig. 22.24a. The roughness amplitude, σ, varies from as low as 0.01 μm in glass to as high as 30 μm in tree bark. Similarly, the correlation length varies from 2 to 300 μm. The scale dependency of the surface parameters is illustrated in Fig. 22.24b. As the scan length of the profile increases, so too does the roughness amplitude and correlation length. Table 22.5 summarizes the scale-dependent factors σ and β^* for all seven sampled surfaces. At a scale length of 80 μm (size of seta), the roughness amplitude does not exceed 5 μm, while at a scale

Table 22.5. Scale dependence of surface parameters σ and β^* for rough surfaces at scan lengths of 80 and 2000 μm

Surface	Scan length			
	80 μm		2000 μm	
	σ (μm)	β^* (μm)	σ (μm)	β^* (μm)
Sycamore tree bark	4.4	17	27	251
Silitstone	1.1	4.8	11	268
Painted drywall	1	11	20	93
Wood laminate	0.11	18	3.6	264
Polished steel	0.07	12	0.40	304
Polished 2024 aluminum	0.40	6.5	0.50	222
Glass	0.01	2.2	0.02	152

length of 2000 μm (size of lamella), the roughness amplitude is as high as 30 μm. This suggests that setae are responsible for adapting to surfaces with roughness on the order of several micrometers, while lamellae must adapt to roughness on the order of tens of micrometers. Greater roughness values would be adapted to by the skin of the gecko. The spring model of Bhushan et al. [15] verifies that setae are only capable of adapting to roughness of a few microns and suggests that lamellae are responsible for adaptation to rougher surfaces.

References

1. Anonymous (2002) Self-assembly, Merriam-Webster's Medical Dictionary, Merriam-Webster, New York
2. Aristotle (1918) Historia Animalium. Book IX, Part 9, translated by Thompson DAW http://classics.mit.edu/Aristotle/history_anim.html
3. Arzt E, Gorb S, Spolenak R (2003) From micro to nano contacts in biological attachment devices. Proc Natl Acad Sci USA 100:10603–10606
4. Autumn K (2006) How gecko toes stick. Am Sci 94:124–132
5. Autumn K, Peattie AM (2002) Mechanisms of adhesion in geckos. Integr Comp Biol 42:1081–1090
6. Autumn K, Liang YA, Hsieh ST, Zesch W, Chan WP, Kenny TW, Fearing R, Full RJ (2000) Adhesive force of a single gecko foot-hair. Nature 405:681–685
7. Autumn K, Sitti M, Liang YA, Peattie AM, Hansen WR, Sponberg S, Kenny TW, Fearing R, Israelachvili JN, Full RJ (2002) Evidence for van der Waals adhesion in gecko setae. Proc Natl Acad Sci USA 99:12252–12256
8. Bergmann PJ, Irschick DJ (2005) Effects of temperature on maximum clinging ability in a diurnal gecko: evidence for a passive clinging mechanism? J Exp Zool 303A:785–791
9. Bertram JEA, Gosline JM (1987) Functional design of horse hoof keratin: the modulation of mechanical properties through hydration effects. J Exp Biol 130:121–136
10. Bhushan B (1996) Tribology and Mechanics of Magnetic Storage Devices. 2nd edn, Springer, Berlin, Heidelberg, New York
11. Bhushan B (1999) Principles and Applications of Tribology, Wiley, New York
12. Bhushan B (2002) Introduction to Tribology, Wiley, New York
13. Bhushan B (2005) Introduction to Nanotribology and Nanomechanics. Springer, Berlin, Heidelberg, New York

14. Bhushan B (2006) Springer Handbook of Nanotechnology. 2nd edn, Springer, Berlin, Heidelberg, New York

15. Bhushan B, Peressadko AG, Kim TW (2006) Adhesion analysis of two-level hierarchical morphology in natural attachment systems for smart adhesion. J Adhesion Sci Technol (in press)

16. Bikerman JJ (1961) The Science of Adhesive Joints, Academic, New York

17. Chui BW, Kenny TW, Mamin HJ, Terris BD, Rugar D (1998) Independent detection of vertical and lateral forces with a sidewall-implanted dual-axis piezoresistive cantilever. Appl Phys Lett 72:1388–1390

18. Davies DK (1973) Surface charge and the contact of elastic solids, J Phys D Appl Phys 6:1017–1024

19. Dellit WD (1934) Zur Anatomie und Physiologie der Geckozehe, Jena. Z Naturwiss 68:613–658

20. Derjaguin BV, Krotova NA, Smilga VP (1978) Effect of contact deformations on the adhesion of particles. J Colloid Interface Sci 53:314–326

21. Fan PL, O'Brien MJ (1975) Adhesion in deformable isolated capillaries. In: Lee LH (ed) Adhesion Science and Technology, Vol 9A, Plenum, New York, p 635

22. Federle W, Riehle M, Curtis ASG, Full RJ (2002) An integrative study of insect adhesion: mechanics of wet adhesion of pretarsal pads in ants. Integr Comp Biol 42:1100–1106

23. Full RJ, Fearing RS, Kenny TW, Autumn K (2004) Adhesive microstructure and method of forming same. US Patent 6,737,160

24. Gao H, Wang X, Yao H, Gorb S, Arzt E (2005) Mechanics of hierarchical adhesion structures of geckos. Mech Mater 37:275–285

25. Geim AK, Dubonos SV, Grigorieva IV, Novoselov KS, Zhukov AA, Shapoval SY (2003) Microfabricated adhesive mimicking gecko foot-hair. Nat Mater 2:461–463

26. Gennaro JGJ (1969) The gecko grip. Nat Hist 78:36–43

27. Glassmaker NJ, Jagota A, Hui CY, Kim J (2004) Design of biomimetic fibrillar interfaces: 1. Making contact. J R Soc Interface 1:23–33

28. Glassmaker NJ, Jagota A, Hui CY (2005) Adhesion enhancement in a biomimetic fibrillar interface. Acta Biomater 1:367–375

29. Hamaker HC (1937) London van der Waals attraction between spherical bodies. Physica 4:1058

30. Han D, Zhou K, Bauer AM (2004) Phylogenetic relationships among gekkotan lizards inferred from C-mos nuclear DNA sequences and a new classification of the Gekkota. Biol J Linn Soc 83:353–368

31. Hanna G, Barnes WJP (1991) Adhesion and detachment of the toe pads of tree frogs. J Exper Biol 155:103–125

32. Hansen WR, Autumn K (2005) Evidence for self-cleaning in gecko setae. Proc Natl Acad Sci USA 102:385–389

33. Henry PSH (1953) The role of asymmetric rubbing in the generation of static electricity. Br J Appl Phys S2:S31–S36

34. Hiller U (1968) Untersuchungen zum Feinbau und zur Funktion der Haftborsten von Reptilien. Z Morphol Tiere 62:307–362

35. Hinds WC (1982) Aerosol Technology: Properties, Behavior, and Measurement of Airborne Particles, Wiley, New York

36. Hora SL (1923) The adhesive apparatus on the toes of certain geckos and tree frogs. J Proc Asiat Soc Beng 9:137–145

37. Houwink R, Salomon G (1967) Effect of contamination on the adhesion of metallic couples in ultra high vacuum. J Appl Phys 38:1896–1904

38. Huber G, Gorb SN, Spolenak R, Arzt E (2005a) Resolving the nanoscale adhesion of individual gecko spatulae by atomic force microscopy. Biol Lett 1:2–4

39. Huber G, Mantz H, Spolenak R, Mecke K, Jacobs K, Gorb SN, Arzt E (2005b) Evidence for capillarity contributions to gecko adhesion from single spatula and nanomechanical measurements. Proc Natl Acad Sci USA 102:16293–16296

40. Hui CY, Jagota A, Lin YY, Kramer EJ (2002) Constraints on micro-contact printing imposed by stamp deformation. Langmuir 18:1394–1404

41. Irschick DJ, Austin CC, Petren K, Fisher RN, Losos JB, Ellers O (1996) A comparative analysis of clinging ability among pad-bearing lizards. Biol J Linn Soc 59:21–35

42. Israelachvili JN, Tabor D (1972) The measurement of Van der Waals dispersion forces in the range of 1.5 to 130 nm. Proc R Soc Lond Ser A 331:19–38

43. Israelachvili JN (1992) Intermolecular and Surface Forces, 2nd edn, Academic, San Diego

44. Jaenicke R (1998) Atmospheric aerosol size distribution, In: Harrison RM, van Grieken R (eds) Atmospheric Particles. Wiley, New York, p 1–29

45. Jagota A, Bennison SJ (2002) Mechanics of adhesion through a fibrillar microstructure. Integr Comp Biol 42:1140–1145

46. Johnsen A, Rahbek K (1923) A physical phenomenon and its applications to telegraphy, telephony, etc. J Inst Electr Eng 61:713–725

47. Johnson KL, Kendall K, Roberts AD (1971) Surface energy and the contact of elastic solids. Proc R Soc Lond Ser A 324:301–313

48. Kluge AG (2001) Gekkotan Lizard taxonomy. Hamadryad 26:1–209

49. Losos JB (1990) Thermal sensitivity of sprinting and clinging performance in the tokay gecko (Gekko gecko). Asiat Herp Res 3:54–59

50. McFarlane JS, Tabor D (1950) Adhesion of solids and the effects of surface films. Proc R Soc Lond Ser A 202:224–243

51. Menon C, Murphy M, Sitti M (2004) Gecko inspired surface climbing robots, In: IEEE International Conference on Robotics and Biomimetics, August 22–26, p 431–436

52. Northen MT, Turner KL (2005a) A batch fabricated biomimetic dry adhesive. Nanotechnology 16:1159–1166

53. Northen MT, Turner KL (2005b) Multi-scale compliant structures for use as a chip scale dry adhesive. Transducers 2:2044–2047

54. Pan B, Gao F, Ao L, Tian H, He R, Cui D (2005) Controlled self-assembly of thiol-terminated poly(amidoamine) dendrimer and gold nanoparticles. Colloids Surf A 259:89–94

55. Peressadko A, Gorb SN (2004) When less is more: experimental evidence for tenacity enhancement by division of contact area. J Adhes 80:247–261

56. Persson BNJ (2003) On the mechanism of adhesion in biological systems. J Chem Phys 118:7614–7621

57. Persson BNJ, Gorb S (2003) The effect of surface roughness on the adhesion of elastic plates with application to biological systems. J Chem Phys 119:11437–11444

58. Phipps PB, Rice DW (1979) Role of water in atmospheric corrosion, ACS Symposium Series No. 89, American Chemical Society, Washington

59. Ruibla R, Ernst V (1965) The structure of the digital setae of lizards. J Morph 117:271–294

60. Russell AP (1975) A contribution to the functional morphology of the foot of the tokay, Gekko gecko. J Zool Lond 176:437–476

61. Russell AP (1986) The morphological basis of weight-bearing in the scansors of the tokay gecko. Can J Zool 64:948–955

62. Schäffer E, Thurn-Albrecht T, Russell TP, Steiner U (2000) Electrically induced structure formation and pattern transfer. Nature 403:874–877

63. Schleich HH, Kästle W (1986) Ultrastrukturen an Gecko-Zehen. Amphib Reptil 7:141–166

64. Schmidt HR (1904) Zur Anatomie und Physiologie der Geckopfote. Jena Z Naturwiss 39:551

65. Shah GJ, Sitti M (2004) Modeling and design of biomimetic adhesives inspired by gecko foot-hairs. In: IEEE International Conference on Robotics and Biomimetics. August 22–26, p 873–878
66. Shaw PE (1923) Electrical separation between identical solid surfaces. Proc Phys Soc 39:449–452
67. Simmermacher G (1884) Untersuchungen über Haftapparate an Tarsalgliedern von Insekten. Zeitschr Wiss Zool 40:481–556
68. Sitti M (2003) High aspect ratio polymer micro/nano-structure manufacturing using nanoembossing, nanomolding and directed self-assembly, In: Proceedings of the IEEE/ASME Advanced Mechatronics Conference, July 20–24, Vol 2, p 886–890
69. Sitti M, Fearing RS (2002) Nanomodeling based fabrication of synthetic gecko foot-hairs, In: Proceedings of the IEEE Conference on Nanotechnology. August 26–28, p 137–140
70. Sitti M, Fearing RS (2003a) Synthetic gecko foot-hair for micro/nano structures as dry adhesives. J Adhes Sci Technol 18:1055–1074
71. Sitti M, Fearing RS (2003b) Synthetic gecko foot-hair for micro/nano structures for future wall-climbing robots, In: Proceedings of the IEEE International Conference on Robotics and Automation. September 14–19, Vol 1, p 1164–1170
72. Skinner SM, Savage RL, Rutzler JE (1953) Electrical phenomena in adhesion I: electron atmospheres in dielectrics. J App Phys 24:438–450
73. Stork NE (1980) Experimental analysis of adhesion of Chrysolina polita on a variety of surfaces. J Exp Biol 88:91–107
74. Stork NE (1983) J Nat Hist 17:829–835
75. Timoshenko SP, Gere JM (1961) Theory of Elastic Sstability, McGraw-Hill, New York
76. Tinkle DW (1992) Gecko. Encyl Am 12:359
77. Van der Kloot WG (1992) Molting. Encyl Am 19:336–337
78. Wagler J (1830) Naturliches System der Amphibien, Cotta'schen, Munich
79. Wahlin A, Backstrom G (1974) Sliding electrification of teflon by metals. J Appl Phys 45:2058–2064
80. Williams EE, Peterson JA (1982) Convergent and alternative designs in the digital adhesive pads of scincid lizards. Science 215:1509–1511
81. Yurdumakan B, Raravikar NR, Ajayan PM, Dhinojwala A (2005) Synthetic gecko foot-hairs from multiwalled carbon nanotubes. Chem Commun 3799–3801
82. Zimon AD (1969) Adhesion of Dust and Powder, translated from Russian by Corn M. Plenum, New York
83. Zisman WA (1963) Influence of constitution on adhesion. Ind Eng Chem 55(10):18–38.

23 Novel AFM Nanoprobes

Horacio D. Espinosa · Nicolaie Moldovan · K.-H. Kim

23.1
Introduction and Historic Developments

The atomic force microscope (AFM) is a member of the family of scanning probe microscopes, which makes use of specialized probes to scan a sample surface to produce maps of topography, conductivity or friction among many others. The resolution of the technique is highly dependent on the probe quality and sharpness. To appreciate this, let us start with some historic developments. Scanning probe microscopy started in 1981 with the discovery of the scanning tunneling microscope (STM) by Binnig and Rohrer, who were awarded the Nobel Prize in Physics in 1986 for this invention. The STM operates by keeping constant a tunneling current established between a sharp conductive probe and a conductive sample, while scanning the sample using a high-accuracy xyz piezoactuator. Although the first STMs had relative modest probe sharpness (approximately 1-μm tip radius), atomic resolution could be obtained with remarkable simplicity in view of the fact that the tunneling current was established between one or a few atoms of the probe tip and one or a few atoms of the substrate. To keep the tunneling current constant, a constant tunneling gap had to be maintained during the lateral motion of the probe. This was achieved through feedback by amplifying the tunneling current and coupling the signal to a z-motion actuator to which the probe was attached. While the feedback acts to keep the tunneling current constant, it provides a voltage signal proportional to the topographic height information on the scanned surface, which is recorded.

Soon after the invention of the STM, the challenge was to extend the technique to nonconductive samples. Experiments showed that whenever a probe tip was close enough to a sample to establish a tunneling current, there were also small but significant forces acting on the probe. With cantilevered probes, these forces could be used to retrieve the topographic information by monitoring the cantilever deflection. This latter task was accomplished by detecting the position of a laser beam reflected on the cantilever, forming a so-called *optical lever*. The signal provided by the light detector was coupled via an amplifier to the z-motion actuator of the probe relative to the substrate to achieve feedback control. This constituted the AFM, as invented in 1986 by Binnig and developed in the same year by Binnig, Quate, and Gerber (1986). Modern AFMs have a four-quadrant photodiode (Fig. 23.1) to detect both vertical and lateral displacements of the laser spot. The latter is used to record probe tilt, which is due to lateral forces (friction) and constitutes the basis for lateral force microscopy (LFM). Maps of the friction coefficient between the probe and sample

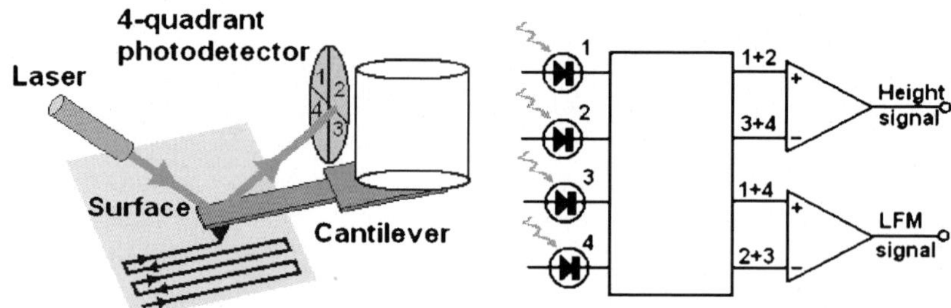

Fig. 23.1. Optical lever scheme of an atomic force microscope (AFM) and signal processing from a four-quadrant photodiode; the height signal is used as a feedback to keep the cantilever deflection constant, while the lateral force signal is processed for lateral force microscopy (LFM) measurements

can be obtained using this scheme. For a detailed history of the development of the AFM and other scanning probe techniques, we refer the reader to [1,2].

Over the years, many types of atomic force microscopy techniques were developed. Among them, we distinguish those operating with the probes in contact or noncontact mode, lateral force probes [3], vibrating probe techniques [4], AFMs operating in different environments (vacuum [5,6], air, gas[1] or immersed in liquids [7,8]), probes sensitive to other type of forces (electrostatic [9], magnetic [10]) or signals (electrical [11,12], capacitive Kelvin probes [13–15], temperature [16,17], light [18], chemical [19]), and probes integrating electrostatic [20,21], piezo-, thermal or magnetic actuators for their z-motion or vibration [22,23]. During the last decade, more complex probes integrating different types of sensors or even microfluidics were developed [24–28]. Novel AFM techniques were also developed in which the z-motion feedback is achieved with other types of sensors independently of the optical lever (e.g., piezoresistive [29], capacitive [30], complementary metal oxide superconductor, CMOS, integrated deflection sensors [31–33], interferential optical schemes, and membrane deflection [34]).

AFM-related techniques were further developed during the last decade for actively producing nanoscale changes on the sample surface. This latter research domain showed a strong development leading to specialized probes and techniques for various purposes especially in biochemistry and the life sciences. New tools were developed, along with methods and software. Ultimately, entire companies emerged, specializing in their production. The same probes may be capable of multitasking by combining different modes of operation and sensing [35,36].

The methods for local surface modification with AFM probes range from simply mechanically scratching the surface [37–39] or nanomanipulation of nanoscale objects on surfaces for the purpose of building structures [40–42], to chemical

---

[1] Several companies produce environmental chambers for SPM. Some of them are EnviroView 1000, Nanonics product sheet, http://www.nanonics.co.il; EnviroScope atomic force microscope, Digital Instruments, Veeco product sheet, http://www.veeco.com/appnotes/DS48_EScope_Compressed.pdf; Environmental control chamber, Molecular Imaging, product sheet, http://www.molec.com/products_options.html#env_ctrl

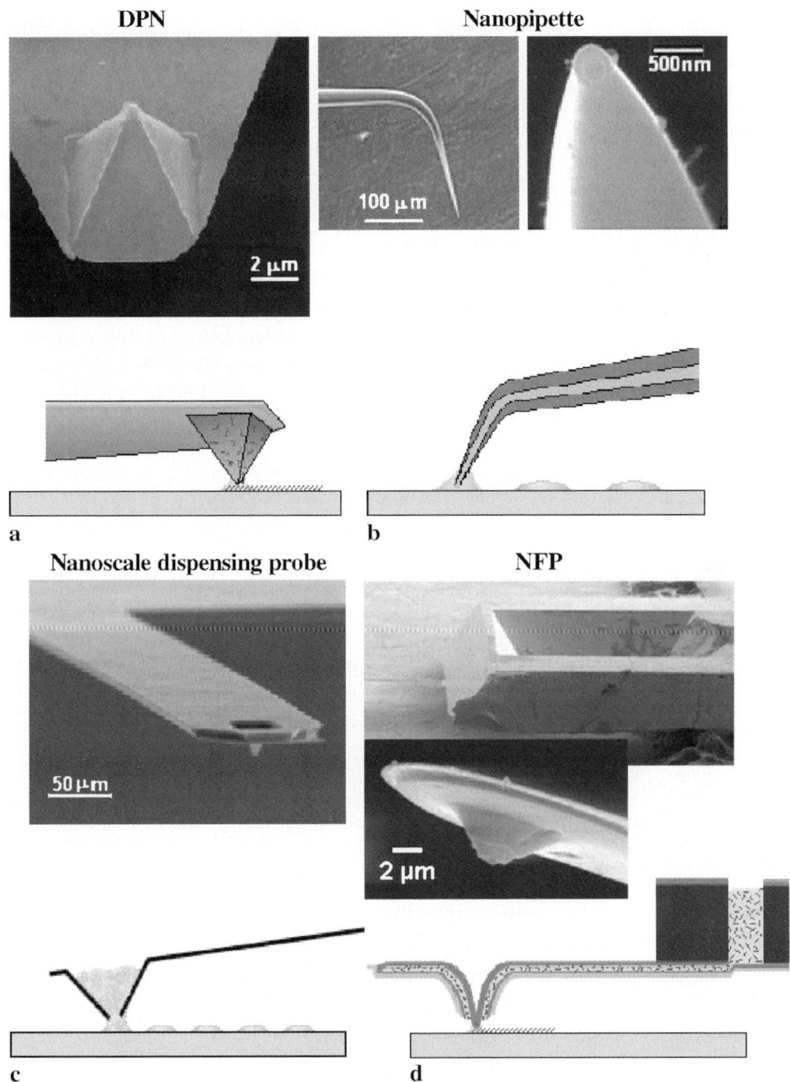

Plate 23.1. Comparison of different atomic force microscopy nanopatterning tools. (**a**) dip-pen nanolithography – DPN [50–54]; (**b**) nanopipettes [24]; (**c**) apertured probes [28]; (**d**) nanofountain probe – NFP [58]

or electrochemical-induced modifications, such as oxidation of surfaces [43–46], etching (e.g., Cr for mask repair [24]), local chemical vapor deposition (CVD) of metal nanowires [47, 48] and electrodeposition [49]. A distinct class with a variety of direct deposition methods emerged from a technique first proposed in 1999 by Mirkin [50–54], dip-pen nanolithography (DPN). This uses AFM probes coated with certain molecular ink species to write patterns on substrates, as will be detailed in Sect. 23.2.

DPN techniques for writing with a large variety of inks on suited substrates were developed, but they all exhibit a common feature: DPN patterning, owing to its serial nature and diffusion-controlled mechanism, is slow (a few microns per second writing speed). Likewise, interruptions for reinking, changing molecular inks and re-alignment as well as for changing the probes owing to wear or particle contamination are customary. Several techniques and specialized probes were developed to alleviate these inconveniences. Among them are nanofountain probe (NFP) lithography (in which ink is delivered to the probe in liquid form) [24,55–59], electro-pen lithography (in which, additional to inking, a voltage is applied between the probe and the substrate, producing oxidation of the substrate and an increase in writing speed) [60], nanopipette writing [24], and quills-based techniques [26, 27] with applications mainly in biomolecular chemistry. In Plate 1, a comparison of these techniques is presented. Each of them has its advantages and disadvantages. DPN (a) is characterized by simplicity, high writing resolution (smaller than 100 nm) and a large amount of literature available, but is the slowest of the techniques; apertured probes (b, c) can store and dispense larger amounts of ink molecules, but writing speed is not increased substantially and the writing resolution (regular approximately 300 nm, best reported 70-nm-diameter spots) is inferior to DPN. Moreover, because of the ink supply into the probe cavities, it cannot be easily handled and integrated. Pulled-glass nanopipettes (b) have continuous ink delivery, but lower resolution (approximately 1 μm), and cannot be integrated by microfabrication into larger systems; NFP (d) have continuous ink delivery for a long-range writing capability, resolution close to that of DPN, and can be integrated in arrays and systems, although the probes and AFM systems become more complicated and costly. The electro-pen approach was not mentioned separately, since, although developed only for DPN, it may work with any of the other types of probes, provided they are made conductive and a voltage is applied.

The growing complexity of the probes makes them more laborious to manufacture and quite costly. AFM probes, even if complex, remain disposable laboratory accessories owing to inherent demands on avoiding cross-contamination and tip sharpness. However, it becomes more and more desirable to extend their lifetime for applications involving large area patterning or large-scale imaging. In these cases wear of the tip is one of the main limiting factors. Integration of low-wear materials such as diamond into the tips of the probes is one solution we will discuss in Sect. 23.3. Usage of conductive diamond also has advantages regarding the possibility of their use in electro-pen nanolithography (EPN) and other conductive probe AFM techniques while preserving tip sharpness. We will demonstrate this technology in the same section.

Recently, AFM techniques have been used in identifying mechanical properties of nano-objects, ranging from nanowires [61] and carbon nanotubes [62] to single molecules [63–65] (molecular pulling). For these techniques, the probes must be precisely calibrated regarding their spring constant (force spectroscopy) and vibration properties (dynamic force spectroscopy) [66]. For retrieving the properties of the objects investigated, the motion characteristics of the probes must be understood and analyzed coupled with the mechanical properties of the molecules investigated. In many cases, this involves molecular dynamics simulations.

As the ultimate force measuring tools at the nanoscale (down to piconewtons – picoforce meters), AFMs have been involved also in detecting the occurrence of liquid–solid phase transitions at the nanoscale on surface-adsorbed water. The study revealed ice formation at room temperature if voltage in the range 5–10 V is applied between the probe and the substrate [67]. This is particularly interesting and relevant for electrically biased DPN techniques such as electrochemical DPN and EPN, in relation to the formation of a water meniscus between tips and substrates, wear mechanisms and molecular ink transport.

23.2
DPN and Fountain Pen Nanolithography

Tips integrated on microcantilevers may function as pens or quills, to locally deliver molecules previously present on the tip surface. The tip sharpness allows sub-100-nm patterning with various molecular species. The domain known as DPN is relevant for high-resolution patterning of limited areas, such as for life sciences applications (nanoassays). In most cases, the ink molecules produce a local reactive functional-ization of the surfaces, such that specific biochemical adhesion experiments can be conducted at these length scales. In general, the deposited material and the substrate must be paired, such that a chemical reaction occurs upon delivery, or a surface self-assembled monolayer (SAM) is formed, such that a reading is possible using the same AFM technique. In some cases, this reaction or SAM formation is not needed; for example, in the case of simple patterning with fluorescent dyes in which the reading can be achieved optically.

In the DPN technique (Fig. 23.2), the ink molecule species are physisorbed on the tip surface, such that the tip geometry is not significantly altered. When the tip is placed in contact with a target surface, the molecules migrate from the tip onto the surface and the time of surface contact directly correlates with the amount of material transferred. A model of the DPN material transfer was elaborated by Jang et al. [68]. It supposes that the ink molecules form a monolayer spot on the substrate, developing from the area of probe contact. The probe acts as an infinite source of molecules, while the surface diffusion of ink molecules on the existent monolayer transports the molecules to the edge of the spot, where they stick to the surface, extending the spot diameter. The model equations lead to a dependence of spot diameter proportional to the square root of contact time, which is verified experimentally in most cases of practical interest, evidently before the ink exhaust on the tip. The diffusion-driven

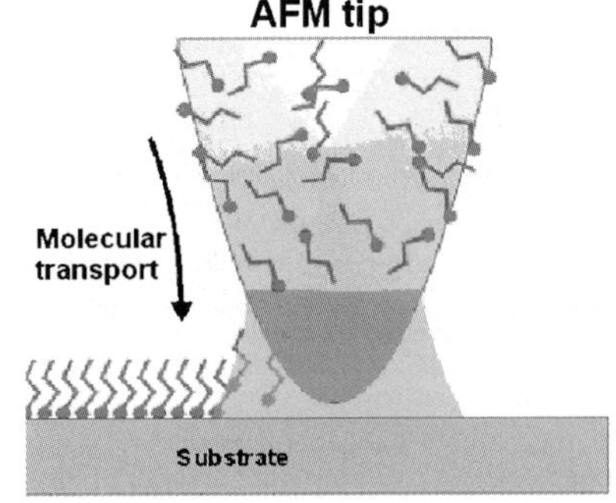

Fig. 23.2. Writing mechanism in dip-pen nano-lithography (DPN) after Piner et al. [75]. Ink molecules from an AFM probe diffuse through a water meniscus formed by capillary condensa-tion and migrate to the substrate, where they form a self-assembled monolayer

writing process also explains the dynamics of line writing and is responsible for the low writing speed in DPN, typically below 1 µm/s. According to the model, the writing speed also depends on the reactivity of the ink with the substrate and the contact radius of the tip. This was confirmed experimentally, although a quantitative verification was not reported.

A water meniscus formed by capillary condensation was suggested as the mechanism of molecular transfer from a tip to the substrate in DPN, and the formation of such a meniscus was recently confirmed using an environmental scanning electron microscopy (SEM) technique (Fig. 23.3) [69]. The formation of this meniscus causes the ink delivery to depend on the humidity, temperature and air flow in the AFM chamber. Because of this, specialized equipment for DPN writing is usually placed in chambers with controlled environmental parameters. Typical values reported for successful writing are 10–100% humidity and 23–33 °C [70]. However, DPN in very dry environments and on carefully dried substrates was also reported [71], leading to a controversy. Our recent investigations on the aspect of ink-coated DPN probes showed that although ink is supplied from solutions in most DPN writing schemes, the ink species dries up during usage, forming a compact, quite irregular coating on the writing AFM probe.

Figure 23.4 presents a SEM view of a standard Si_3N_4 probe before and after dipping it into a 16-mercaptohexanoic acid (MHA) solution and drying. Even when a careful dip was performed using a micromachined ink well and the AFM head motion in order to achieve a controlled and partial coating [72] (Fig. 23.4c), the features of the coating suggest that writing performed with such probes is quite

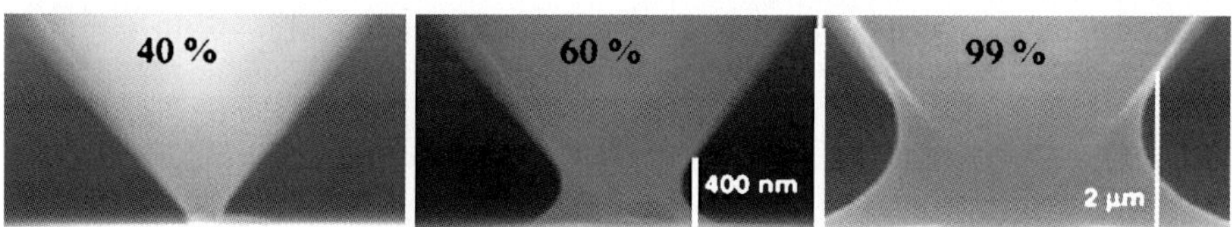

Fig. 23.3. Sequence of images collected in an environmental scanning electron microscope (SEM) at various relative humidity levels, showing meniscus formation between an AFM tip and a substrate. All images were collected at 5 °C, 15.0-kV accelerating voltage, at ×35,000. The humidity was varied by decreasing the pressure of the water vapor from 6.4 Torr down to 1 Torr: 40% relative humidity, 2 Torr; 60% relative humidity, 3.2 Torr; 99% relative humidity 6.4 Torr. (After [69])

Fig. 23.4. SEM images of a standard nitride AFM probe before (**a**) and after (**b**) dipping it in 16-mercaptohexadecanoic acid (MHA) solution and drying. (**c**) Probe dipped carefully, by using a micromachined ink well and the AFM motion for dipping

different from the simplified picture provided by the meniscus-driven model. Even if capillary condensation plays a strong role in the DPN patterning, these images suggest that writing can actually be performed at least in part like with a pencil, rather than with a quill. This may explain why DPN writing was also successfully accomplished in a dry nitrogen atmosphere and with very dry substrates [71] – ultimately, the transfer of molecules should also be possible by direct contact, without the aid of a meniscus. This writing mode continues to support the square-root-of-time dependence of the spot diameter, since the latter reflects only the behavior of ink diffusion to the substrate.

If environmental conditions and ink and substrate chemistry are optimized, patterning small feature sizes is critically dependent on the material and radius of curvature of the AFM probe tips. The most commonly used probes in the DPN technique are made of silicon nitride cantilevers with integrated pyramidal tips, which are typically fabricated using the pyramidal-pit molding technique [73]. For higher-resolution DPN writing, sharper tips produced by oxidation-sharpening [74] are usually employed. The dimensions of the cantilevers have been selected to meet the desired range of spring constants 0.03–0.3 N/m [50].

DPN was originally reported to pattern gold surfaces with a solution of alkanethiols [75]. Its applications have been subsequently extended to patterning surfaces with versatile types of materials such as biomolecules [51, 52, 76], polymers [77], small organic materials [78], sol precursors [79] and metal salts [46]. One of the advantages of the DPN technique is that biomolecules, such as DNAs or proteins, can be patterned both by direct write and indirect assembly [51, 52, 54, 80–82], which can be utilized to build nanoscale biomolecular sensor arrays with higher sensitivity and selectivity requiring much smaller sample volumes. For example, modified single-strand DNAs with a thiol group at one end were patterned on a gold surface and used to capture complementary DNA sequences tagged with gold nanoparticles [51]. In this way, a pattern of gold particles could be assembled (*bottom up*) on the surface, in a completely different way than by thin film deposition and etching, or lift-off (traditional *top-down* techniques). Feature sizes ranging from a few micrometers to less than 100 nm were achieved.

Once a tip is coated with the molecules of interest, patterning is typically controlled by commercially available AFM instruments to precisely deposit desired amounts of molecules at controlled locations. DPN writing has been typically performed with a single probe; however, patterning large areas with a single tip, owing to its serial nature and the limited scan size of the AFM, is very inefficient. In an effort to improve the throughput of the DPN technique, the feasibility of parallel DPN patterning with a commercially available tip array was demonstrated in a commercially available AFM [50].

Furthermore, linear arrays of high-density probes with integrated tips for DPN were microfabricated [83, 84]. Two types of DPN probe arrays were developed using surface micromachining techniques. The first type, or type-1 probe array, was fabricated out of thin-film silicon nitride using the molding technique with a protruding tip, whereas the second type, or type-2 probe array was fabricated from heavily boron doped silicon (Fig. 23.5). The type-1 probe array consists of 32 straight probes in a 1D arrangement, with the space between consecutive probes being 100 μm. The dimensions of an individual cantilever were 400-μm long, 50-μm

Fig. 23.5. SEM micrographs of DPN probe arrays with the type-1 (*left*) and type-2 (*right*) arrays [83]

wide and 0.6-μm thick. The type-2 array had eight probes separated from each other by a spacing of 310 μm, whereas each cantilever was in a multifold configuration. The dimensions were 1400, 15 and 10 μm for the cantilever length, width and thickness, respectively.

23.2.1
NFP Chip Design – 1D and 2D Arrays

As mentioned, a major limitation of the DPN approach is the slowness in writing. There are several causes for the this: the diffusion-driven ink dispersion on the surface; the necessity to reink and reposition the probes if the species delivered are exhausted; the limited response time of the electromechanical scanning systems; and the necessity to scan large areas with a single probe in a serial type of writing. To speed up the process, parallel writing with probe arrays, continuous ink delivery to avoid reinking and realigning interruptions, and speeding up of the ink transfer as much as possible are needed. Parallel writing with probe arrays involves large-scale fabrication processes using microelectromechanical systems (MEMS) techniques, which eliminates the use of pulled-glass nanopipette types of probes; continuous ink delivery means integration of microfluidics into probes – which replaces dip-and-write techniques with fountain-pen-type writing; attempts to speed up the ink transfer can be addressed by optimization of ink solutions, by using electrical biasing of probes, like in EPN. Hidden in the parallel writing requirement lies the necessity of individual control on the positions of the probes in the vertical direction, with independent actuation and independent position readout and, eventually, feedback. Since the primary task for such probes is to write, the optimization of writing is the first to be solved, which leads to a gradual approach for increasing the probe's complexity: first individual probes with fountain-pen-type ink delivery, then arrays of fountain probes, then integration of independent actuation and independent sensing have to be pursued. For the time being, individual fountain probes [57, 58] and arrays of fountain probes have been fabricated, while independent actuation and sensing are being tested separately and can benefit from the experience accumulated from other parallel AFM techniques, such as imaging or data storage/readout [85, 86]. In the following, we present some developments in these directions.

23.2.1.1
Writing Mechanisms with Direct Feeding of Ink

Feeding fluid ink to the AFM probe for writing brings into question if the writing mechanism is or is not the same as in DPN. The answer to this question is rather complex, and requires a deeper insight into the phenomena at the tip-to-substrate transfer of molecular species. Figure 23.6 presents a schematic comparison between writing mechanism of DPN, nanopipettes, apertured AFM probes and NFP.

The "standard" mechanism of DPN starts from an infinite source of ink molecules adsorbed on the probe, which are transferred with the aid of a water meniscus to the substrate surface, where they form a SAM and continue their ride from the tip to the boundary of this layer, as previously discussed. The water meniscus, if present, is formed between the tip and the substrate owing to capillary condensation from the environment, but also by gathering water molecules already existing on the surface itself, as an adsorbed or prewetting layer. The role of this water meniscus (shown also in Fig. 23.3) could be to ease the molecule transfer, but, eventually, the meniscus could be very limited or even missing, while molecules still transfer by direct contact. The ink molecules are usually species with very low solubility in water (MHA, octadecanethiol, ODT, nanoparticles) and, if amphiphilic, as is mostly the case, are supposed to have a preferential orientation on the surface of water, with the hydrophilic end facing the water and the hydrophobic end facing out. The low solubility and supersaturation regime (due to the heavy evaporation of solvent) is favorable for the micellar transition, which means that the amphiphilic molecules will try to connect by the hydrophobic ends through van der Waals forces to form colloidal groups. However, the limited size of the meniscus suggests that the volume properties of solutions at equilibrium, such as concentration and micellation (as well as density, temperature and pressure), cannot describe properly what is happening at this scale, the more suggestive picture being probably a succession of fluctuations with some statistical tendencies. The physical properties of adsorbate layers, especially the coadsorption and codiffusion properties, are also not well known. DPN experiments performed with MHA molecules as ink, with coating of the tips by dipping into solutions of different solvents (ethanol, acetonitrile, dimethylformamide, etc.), and different concentrations, showed different writing dynamics. This suggests that, although not much solvent is left on the tips after drying, it still alters the surface diffusion in the coadsorbed layer.

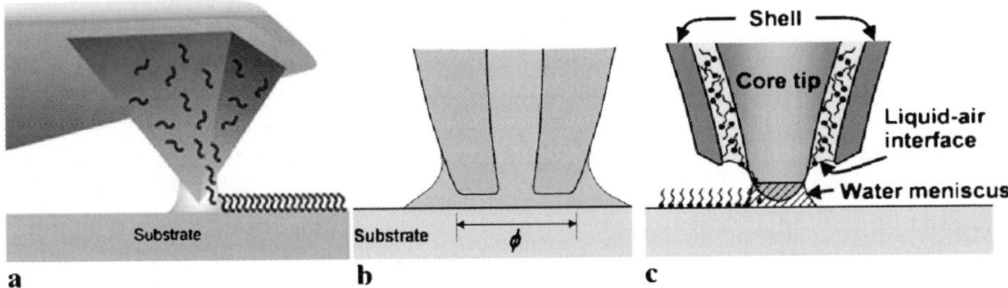

Fig. 23.6. Writing mechanism of DPN (**a**), nanopipettes and apertured probes (**b**) and nanofountain probes (NFP) (**c**)

In case of probes such as nanopipettes [24] and apertured tips [28] (Fig. 23.6b) liquid ink solution is touching the substrate and, owing to capillary action, has the tendency to form an exterior meniscus. The presence of sufficient solution allows this meniscus to grow close to the equilibrium shape, dictated by the pressure balance at the ends of the capillary channel; thus, it is not the inner diameter of the aperture, but the outer diameter of this meniscus that determines the ultimate writing resolution. The best resolution reported with such probes is in the 1-μm range, although capillary probes with outer diameters as small as 20 nm were used. Open channel microspotters [87] essentially share the same meniscus-controlled writing mechanism. Table 23.1 shows a comparison of performances of DPN, NFP and various other types of microfluidic probes.

NFP (Fig. 23.6c) combine both the tip-transfer of ink, similar to DPN, and the presence of liquid ink in close proximity of the tip. One early problem is the transport of the fluid ink through the microchannels, from a reservoir to a special type of tip, with a volcano structure. While in nanopipettes the capillary lumen is large (20–50 μm) and shrinks to a few nanometers (best reported 5 nm [24]) only at the end of the probe, transport by capillary action is very similar to that predicted by classical theory. In the case of NFP, channels with cross sections as small as $0.3-0.5\,\mu$m $\times\ 4-10\,\mu$m are customary, stretching over approximately 1 mm lengths. The surface-to-volume ratio of the liquid is much increased in this case, possibly allowing the formation of an ink concentration gradient along the channel; this is likely to be the case for low-solubility ink molecules which try to occupy the liquid–solid interfaces. The phenomenon is similar to prewetting effects in separation columns, known to work even for mixtures containing highly soluble components. The preferential evaporation of solvent at the tip end creates another source of concentration gradient, this time with higher concentration in the downstream (tip) end of the microchannel. This concentration increase, if not balanced by a continuous supply of solvent by diffusion, may lead to the supersaturation and precipitation of ink molecules, clogging the channel.

Microfluidic simulations performed for the NFP's volcano-shaped tips show different possible scenarios of meniscus formation depending on the contact angles of the materials used for the tips and the shell material (Fig. 23.7). For contact angles, $C_t = 20°$ and $C_s = 65°$, for a tip made of SiO_2 and a shell made of Si_3N_4, a stable fluid–air interface develops. In the case of small contact angles,

Table 23.1. Comparison of different types of micropatterning and nanopatterning probes

	Dip-pen nano-lithography [75]	Nano-pipette [24]	Microspotters SPT (Surface Patterning Tool) [87]	Apertured tip [28]	Nanofountain probe [58]
Resolution	< 100 nm; best 15-nm lines/ 5-nm spaces	~ 1 μm	2–3 μm	~ 1 μm; best 300 nm	< 100 nm; best 40-nm lines
Micro-fabrication	Commercial tips	No	Developed steps	Developed steps	Developed steps
Continuous delivery	No	Yes	Yes	Yes, very limited volume	Yes

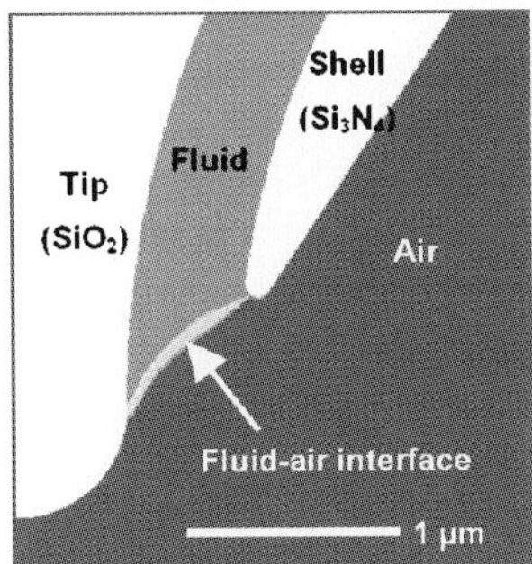

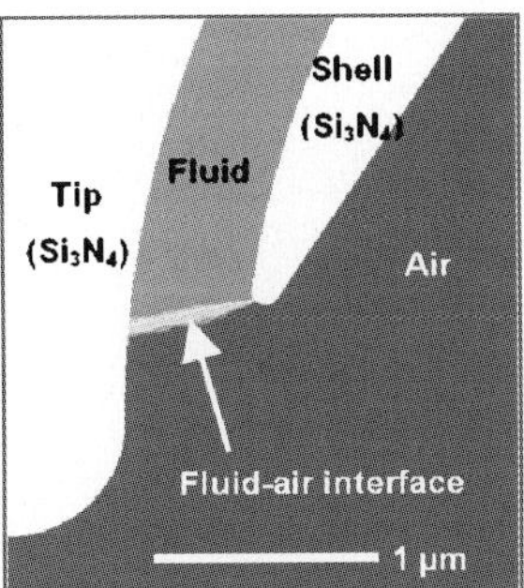

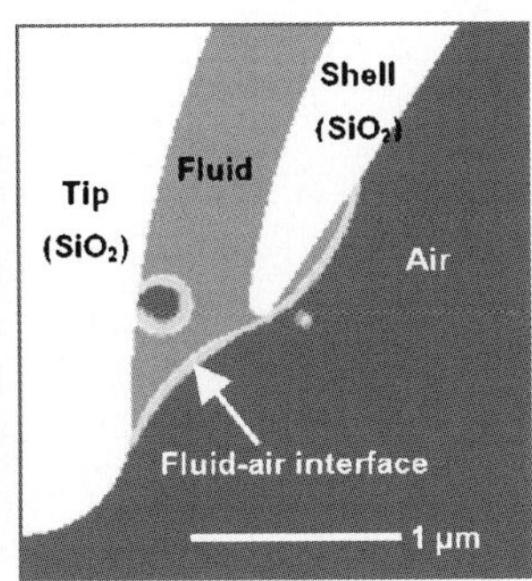

Fig. 23.7. Finite-element simulations (CFD-ACE+) of capillary motion of water through microchannels and a volcano tip, stopping in the final equilibrium position for the meniscus. The contact angles with the tip and shell are, respectively, $C_t = 20°$ and $C_s = 65°$, SiO_2 tip and Si_3N_4 shell (*left*), $C_t = C_s = 65°$, tip and shell both Si_3N_4 (*middle*) and $C_t = C_s = 20°$, tip and shell both SiO_2 (*right*)

$C_t = C_s = 20°$, for a tip and shell both made of SiO_2, the meniscus can surround the shell and lead to unstable flow, potentially affecting the local diffusion processes (Fig. 23.7). While the simulations in Fig. 23.7 were performed for pure SiO_2 and Si_3N_4, crossover or different values for the contact angles can be easily reached by surface oxidation of the nitride, by etching processes, contamination or targeted chemical functionalization. The presence of ink outside the volcano tip is confirmed by SEM images of probes after the ink solution had dried and solidified (Fig. 23.8). However, ink can be present even on quite remote points on the probes, for which meniscus reshaping is a poor hypothesis. A typical case is the ink climbing from the base on long whiskerlike probes, such as those grown on volcano probes by electron-beam-assisted W deposition (Fig. 23.9). Such W nanowires show ink clusters and also performed writing, confirming once again the presence of molecular ink. One possible explanation is that ink molecules are coevaporated with solvent molecules and can recondense on solid surfaces such as the tip and substrate, contributing to the transport of ink by mechanisms other than in the case of DPN. This actually adds to the complexity of the phenomena, since the ink solvent, as viewed from the size scale

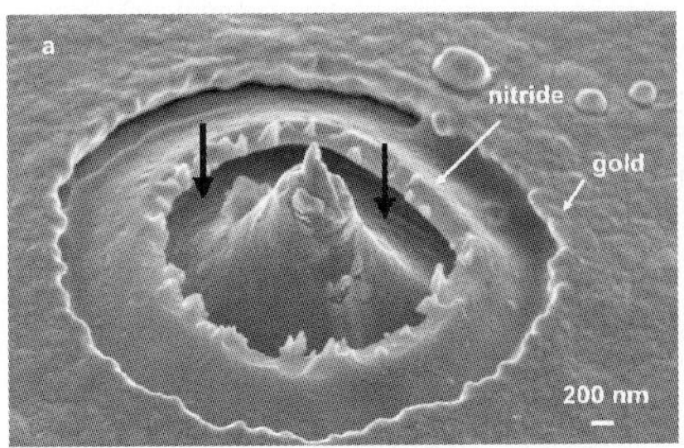

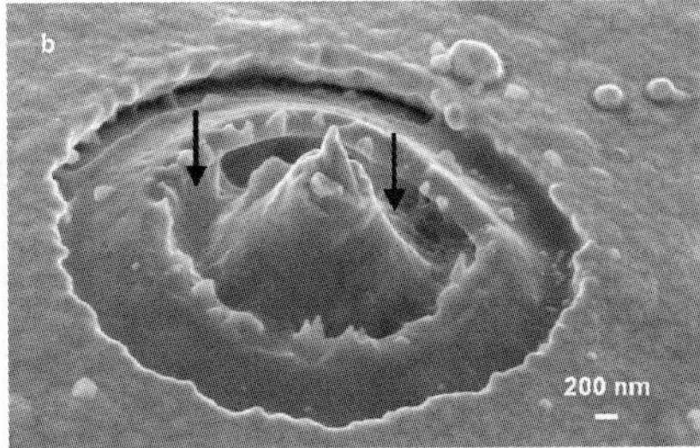

Fig. 23.8. Image of a NFP before (**a**) and after (**b**) ink delivery through the channel. The image corresponds to a first-generation NFP, with the gold sealing layer apparently forming a second volcano shell, although no gap is in reality present between the gold and the nitride

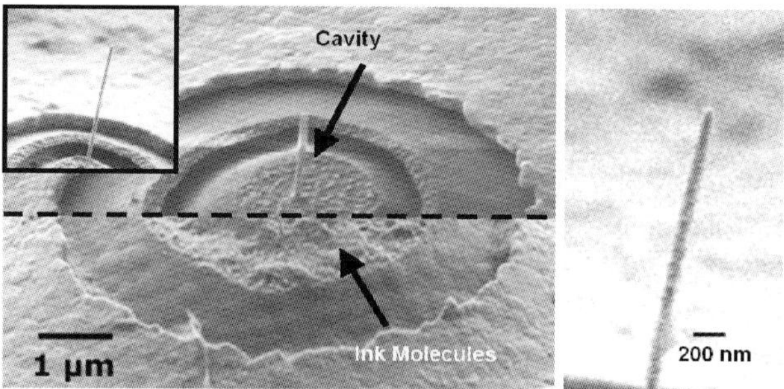

Fig. 23.9. Volcano-shaped probe with a W nanowire. Empty volcano cavity prior to ink supply (*above the line*); volcano filled with molecules after ink supply (*below the line*). Two images are combined for comparison purposes. A nanowire grown by electron-beam-enhanced deposition of W is shown at the center of the volcano. The *inset* and *right image* show the W nanowire with clusters of MHA molecules coating its surface [57]

of the tip, is subjected to heavy evaporation and recondensation. Such phenomena are expected to play an even stronger role in the case of microspotters and open channel devices [85], where they can lead to cross-contamination of probes.

Returning to the microfluidic transport model (Fig. 23.7), another degree of complexity observed in NFP writing is the relative motion of the core tip within the shell. In such architectures, the core tip and shell form a highly compliant micromechanical structure, which could be evidenced during observation of volcano tips with SEM. Owing to charging or possible unhomogeneous heating and thermal expansion, core tips can be seen (live) to move from their nominal center position (Fig. 23.15b). Similar motion is expected to occur when the volcano is filled with liquid. The capillary force on a center pillar in a ring surrounded by a meniscus increases with the off-center displacement of the pillar, leading to mechanical instability. The force opposing this instability is the elastic deformation, which renders the central position of the tip metastable. Thus, during writing, the volcano tip is likely to have an off-center position, touching the shell. The high elastic compliance of the volcano core also allows its vertical motion relative to the shell, resembling a rapidograph writing mechanism. While the vertical motion of the core is favorable (allows rewetting of the tip, "calligraphy" by changing of writing features with the contact force in the AFM cantilever, and can be used as a declogging means), the lateral motion is less desirable, leading to misalignment and decreased position accuracy. This allows the development of different classes of probes: more rigid for high position accuracy, less rigid for calligraphy, etc.

23.2.1.2
1D Fountain Probe Array

The design of a first microfluidic nanoprobe with integrated microfluidics is presented in [58] and was driven by the functionality requirements discussed earlier. A first constraint in the design was imposed by the desire to preserve the high-resolution DPN writing mode. For this reason, the ultimate ink delivery system required an AFM tip, rather than a tube or simple aperture. However, the supply of ink had to be continuous to avoid interruptions for the replenishing of molecules. Since distances along the chip and cantilevers are in the millimeter range, a liquid source was supposed to be the only one providing good molecular transport. In this scenario, the ink molecules must be soluble in the ink solvent (at least to some degree), and the liquid must be delivered through capillary channels close to the tip apex. However, the meniscus must end prior to reaching the apex of the AFM probe to avoid the formation of an outer meniscus as in nanopipette-type devices, which have poor resolution (approximately 1 μm) [24]. To avoid the rapid evaporation of the ink solvent and cross-contamination through coevaporation and remote condensation of the ink species, the channels were preferred to be closed along the path from the reservoir to the dispensing tip. A schematic representation of this design is presented in (Fig. 23.10).

Other constrains on the cantilevers were imposed by the requirement to keep their stiffness low (0.03–0.3 N/m), to achieve both normal (for writing) and LFM (for reading) high resolutions, while the tip material had to be hard and sufficiently hydrophilic to facilitate molecular transport. For instance, silicon nitride was proven to give good results in most DPN experiments. The minimization of processing steps

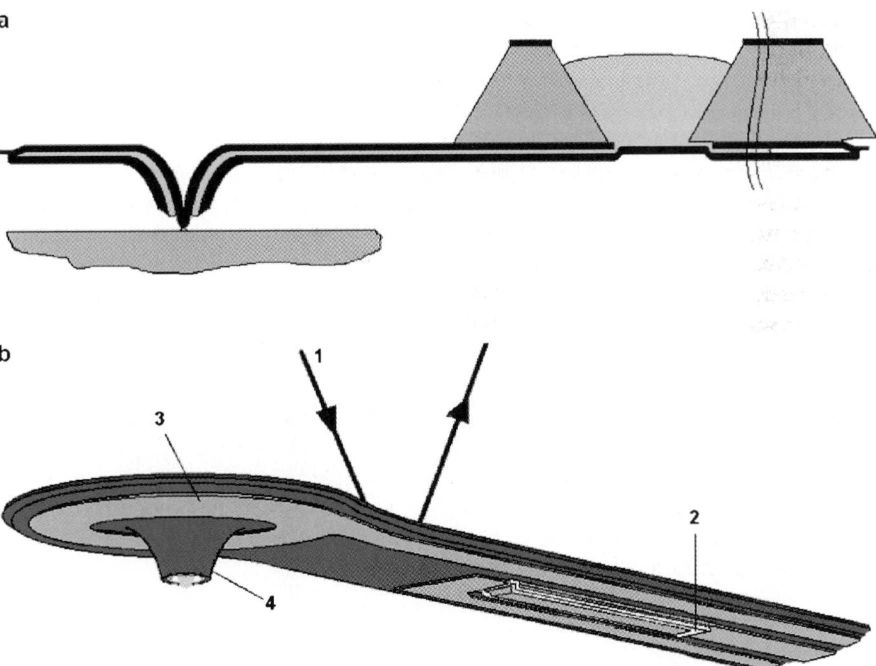

Fig. 23.10. General view of microfluidic nanoprobe: (**a**) cross-sectional view showing microfluidic components, from reservoir to tip; (**b**) 3D schematic view of the cantilever, showing *1* laser beam from the AFM position sensor, *2* piezoactuator, *3* electrode for applying pressure pulses and *4* microfluidic volcano tip. Note that *2* and *3* are in development stages and are discussed here only from a design viewpoint

and the use of common fabrication techniques and materials available in standard microfabrication laboratories were also sought. A versatile design was adopted to permit easy addition of upgrades, such as variation in tip shapes, cantilever stiffness, integration of independent actuation and large-scale arraying capabilities for both the micromechanical parts and the electrical and microfluidic circuitry.

A first exploratory device of this kind, used for testing the writing capability of NFP with ink fed through microchannels as long as 1.35 mm, is presented in Figs. 23.10 and 23.11. The NFP chips had a single on-chip reservoir and five cantilever probes of different lengths, with volcano-shaped ink delivery tips. The chips were designed to use only one probe at a time.

The fabrication of the device used surface micromachining for producing the microfluidic structure, comprising a sacrificial SiO_2 layer for forming the lumen of the microchannels, sandwiched between two low-stress Si_3N_4 layers – the structural material of the cantilevers. The channels were sealed by a bird's beak oxidation to bring the two edges of the nitride together, followed by the deposition of an additional sealing Si_3N_4 layer. The on-chip reservoirs were formed and chips were released by KOH etching.

The first generation of microfabricated NFP chip was capable of writing with sub-100-nm resolution, down to 45-nm line widths, but also revealed fabrication difficulties and possible failure modes. Among them were quite large tip radii in the range 300–500 nm, channel sealing difficulties resulting sometimes in leaking of the channels, easy clogging of the microchannels owing to their small lumen size in some critical portions (0.1–$0.4 \times 5\ \mu m^2$ cross section), a too low cantilever torsional

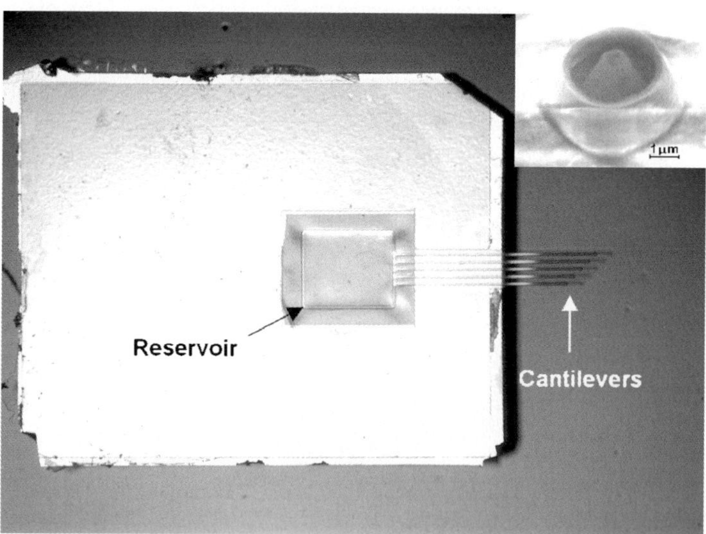

Fig. 23.11. First-generation NFP with on-chip reservoir, microchannels incorporated into cantilevers and volcano-shaped dispensing tips. The *inset* shows a detail of the volcano-shaped tip, at the free end of the cantilever. The dimensions are 2.66 mm × 3.27 mm (chip excluding cantilevers), 810 μm × 891 μm (reservoir) and 400–600 μm (cantilevers). The chip is designed for individual probe writing

stiffness and too small reflective area leading to poor signals in the AFM optical lever detectors, difficulties in achieving the connections between the channels and volcano tips, a large fracture rate of the reservoir membranes during the fabrication owing to the composite nature of the membranes and their different residual stresses, and a quite loose management of wafer real estate owing to the KOH release of the chips.

The decrease of tip radius for this first-generation device was made possible by growing W nanowires through electron-beam-assisted deposition [57]. The deposition of the W nanowire was accomplished when the electron beam was focused at one spot on the tip and a W precursor gas was injected in the SEM chamber (Zeiss 1540XB Crossbeam). For these experiments, configurations with poor volcano-shaped tip geometry were chosen, usually obtained close to the edges of the wafers. SEM observations after ink feeding revealed that transport of molecular inks along the W nanowire is in fact achieved (Fig. 23.9). Molecular precipitation was observed as an increase in nanowire surface roughness.

Although the reinking time could be spared with these devices, the writing remained slow since single-probe patterning is slow as a result of the diffusion-driven and sequential nature of process. This drawback can be solved in part by arraying and parallel writing.

The principle of the first 1D array of NFP fed by two on-chip reservoirs is presented in Fig. 23.12. In the first stage, the 1D array chip should be designed to fit into commercially available AFMs to exploit their scanning and optical lever schemes. By arranging multiple on-chip reservoirs, we should be able to achieve patterning surfaces with different types of inks.

The difficulties of the first generation of NFP were analyzed and overcome by enhancements implemented in a newly designed device and fabrication sequence. The new-generation NFP chip contains a linear array of 12 cantilever probes, with microfluidic channels connecting two on-chip reservoirs to volcano-shaped ink delivering tips. The 12 probes can be used to write in parallel, thus helping to speed

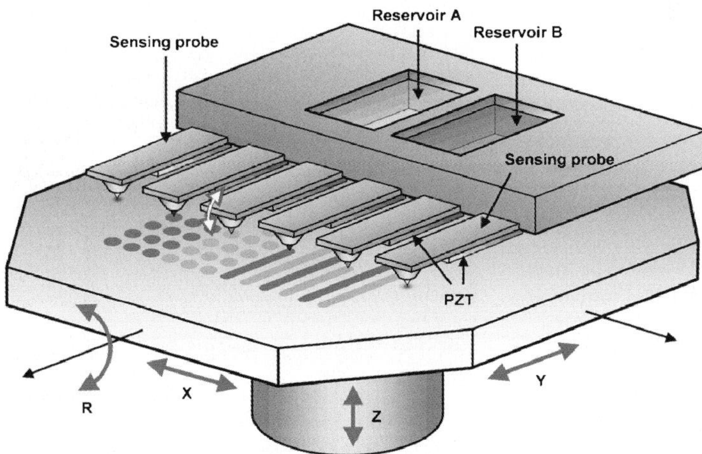

Fig. 23.12. The operation of the 1D array of a NFP with two ink reservoirs. PZT lead zirconate titanate

up the process, but can be also fed with two different inks. A similar array of probes and reservoirs is placed on both sides of the chip to increase the success rate and usage of each chip, which was not possible with the old version of the NFP.

A general view of the microfabricated second-generation NFP device can be seen in Fig. 23.13. The chip has an overall size of 1.8 mm × 3.2 mm (excluding the cantilever lengths), to fit easily into commercial AFM equipment. The cantilever lengths are 630 and 520 µm, respectively, on the two sides of the chip. The different lengths account for a longitudinal bending stiffness of 0.175 and 0.312 N/m, respectively, to provide a choice for different applications. Otherwise, the chip has a fourfold symmetry, and uniform and repetitive architecture, to help in reducing the burden of optimizing the fabrication processes. The reservoirs feed six microchannels each. The microchannels are in part embedded in the silicon chip body and are in part embedded along the length of the silicon nitride cantilevers. The total lengths of the microchannels are 1110 (1000), 1175 (1055) and 1310 (1200) µm, respectively, for the three types of channels surrounding symmetrically the reservoir and the two sides of the chip. Figure 23.14 shows a zoom-in view of a quarter of the chip, where the different components can be identified.

The cross section of the channels is rectangular, with a height of 0.5 µm and a width of 12 µm. The reservoirs consist of a double silicon nitride membrane and a cylindrical well of 160-µm diameter, which is etched from the backside of the chip, to reach this membrane. This cylindrical well gives the volume of the on-

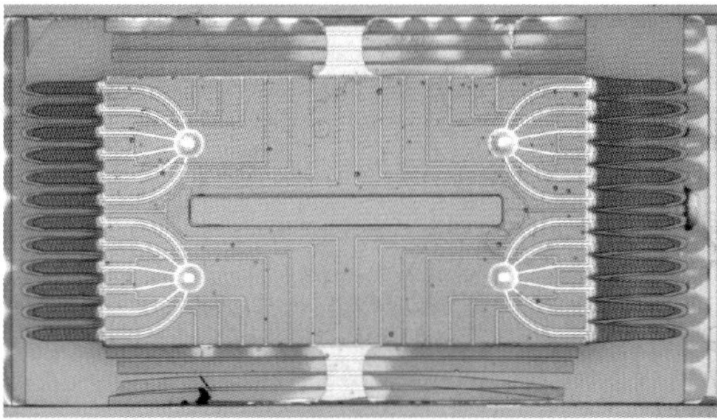

a

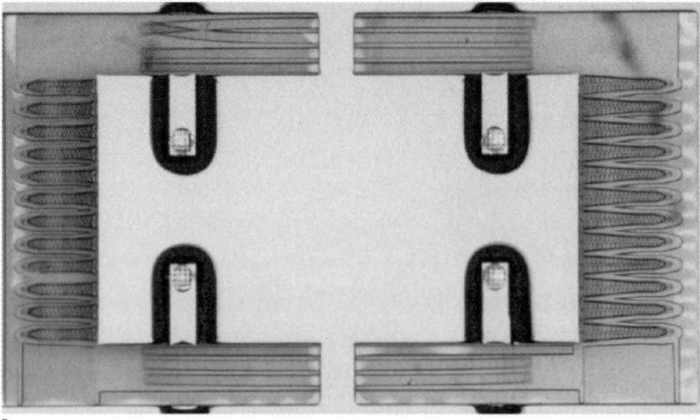

b

Fig. 23.13. Overview of a second-generation NFP chip: (**a**) front-side view; (**b**) backside view. The chip is shown attached to the Si wafer by a pair of Si bridges with notches at their base, for controlled breaking. The long cantilever strips attached to the Si bridges are Si₃N₄ dummy structures introduced for better control of etching and mechanical tests

chip reservoir, which is approximately 60 nl. The well opens at the backside of the chip into a trapezoidal trench of 40-μm depth and 220-μm width (at the wafer surface), extending from the reservoir to the margin of the chip over a total length of 965 μm. This trench is designed to host a 100-μm-diameter capillary which can be used to connect the on-chip reservoir to an even larger, remote reservoir (Figs. 23.13b and 23.14). In this case, the capillary needs to be glued and capped with a thin counterplate to cover the whole backside of the chip. The other side of the

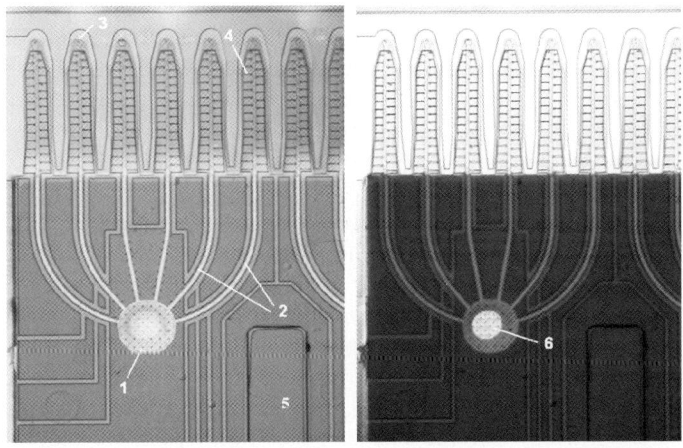

Fig. 23.14. Zoom-in view on a quarter of the NFP chip in combined reflected and transmitted light optical microscopy: (**a**) top view with more reflected light; (**b**) top view with more transmitted light, showing the reservoir well. *1* Reservoir membrane, *2* microchannels, *3* tip, *4* stabilization beam, *5* central spacing post, *6* reservoir well

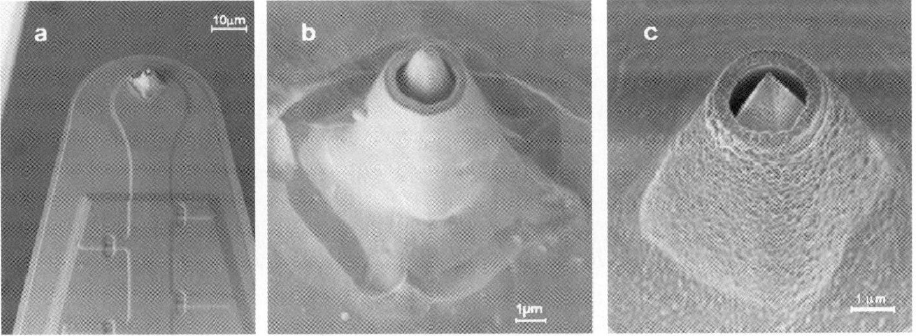

Fig. 23.15. Volcano ink dispensing tips: perspective on the cantilever end (**a**) and tip details (**b**),(**c**). The radius of the tips varies between the thickness of the lower thin nitride film (250 nm) molded around the convex Si precursor (**b**), to tens of nanometers (**c**). The latter was achieved by a sharpening process that occurred with prolonged CF_4 reactive ion etching (RIE) after removal of the top nitride. The off-center position of the core tip is an effect of reversible deformation of the structures that has been regularly observed live during the charging produced by the SEM imaging. The highly compliant structure of the tip is beneficial to the writing process and may lead to a mechanism similar to that of the rapidograph writing nibs

microchannels opens into a volcano-shaped dispensing tip with a height of 7–8.5 μm at the end of the cantilever (Fig. 23.15). The front side of the chip is provided with a central rectangular post of 9–10 μm thickness, to prevent the damaging of the reservoir membranes when mounting the chip under the spring clamps of AFM heads. In case electrodes are present on the chip, as planned for future applications such as EPN or actuated probes, this prevents also their shortcut.

The fabrication of the novel, second-generation NFP chip is presented in Sect. 23.2.2. Various tests performed with NFP chips to test their functioning and performances are presented and discussed in Sect. 23.2.4, where we give also an outlook to the future applications and device developments. The quest for integrating piezoactuators on these types of cantilevers is presented in Sect. 23.2.3.

23.2.2
Microfabrication of the NFP

The fabrication of the device in Fig. 23.16 starts with the formation of silicon tips on top of a silicon wafer, by underetching SiO_2 precursor caps in KOH to form {114}-faceted pyramids. These are further etched by a hydrofluoric acid–nitric acid–acetic acid isotropic etchant for detaching the caps and sharpening. This process is identical to the one used and optimized for the first-generation NFP [58]. The etching for the formation of the tips also produces the central mesa structure used as a spacer in the AFM mount, and recesses the active chip surface about 9 μm below the wafer surface, which prevents damage of the tips in the later lithographic processes and wafer handling. This necessitated the performance of all subsequent optical lithography processes using relatively thick photoresist (7–12 μm of Shipley SPR 220-7), with special care for high aspect ratio structures and side wall profiles.

A first layer of low-stress Si_3N_4 (low-pressure CVD, LPCVD) of 300–350-nm thickness is then deposited and patterned on the backside of the wafer, to form the rectangular windows for etching the trapezoidal trenches for the attachment of the capillaries (trenches can be seen in Fig. 23.13). The process flow in a cross section of the reservoir area can be followed in Fig. 23.16. A KOH etching of these trenches is subsequently performed for a depth of approximately 40 μm. The front-side nitride is then patterned (CF_4 reactive ion etching, RIE) outlining the cantilevers and the interchip. The pattern also contains an array of 3-μm-diameter holes at the place of the reservoir, forming a sieve to later connect the microchannels with the reservoir wells. This bottom nitride layer makes a conformal covering of the Si pyramids and forms the tip material. A SiO_2 layer is then deposited (500 nm, low-temperature plasma-enhanced, PECVD, at 200 °C) and patterned, to form the microchannel core sacrificial layer. Besides the path of the microchannel lumen and reservoir, the SiO_2 pattern contains lateral beams all along the microchannels (details visible in Fig. 23.17). These beams have a twofold role: on one hand, they provide the place from where the sacrificial material will later be contacted and removed while minimizing the sealing perimeter; on the other hand, they stiffen the cantilevers for lateral bending or torsion, while keeping the longitudinal bending stiffness low. This solution was adopted to improve upon the former-generation NFP, while keeping the cantilever stiffness in the desired range. The reflective area of the cantilevers was also increased, to provide a stronger reflected light signal for the optical lever.

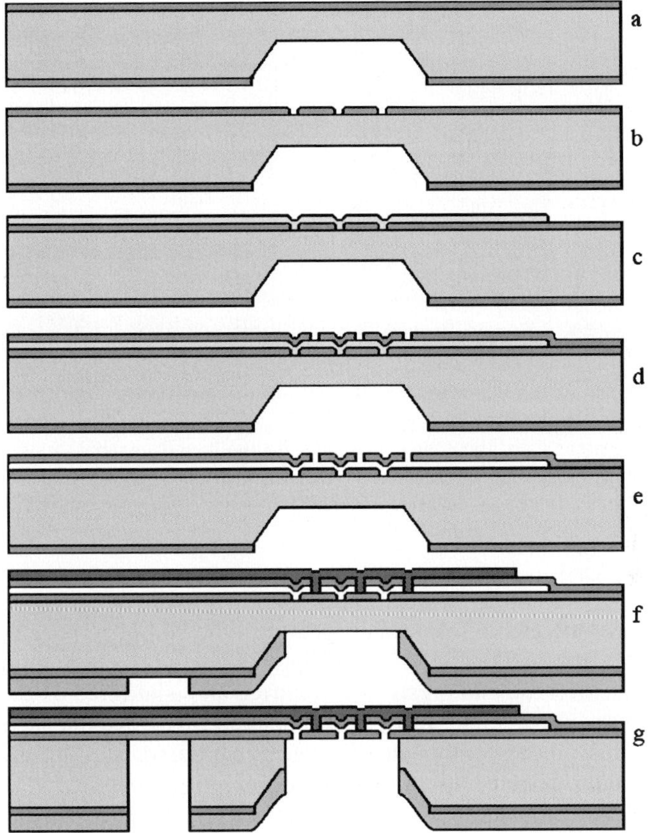

Fig. 23.16. Fabrication sequence of the NFP chip after the formation of the Si tips – the reservoir area. (**a**) Si_3N_4 deposition, backside lithography, and backside KOH trench formation; (**b**) front-side lithography for cantilever and connection holes delineation; (**c**) plasma-enhanced chemical vapor deposition (PECVD) of SiO_2 and patterning for channel core and reservoir delineation; (**d**) low-pressure chemical vapor deposition of low-stress nitride and patterning; resist deposition on front side and etching of protruding tips (not shown); (**e**) etching of SiO_2 sacrificial layer; (**f**) PECVD of SiO_2 sealing layer on the front side and thick resist lithography on the backside of the wafer; (**g**) formation of reservoir wells and chips by deep RIE of Si

The deposition of the top Si_3N_4 layer (500 nm, low-stress LPCVD) continues with the lithographic patterning to enclose the formerly delineated cantilevers, and the SiO_2 pattern. Small openings ($2\,\mu m \times 3\,\mu m$) are provided across the base of the earlier-mentioned SiO_2 beams, and an array of 3-μm-diameter holes is also formed in the reservoir area, symmetrically intercalated with the similar holes made in the first nitride layer. Etching of these features in the top Si_3N_4 layer was done by CF_4 RIE through a thick (10-μm) photoresist mask, preventing the attack upon the tips. This photoresist was removed and replaced by a 5-μm-thick photoresist, through which the tips were this time protruding for about 2–$3\,\mu$m. A $CF_4 + O_2$ RIE was performed to remove the nitride from the top part of the tips, exposing

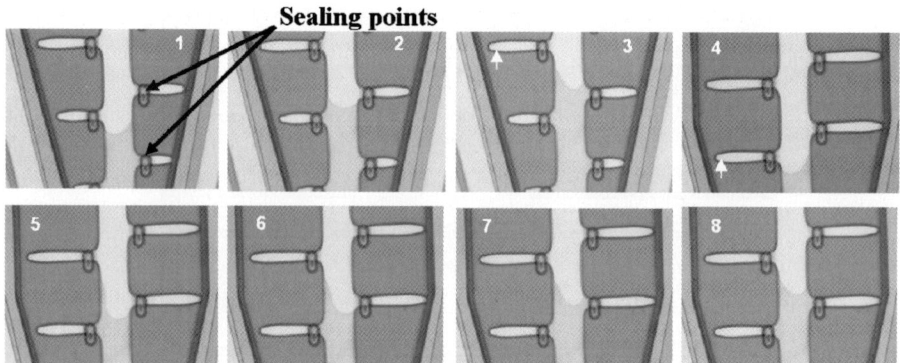

Fig. 23.17. Snapshots of a movie showing the motion of water (*darker tint*) inside the sealed microchannels. *Arrows* point to sealing points and lateral channels. One can notice there is no penetration of water from the central channel into the lateral channel, a sign of a correct sealing. After snapshot *3*, the sample was moved such that the meniscus was still observable; the *upper-left channel* in images *1–3* becomes the *lower-left channel* in images *4–8*. The time spacing between the frames is approximately 1 s

the SiO_2 layer. In variations of this process, leading to sharp tip geometries such as in Fig. 23.15c, the etching was prolonged until the complete penetration of the SiO_2 layer at the tips, and continued for a slight etching of the bottom nitride layer. After these processes, the photoresist was removed and a buffered oxide etch (BOE 10:1) was performed for about 400 min, to completely remove the sacrificial SiO_2 layer. During this process, although not clearly measured, a slight attack on the Si_3N_4 is believed to have occurred, leading to thinning of the nitride especially in the portions where the material was exposed on both sides and for longer time, such as the channels and the tips. This may have contributed to an additional sharpening of the tips. To prevent the collapse of the double nitride membrane in the reservoir region, and the channels in general, supercritical CO_2 drying was used after every wet process following from this step.

The fabrication continued by depositing a SiO_2 layer (approximately 1.5 μm, low-temperature PECVD) to completely seal the holes through which the channels and the reservoirs were cleared from the sacrificial material. Since minimal deposition conformity was desired to ensure that the channel lumen would not be reduced, the temperature in the PECVD process was reduced as low as 150 °C, which produced a material with residual stress similar to that of low-stress silicon nitride (approximately 180 MPa tensile), thus causing the cantilevers to remain reasonably straight after release. The sealing film was patterned lithographically by wet etching (BOE), to remove it from certain areas, including the tip region, such that a complete sealing along the channels and reservoirs was preserved while making the tip orifice communicate with the channels. At this step, the tip–channel–reservoir communication could be proven by watching the water motion and evaporation under the microscope, as shown in Fig. 23.17 and explained later.

The next step was the deep RIE (DRIE, Bosch process) of the backside of the wafer, through a 9-μm-thick photoresist layer patterned with the reservoir wells and the chip delineation pattern. For this process, the photoresist was UV-hardened and baked for 9 h at 80 °C, to prevent the reflow, especially critical in the well region, since the wells are in the deeply recessed area of the trapezoidal trenches for the capillary connections. This fact imposed an optimization of the lithographic process. Since the etching rate in the DRIE process is highly sensitive to the area of the features, simultaneous opening of the reservoir well with reaching the cantilevers from the backside in the interchip required special attention. A preliminary study with dummy wafers showed that, if starting from a flat surface, the interchip complete opening was reached while the reservoir well remained with about 40 μm to be etched. This result obviously depended on the particular wafer size (in this case 3-in. wafers of 380–425-μm thickness), the exposed-to-unexposed area ratio, the relative positioning of features, the equipment used (Unaxis SLR 770), etc. The 40-μm "handicap" for the large area proved reasonable for the thickness variation of wafers in our case.

For the etching, the 3-in. wafers were glued on handling 4-in. wafers using a ZnO-containing Dow Corning 340 silicone heat sink compound, which was applied manually on the front side of the NFP wafer, mainly on the Si frame area and especially avoiding the reservoir areas. This mounting also prevented the stopping of the process at the first punch-through events on the NFP wafer. The DRIE process was carried on in sequentially decreasing steps, by carefully measuring the initial thickness for each individual wafer and the etching depths achieved at each step. The rigorous control was needed since the Si_3N_4, although is has a lower etching rate than silicon in the Bosch process, is not an effective etch stop material and could have led to damages at the rear side of the cantilever channels. The DRIE process was stopped when the reservoir well reached the double Si_3N_4 membrane. Fine removal of Si for clearing completely the backside of the cantilevers was possible in some cases using XeF_2 etching, but for a limited time, since after approximately 30 min of etching, damages on the Si_3N_4 could be noticed. However, in most cases, this process was not necessary.

The choice of DRIE over KOH etching for the chip release and reservoir formation was motivated by the small amount of space available in the chip to fit two reservoirs, given the fact that the NFP chip should be of standard size for conventional AFM heads. Another advantage of DRIE is the increase in the number of chips fabricated per wafer, owing to the elimination of tilted walls and convex corner compensation beams in the interchip, required by the KOH release process [58].

After completing the etching, the NFP wafers were detached from the handling wafers in hot Nanostrip solution (Cyantek) by careful lateral sliding and were kept in solution until complete removal of the silicon heat sink compound was achieved. Deionized water and methanol rinsing followed and supercritical CO_2 drying completed the process. At this step, the NFP chips remained attached to a Si frame for easy handling, as shown in Fig. 23.13. The NFP wafers later underwent a gold sputtering (approximately 20 nm) on the backside, to enhance the reflectivity of the cantilevers (not shown in Fig. 23.13).

The connectivity between the reservoir, channels and volcano dispensing tips could be checked, after etching the sacrificial oxide layer and performing the sealing step, by imaging the chip under an optical microscope. In this experiment, a water was fed into the channels by placing a droplet onto a volcano tip using a micropipette.

The motion of the water meniscus, due to capillarity and evaporation, could then be clearly observed and recorded (Fig. 23.17). After releasing the chips, the connectivity between the reservoir and microchannels was also tested by feeding water with a micropipette into the reservoir well. Proper sealing could also be tested in this manner, by observing the color change and meniscus motion, not proceeding beyond the sealing points and into the lateral channels. The connectivity from the reservoir up to the dispensing tips could also be substantiated from writing tests performed with various inks.

Figure 23.18a shows a SEM image of a volcano-shaped tip after feeding the reservoir with DNA and buffer solution. The image shows evidence of unstable flow with molecular ink around the tip and the volcano outside wall. As discussed in the context of Fig. 23.7c, depending on the wetting angles, which in turn are a function of the buffer chemistry for a given microchannel material, unstable flow may result. In this case it is clear that such unstable flow took place in agreement with the numerical predictions. The experiment was conducted to demonstrate the capability of the NFP to deliver copious amounts of ink if desired.

Evidence of fluid ink persistence in the reservoir and channels during sufficient periods of time, such as those required for performing writing tests, was also obtained. Simple timing measurements performed under a stereomicroscope revealed that solutions of water and alcohol persisted in liquid form for 10–30 minutes in the well even in the absence of capping. It was observed that the capillarity dynamics keeps the channels filled with liquid during all this time and that the channels are the last to dry out.

Despite these observations, there is also more direct evidence of fluid flow in the NFP system during writing. For instance, in tests performed with a solution containing fluorescein isothiocyanate labeled DNA there were instances in which a 10-μm-diameter spot formed rapidly as a result of the unstable flow previously described (Fig. 23.18b). We found that by controlling the buffer chemistry, stable flow could be achieved. We will illustrate the NFP direct patterning of DNA molecules with a resolution of 200 nm in Sect. 23.2.4.

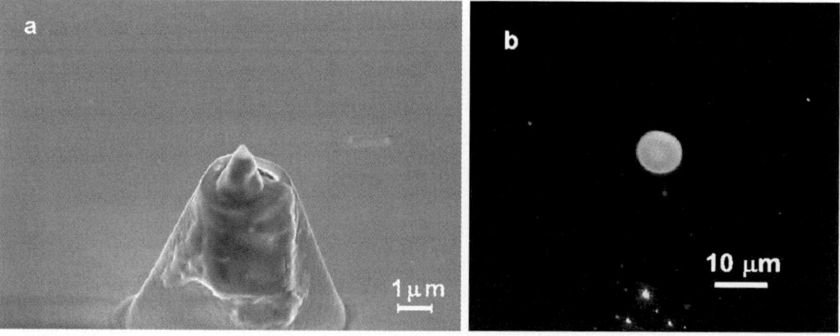

Fig. 23.18. Evidence of molecular ink coating and unstable flow during the writing process. (a) SEM micrograph showing unstable flow from a volcano tip, after feeding a solution of DNA and buffer into the reservoir. (b) A 10-μm-diameter fluorescent dot of fluorescein isothiocyanate labeled DNA on a gold substrate formed as result of unstable flow (see also Fig. 23.7c and its discussion)

23.2.3
Independent Lead Zirconate Titanate Actuation

Independent actuation of probes and position/contact sensing are important aspects in the modern operation of AFMs. Normally, an AFM probe is moved in the z-direction by a piezocrystal mounted in the AFM head. The motion in the x–y-directions relative to the sample surface is achieved either by the x–y motion of the sample-holding stage, or by the x–y motion of the AFM head. In both cases, thermal or piezoactuation can be employed. In most modern AFM equipment, the x–y (scanning) motion of the AFM head is performed with a piezoelectric crystal of a special geometry, allowing the independent control of the z-motion and the x–y scanning. This is achieved at the expense of increasing the size of the AFM head (usually approximately 2–4 cm in diameter). The optical z-motion detector may also be included in this head. In the case of arrays of cantilevers as AFM probes, AFM heads can provide only one degree of freedom in the z-direction, i.e., a global motion of the entire probe chip. In the specific case of the "millipede" approach developed by IBM [86, 88], the z-motion of each cantilever in a 32×32 array is controlled by a thermal actuator. The device works in contact mode, such that a continuous contact is necessary during scanning. A global leveling scheme is used to keep the probe array parallel and in contact with a flat thermoplastic polymer surface. For this, four cantilevers at the corners of the 2D probe array are feedback-controlled to keep the probe array at approximately 1 μm from the substrate. Except for the four feedback-controlled cantilevers, all the others are moving without control. The method was previously used by Lutwyche et al. [89] to perform parallel imaging with a 5×5 array of AFM probes.

Multispot optical sensors for measuring the positions of several cantilevers at the same time (or sequentially, but at a fast rate) have not yet been developed. It is obvious then that independent actuation and control of probes in multitip AFM devices has to pursue a different path. Several methods for independent actuation were experimented.

Thermal actuation in the context of the DPN technique was pursued by Bullet et al. [90]. In this implementation, a general z-motion control for a linear array of probes is provided by the AFM head with feedback on one cantilever, while the others are independently actuated via bimorph thermal actuators with contact only assumed by the parallel placement of the probe array with regard to the substrate.

Piezoelectric actuation of AFM cantilevers was successfully pursued and is documented in the literature [91–93]. Two piezomaterials were employed: ZnO (single cantilevers with ZnO actuation are commercially available [94]) and lead zirconate titanate (PZT) [95, 96]. These materials can be used both for actuation (direct piezoelectric effect) and in sensing (inverse piezoelectric effect). Sensing can also be achieved on the basis of the piezoresistive effect in thin films or doped regions in silicon cantilevers. Complicated 3D structures for piezoresistive sensing both in the z-direction and in the lateral direction (y) were developed [97] and capacitive sensing was also reported [98]. The read-out circuitry, to measure sensor deflection, can be integrated on the chip, using a standard CMOS process [99], in which case the cantilever definition has to be a postprocessing step. The presence of an actuator on the AFM cantilever can be used also to control actively the Q

factor of the cantilever, with benefits in increased scanning speed. For this, a force proportional to the deflection speed can be applied on the actuator, mimicking an increased damping factor.

The major difficulty in piezoelectric actuation is the integration of the piezoelectric material deposition, thermal curing and patterning in the general probe fabrication process. Piezoelectric materials are generally high thermal budget materials: they require high-temperature deposition or a thermal treatment for proper crystal formation. Adhesion is also a problem; hence, well-tailored electrodes are necessary such as Al with SiO_2 protection for ZnO [91] or a stack consisting of 0.25 μm Ti/Au, 3.5 μm ZnO, 0.25 μm Ti/Au [94]. PZT usually requires sputtered Pt thin films as electrodes and as adhesion layers, which also demand special attention in their patterning. Additionally, ZnO and PZT have a poor chemical resistance to most acids and bases used in the lithographic processing of the cantilevers, for which the piezomaterial needs to be protected during wet etching processes (such as KOH etching of Si or buffered HF for etching of SiO_2). RIE of ZnO or PZT is based on chlorine or fluorine chemistries, which also etch Si, Si_3N_4 and SiO_2. This low selectivity with regard to metals and Si-based materials makes it necessary to deposit additional protective layers (e.g., polyimide, wax, Teflon AF), while patterning of the top electrodes can be achieved only by lift-off processes [85] or by processes that etch both the metal and the piezomaterial [91]. While PZT has better piezoelectric properties than ZnO, it contains lead, which presents contamination problems. As a result, only postprocessing schemes or dedicated equipment can be used for patterning (such as RIE, DRIE, furnaces). Both ZnO and PZT are brittle materials and as such they raise challenges related to cracking, delaminating or bending induced by thermal mismatch or intrinsic stress. For DPN or NFP patterning, the desired cantilever stiffness is in the range 0.03–0.30 N/m; hence, the significant increase in stiffness resulting from the presence of a piezoelectric film is somewhat undesirable and needs to be taken into account in the cantilever design. Because of this and the inherent presence of electrode materials, low-stiffness AFM probes can be achieved only at the expense of increasing the cantilever lengths (700–1000 μm) [91]. Depending on the deposition method, the piezoelectric materials for cantilever actuators have thicknesses in the range 0.2–5 μm. Although better piezoelectric properties are expected from thinner films [100], the necessary thickness of the material has to be determined from constraints in operating voltage and required cantilever end deflection. The calculation method is presented in [85]. Thickness and piezoelectric constants are dictated by the type of material (ZnO or PZT) and deposition method: 0.5–2 μm of ZnO films can be deposited by direct or reactive sputtering (sputtering from a Zn target in an O_2 atmosphere), while PZT can be deposited by metal organic CVD (MOCVD), sputtering [101], reactive RF sputtering [100] or sol–gel methods [102]. A thermal treatment at 650 °C is usually necessary for sintering the PZT to form the perovskite phase, while a Pt film on the substrate (also used as the electrode) helps in texturing the grains with the c-axis (which presents the highest transversal piezoelectric effect) normal to the substrate. To exploit this orientation of the c-axis, the PZT has to be sandwiched between the electrodes (rather than forming interdigital electrodes with PZT on top).

The integration of ZnO and PZT films on special AFM cantilevers of the type used in the NFP chip was experimented. The PZT material proved superior in terms of general process compatibility and thickness requirements to achieve the same bending effect. The PZT deposition and patterning processes (Fig. 23.19) were studied and optimized to avoid destructive interferences in the fabrication processes. Successful cantilever array actuation was achieved by depositing a 250-nm-thick PZT

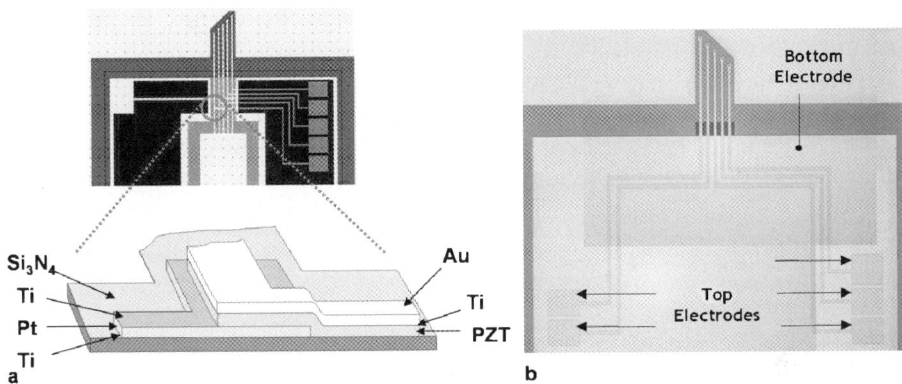

Fig. 23.19. Fabrication scheme (**a**) and optical image of cantilevers with PZT actuators and electrodes prior to release (images were taken portion by portion owing to the limited magnification of the microscope used, and were subsequently combined). Cantilevers were released afterwards by XeF$_2$ etching and were successfully actuated by applying square waves through the electrodes

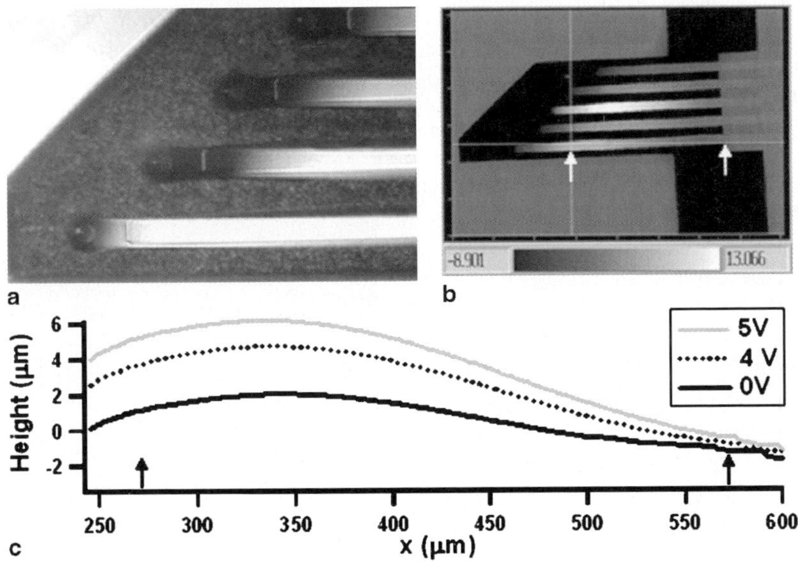

Fig. 23.20. (**a**) Optical micrograph showing released cantilevers of the first-generation NFP with PZT actuation, with the lower cantilever actuated (raised) and the three upper cantilevers unactuated (down); actuation movie available at [103]. (**b**) Phase-shift interferometric microscopy image for actuation studies. (**c**) *Curves* showing the cantilever longitudinal profile upon applying actuation voltages from 0 to 5 V

layer by MOCVD on patterned Pt electrodes (lift-off). After sintering at 650 °C, a Pt top electrode was patterned on the top of the PZT material by lift-off. The Pt electrode was also used as a masking layer to etch the PZT in a CH_2FCl RIE plasma (100 standard cm^3/min, 50 mTorr, 200 W). The cantilevers were released by XeF_2 etching of the underlying Si, a process that did not interfere much with the NFP cantilever structures (although a slight attack on the Si_3N_4 is known to result during long exposure times). Released cantilevers are shown in Fig. 23.20a. Employment of a phase-shift interferometric microscope allowed independent actuation of the cantilever to be assessed (Fig. 23.20b,c) [103]. The bending effect resulted in a total cantilever end displacement of 4 μm for a 500-μm-long cantilever at an applied voltage of 5 V. Repeated bending by application of a 5-V peak-to-peak square wave with a frequency of 100 Hz for 10 h showed no change in the electromechanical response.

23.2.4
Applications

23.2.4.1
Molecular Ink Writing with Continuous Ink Feeding

Several types of inks were tested with the NFP devices. To take advantage of the experience accumulated from DPN, ODT and MHA inks in conjunction with gold substrates were first tested. Patterns with line widths as small as 40 nm have been successfully written using "normal" volcano tips (without a W nanowire). In these experiments, an ink – a solution of MHA in ethanol at a concentration of 1 mM – was supplied via the on-chip reservoir to the dispensing tip by capillary action. After the tip had been brought into contact with a gold substrate, the molecules were transported to the substrate, creating the desired pattern as the tip was laterally

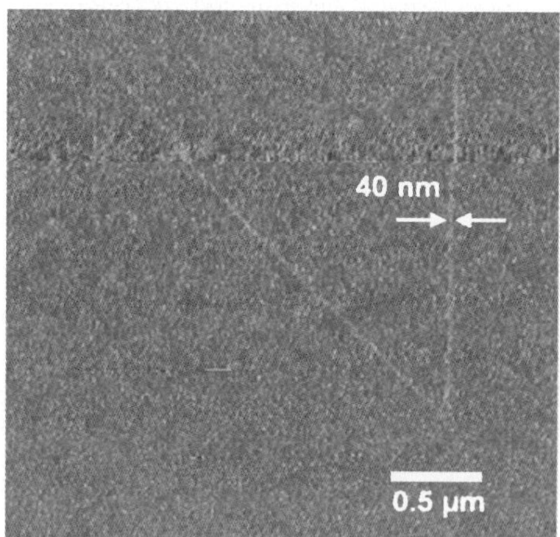

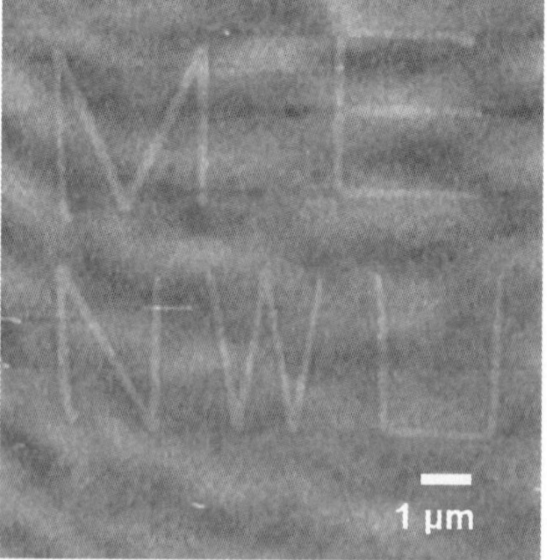

Fig. 23.21. Lateral force AFM image of MHA deposited in the shape of the letter N [57] (**a**) and text (**b**) [58] onto a gold substrate by the volcano tip with first-generation NFP. Both writing and imaging were performed by the same tip

moved along a preprogrammed path (Fig. 23.21). The pattern was obtained with a sweeping rate of 0.05 μm/s, at room temperature and a relative humidity of 60%. The writing process was as repeatable as the DPN writing mode.

Taking into consideration that the microfabricated tip radius is about 3 times the patterned line width, we find that the resolution is somewhat surprising. We attributed this feature to the tip roughness, resulting from the RIE step during its microfabrication, which leads to a much smaller contact area than the one that would occur if the tip were perfectly smooth.

For a given protocol, writing with MHA produced repeatable and uniform line patterns (Figs. 23.21b and 23.22). The reported writing tests were generally performed 5–10 min after feeding the ink into the reservoir. Observation of the ink drying process in the reservoir of unmounted chips showed that the ink remains in the liquid state for about 12 min. Tips with longer delaying times before writing also performed well (dried ink).

Several writing protocols using thiol-based inks in conjunction with gold substrates and acetonitrile as solvent were performed with the second-generation NFP with similar results. More interestingly, the device was able to achieve parallel and dual ink writing. Simultaneous patterning with two types of inks was performed using solutions of different thiolates. MHA and $1H,1H,2H,2H$-perfluorododecane-1-thiol (PFT) were selected for testing since PFT produces dark contrast in AFM friction images while MHA gives bright contrast. We prepared saturated solutions in acetonitrile of both materials. One reservoir was filled with a droplet of MHA solution using a micropipette. Subsequently, with the micropipette tip replaced, the other reservoir was filled with the PFT solution. This feeding process was performed under a stereomicroscope to confirm that the droplets of both solutions did not cross-contaminate. Following the feeding procedure, the chip was mounted in an AFM instrument (Thermomicroscopes CP). Since only four cantilevers could be

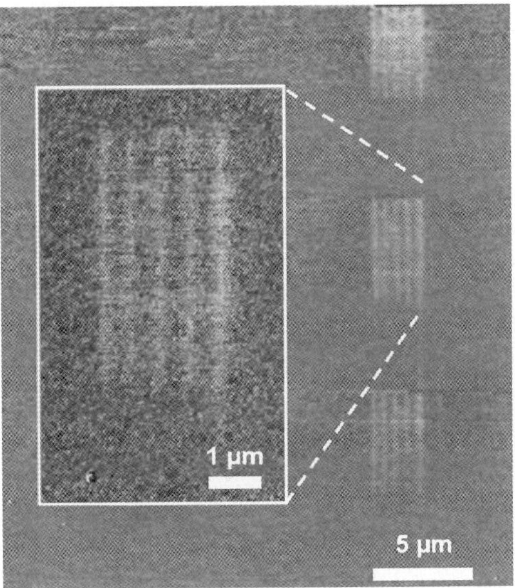

Fig. 23.22. Lateral force image of arrays of lines written with MHA on gold, using the second-generation NFP device. The line widths in this picture are approximately 150 nm, but lines as fine as 40 nm were obtained [57]

seen by the CCD camera of the AFM, the viewing field was adjusted to capture the four cantilevers from the middle of the 12-cantilever array. While the deflection of only one of the four cantilevers was monitored by the AFM optical lever, the others were passively operated and supposed to touch the substrate roughly in the same way as the monitored one. Once the tips had been brought into contact with a gold surface, customized software was used to move the tips along the surface to pattern dot arrays. For characterization purposes and to avoid distortion of the dot patterns by depositing new ink, the patterns formed were examined with a different probe, a commercially available silicon nitride imaging tip. With a reference mark on the surface, the imaging tip was located first on an area where a MHA-fed tip was operated. The surface was scanned to obtain a LFM map (Fig. 23.23a). Subsequently, the tip was moved to an area where PFT was patterned, followed by imaging the area (Fig. 23.23b). Similar patterns were recorded for arrays of dots written with different probes of the array. The two molecular inks were chosen on purpose to have similar writing dynamic. For instance, independent tests demonstrated the writing capability of the NFP with both MHA–acetonitrile and ODT–ethanol solutions; however, the much lower writing speed of ODT made the simultaneous writing impractical.

In separate experiments, NFP chips were tested for patterning lines, using MHA solution in acetonitrile as the ink at various scanning speeds. Writing proved reliable up to a writing speed of about 15 μm/min. The minimum line width obtained was approximately 78 nm. (Fig. 23.24).

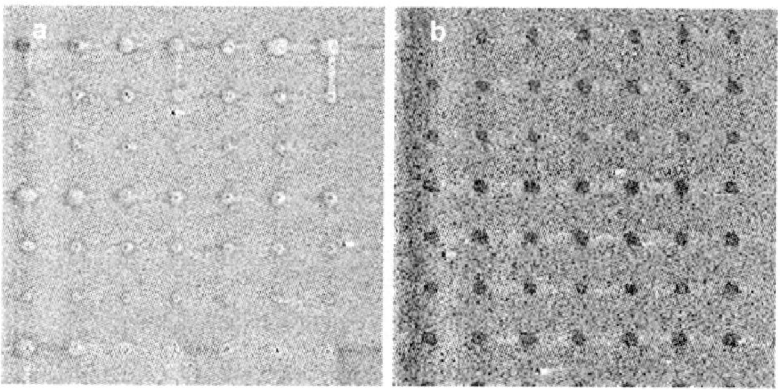

Fig. 23.23. Dot calligraphy test with the NFP for simultaneous writing with two different inks. LFM images of areas (15 μm × 15 μm) patterned with MHA (**a**) and 1*H*,1*H*,2*H*,2*H*-perfluorododecane-1-thiol (**b**). The difference in the sizes of the dots is deliberate: contact times were 6, 4 and 2 s for alternating rows, respectively. Relative humidity was maintained at 70% ± 5

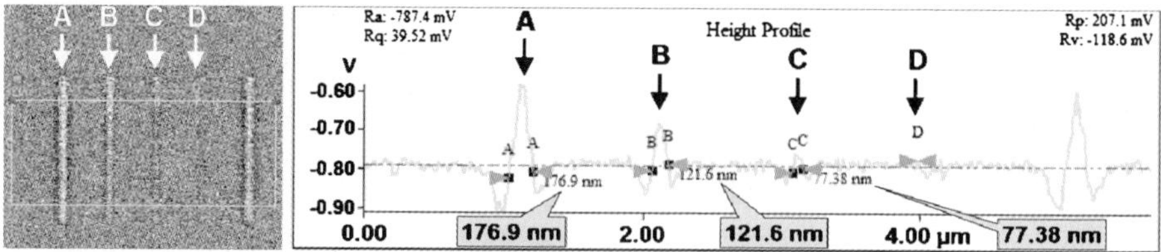

Fig. 23.24. Line patterns written with the NFP on a gold substrate with MHA ink. The LFM image was used for section analysis. The sweep rates for the unit set of lines (*A*, *B*, *C* and *D*) were 3.2, 6.4, 12.8 and 25.6 μm/s, respectively

The patterning capabilities of NFP were also tested with other type of inks, such as ODT, hexamethyldisilazane, diamond nanoparticles, DNA and proteins. The last two are of particular significance for making the NFP a specialized tool for biopatterning, as presented in the next section.

23.2.4.2
DNA and Protein Arrays

Precise placement and control of biological materials down to a few molecules per surface spot is crucial in developing bioassays and novel biomolecular recognition techniques, such as sequencing, protein identification and combinatorial drug testing. Reducing the size of the active assay area is essential for maximizing the probability of interaction between molecular species, when they randomly diffuse over a nanoarray area, especially when the number of molecules to be identified is limited. While decreasing the number of reacting molecules, there is, however, an increase in the relative weight of fluctuation in the number of occurrences per spot, which makes an assay significant statistically only if the experiment involves a high number of spots. Thus, arrays of high density of spots are needed and are highly desirable in biomolecular applications.

Since the introduction of DNA or protein microarrays, which consist of a dot matrix of such molecules, developments in the life sciences have been expedited. Such controlled deposition is the key leading to the development of many surface patterning and advanced cantilever-based patterning tools [24,27,28,57,87,104,105]. A few micron-diameter arrays of dots are currently produced with microspotters, but the challenge is to produce submicron dot arrays, with sub-100-nm diameters, for which DPN and NFP are a sensible technical solutions.

DPN was used to pattern modified oligonucleotides on surfaces of gold and modified silicon oxide [51]. In typical DNA patterning by the DPN technique, silicon nitride tips were coated with 3-aminopropyltrimethoxysilane (APTMS) in order to improve the coating effectiveness of DNA. Micropipettes were also employed to deliver such molecules. A solution contained in a pipette was ejected by electrochemical current on a glass surface [106]. Likewise, cantilevered nanopipettes were employed to locally deliver proteins on surfaces [107]. The resolution of the micropipette-based technique is critically determined by the outer diameter of the tip as the solution tends to flood out of the aperture when brought into contact with surfaces – especially hydrophilic surfaces. The feature size is usually above 1 µm and the minimum feature size achieved so far is approximately 250 nm. To achieve this resolution, it is pivotal to fabricate tips with a small aperture. Unfortunately, reproducibility in micropipette or nanopipette fabrication is low; hence, micromachining techniques have attracted significant interest. For example, cantilevers were micromachined to mimic conventional pin-type microspotters. A buffer solution is transported via open microchannels to the gap between cantilever ends and deposited on a substrate by direct contact. Miniaturized gaps were produced by improved design to achieve patterned feature sizes as small as 2–3 µm [27]. However, it still remains a challenge to accomplish submicron patterning with such microspotters because the feature size is determined by the gap size, which is about 1 µm when conventional UV lithography is employed. The so-called NFP provides a much better resolution for the reasons elaborated in a previous section. Continuous molecular ink-feeding in closed microfluidic systems and high-resolution writing, as small as approximately 40 nm, were achieved with this technique [57]. Direct submicron pattering of DNA solutions, continuously supplied to the volcano tip rather than coated by dipping and *without* the need for APTMS priming, was also demonstrated. Capillary feeding was utilized to continuously supply ink to the tip from an on-chip reservoir. As presented in Fig. 23.25, the NFP device was used to pattern gold surfaces with alkanethiol-modified oligonucleotides. The patterned DNA spots were subsequently hybridized with complementary DNA functionalized Au nanoparticles (approximately 15 nm in diameter).

Patterning was performed with a commercial AFM (Thermomicroscopes CP), with one cantilever selected for deflection detection, while the other cantilevers passively made contact with the surface. The AFM was placed in an environmentally controlled glove box at a relative humidity of $70 \pm 5\%$ and room temperature.

The ink used was a solution of 6.60 µl phosphate buffered silane, 3.06 µl dimethylformamide and 0.34 µl water containing 0.3 M $MgCl_2$ and 10 mM hexanethiol-modified oligonucleotides. The solution was fed into the on-chip reservoir using a micropipette, but direct feeding via a capillary tube adapted to the chip is possible for large-scale applications. Immediately after the feeding, the chip was mounted into the AFM. Once the tip had been brought into contact with the gold surface, customized software was used to laterally move the tips. An array of dots was patterned with three different contact times of 2, 1 and 0.5 s, for each alternating row of the array. After the patterning, the surface was passivated by immersing the substrate in a 1 mM C6 hexanethiol solution of ethanol for 5 min. Subsequently, the substrate was rinsed with ethanol followed by distilled water, and then blown dry with nitrogen gas.

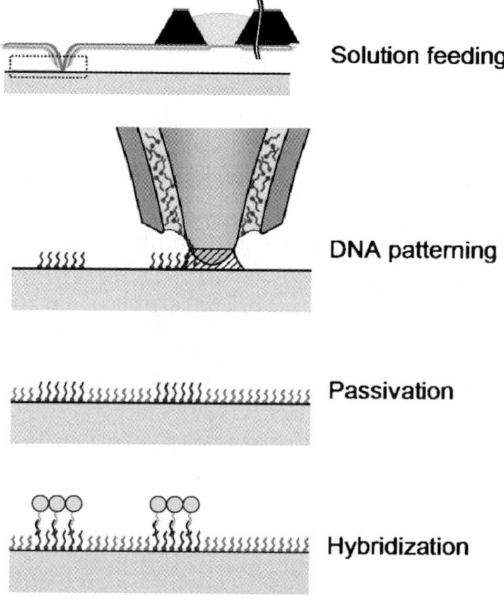

Fig. 23.25. DNA patterning by NFP

In order to examine the patterned DNA spots, complementary DNA functionalized Au nanoparticles (approximately 15 nm in diameter) were applied. A three-DNA-strand system was used, consisting of capture, target and probe strands [108, 109]. While the patterned DNA worked as a capture strand, target and probe strands were applied after the patterning and the C6 passivation. The Au nanoparticles had been functionalized with the probe strand and the target strand was used to link the capture and probe strands. The substrate with the patterned DNA array was immersed in a solution containing both linker and probe strands. After approximately 40 min of hybridization at 37 °C, the substrate was rinsed with ammonium acetate buffer to remove nonspecifically bound Au nanoparticles.

The hybridized spot array was imaged by dark-field optical microscopy and AFM topography (Fig. 23.26). In the dark-field optical micrograph shown in Fig. 23.26a, the assembled nanoparticles produced Rayleigh scattering. The larger dots in rows 1 and 4, produced by 2 s of contact time, are easily visible. Some of the smallest dots in rows 3 and 6, produced by 0.5 s of contact time, are less clear. The AFM topography was obtained in tapping mode, and the measured height was approximately 20 nm. All the dots are clearly defined except for a few missing dots.

The typical spot size routinely produced with this technique was about 200–300 nm in diameter, depending on the contact time. The oblong shape of the spots is directly related to the tip shape and, eventually, to its mechanical deformation during the contact time as a result of the tip compliance within the volcano shell. The resolution achieved is much better than that obtained in previously reported cantilever-based tools. In contrast to micropipette-based devices, it has the advantage of scalability. With the simultaneous multi-ink patterning combined with the DNA patterning and scalability to a 2D array, the NFP has the potential of being the tool of choice to mass-produce microarrays of DNA and proteins. Since it does not need pretreatment of the tip prior to DNA patterning, it would require fewer preparatory steps and fewer opportunities for contamination or damage in the array of tips.

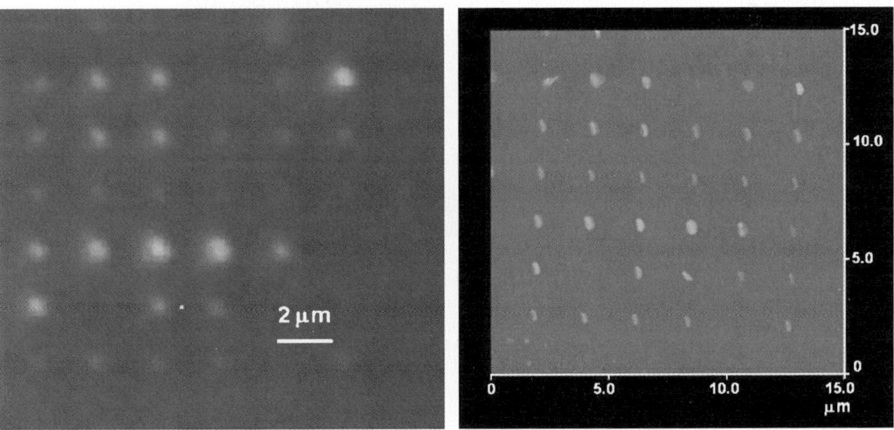

Fig. 23.26. Optical dark field image (*left*) and AFM topography (*right*) of the DNA dot array patterned by second-generation NFP. The contact times were 2, 1 and 0.5 s, respectively, for each alternating row of the array

23.2.5
Perspectives of NFP

The future of NFP will likely consist of the development of 2D arrays of probes capable of writing with several molecular species, with independent actuation, and eventual independent sensing. The developments are supported strongly on one hand by the need to increase writing speed and on the other by the capability to fabricate by standard micromachining processes probes of high complexity. Figure 23.27 illustrates a possible architecture for such a 2D NFP array.

When expanded to 2D arrays, the use of a scanning stage for the motion of the sample will be necessary because the z motion of the piezotubes in commercial AFMs is coupled with the x–y scanning and also because the mounting angle of a probe is usually not parallel to the scanning substrate (probes are mounted with tilt angle of approximately 15° to the substrate). X–Y scanning stages with feedback control, nanometer resolution and five degrees of freedom (x, y, z, θ_x, θ_y) are commercially available [110]. Such stages can be used for the alignment and scanning of the 2D probe array. In order to align the substrate to the NFP array, avoiding independent sensing on each probe, three microfabricated deflection sensors placed on the periphery of the array can be used to control the plane of the NFP array. Since PZT is capable of both actuation and sensing, probes at three locations on the periphery of the array can be designed to work as sensors that could send feedback contact information to the corresponding stage actuators. Using feedback-control algorithms, the stage actuators can, in turn, adjust the substrate (stage) parallel to the array. With the distance between the array and stage maintained constant while scanning, the writing status of individual cantilevers can be turned on or off by applying an activation voltage to the independent PZT probe actuators. Ideally, each cantilever requires two electrical connections. When a 2D array becomes massively parallel for

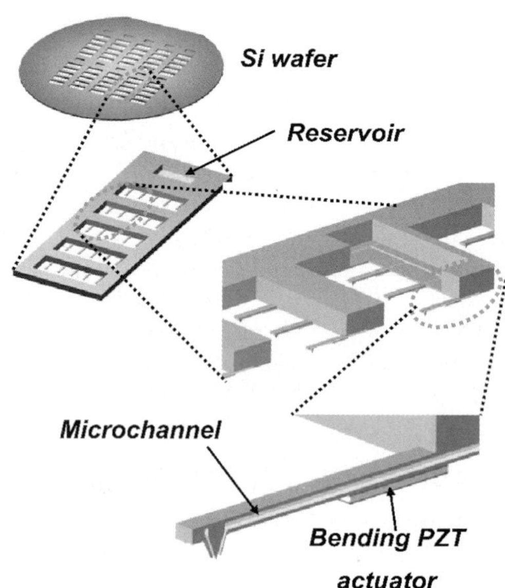

Fig. 23.27. 2D array of NFP including multiple reservoirs and independent PZT actuation of cantilevers

higher throughput, the area needed for electrical connections and pads proportionally becomes massive. In order to reduce the complexity, pads can be shared row-by-row and column-by-column with use of an $N \times N$ multiplexed addressing schemes as in random access memories (RAM).

For future development of PZT actuated probes, special attention will be needed to minimize the hysteretic effects in the piezoelectric materials. The deflections presented in Fig. 23.20c were obtained for a monotonic increase in the applied voltage starting from a given material history. With further application of square waves, slightly different deflections were measured. This problem can be alleviated by optimizing the deposition of the piezoelectric material such that hysteresis is minimized. Fatigue and exhaustion of the piezoelectric response can also limit the function and life time of independently actuated probes. This and many other challenges remained to be addressed in the design of systems for massively parallel patterning of a variety of molecular species. One of them, tip wear, is addressed in the following section.

23.3
Ultrananocrystalline-Diamond Probes

Wear resistance and long functional life time become paramount when designing complex 1D and 2D arrays of AFM probes, especially for contact mode techniques such as DPN and NFP. Clearly the cost of fabricating such probes is high because of the number of processing steps. Hence, materials exhibiting low wear should be employed to maximize the life of the array and as a result produce more cost-effective probes. Wear resistance can be defined in terms of scanning distance and wear rate (i.e., amount of material removed per unit mechanical work). The materials in contact, the force applied on the probe, the environmental parameters (temperature, humidity, gas composition in the AFM chamber, contamination species) and the ink types are all important parameters controlling the wear rate. Even though the theory of friction in AFM applications has gained some attention, leading to interesting concepts such as wear or friction-free mechanical motion or nanoscale lubrication, systematic experimental measurements have not yet been reported.

Engineering new materials and developing simple fabrication processes is one way of addressing this problem. Both aspects will be discussed here. We also focus on probes which may be attractive for integrating new functions, such as independent sensing and actuation, arraying, multiple probing techniques, microfluidics and/or scanning probe machining. The life time of probes is typically shortened by mechanical failure in operation and handling, pickup of material and particles from the samples, and wear. While the former can be enhanced by more careful procedures, the latter is especially important in contact mode techniques, such as contact mode AFM imaging [111], scanning spreading resistance microscopy [112], atomic scale potentiometry [113], scanning thermal microscopy [114] and lithography [71, 115], for example, DPN and fountain-pen nanolithography [46, 57, 68, 69, 71]. To reduce probe wear, hard materials are typically employed, among which diamond is the obvious material of choice. Furthermore, diamond possesses surface and bulk properties that are ideal for probes: very low chemical reactivity, a low work function when the surface is chemically conditioned, no oxide layer formation, tunable electrical conductibility by doping, and thermal conductivity ranging from relatively low (approximately 10 W/K m) for ultrananocrystalline diamond (UNCD) to extremely high (approximately 2000 W/K m) for single-crystal diamond [116].

In previous work, several species of diamond films have been employed in probe fabrication by different groups [117–122], differing mainly in the degree of crystallinity of the diamond. Initial attempts at producing conductive diamond probes for scanning tunneling microscopy involved boron-implanted macroscopic diamond crystals, which were machined by polishing and were mechanically assembled into AFM cantilevers [117]. Later attempts used boron-doped epitaxial CVD layers grown on natural diamond [120]. These approaches showed the feasibility of employing conductive diamond in probe manufacturing, but are obviously not well suited for integration. Microcrystalline diamond (MCD) or nanocrystalline diamond (NCD) films [121] have superior mechanical characteristics (wear, hardness) with respect to amorphous or diamond-like carbon (DLC) materials [122], but have higher surface roughness than the latter. DLC films are smoother and easier to integrate in more complex fabrication schemes but cannot be made highly conductive. Because of this, micromachining techniques based on molding methods [118, 119, 123] were developed to minimize the major problem of crystalline diamond films, i.e., their surface roughness when used as coatings. An alternative to molding is to use thin conformal films to cover probes made of other materials. The latter technique has the disadvantage of increasing the initial tip radius of the probes (10–20 nm) by the thickness of the diamond film (typically 70–100 nm to achieve full coverage of the substrate); thus, resulting in much lower tip sharpness. Typical commercial diamond-coated tips have radii in the 100–200-nm range. In particular cases, nanoroughness features can improve the radius of the contact area, but the general shape and aspect ratio of the probes is compromised. Molding of crystalline diamond works reasonably well, but leaves the growing surface of the diamond very rough, improper to continue the integration with other, later-deposited layers and further processes.

UNCD films [124], with grain sizes in the 2–5-nm range, retain most of the surface and bulk properties of crystalline diamond as well as the smoothness of the substrate [125, 126]. The material is deposited by microwave PECVD (MPCVD) from an Ar–CH_4 (99%:1%) gas mixture. Table 23.1 shows some of the remarkable properties of UNCD, compared with other forms of diamond. The term UNCD is used to distinguish this material from MCD [121], NCD [127] and DLC [122]. Owing to the small size of the UNCD grains, the ratio of grain boundary atoms (which consist of a mixture of sp^2, sp^3 and other forms of carbon bonding) to bulk atoms (sp^3) is high, leading to interesting material properties, like a predictable fracture strength equal to or higher than that of single-crystal diamond and the ability to incorporate nitrogen into the grain boundaries, which gives rise to greatly increased (up to 250/Ωcm) room-temperature n-type conductivities. A comparison between these species of diamond is given in Table 23.2.

The remarkable hardness of UNCD makes it the material of choice for contact mode nanoprobe tips. Erdemir et al. [127] measured wear rates on MCD films using a pin (SiC)-on-disc tribometer measurement technique. They found that MCD films exhibit wear rates from 0.48×10^{-6} to 55.0×10^{-6} mm^3/N m. By contrast, UNCD

Table 23.2. Characteristics of different diamond film species

	Microcrystalline diamond	Nanocrystalline diamond	Ultrananocrystalline diamond	Diamond-like carbon ta-C	ta-H:C
Growth species	CH_3^* (H°)	CH_3^* (H°)	C_2	C	C
Crystallinity	Columnar	Mixed diamond and non-diamond	Equiaxed diamond	Mixed diamond and amorphous	Amorphous
Grain size	$\sim 0.5-10\,\mu m$	$50-100\,nm$	$2-5\,nm$	Variable	–
Surface roughness	$400\,nm-1\,\mu m$	$50-100\,nm$	$20-40\,nm$	$5-100\,nm$	$1-30\,nm$
Electronic bonding character	sp^3	Up to 50% sp^2 (secondary phase)	2–5% sp^2 (grain boundary)	Up to 80% sp^3	Up to $\sim 40\%\ sp^3$
Hydrogen content (%)	< 1	< 1	< 1	< 1	15–60

films exhibit a wear rate as low as $0.018 \times 10^{-6}\,mm^3/N\,m$. It was also found that the as-grown UNCD films have friction coefficients roughly 2 orders of magnitude lower than those of MCD films of comparable thickness. The wear rate of a SiC pin rubbed against an UNCD film was found to be approximately 4000 times lower than that of a SiC pin rubbed against an as-deposited MCD film.

Next we report the fabrication of UNCD probes for AFM, integrating tips and cantilevers made entirely of this material, both in nonconducting (undoped) and in conducting (nitrogen-doped) states. The probes were characterized by electron microscopy and resonance measurements, and their performances were tested in imaging, DPN writing and conductive modes. Wear tests were also conducted to demonstrate the superior behavior of the microfabricated UNCD tips compared with that of commercial silicon nitride tips.

23.3.1
Chip Design

Two types of cantilevers with different designs were fabricated. One featured a high-stiffness triangular cantilever and the other a low-stiffness arrow-shaped cantilever. The length of both cantilevers was 170 μm, while the width of the arrow-shaped cantilever was 12 μm and the width of the arms of the triangular cantilevers was 18.8 μm. The thickness of the diamond (UNCD) was between 0.8 and 1.4 μm, depending on batch and wafer-level deposition nonuniformities. The arrow-shaped cantilever was chosen instead of a simple rectangular cantilever to increase the reflective area of the probe near the tip while keeping the stiffness at a minimum. The chip body was of rectangular shape with dimensions 1.6 mm × 3.6 mm. Four cantilevers of the same type were placed on each chip, one pair spaced 300 μm and the other pair spaced 600 μm on opposite sides of the chips. This design follows roughly the geometry employed in commercial tips (Veeco, Ultralever). A view of the two types of cantilevers is seen in Fig. 23.28a and b, while a zoom-in of the tip apex is shown in Fig. 23.28c.

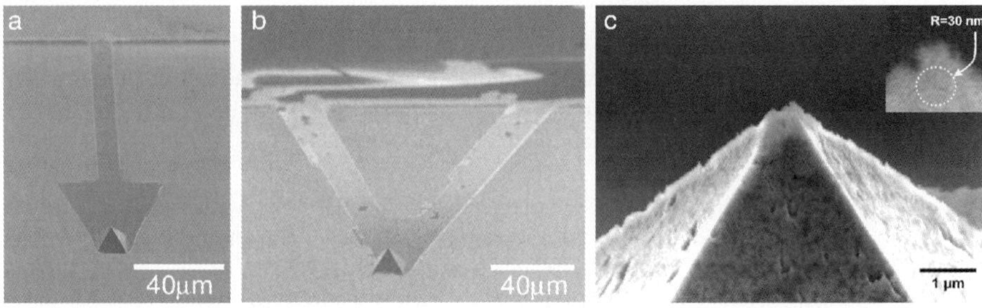

Fig. 23.28. SEM pictures of the two types of ultrananocrystalline diamond (UNCD) cantilevers with tips: (**a**) arrow-shaped cantilever, (**b**) triangular cantilever; (**c**) detail of a tip, with the *inset* showing the apex and a 30-nm-radius circle for reference. The 2–5-nm-scale grain size makes us believe that ultimately the tip vertex is much sharper

23.3.2
Molding and Other Fabrication Techniques

Molding is well known as a fabrication method for ultrasharp tips of a large variety of materials [128,129], including diamond [130], for which tip radii of 30 nm have been reported [131]. However, the reports on diamond molding did not include oxidation sharpening as an option for increased sharpness of the probes. Tip radii were limited by the geometrical precision of the pyramidal pit etched into silicon, and by the diamond deposition and seeding parameters. The ultimate shape of such a pyramidal pit in Si(100) is given by many factors, including the accuracy of alignment with respect to the crystallographic orientation, and the lithographic performances in providing optimum geometries for windows in the masking layer used for pyramidal etching. A slight increase in the window size in one direction may result in formation of a line-edge probe rather than a point-tip probe. Since the alignment, lithography and etching processes are never perfect on the nanometer scale, one expects line-edge probes to be always obtained, depending on how much magnification is used in observing the tip end. In the approach presented, besides a sufficiently rigorous lithography (±0.1 μm) and care in alignment (better than ±1° for both flat-to-crystal and mask-to-flat alignment), an oxidation sharpening step was added (Fig. 23.29). This step, besides the additional sharpening due to constraints in the oxide growth in pyramidal pits [128], performed well in leading to single-point tip geometries.

The molding method has the general inconvenience that the tip is fabricated facing towards the substrate; thus some microfabrication effort is required to reverse the probe cantilevers with respect to the handling chip body. Several methods were reported for reversing the diamond tips: (1) building chip bodies by micromachining on separate silicon or glass wafers and gluing them onto the tip-fabrication wafer [123]; (2) fabricating tips and portions of the cantilevers on one wafer and gluing them onto cantilevers fabricated on other wafers (eventually, made of other materials), followed by releasing [132]. All these methods require aligned bonding procedures and a good resistance of the glue joint during the chip release and

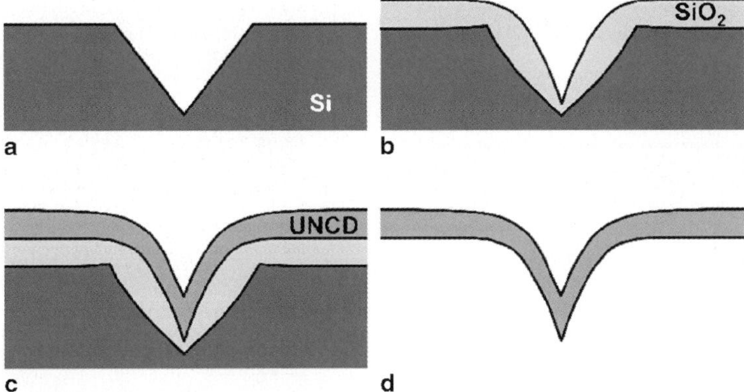

Fig. 23.29. Molding of UNCD by (**a**) forming a pyramidal pit in Si(100) by orientation-dependent etching, (**b**) oxidation sharpening, (**c**) deposition of UNCD and (**d**) removal of the Si and SiO_2 mold by chemical etching

operation processes. In order to simplify this sequence, we investigated the use of metal electroforming to build a chip body within an SU-8 photoresist mold, followed by release through dissolution of the silicon substrate. Although bonding of a complementary wafer for the chip body has the advantage of offering a medium to develop electronic circuitry without interfering much with the tip fabrication, it is an unnecessary effort in the case of simple conductive probes. In our approach, simple sensors/actuators in film form, such as piezoactuators or piezoresistors, can be still integrated on top of the diamond by proceeding from the mold-side silicon wafer. Additional precautions in protecting them during the silicon mold removal may be required. Less harsh methods to remove the silicon substrate can be also considered, such as XeF_2 etch or sequential combinations of partial etch with KOH solution and finalizing with XeF_2.

The processing steps employed in fabricating molded UNCD probes are summarized in Fig. 23.30. The fabrication starts by forming an oxide mask (thermal oxidation, 500 nm), which was patterned lithographically by mask M1 with square openings (12 μm × 12 μm), followed by KOH (30%, 80 °C) etching of pyramidal pits in the Si(100) wafer. Several groups of different sized squares and rectangles were fabricated simultaneously with mask M1, corresponding to different alignment marks, such that a large variety of tip geometries could be obtained. A thermal oxidation sharpening process at 900 °C followed, which resulted in a SiO_2 layer greater than 1 μm in thickness on the {100} surface of the Si wafer. Figure 23.31a shows one sharpened pyramidal mold, exhibiting a central black area corresponding to the recessed (sharpened) zone. An ultrasonic seeding procedure was applied with a 4–6-nm grain diamond powder suspended in methanol (5 mg/l), to which the wafers were exposed for 30 min and rinsed with 2-propanol, then ultrasonically cleaned in methanol for 5 min and dried. Growth of an UNCD layer (0.5–1-μm thick) was achieved by MPCVD in a methane–argon gas mixture, containing also nitrogen in the case of the n-doped films [124]. Next, an Al mask (80 nm) was deposited by electron beam evaporation and patterned with mask M2, defining the cantilevers. The pattern was transferred into the UNCD by RIE, using oxygen

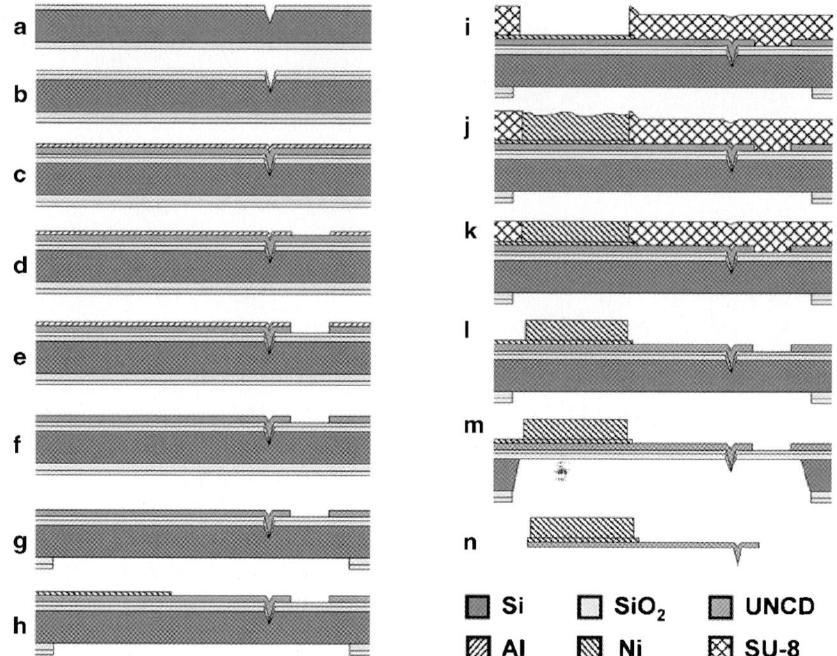

Fig. 23.30. Fabrication of UNCD probes – processing sequence. (**a**) Oxide growth followed by lithography with mask M1, and then etching in KOH solution. (**b**) Oxidation sharpening. (**c**) UNCD and Al deposition. (**d**) Lithography (mask M2) followed by Al etching. (**e**) UNCD etching using RIE. (**f**) Al removal. (**g**) After lithography (mask M3), oxide etching on the backside. (**h**) Ti/Ni deposition followed by patterning (mask M4). Ti layer (not shown in the figure) used as an adhesion layer between UNCD and Ni. (**i**) SU-8 spin-coating and patterning (mask M5). (**j**) Ni electroforming. (**k**) Lapping/polishing. (**l**) SU-8 removal. (**m**) Etching in KOH solution. (**n**) Oxide etching

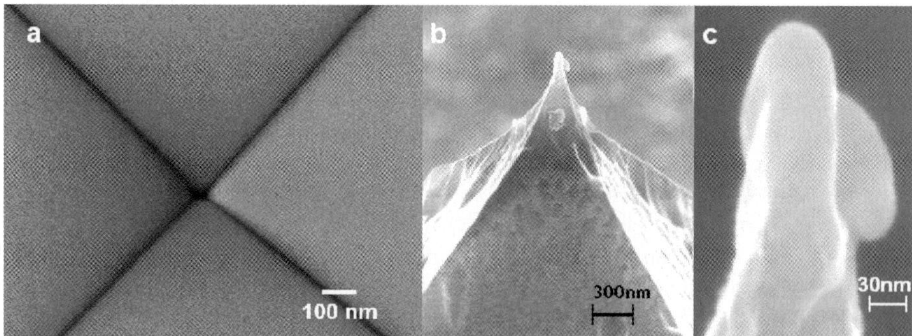

Fig. 23.31. (**a**) An anisotropically etched pyramidal hole, after sharpening by thermal oxidation. (**b**) Molded UNCD probe. (**c**) Zoom-in showing the probe apex with approximately 30 nm radius. A more exact calculation of the radius is presented in Sect. 4.3.1

plasma (30 mTorr, 50 standard cm^3/min, 200 W), according to a process described in [133], after which the Al mask was removed by wet chemical etching. The backside oxide was patterned with mask M3, for subsequent release of the structures. A top metal was then deposited by electron beam evaporation and patterned on the front side, forming a plating base for the subsequent electroforming of the chip body.

The plating base metal must suit the electroforming metal, for example, Ni 150-nm plating base in the case of Ni electroforming, or Ti 10 nm/Au 200 nm for electroforming of gold. The choice of the electroforming metal is based on compatibility with the release procedure. Patterning of the plating base layer (mask M4) is required for avoiding the presence of the metal in the area adjacent to the cantilevers, where it may result in metal debris hanging around the cantilevers after their release. At the same time, the plating base pattern has to be continuous, for providing electrical contact to all the areas that need to be plated. Optionally, the plating base can be left on the cantilevers, to act as a reflective coating, provided no electroplating will be performed there.

In our process, a negative-tone SU-8 photoresist (MicroChem, 330-μm thickness) layer was deposited and patterned with mask M5, to form a mold for the metal handling chip body. For the chip discussed here, we used gold electroforming, but other metals may be considered for cost reduction (Ni, Cr) or for special applications including piezoactuators where Pt would be the metal of choice to subsequently grow a piezoelectric layer on top. The electroforming of the gold layer was performed in two steps: it started with the deposition of 3 μm of gold with a neutral pH Techni-gold 25E (Technic) sulfamate solution [134], to provide a good adhesion, and continued with the deposition of the thick chip body from a mildly acid pH Neutronex 309 (Enthone-OMI) bath. Both processes were performed at 40 °C, with a current density of 1 mA/cm^2. The metal thickness ranged between 250 and 300 μm. The gold roughness and incidental side wall overplating were removed by mechanical polishing. The SU-8 resist mold was then removed in a piranha solution. The removal of the Si substrate was performed by KOH etching (30%, 80 °C), and the remaining oxide was removed by a buffered HF solution. After the removal of the sacrificial silicon mold, the chips remained suspended on a Si and gold frame, each supported by two bridges from where they could be easily clipped off using tweezers.

23.3.3
Performance Assessment and Wear Tests

The tips were characterized and sorted using field-emission SEM. The high conformity of the UNCD deposition enabled a good coverage of the pyramidal pits used as molds, resulting in tips with a nanometer-scale apex. In most cases very sharp tips were achieved from protruding features (Fig. 23.31). Owing to the nanograin size (2–5 nm) of the material, the ultimate tip radius could be in the 2–5-nm range.

Examination of the UNCD surface morphology revealed that film growth is achieved from seeding nanoparticles, which leads to clustering. As discussed in the context of UNCD strength [135,136] a large number of grains is present in each cluster and imperfections between clusters were observed in the form of voids [135,136]. In the pit area, larger voids are clearly visible between grain clusters (Figs. 23.28c inset, 23.33). Growth of UNCD on ultrasonically seeded SiO_2 is more challenging than growth on Si; hence, cluster and void sizes generally tend to increase. Similar structures, but made of crystalline grains, have been observed in MCD films grown on side walls of pyramidal pits in Si [131]. The nucleation of MCD grains and intergrain gap formation on tilted surfaces were linked by Scholtz et al. [131] to the size of the diamond particles used in the ultrasonic seeding process. In their experiments, ultrasonic abrasion with 4-µm-diameter diamond particles produced minimal intergrain gaps on flat surfaces, while for pyramidal holes, the optimum diamond particle size for uniform coverage was found to be 1 µm. In the case of UNCD, the initiation of the film growth is done mostly by the nanometer-size diamond nanoparticles used in the seeding step, but the nucleation and growth obeys similar rules. The inset of Fig. 23.28c is a zoom-in view of one of the top clusters. A coral-like surface morphology is observed.

The surface roughness of UNCD films can be minimized to values in the 4–7-nm root-mean-square range by depositing a very thin (5–10-nm thick) metallic layer, for example, W, Mo or other carbide-forming materials, as a nucleation promoter. We have found that SiO_2 is a more difficult nucleation medium than Si, resulting in poor UNCD film adhesion, especially in the case of doped UNCD. Nonetheless, it is indispensable for maximum sharpening. In a variant of the processing sequence, we used an additional lithography step to selectively remove the oxide prior to the UNCD deposition from all areas, except the pyramidal pits. For this purpose, mask M1 was realigned, but using a reversible (negative) photoresist (Shipley AZ5214E), through which the oxide was removed in BOE.

A vibration test on the arrow-shaped UNCD cantilevers was performed using a Polytec vibrometer in differential interferometric mode (the reference beam on the chip body and the measuring beam on the cantilever). The first three resonant frequencies were measured at 32.2, 52.31 and 53.94 kHz, respectively (movies are provided as supporting materials [137]). The first and the last vibration frequencies correspond to vibration modes in which the pyramid axis oscillates with rotation around the perpendicular to the chip surface. The intermediate frequency corresponds to a motion in which the pyramid axis oscillates without rotation, i.e., it remains perpendicular to the chip surface during the whole motion. This vibration mode is well suited for AFM techniques involving cantilever oscillations perpendicular to the surface being scanned (tapping mode) [4]. Note that the high stiffness achievable with UNCD cantilevers makes it possible to increase the working frequencies, with obvious benefits in sensitivity and resolution.

The stiffness of the UNCD cantilevers was measured by the cantilever deflection method, using an AFM probe after calibration with a reference cantilever of known stiffness. We measured a stiffness value of 3.6 N/m for a 1.2-µm-thick triangular cantilever, and a stiffness of 0.94 N/m for a 0.9-µm-thick arrow-shaped cantilever. Knowing the geometry of the cantilever, we could deduce a value of 923 ± 50 GPa for Young's modulus of UNCD, which correlates well with the value of 960 ± 60 GPa measured by membrane deflection experiments [125, 126]. The stiffness of the arrow-shaped cantilever is close to the stiffness of one of the triangular nitride cantilevers of a commercial nitride probe (tip A of Veeco Si_3N_4 contact mode probes, 0.58 N/m). This information was used to perform a comparative test on wear resistance of the tips. For this purpose, a Digital Instruments 3100 AFM was used for repeated scanning in contact mode. Imaging of an UNCD surface, $10\,\mu m \times 10\,\mu m$ in area, was performed with a constant force of 30 nN. The scanning of the UNCD tip was performed on the surface of an UNCD film deposited on Si for the purpose of accelerating the wear tests. The tests were run for UNCD and Si_3N_4 tips. The tips were imaged before and after the test using a field-emission SEM (LEO Gemini 1525). After 1 h of scanning, the nitride tips showed visible wear, while the UNCD tips showed no appreciable change (Fig. 23.32). The degrading of the silicon nitride tip could be detected also in the quality of topographic AFM images, while the images recorded with the UNCD tip showed no appreciable change.

In an attempt to look for the dominant wear mechanisms in UNCD, we looked for changes in the diamond tip after long scans performed for 3 h with an increased constant force of 50 nN using the triangular-shaped UNCD cantilever. Figure 23.33 presents a view of the tip apex before and after the wear experiment. One can notice the absence of a relatively large portion of the tip apex and debris gathered on the side walls of the pyramid. The geometry of the worn tip suggests that the observed wear is due to the dislodgment of a cluster of UNCD grains as a result of intercluster cracking, rather than a gradual atom-by-atom removal. This finding shows that when increasing the wear resistance of UNCD probes, one has to optimize the intercluster strength or eliminate the formation of clusters. The role of seeding in UNCD surface morphology and strength is discussed in [125, 126, 136]. Elimination of clustering

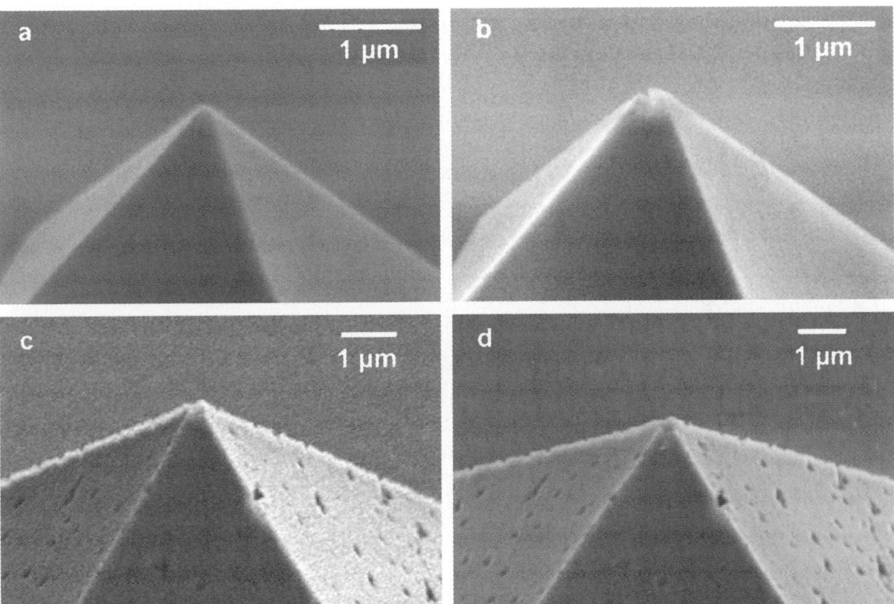

Fig. 23.32. Si_3N_4 tip before (**a**) and after (**b**) the wear test, showing damage. UNCD tip before (**c**) and after (**d**) the same test, showing no visually detectable damage

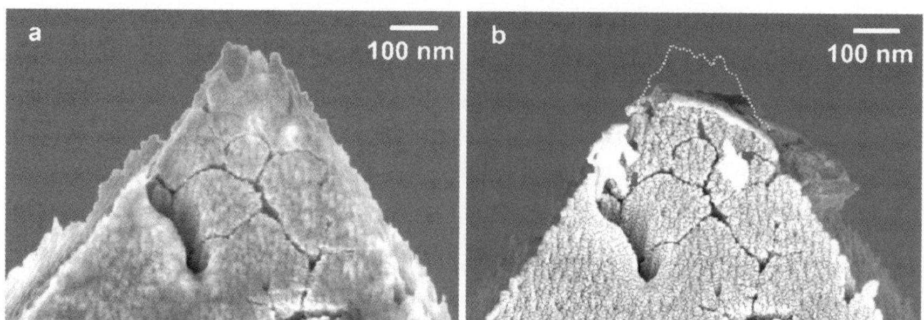

Fig. 23.33. High-resolution SEM image of an UNCD tip before (**a**) and after (**b**) a prolonged wear test. The absence of portions of the material roughly following the *intercluster contour line* suggests failure happens mainly by detaching of clusters, rather than a continuous smooth wear

would also have a positive impact in obtaining smoother, high-conformity filling of the mold template with diamond, resulting in sharper tips with controlled geometry at the nanoscale. Research on the optimization of seeding and deposition parameters was performed. In order to decrease the agglomeration of the powder in the solution, a couple of droplets of surfactant (sodium dodecyl sulfate, Fluka) were also added using a regular glass pipette.

23.3.4
Applications

Besides wear resistance, the performance of the UNCD AFM probes resides in their low work function, low chemical reactivity and high conductivity when doped. These features enable advanced conductive AFM techniques such as AFM potentiometry and nanoelectrochemistry. Molecular writing and lateral force imaging with UNCD probes were also demonstrated. Hence, combinations of conductive and writing techniques such as electrochemical DPN [46] or EPN [67] are possible. These applications are discussed next.

23.3.4.1
AFM Potentiometry

UNCD can be doped with nitrogen to enhance its electrical conductivity [138]. Since a very high concentration of dopant results in the weakening of bond strength at the grain boundaries, there is a trade-off between mechanical strength and electrical conductance of the material. We have found that a good balance between strength and conductivity can be achieved when UNCD is grown from gas mixtures containing 10% nitrogen, which leads to a material with a conductivity of $30/\Omega cm$ [139]. This material was used to prepare AFM cantilevers for various applications, following the fabrication method presented in the previous section.

To theoretically evaluate the contact resistance of a doped UNCD probe with a highly conductive substrate (such as a gold substrate), we began by computing the electrical resistivity of the probe on the basis of the geometry described in Figs. 23.34

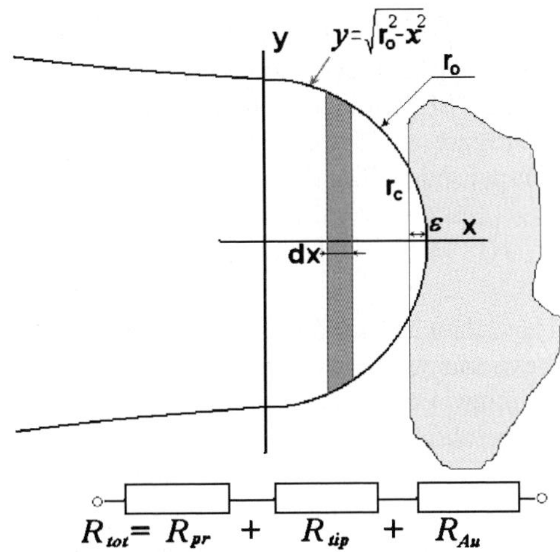

Fig. 23.34. Probe in contact with a gold substrate and equivalent electrical circuit

and 23.35. The electrical scheme of the total resistance of the probe can be viewed as composed of the resistance of the approximate pyramidal body of the probe, the resistance of the tip end (approximately a hemisphere of radius r_o) and the resistance of the gold substrate. The resistance of the probe cantilever was neglected, being approximately 4 Ω, owing to a gold coating of approximately 50 nm, on the backside of the cantilever, deposited to improve the reflectivity. This layer is present also in the inner side of the pyramid body, as schematically depicted in Fig. 23.34.

For the calculation of the resistance of the pyramid body, an exact shape of the body must be considered, since the main contribution to the resistance comes from the vortex of the probe, where an important geometrical change is induced by the oxidation sharpening. The exact shape of the probe can be determined from a digitized SEM frontal image of the probe, taking into account the tilts. The tilt calibration can be avoided by knowing the angle of the probe pyramid from the

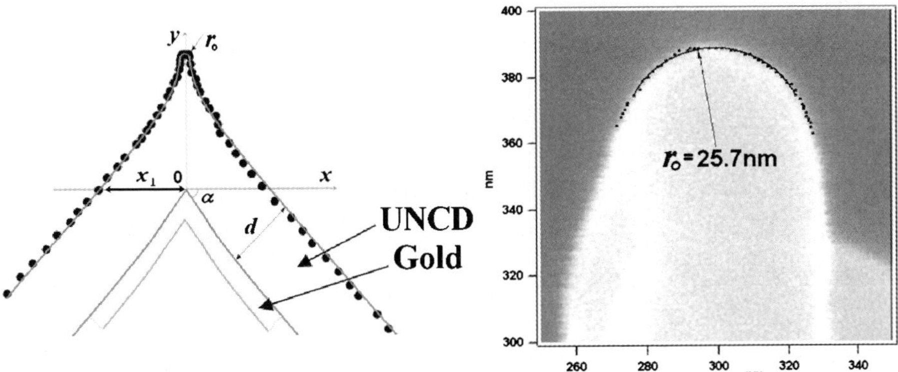

Fig. 23.35. (a) Geometry of the tip defined by digitizing an SEM image of the probe shown in Fig. 23.31b; the drawing is not at scale. (b) Same procedure applied at a smaller scale for the tip end radius, based on the image shown in Fig. 23.31c

crystallographic orientation of its facets [(111) planes]. Figure 23.35 shows the digitized shape corresponding to a transversal section through the probe, derived from an SEM image similar to that in Fig. 23.31b.

Figure 23.35 shows the profile resulting from a digitized SEM image after tilt compensation. The data points were fit by the least-squares method using the function

$$y(x) = y_0 - |x| \tan \alpha + \frac{a}{|x|^\mu} \,. \tag{23.1}$$

The values identified for y_0, a and μ are those listed in Table 23.2. The origin of the x-axis was considered at the tip point. The radius r_0 of the tip was calculated in a similar way from a high-magnification SEM image (Fig. 23.35b). Alternatively, it can be obtained by deconvolving scans of spherical beads of known diameter [139]. The tip radius was used to calculate the contact area, which in turn was employed in computing the contact resistance.

In the following, the pyramidal part of the probe will be considered to fit the hemisphere of radius r_0 forming the tip, although is it evident that one has a square cross section, while the other has a circular base. Better approximations can be eventually adopted, but more important errors are supposed to come from the value of the resistivity and other geometrical imperfections.

For an estimation of the pyramid body resistance, note that the material under the horizontal line in Fig. 23.35a (in contact with the gold layer) has a small contribution to the total resistance. This will become more evident in the following developments. Thus, a good approximation of the contact resistance can be obtained by calculating the resistance of the pyramid trunk extending from point $y(r_0)$ to $y(x_1)$ in the vertical direction, where the origin of the axes was chosen at the apex of the gold layer, and x_1 is the coordinate at level $y = 0$.

The thickness of the UNCD layer was measured as $d = 0.9\,\mu$m and the thickness of the gold film (for increase of reflectivity) as 50 nm. From the geometry of the silicon v-grove, $\tan \alpha = \sqrt{2}$ and the half-width of the probe at the point of the gold layer can be calculated as

$$x_1 = \frac{d}{\sin \alpha} = d\sqrt{\frac{3}{2}} \,. \tag{23.2}$$

For an elementary segment of thickness dy and cross section $(2x)^2$ in the probe body region, the resistance is given by $dR_{\mathrm{pr}} = \rho \frac{dy}{(2x)^2}$, where ρ is the resistivity of UNCD. For the total resistance, we have to integrate over x from x_1 to r_0:

$$\begin{aligned}
R_{\mathrm{pr}} &= \int_{x_1}^{r_0} \rho \frac{dy}{(2x)^2} = \frac{\rho}{4} \int_{x_1}^{r_0} \left(\sqrt{2} + \frac{a}{x^{\mu+1}} \right) \frac{dx}{x^2} \\
&= \frac{\rho}{4} \left[\sqrt{2} \left(\frac{1}{r_0} - \frac{1}{x_1} \right) + \frac{a\mu}{(\mu+2)} \left(\frac{1}{r_0^{\mu+2}} - \frac{1}{x_1^{\mu+2}} \right) \right] ,
\end{aligned} \tag{23.3}$$

where (23.1) for the function $y(x)$ was taken into consideration and the absolute value was omitted since the calculation is performed in the positive half plane. This expression shows that, since $r_0 \ll x_1$, ($r_0 \sim 25.7$ nm, $x_1 \sim 1200$ nm) the terms in x_1 accounting for the larger side of the pyramid trunk considered are not significant. This supports the fact that the approximations done on that side of the object are of minor significance. For the values being considered (Table 23.3) the resistance of the probe body as function of the tip radius r_0 is given in Fig. 23.36.

From this graph, the resistance of the pyramidal part for the case being considered, $r_0 = 25.7 \pm 1.5$ nm, is $R_{pr} = 560 \pm 51\ \Omega$, where the error in R_{pr} stems mainly from the accuracy of the resistivity (5% error) and the error in the tip radius calculation (Fig. 23.35b).

The resistance of the hemispherical tip can be calculated in a similar way, i.e., by considering the elementary resistance dR_{tip} of a layer of circular cross section and infinitesimal thickness dx (Fig. 23.34). Integrating over the domain $x = 0$ to $r_0 - \varepsilon$, where ε is the indentation depth, we obtain

$$dR_{tip} = \frac{\rho\,dx}{\pi y^2} \tag{23.4}$$

and

$$R_{tip} = \int_0^{r_0-\varepsilon} \frac{\rho\,dx}{\pi y^2} = \frac{\rho}{\pi} \int_0^{r_0-\varepsilon} \frac{dx}{r_0^2 - x^2} = \frac{\rho}{2\pi r_0} \int_0^{r_0-\varepsilon} \left(\frac{1}{r_0 - x} + \frac{1}{r_0 + x} \right) dx$$

$$= \frac{\rho}{2\pi r_0}\left[\ln(r_0 + x) - \ln(r_0 - x) \right]\big|_0^{r_0-\varepsilon} = \frac{\rho}{2\pi r_0} \ln\left(\frac{2r_0 - \varepsilon}{\varepsilon} \right). \tag{23.5}$$

The contribution to the total resistance from the gold side can be evaluated from the formula of a circular contact of radius r_c on a semi-infinite conductor of resistivity ρ_{Au}, for our case [140]:

$$R_{Au} = \frac{\rho_{Au}}{4r_c} = \frac{\rho_{Au}}{4\sqrt{\varepsilon(2r_0 - \varepsilon)}}. \tag{23.6}$$

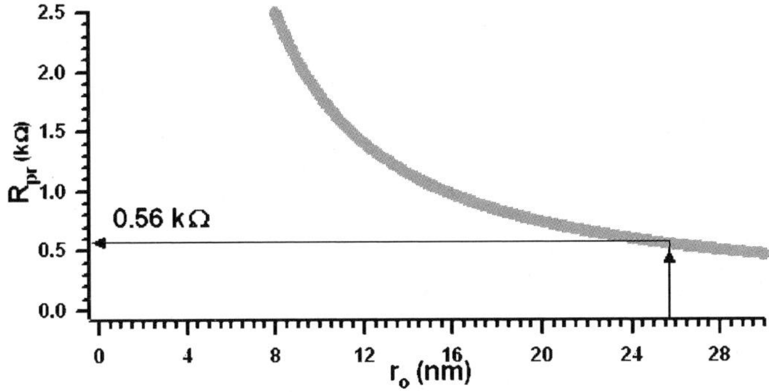

Fig. 23.36. Resistance of the probe body as a function of the tip radius, calculated from the *fitted curve* representing the shape of the tip, with the parameters given in Table 23.3. The *thickness of the line* corresponds to the error domain

Table 23.3. Values of the parameters used to obtain the plot shown in Fig. 23.36

d	1000 ± 100 nm
x_1	1225 ± 125 nm
ρ	3.85×10^{-5} Ωm
μ	0.588 ± 0.08
a	80 ± 20 nm$^{\mu+1}$
y_0	3840 ± 25 nm

Combining (23.3), (23.5) and (23.6), we can write

$$
R_{\text{tot}} = \frac{\rho}{4} \left[\sqrt{2} \left(\frac{1}{r_0} - \frac{1}{x_1} \right) + \frac{a\mu}{(\mu + 2)} \left(\frac{1}{r_0^{\mu+2}} - \frac{1}{x_1^{\mu+2}} \right) \right]
$$

$$
+ \frac{\rho}{2\pi r_0} \ln \left(\frac{2r_0 - \varepsilon}{\varepsilon} \right) + \frac{\rho_{\text{Au}}}{4\sqrt{\varepsilon(2r_0 - \varepsilon)}} \, . \tag{23.7}
$$

Neither the tip resistance nor the gold substrate resistance can be evaluated without knowing the indentation depth ε. To gain insight into the working of the tip-to-substrate contact for the case of UNCD probes in contact with highly conductive metal substrates, we will estimate the indentation depth from a mechanical calculation based on the applied force and the constitutive behaviors of the materials.

For low contact forces, the radius of the contact spot is a function of the applied force. The substrate material in this experiment was chosen to be gold such that plastic indentations would develop at relatively low contact forces. The force applied to the tip in the experiments was evaluated to be 0.20 ± 0.08 μN. The evaluation was based on the deflection sensitivity of the piezotube (75 nm/V) and a prescribed displacement equivalent to 3 V. The stiffness of the cantilever was computed to be 0.88 ± 0.32 N/m. This cantilever stiffness was determined from its geometry and confirmed by load-deflection measurements performed with an X-ray photoelectron spectroscopy nanoindenter.

To estimate the indentation depth ε, using contact mechanics solutions, we can begin by using the Hertzian elastic solution:

$$
a = \left(\frac{3Pr_0}{4E^*} \right)^{1/3} , \tag{23.8}
$$

$$
\varepsilon = r_0 - \sqrt{r_0^2 - a^2} , \tag{23.9}
$$

where P is the applied force, E^* is the reduced modulus of the UNCD–gold interface and a is the contact radius. For a contact force of 0.20 μN and tip radius of 25.7 nm, we compute a penetration depth of 0.256 nm and a contact stress of 7.3 GPa, which is much higher than the yield stress of gold. Therefore, an elastoplastic solution needs to be considered to estimate the actual penetration depth. A model proposed by Zheng and Cuitino [141] corresponding to the contact of two spheres pushing against each other was used. One sphere of radius r_0 is considered for the UNCD tip, while the other, of radius $r \to \infty$ is considered for a flat gold surface. The penetration depth can then be calculated from the following expression [141]:

$$\frac{P}{\pi r_\text{o}^2 \Gamma} = B(m)\left[2c^2(m)\right]^{1+\frac{1}{2m}} k(m) \left(\frac{\varepsilon}{r_\text{o}}\right)^{1+\frac{1}{2m}}, \tag{23.10}$$

where m is the hardening exponent defining the plastic regime of gold, which is modeled to follow a power law. $B = 0.74 + 0.26^{1/m}$ ($m < 3$), $c^2 = 1.5 - (1/m)^{0.5}$ and $k = 3.07 \times 0.16^{1/m}$ are constants depending on the value of m. Γ is the equivalent reference stress calculated on the basis of the yield stress values for gold and UNCD, i.e., $(1/\Gamma)^m = (1/\sigma_\text{Au})^m + (1/\sigma_\text{UNCD})^m$; with $\sigma_i = \sigma_y^{1-1/m} E_i^{1/m}$. The value of m and the yield stress of gold were obtained from the tensile testing of gold thin films reported by Espinosa and Prorok [142]. A yield stress of 165 MPa and a hardening exponent $m = 1.664$ were identified from the reported stress–strain curves. For a contact force of $0.20 \pm 0.08\,\mu$N, the penetration depth was then calculated to be $\varepsilon = 0.7 \pm 0.21$ nm, which corresponds to a contact radius $a = 6.0 \pm 0.8$ nm.

Substituting this value for the penetration depth, along with the other parameters listed in Table 23.3, into (23.7), we obtain an estimated value for the total resistance of approximately $1.04\,$kΩ. It is interesting to consider the various contributions to the total resistance $R_\text{tot} = 2.1\,$kΩ as predicted from this calculation. The pyramidal resistance is $R_\text{pr} = 0.56\,$kΩ, the tip resistance is $R_\text{tip} = 1.04\,$kΩ and the gold substrate resistance is only $R_\text{Au} = 0.5\,\Omega$. This shows that the small volume at the tip apex, with its possible deviations from the considered geometrical model, is responsible for most of the probe resistance.

If a simpler model consisting of a contact region with radius a and a half space is used for the contact resistance, the formula for the contact resistance is $R_\text{c} = \rho/4a$, with ρ being the average resistivity. Using the values of ρ for UNCD and gold, and a from the mechanical analysis, we obtain $R_\text{c} = 2.09\,$kΩ.

For an experimental evaluation of the total contact resistance R_tot, the UNCD probe (previously characterized in terms of shape) was brought into contact with a gold thin film, deposited on a silicon substrate, by applying controlled forces. Figure 23.37a shows the electrical circuit used in the measurement. A sweep voltage varying between -50 and $50\,$mV was applied using a Keithley SCS instrument and the current flowing through the circuit was simultaneously measured. An I–V curve is plotted in Fig. 23.37b, from which the effective resistance of the circuit can be

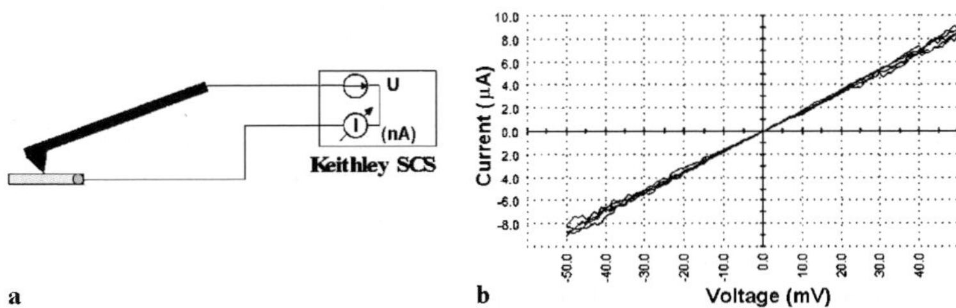

Fig. 23.37. (a) The electrical circuit used for getting the contact resistance. The substrate shown is a gold surface. (b) I–V curves obtained showing ohmic behavior of the contact

calculated. The same experiment was repeated several times to ensure the consistency of the measurements. A total resistance of $R_{tot} = 6.25 \pm 0.05\,\mathrm{k\Omega}$ was obtained, which is larger than the value calculated theoretically. This shows that the tip-to-substrate contact contains a resistance (approximately $4\,\mathrm{k\Omega}$) that can be attributed neither to the probe nor to the gold substrate or electrodes; thus, it must be a material of high resistivity entrapped between the probe and the gold. This can be a thin oxide layer, a contamination layer, an alteration of the diamond tip owing to nanoscale discharges and heat generated during the conductive measurements or, more likely, a combination of these features. It is worth mentioning that the values enabling the comparison stem from measurements under a relatively high contact force. At lower contact forces, the electrical contact area is not formed simply by the geometrical indentation footprint, but involves contamination layers (oxides, adsorption layers) and roughness features of the two surfaces in contact. These features change both the electrical and the mechanical models. It may involve tunneling or dielectric breakdown and the mechanical response is more size-dependent. From a practical viewpoint, at low contact forces, the contact I–V characteristics of the type shown in Fig. 23.37b are history-dependent and become unreliable owing to noise and deviations from the ohmic behavior.

To evaluate the UNCD probe capabilities for AFM conductivity measurements, a thin film of gold deposited on a silicon substrate was scanned using the conductive

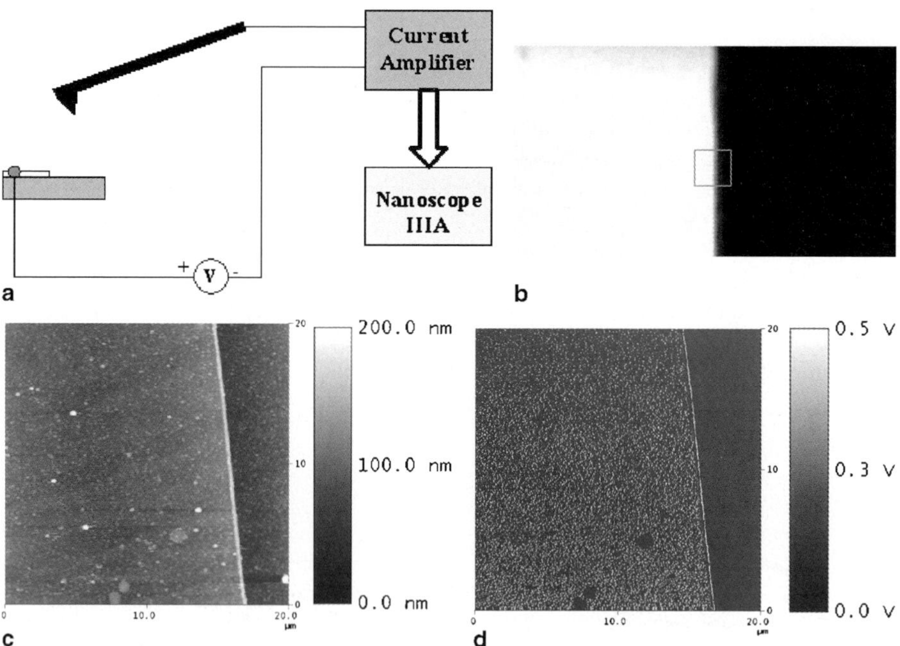

Fig. 23.38. (a) Electrical scheme used for getting a conduction map of the sample shown (b) A CCD image of the scanned region. The *bright portion* is gold film deposited on Si substrate (*dark region*). (c) Topography of the scanned region. (d) Conduction map, showing spikes of voltages in the conductive region

UNCD probe. A small anodic voltage V (of the order of millivolts) was applied to the gold surface with the UNCD probe grounded. The current in the circuit was amplified using a preamplifier, the output of which was inputted to the AFM Nanoscope III controller. Figure 23.38a shows the schematic of the electrical setup used during scanning and Fig. 23.38b shows the scanned region. When the tip comes in contact with the conductive substrate, a current is established in the circuit. This current is amplified by the preamplifier, and converted into a voltage signal with a known I–V sensitivity. The voltage plots were recorded simultaneously with the topographic and lateral force images using the AFM Nanoscope software. Figure 23.38c and d shows the topography and the conduction map of the scanned region.

The direct measurement of current and voltage through the conductive probe has the limitation that the probe continuously perturbs the voltage distribution on the sample. If this is to be avoided, the AFM probe can also be used in a potentiometric scheme (Fig. 23.39a). If the current through the probe I_p is kept sufficiently low, the overall voltage distribution on the sample is not affected by the measurement, provided the current through the main sample circuit is much higher ($I_s \gg I_p$). The example shown in Fig. 23.39 used a thin Cr film resistor of resistivity $0.635\,\Omega/\mathrm{cm}$ and a total resistance of $100\,\mathrm{k}\Omega$ with an applied voltage of $10\,\mathrm{V}$. A resistor, $R_p \gg R$,

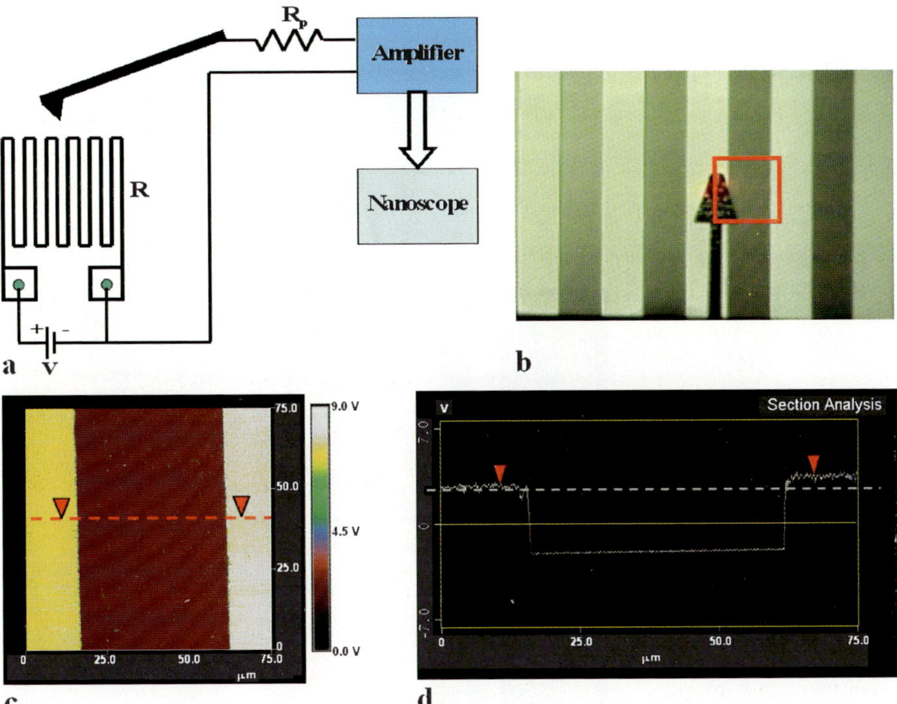

Fig. 23.39. (**a**) Electric circuit used for AFM potentiometry. (**b**) CCD image of the scanned region, where the *bright lines* are the metal traces and the *dark lines* are the spaces. (**c**) Conduction map of the two fins shown. (**d**) Cross-sectional profile along the line shown

was connected in series with the tip to limit the probing current. A CCD image of the scanned region is shown in Fig. 23.39b. Two fins of the resistor are scanned and a potential difference between the voltages at the two fins is observed. Figure 23.39c shows the corresponding conduction map and Fig. 23.39d shows a cross section of the voltage profile along the line shown (in red). For the plots shown, an amplification factor of 10^6 is used and $R_p = 1\,M\Omega$. These values correspond to a voltage difference of approximately 0.6 V between the two fins. The voltage applied across the fins is 10 V and the total length of the resistor is around 40 cm, which means a voltage gradient of 25 V/m. Therefore, a voltage difference of 0.6 V means a difference of around 1.5 cm between the locations of the two fins. The distance between the centers of the two locations is measured to be around 1.35 cm, which matches well with the electrically calculated value.

If the previous examples were obtained with relative large samples, the true advantage of AFM potentiometry is demonstrated by nanometer-scale measurements. Figure 23.40 shows an example of scanning with the potentiometric setup of a Pt

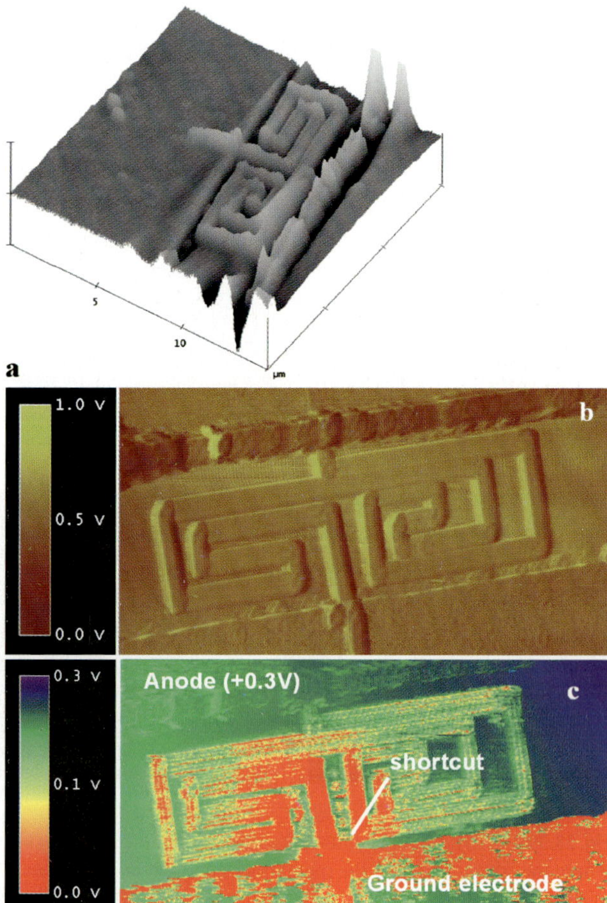

Fig. 23.40. Topographic image of a focused ion beam (FIB) deposited Pt nanowire (**a**), LFM image of the device (**b**) and AFM potentiometry map (**c**), showing that the voltage does not follow a uniform linear decrease along the nanowire resistor, owing to small shortcuts between the elements of the resistor, which occurred during the FIB deposition

nanowire, deposited by a focused ion beam between two electrodes, through which a current $I = 33\,\mu A$ was established by applying a voltage of 50 mV. Figure 23.40a is a topographic image of the Pt nanowire, while Fig. 23.40b is a lateral force image. The AFM potentiometric map is shown in Fig. 23.40c.

In the potentiometric map, it can be observed that the resistor is short-circuited at some places, a feature that is not evident in the topographical image. By contrast, the conduction map reveals that all the dark regions are electrically connected and are therefore at the same potential.

23.3.4.2
Molecular Writing

For DPN, the UNCD tip was dipped into a 10% MHA solution in acetonitrile, dried and mounted on the AFM head (Digital Instruments 3100). Writing was performed on the surface of a freshly deposited thin film of gold (100-nm Au electron-beam-evaporated onto a microscope glass slide). Writing of the letters U-N-C-D with the molecular ink was achieved with a line width of about 200 nm (Fig. 23.41).

The writing speed, which is controlled by the rate of molecular diffusion, was compared between the UNCD arrow-shaped and commercial nitride tips by performing static contact depositions of MHA dots on gold, with contact times of 5, 10 and 20 s. Both experiments were performed in similar conditions (contact force 50 nN, humidity $50 \pm 5\%$, temperature 23 °C). The results are shown in Figs. 23.42 and 23.43. The UNCD tips showed a slightly faster deposition rate, but within the limits of variability encountered among different nitride tips. This is consistent with the wetting properties of diamond, which is slightly more hydrophilic than silicon nitride. The smallest features that could be written with UNCD tips, using the DPN technique, were 80-nm-diameter dots (Fig. 23.42b).

The imaging of the DPN-deposited dots and letters was performed in all cases with the same tips used for deposition by running the instrument in frictional force mode. This also demonstrates that the UNCD probes developed produce high-quality frictional force AFM images.

Experiments also showed that the UNCD tips can be used to write by scratching the relatively soft gold surface. In this continuous writing mode, the tip is moved between a discrete number of points while keeping it in contact mode at relatively high contact force. The result of such a scratch test is shown in Fig. 23.44.

Fig. 23.41. "UNCD" written with MHA on a freshly deposited gold film, using a UNCD triangular-shaped probe. The line width is 200 nm. Imaging of the MHA pattern was performed with the same UNCD tip in lateral force mode

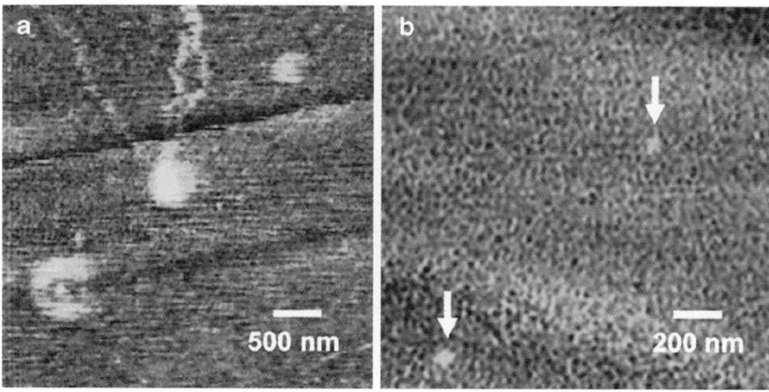

Fig. 23.42. (**a**) Three MHA dots on gold, obtained by contacting the surface with the arrow-shaped UNCD probe for 5, 10 and 20 s (from *top right* to *bottom left*, respectively). The image was recorded with the same probe in lateral force imaging mode. The diameters of the dots are 360, 540 and 710 nm, respectively. (**b**) Dots of 80-nm diameter are the best resolution features obtained so far

Fig. 23.43. The diameter of the MHA dots obtained with UNCD and Si_3N_4 tips in similar mode, as function of contact time. *Continuous fits* were done with square-root functions $Y = At^{1/2}$ showing good agreement with the model of Jang et al. [68]

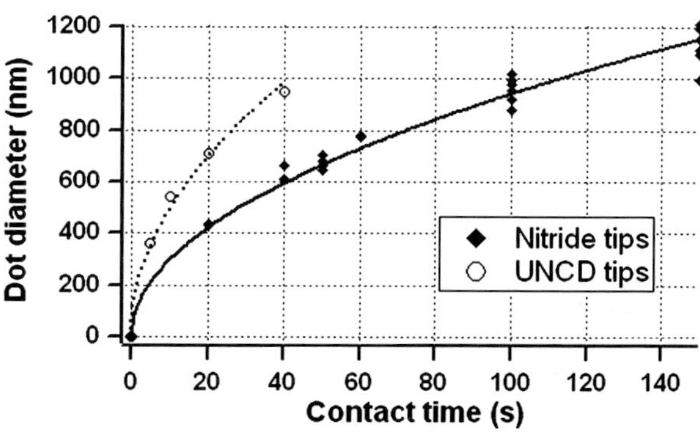

Fig. 23.44. AFM topography image of a scratch on a gold surface obtained by translating an UNCD tip in contact mode between two specified points

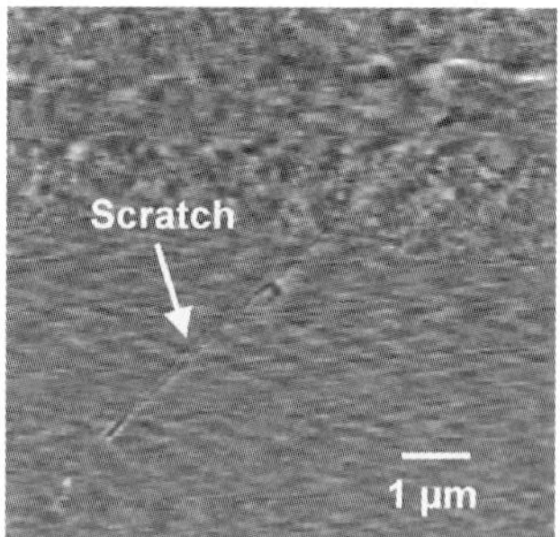

23.3.5
Perspectives for Diamond Probes

Diamond is expected to play an important role in future implementations of complex scanning probe systems owing to its excellent wear resistance, biocompatibility and the potential to make it conductive by doping. Complete diamond probes or probes with diamond forming only the probe tips are envisioned. For such applications, the use of diamond becomes a matter of integration, i.e., microfabrication processes and materials compatibility. Diamond in its various forms is still difficult to integrate. Problems related to adhesion, particulate and impurities contamination, film uniformity, reproducibility and selectivity in dry etching relative to some of the standard MEMS/nanoelectromechanical systems (NEMS) materials, for example, Si, SiO_2 and Si_3N_4, need to be addressed.

A class of applications for which diamond will be extremely useful is AFM-based microelectrochemistry or nanoelectrochemistry. In such cases, a critical aspect of the experimental measurement is the choice of material for the conducting AFM tip. In air, highly doped Si tips suffer from the growth of a native oxide layer that quickly deteriorates the tip conductivity. Metal-coated tips also fail to provide a continually low contact resistance as the metal wears from the tip apex during scanning [143, 144]. One approach to minimize metal wear is to operate the AFM predominantly in tapping mode to minimize lateral forces on the tip. An alternative solution is the use of diamond-coated Si tips [145], or the much sharper molded diamond tips presented before.

Although doped diamond can serve as an electrode for electrochemistry and conductive scanning probe techniques, including eventually EPN, it is likely that its superior wear resistance properties will not last under all possible environmental and electrical conditions employed in probing or writing. Recent conductive scanning probe microscope studies performed with metal probes on various diamond surfaces have shown that diamond can be etched [146], oxidized [147] or its surface modified [147, 148], thus providing a good substrate for AFM lithography and the formation of surface nanoelectronic circuits [148]. Oxidation of conductive diamond happens especially at anodic biases of a few volts (4–10 V) in air, which supplies the oxygen for the reaction. Oxidation occurs, likely with an initial swelling of the diamond surface [146], and then a surface recess is formed in a few minutes. However, swelling was observed even while performing the experiments in a vacuum, and persisted as long as the vacuum was maintained. Thus, the swelling may be a form of local carbon atom disorder, different from diamond, with high oxygen reactivity, or the accumulation of water molecules on oxidized carbon sites, desorbed later and leaving behind patterns of oxidized or etched diamond [147]. The low wear rate of diamond species is related also to the low friction coefficient, owing to the presence of hydrogen-terminated surface bonds. Local removal of hydrogen atoms during conductive techniques may thus increase the friction coefficient and wear. To gain insight into these mechanisms, to define the range of optimal material applications and to develop new diamond probes, extensive research is needed. Clearly, like any material, diamond is the right choice only for applications that exploit its superior properties. Among them, certainly are the probes of various future scanning probe microscopes, nanolithography, molecular pulling and data storage techniques.

References

1. Bonnell D (ed) (2001) Scanning probe microscopy and spectroscopy: theory, techniques, and applications. Wiley-VCH, Weinheim

2. Colton RJ, Engel A, Frommer JE, Gaub HE, Gewirth A, Guckenberger R, Rabe J, Heckl WM, Parkinson B et al (ed) (1998) Procedures in scanning probe microscopies. Wiley, New York

3. Colchero J, Bielefeldt H, Ruf A, Hipp M, Marti O, Mlynek J (1992) Phys Status Solidi A 131:73

4. Giessibl FJ (2003) Rev Mod Phys 75:949

5. Gilsinn J, Zhou H, Damazo B, Fu J, Silver R (2004) In: Technical Proceedings of the 2004 NSTI Nanotechnology Conference and Trade Show. Boston USA, Vol 3, p 456–459

6. Nanonics High vacuum environmental NSOM/SPM systems product sheet http://www.nanonics.co.il

7. Xu S, Arnsdorf MF (1995) Proc Natl Acad Sci USA 92:10384

8. Sulchek T, Hsieh R, Adams JD, Minne SC, Quate CF, Adderton DM (2000) Rev Sci Instrum 71:2097

9. Gil A, de Pablo PJ, Colchero J, Gómez-Herrero J, Baró AM (2002) Nanotechnology 13:309–313

10. Martin Y, Wickramasinghe HK (1987) Appl Phys Lett 50:1455

11. Trenkler T, Hantschel T, Stephenson R, De Wolf P, Vandervorst W, Hellemans L, Malavé A, Büchel D, Oesterschulze E, Kulisch W, Niedermann P, Sulzbach T, Ohlsson O (2000) J Vac Sci Technol B 18:418

12. Kinser CR, Schmitz MJ, Hersam MC (2005) Nano Lett 5:91

13. Nonnenmacher M, O'Boyle MP, Wickramasinghe HK (1991) Appl Phys Lett 58:2921

14. Grover R, McCarthy B, Zhao Y, Jabbour GE, Sarid D, Laws GM, Takulapalli BR, Thornton TJ, Gust D (2004) Appl Phys Lett 85:3926

15. Chu LL, Takahata K, Selvaganapathy PR, Gianchandani YB, Shohe JL (2005) J Microelectromech Syst 14:691

16. Shi L, Majumdar A (2001) Microscale Thermophys Eng 5:251–265

17. Dekhter R, Khachatryan E, Kokotov Y, Lewis A, Kokotov S, Fish G, Shambrot Y, Lieberman K (2000) Appl Phys Lett 77:4425

18. Betzig E, Trautman JK (1992) Science 257:189

19. Kueng A, Kranz C, Lugstein A, Bertagnolli E, Mizaikoff B (2005) Angew Chem Int Ed Engl 44:3419

20. Miller SA, Turner KL, MacDonald NC (1997) Proc SPIE 3009:72

21. Bullen D, Liu C (2006) Sens Actuators A 125:504

22. Heller MJ (2002) Annu Rev Biomed Eng 4:129

23. Offereins HL, Sandmaier H, Marusczyk K, Kuhl K, Plettner A (1992) Sens Mater 3:127

24. Lewis A, Kheifetz Y, Shambrodt E, Radko A, Khatchatryan E, Sukenik C (1999) Appl Phys Lett 75:2689

25. Ionescu RE, Marks RS, Gheber LA (2003) Nano Lett 12:1639–1641

26. Xu J, Lynch M, Nettikadan S, Mosher C, Vegasandra S, Henderson E (2006) Sens Actuators B 113:1034

27. Xu J, Lynch M, Huff JL, Mosher C, Vengasandra S, Ding G, Henderson E (2004) Biomed Microdevices 6:117

28. Meister A, Jeney S, Liley M, Akiyama T, Staufer U, de Rooij NF, Heinzelmann H (2003) Microelectron Eng 67–68:644

29. Barth W, Debski T, Abedinov N, Ivanov T, Heerlein H, Volland B, Gotszalk T, Rangelow IW, Torkar K, Fritzenwallner K, Grabiec P, Studzinska K, Kostic I, Hudek P (2001) Microelectron Eng 57–58:825

30. Mueller-Falcke C, Song Y-A, Kim S-G (2004) Industrial Optical and Robotic Systems Technology and Applications. Optics East, SPIE, Philadelphia, p 5604

31. Ono M, Lange D, Brand O, Hagleitner C, Baltes H (2002) Ultramicroscopy 91:9

32. Lange D, Akiyama T, Hagleiter C, Tonin A, Hidber HR, Niedermann P, Staufer U, de Rooij NF, Brand O, Baltes H (1999) In: Proceedings of IEEE Micro Electro Mechanical Systems (MEMS), p 447
33. Hagleiter C, Lange D, Akiyama T, Tonin A, Vogt R, Baltes H (1999) Proc oSPIE Smart Struct Mater 3673:240
34. Onaran AG, Balantekin M, Lee W, Hughes WL, Buchine BA, Guldiken RO, Parlak Z, Quate CF, Degertekin FL (2006) Rev Sci Instrum 77:023501
35. Golszalk T, Grabiec P, Shi F, Dumania P, Hudekd P, Rangelow IW (1998) Microelectron Eng 41–42:477
36. Kley VB (2002) USt Patent 6,353,219 B1, March 5
37. Kunze U, Klehn B (1999) Adv Mater 11:1473
38. Popp V, Kladny R, Schimmel T, Küppers J (1998) Surf Sci 401:105
39. Liu GY, Xu S, Qian YL (2000) Acc Chem Res 33:457
40. Falvo MR, Clary GJ, Taylor RM, Chi V, Brooks FP, Washburn S, Superfine R (1997) Nature 389:582
41. Wong EW, Sheehan PE, Lieber CM (1997) Science 277:1971
42. Harfenist SA, Cambron SD, Nelson EW, Berry SM, Isham AW, Crain MM, Walsh KM, Keynton RS, Cohn RW (2004) Nano Lett 4:1931
43. Sugimura H, Nakagiri NJ (1996) J Vac Sci Technol A 14:1223
44. Legrand B, Stievenard D (1999) Appl Phys Lett 74:4049
45. Vaccaro PO, Sakata S, Yamaoka S, Umezu I, Sugimura A (1998) J Mater Sci Lett 17:1941
46. Li Y, Maynor BW, Liu J (2001) J Am Chem Soc 123:2105
47. Möller D, Rüetschi M, Haefke H, Güntherodt H-J (1999) Surf Interface Anal 27:525
48. Möller D, Eckert R, Haefke H, Güntherodt H-J (2000) J Vac Sci Technol B 18:644
49. Rossiter C, Suni II (1999) Surf Sci 430:L553–L557
50. Hong S, Mirkin CA (2000) Science 288:1808
51. Demers LM, Ginger DS, Park S-J, Li Z, Chung S-W, Mirkin CA (2002) Science 296:1836
52. Lee K-B, Park S-J, Mirkin CA, Smith JC, Mrksich M (2002) Science 295:1702
53. Wilson DL, Martin R, Hong S, Cronin-Golomb M, Mirkin CA, Kaplan DL (2001) Proc Natl Acad Sci USA 98:13660
54. Lee K-B, Lim J-H, Mirkin CA (2003) J Am Chem Soc 125:5588
55. Kim K-H, Ke C, Moldovan N, Espinosa HD (2003) In: Proceedings of the Society of Experimental Mechanics Annual Conference, Charlotte, p 235
56. Deladi S, Tas NR, Berenschot JW, Krijnen GJM, de Boer MJ, de Boer JH, Peter M, Elwenspoek MC (2004) Appl Phys Lett 85:5361
57. Kim K-H, Moldovan N, Espinosa HD (2005) Small 1:632
58. Moldovan N, Kim K-H, Espinosa HD (2006) J Microelectromech Syst 15:204
59. Moldovan N, Kim K-H, Espinosa HD (2006) Journal of Micromechanics and Microengineering (submitted)
60. Cai Y, Ocko BM (2005) J Am Chem. Soc 127:16287
61. Marszalek PE, Greenleaf WJ, Li HB, Oberhauser AF, Fernandez JM (2000) Proc Natl Acad Sci USA 97:6282
62. Yu MF, Dyer MJ, Skidmore GD, Rohrs HW, Lu XK, Ausman KD, Von Ehr JR, Ruoff RS (1999) Nanotechnology 10:244
63. Rief M, Gautel M, Oesterhelt F, Fernandez JM, Gaub HE (1997) Science 276:1109–1112
64. Fantner GE, Oroudjev E, Schitter G, Golde LS, Thurner P, Finch MM, Turner P, Gutsmann T, Morse DE, Hansma H, Hansma PK (2006) Biophys J 90:1411–1418
65. Oberhauser AF, Marszalek PE, Carrion-Vazquez M, Fernandez JM (1999) Nat Struct Biol 6:1025
66. Schirmeisen A, Holscher H, Anczykowski B, Weiner D, Schafer MM, Fuchs H (2005) Nanotechnology 16:S13

67. Choi E-M, Yoon Y-H, Lee S, Kang H (1995) Phys Rev Lett 95:085701
68. Jang J, Hong S, Schatz GC, Ratner MA (2001) J Chem Phys 115:2721
69. Weeks BL, Vaughn MW, DeYoreo JJ (2005) Langmuir 21:8096
70. Rozhok S, Piner R, Mirkin CA (2003) J Phys Chem B 107:751
71. Sheehan PE, Whitman LJ (2002) Phys Rev Lett 88:156104
72. Rosner B, Eby R (2004) Nanoink technical note. http://www.nanoink.net/docs/appnotes/
    appnotes_multiprobe.pdf
73. Albrecht TR, Akamine S, Carver TE, Quate CF (1990) J Vac Sci Technol A 8:3386
74. Akamine S, Quate CF (1992) J Vac Sci Technol B 10:2307
75. Piner RD, Zhu J, Xu F, Hong S, Mirkin CA (1999) Science 283:661
76. Wilson DL, Martin R, Hong S, Cronin-Golomb M, Mirkin CA, Kaplan DL (2001) Proc
    Natl Acad Sci 98:13660
77. Maynor BW, Filocamo SF, Grinstaff MW, Liu J (2002) J Am Chem Soc 124:522
78. Ivanisevic A, Mirkin CA (2001) J Am Chem Soc 123:7887
79. Su M, Dravid VP (2002) Appl Phys Lett 80:4434
80. Lim J-H, Ginger DS, Lee K-B, Heo J, Nam J-M, Mirkin CA (2003) Angew Chem Int Ed
    Engl 42:2309
81. Agarwal G, Sowards LA, Naik RR, Stone MO (2003) J Am Chem Soc 125:580
82. Agarwal G, Naik RR, Stone MO (2003) J Am Chem Soc 125:7408
83. Zhang M, Bullen D, Chung S-W, Hong S, Ryu KS, Fan Z, Mirkin CA, Liu C (2002)
    Nanotechnology 13:212
84. Bullen D, Wang X, Zou J, Hong S, Chung S, Ryu K, Fan Z, Mirkin C, Liu C (2003)
    in: 16th IEEE international micro electro mechanical systems conference, MEMS 2003,
    Kyoto, Japan
85. Minne SC, Manalis SR, Quate Calvin F (1999) Bringing scanning probe microscopy up
    to speed. Kluwer, Boston
86. Vettiger P, Cross G, Despont M, Drechsler U, Durig U, Gotsmann B, Haberle W, Lantz MA,
    Rothuizen HE, Stutz R, Binnig GK (2002) IEEE Trans Nanotechnol 1:39
87. Belaubre P, Guirardel M, Leberre V, Pourciel JB, Bergaud C (2004) Sens Actuators
    A 110:130
88. Vettiger P, Despont M, Drechsler U, Durig U, Haberle W, Lutwyche MI, Rothuizen HE,
    Stutz R, Widmer R, Binnig GK (2000) IBM J Res Dev 44:323
89. Lutwyche M, Andreoli C, Binnig G, Brugger J, Drechsler U, Haberle W, Rohrer H,
    Rothuizen H, Vettiger P, Yaralioglu G, Quate C (1999) Sens Actuators A 73:89
90. Bullen D, Wang X, Zou J, Chang S, Mirkin C, Liu C (2004) JMEMS 13:594
91. Indermuhle PF, Schurmann G, Racine GA, deRooij NF (1997) Sens Actuators A 60:186
92. Sulchek T, Minne SC, Adams JD, Fletcher DA, Atalar A, Quate CF, Adderton DM (1999)
    App Phys Lett 75:1637
93. Rogers B, York D, Whisman N, Jones M, Murray K, Adams JD, Sulchek T, Minne SC
    (2002) Rev Sci Instrum 73:3242
94. Minne SC, Yaralioglu G, Manalis SR, Adams JD, Zesch J, Atalar A, Quate CF (2000)
    Appl Phys Lett 76:1473–1475
95. Itoh T, Suga T (1993) Nanotechnology 4:218
96. Miyahara Y, Fujii T, Watanabe S, Tonoli A, Carabelli S, Yamada H, Bleuler H (1999) Appl
    Surf Sci 140:428
97. Chui BW, Kenny TW, Mamin HJ, Terris BD, Rugar D (1998) Appl Phys Lett 72:1388
98. Brugger J, Buser RA, de Rooij NF (1992) J Micromech Microeng 2:218
99. Lange D, Brand O, Baltes H (2002) CMOS cantilever sensor systems: atomic force mi-
    croscopy and gas sensing applications. Springer, Berlin, Heidelberg, New York
100. Buhlmann S, Dwir B, Baborowski J, Muralt P (2002) Appl Phys Lett 80:3195
101. Watanabe S, Fujiu T, Fujii T (1995) Appl Phys Lett 66:1481

102. Polla DL, Francis LF (1998) Rev Mater Sci 28:563–597
103. http://clifton.mech.northwestern.edu/~nfp/
104. Belaubre P, Guirardel M, Garcia G, Pourciel JB, Leberre V, Dagkessamanskaia A, Trevisiol E, Francois JM, Bergaud C (2003) Appl Phys Lett 82:3122
105. Moldovan N, Kim K-H, Espinosa HD (2006) J Microelectromech Syst 15:204–213
106. Bruckbauer A, Ying LM, Rothery AM, Zhou DJ, Shevchuk AI, Abell C, Korchev YE, Klenerman D (2002) J Am Chem Soc 124:8810
107. Taha H, Marks RS, Gheber LA, Rousso I, Newman J, Sukenik C, Lewis A (2003) Appl Phys Lett 83:1041
108. Zhang H, Li Z, Mirkin CA (2002) Adv Mater 14:1472
109. Chung SW, Ginger DS, Morales MW, Zhang ZF, Chandrasekhar V, Ratner MA, Mirkin CA (2005) Small 1:64
110. http://www.pi.ws/
111. Meyer E, Hug HJ, Bennewitz R (2004) Scanning probe microscopy: the lab on a tip. Springer, Berlin Heidelberg New York
112. Lu RP, Kavanagh KL, Dixon-Warren SJ, Spring-Thorpe AJ, Streater R, Calder I (2002) J Vac Sci Technol B 20:1682
113. Hersam MC, Hoole ACF, O'Shea SJ, Welland ME (1998) Appl Phys Lett 72:915
114. Veeco (2003) Application modules: dimension and multimode manual, Chap. 2 Veeco, Santa Barbara, CA
115. Basu AS, McNamara S, Gianchandani YB (2004) J Vac Sci Technol B 22:3217
116. Angadi MA, Watanabe T, Bodapati A, Xiao X, Auciello O, Carlisle JA, Eastman JA (2005) J Appl Phys 99:114301
117. Kaneko R, Oguchi S (1990) Jpn J Appl Phys Part 1 29 1854
118. Hantschel T, Niedermann P, Trenkler T, Vandervorst W (2000) Appl Phys Lett 76:1603
119. Niedermann P, Hanni W, Blanc N, Christoph R, Burger J (1996) J Vac Sci Technol 14:1233
120. Visser EP, Gerritsen JW, Vanenckevort WJP, Vankempen H (1992) Appl Phys Lett 60:3232
121. Rameshan R (1999) Thin Solid Films 340:1
122. Friedmann TA, Sullivan JP, Knapp JA, Tallant DR, Follstaedt DM, Medlin DL, Mirkarimi PB (1997) Appl Phys Lett 71:3820
123. Mihalcea C, Scholz W, Malave A, Albert D, Kulisch W, Oesterschulze E (1998) Appl Phys A 66:S87
124. Krauss AR, Auciello O, Gruen DM, Jayatissa A, Sumant A, Tucek J, Mancini DC, Moldovan N, Erdemir A, Ersoy D, Gardos MN, Busmann HG, Meyer EM, Ding MQ (2001) Diamond Relat Materi 10:1952
125. Espinosa HD, Prorok BC, Peng B, Kim KH, Moldovan N, Auciello O, Carlisle JA, Gruen DM, Mancini DC (2003) Exp Mech 43:256
126. Espinosa HD, Peng B, Prorok BC, Moldovan N, Auciello O, Carlisle JA, Gruen DM, Mancini DC (2003) J Appl Phys 94:6076
127. Erdemir A, Bindal C, Fenske GR, Zuiker C, Csencsits R, Krauss AR, Gruen DM (1996) Diamond Films Technol 6:31
128. Minh PN, Takahito O, Masayoshi E (2002) Fabrication of silicon microprobes for optical near-field applications. CRC, Boca Raton
129. Zou J, Wang XF, Bullen D, Ryu K, Liu C, Mirkin CA (2004) J Micromech Microeng 14:204
130. Okano K, Hoshina K, Iida M, Koizumi S, Inuzuka T (1994) Appl Phys Lett 64:2742
131. Scholtz W, Albert D, Malave A, Werner S, Mihalcea C, Kulisch W, Oesterschulze E (1997) Proc SPIE 3009:61
132. Niedermann P, Hanni W, Morel D, Perret A, Skinner N, Indermuhle PF, de Rooij NF, Buffat PA (1998) Appl Phys A 66:S31

133. Moldovan N, Auciello O, Sumant AV, Carlisle JA, Divan R, Gruen DM, Krauss AR, Mancini DC, Jayatissa A, Tucek J (2001) Proc SPIE 4557:288
134. Technic Techni-gold 25 E. Technic, Cranston. http://www.technic.com
135. Peng B, Espinosa HD (2004) In: Proceedings of 2004 ASME international mechanical engineering congress, CA
136. Espinosa HD, Peng B, Prorok BC, Moldovan N, Auciello O, Carlisle JA, Xiao X, Gruen DM, Mancini DC (2003) J Appl Phys 94:6076–6084
137. http://clifton.mech.northwestern.edu/~espinosa/publications/movielink.htm
138. Bhattacharya S, Auciello O, Birrell J, Carlisle JA, Curtiss LA, Goyette AN, Gruen DM, Krauss AR, Schlueter J, Sumant A, Zapol P (2001) Appl Phys Lett 79:1441
139. Vesenka J, Manne S, Giberson R, Marsh T, Henderson E (1993) Biophys J 65:992
140. Holm R (1958) Electric contacts handbook, 3rd edn. Springer, Berlin Heidelberg, New York
141. Zheng S, Cuitino AM (2002) KONA Powder and Particle 20:168–177
142. Espinosa HD, Prorok BC (2003) J Mater Sci 38:4125–4128
143. O'Shea SJ, Atta RM, Welland ME (1995) Rev Sci Instrum 66:2508
144. Thomson RE, Moreland J (1995) J Vac Sci Technol B 13:1123
145. Niedermann P, Hänni W, Blanc N, Christoph R, Burger J (1996) J Vac Sci Technol A 14:1233
146. Myhra S (2005) Appl Phys A 80:1097–1104
147. Rezek B, Sauder C, Garrido JA, Nebel CE, Stutzmann M, Snidero E, Bergonzo P (2003) Appl Phys Lett 82:3336–3338
148. Tachiki M, Seo H, Banno T, Sumikawa Y, Umezawa H, Kawarada H (2002) Appl Phys Lett 81:2854–2856

24 Nanoelectromechanical Systems – Experiments and Modeling

Horacio D. Espinosa · Changhong Ke

24.1
Introduction

Nanoelectromechanical systems (NEMS) are made of electromechanical devices that have critical dimensions from hundreds down to a few nanometers. By exploring nanoscale effects, NEMS present interesting and unique characteristics, which deviate greatly from their predecessor microelectromechanical systems (MEMS). For instance, NEMS-based devices can have fundamental frequencies in the microwave range (approximately 100 GHz) [1]; mechanical quality factors in the tens of thousands, meaning low energy dissipation; active mass in the femtogram range [2]; force sensitivity at the attonewton level [3]; mass sensitivity up to attogram [4] and subattogram [5] levels; heat capacities far below a "yoctocalorie" [6]; power consumption in the order of 10 aW [7]; and extremely high integration levels, approaching 10^{12} elements per square centimeter [1]. All these distinguished properties of NEMS devices pave the way for applications such as force sensors, chemical sensors, biological sensors, and ultrahigh frequency resonators.

The interesting properties of the NEMS devices typically arise from the behavior of the active parts, which, in most cases, are in the form of cantilevers or doubly clamped beams with dimensions at the nanometer scale. The materials for these active components include silicon and silicon carbide, carbon nanotubes (CNTs), gold, and platinum, to name a few. Silicon has been the basic material for integrated circuit technology and MEMS during the past few decades, and is widely used to build NEMS as well. Figure 24.1 is a scanning electron microscopy (SEM) image of a double-clamped resonator fabricated from a bulk, single-crystal silicon substrate [8]. However, ultrasmall silicon-based NEMS fail to achieve desired high quality factors owing to the dominance of surface effects, such as surface oxidation

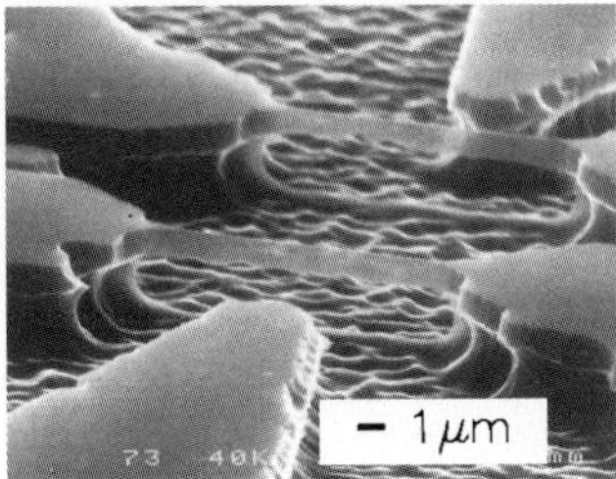

Fig. 24.1. Scanning electron microscope (SEM) image of an undercut Si beam, with length of 7.7 μm, width of 0.33 μm and height of 0.8 μm. (Reprinted with permission from Cleland et al. [8]. Copyright 1996, American Institute of Physics)

and reconstruction, and thermoelastic damping. Limitations in strength and flexibility also compromise the performance of silicon-based NEMS actuators. Instead, CNTs can well represent the ideas of NEMS given their nearly one-dimensional structures with high aspect ratio and perfectly terminated surfaces and excellent electrical, mechanical, and chemical properties. Owing to significant advances in growth, manipulation, and knowledge of their electrical and mechanical properties, CNTs have become the most promising building blocks for the next generation of NEMS. Several CNT-based functional NEMS devices have been reported so far [1, 3, 9–14]. Similar to CNTs, nanowires are another type of one-dimensional novel nanostructures for building NEMS because of their size and controllable electrical properties.

The purpose of this chapter is to provide a comprehensive review of NEMS devices to date and to summarize the modeling currently being pursued to gain insight into their performance. This chapter is organized as follows. In the first part, we review CNTs and CNT-based NEMS. We also discuss nanowire-based NEMS. In the second part, we present the modeling of NEMS, including multiscale modeling and continuum modeling.

24.2
Nanoelectromechanical Systems

24.2.1
Carbon Nanotubes

CNTs exist as a macromolecule of carbon, analogous to a sheet of graphite rolled into a cylinder. They were discovered by Iijima [15] in 1991 and are a subset of the family of fullerene structures. The properties of the nanotubes depend on the atomic arrangement (how the sheets of graphite are rolled to form a cylinder), their diameter, and length. They are light, stiff, flexible, thermally stable, and chemically inert. They have the ability to be either metallic or semiconducting depending on the "twist" of the tube, which is called the "chirality" or "helicity". Nanotubes may exist as either single-walled or multiwalled structures. Multiwalled CNTs (MWNTs) (Fig. 24.2b) are simply composed of multiple concentric single-walled CNTs (SWNTs) (Fig. 24.2a) [16]. The spacing between neighboring graphite layers in MWNTs is approximately 0.34 nm. These layers interact with each other via van der Waals forces.

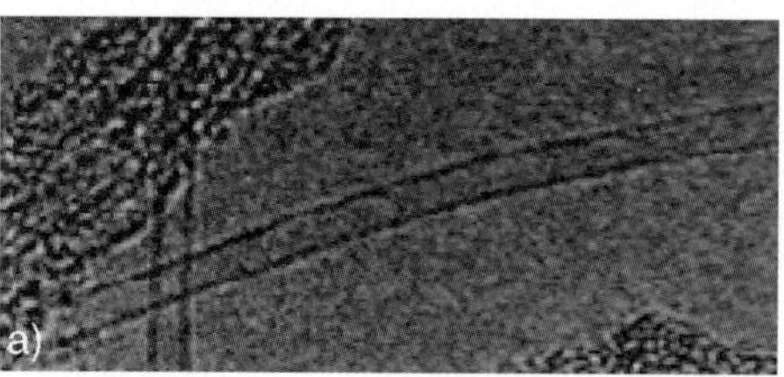

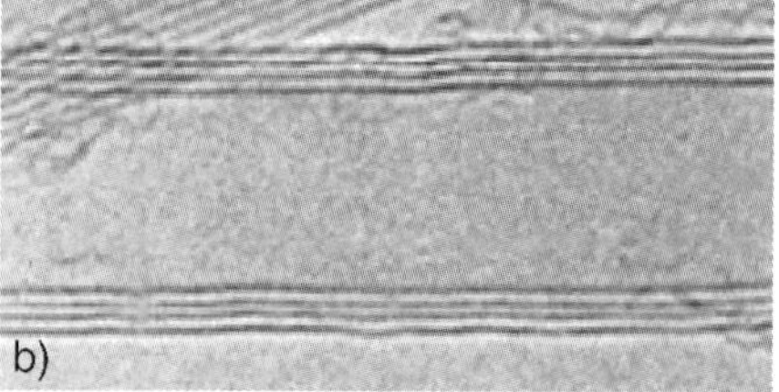

Fig. 24.2. High-resolution transmission electron microscopy image of typical single-walled carbon nanotubes (SWNT) (**a**) and multiwalled carbon nanotubes (MWNT) (**b**). (Reprinted with permission from Ajayan [16] Copyright 1999, American Chemical Society)

The methods used to synthesize CNTs include electric arc discharge [17, 18], laser ablation [19], and catalytic chemical vapor deposition (CVD) [20]. During synthesis, nanotubes are usually mixed with residues including various types of carbon particles. For most applications and tests, a purification process is required. In one of the most common approaches, nanotubes are ultrasonically dispersed in a liquid (e.g., 2-propanol) and the suspension is centrifuged to remove large particles. Other methods, including dielectrophoretic separation, are being developed to provide improved yield.

The mechanical and electrical properties of CNTs have been under intensive study during the past decade. Qian et al. [21] contributed a comprehensive review article from the perspective of both experimentation and modeling. The electronics of CNTs was extensively reviewed by McEuen et al. [22]. Besides, the study of the coupled electromechanical properties, which are essential to NEMS, is rapidly progressing. Some interesting results have been reported regarding the fact that the electrical properties of CNTs are sensitive to the structure variation and can be changed dramatically owing to the change of the atomic bonds induced by mechanical deformations. CNTs can even change from metallic to semiconducting when subjected to mechanical deformation [23–25].

24.2.2
Fabrication Methods

The fabrication processes of NEMS devices can be categorized according to two approaches. *Top-down* approaches, which evolved from manufacturing of MEMS structures, utilize submicron lithographic techniques, such as electron-beam lithography, to fabricate structures from bulk materials, either thin films or bulk substrates. *Bottom-up* approaches fabricate the nanoscale devices by sequential assembly of atoms and molecules as building blocks. Top-down fabrication is size-limited by factors such as the resolution of the electron-beam lithography, etching-induced roughness, and synthesis constraints in epitaxially grown substrates. Significant interest has been shown in the integration of nanoscale materials such as CNTs and nanowires, fabricated by bottom-up approaches, to build nanodevices. Most of the nanodevices reported so far in the literature are obtained by "hybrid" approaches, i.e., combination of bottom-up (self assembly) and top-down (lithographic) approaches [26].

One of the key and most challenging issues of building CNT-based or nanowire-based NEMS is the positioning of nanotubes or nanowires at the desired locations with high accuracy and high throughput. Reported methods of manipulation and positioning of nanotubes are briefly summarized in the following section.

24.2.2.1
Random Dispersion Followed by Electron-Beam Lithography

After purification, a small aliquot of a nanotube suspension is deposited onto a substrate. The result is nanotubes randomly dispersed on the substrate. Nanotubes on the substrate are imaged inside a scanning electron microscope and then this image is digitized and imported to mask-drawing software, where the mask for the subsequent electron-beam lithography is designed. In the mask layout, pads are designed to superimpose over the ends of the CNTs. Wet etching is employed to remove the material under the CNTs to form freestanding nanotube structures. This process requires an alignment capability in the lithographic step with an accuracy of 0.1 μm or better. This method was first employed to make nanotube structures for mechanical testing [27, 28]. The reported NEMS devices based on this method include nanotube-based rotational actuators [10] and nanowire-based resonators [26].

24.2.2.2
Nanomanipulation

Manipulation of individual CNTs using piezo-driven manipulators inside electron microscope chambers is one of the most commonly used methods to build NEMS [9] and structures for mechanical and electrical testing [29–34]. In general, the manipulation and positioning of nanotubes is accomplished in the following manner: (1) a source of nanotubes is positioned close to the manipulator inside the microscope; (2) the manipulator probe is moved close to the nanotubes under visual surveillance of the microscope monitor until a protruding nanotube is attracted to the manipulator owing to either van der Waals forces or electrostatic forces; (3) the free end of the

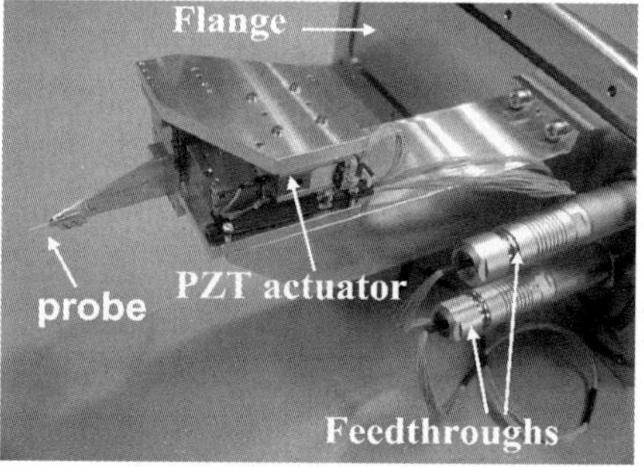

Fig. 24.3. A customized SEM flange holding a three-dimensional nanomanipulator manufactured by Klocke Nanotechnik with two electric feedthroughs, a three-axis piezoelectric transducer (PZT) actuator, and a tungsten probe mounted in the manipulator's probe holder

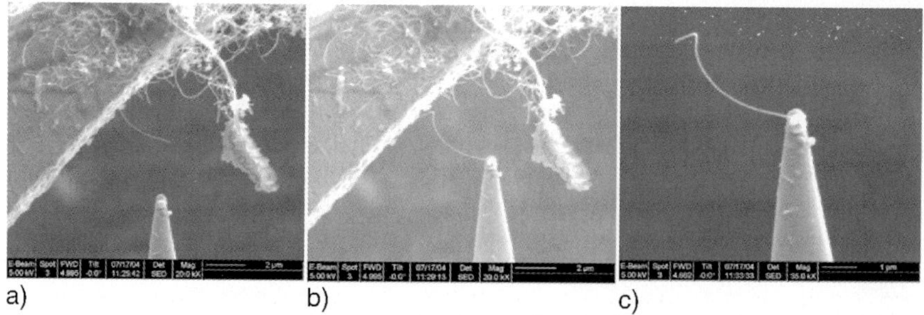

a)   b)   c)

Fig. 24.4. SEM images of the manipulation of carbon nanotubes (CNT) using the 3D Klocke Nanotechnik nanomanipulator. (**a**) Manipulator probe is approaching a protruding nanotube. The sample is a dried nanotube solution on top of a transmission electron microscope (TEM) copper grid. (**b**) Manipulator probe makes contact with the free end of the nanotube and the nanotube is welded to the probe by electron-beam-induced deposition of platinum. (**c**) A single nanotube mounted in the manipulator probe

attracted nanotube is "spot welded" by electron-beam-induced deposition (EBID) of hydrocarbon [9, 34] or metals, like platinum [31], from adequate precursor gases.

Figure 24.3 shows a 3D nanomanipulator (Klocke Nanotechnik, Germany) mounted in a custom-designed scanning electron microscope flange with two electrical feedthroughs, having the capability of moving in X, Y, and Z directions with nanometer-displacement resolution. The manipulation process of an individual CNT is illustrated in Fig. 24.4.

24.2.2.3
External Field Alignment

DC/AC electric fields have been successfully used in the manipulation of nanowires [35], nanotubes [36, 37], and bioparticles [38–40]. Microfabricated electrodes are typically used to create an electric field in the gap between them. A droplet containing CNTs in suspension is dispensed into the gap with a micropipette. The applied electric field aligns the nanotubes, owing to the dielectrophoretic effect, resulting in the bridging of the electrodes by a single nanotube. The voltage drop that arises when the circuit is closed (DC component) ensures the manipulation of only one nanotube. Besides, AC dielectrophoresis has been employed to successfully separate metallic from semiconducting SWNTs in suspension [41]. NEMS devices fabricated using this method include nanotube-based nanorelays [42].

Huang et al. [43] demonstrated another method for aligning nanowires. A laminar flow was employed to achieve preferential orientation of nanowires on chemically patterned surfaces. This method was successfully used in the alignment of silicon nanowires. Magnetic fields have also been used to align CNTs [44].

24.2.2.4
Direct Growth

Rather than manipulating and aligning CNTs after their manufacture, researchers have also examined methods for controlled direct growth. Huang et al. [45] used the microcontact printing (μCP) technique to directly grow aligned nanotubes vertically. Dai et al. [46–49] reported several patterned growth approaches developed in his group. The idea is to pattern the catalyst in an arrayed fashion and control the growth of CNTs from specific catalytic sites. The authors successfully carried out patterned growth of both MWNTs and SWNTs, and exploited methods including self-assembly and external electric field control. Figure 24.5 shows a SEM image of suspended SWNTs grown by an electric-field-directed CVD method [48]. The CNT-based tunable oscillators, reported in [3], were fabricated using this method. Recently, this technique was extended to grow nanotubes in two dimensions to make more complicated nanotube structures, such as nanotube crosses [50].

24.2.2.5
Self-Assembly

Self-assembly is a method of constructing nanostructures by forming stable bonds between the organic or nonorganic molecules and the substrate. Recently, Rao

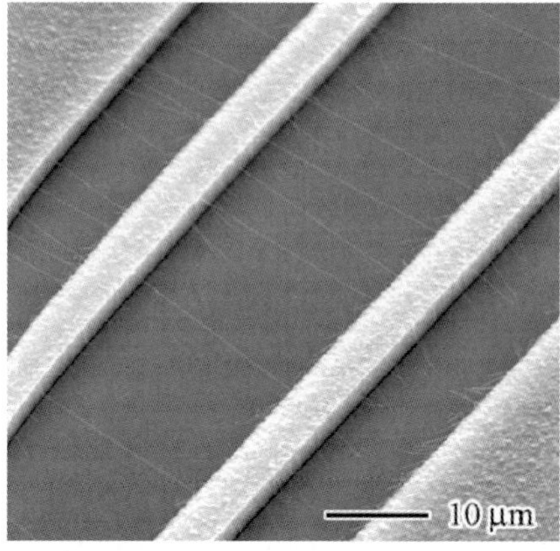

Fig. 24.5. Electric-field-directed freestanding single-walled nanotubes. (Reprinted with permission from Zhang et al. [48]. Copyright 2001, American Institute of Physics)

et al. [51], reported an approach in large-scale assembly of CNTs with high-throughput. Dip-pen nanolithography (DPN), a technique invented by Mirkin's group [52], was employed to functionalize the specific surface regions either with polar chemical groups such as amino ($-NH_2/-NH_3^+$) or carboxyl ($-COOH/-COO^-$), or with nonpolar groups such as methyl ($-CH_3$). When the substrate with functionalized surfaces was introduced into a liquid suspension of CNTs, the nanotubes were attracted towards the polar regions and self-assembled to form pre-designed structures, usually within 10 s, with a yield higher than 90%. The reported method is scalable to large arrays of nanotube devices by using high-throughput patterning methods such as photolithography, stamping, or massively parallel DPN.

24.2.3
Inducing and Detecting Motion

For nanostructures, both inducing and detecting motion are challenging. Some of the methods routinely used in MEMS face challenges when the size shrinks from microscale to nanoscale. For example, optical methods, such as simple beam deflection schemes or more sophisticated optical and fiber-optical interferometry – both commonly used in scanning probe microscopy to detect the deflection of the cantilevers – generally fall beyond the diffraction limit, which means these methods cannot be applied to objects with cross section much smaller than the wavelength of light [53]. During the past decade, significant progress has been made in nanoscale actuation and detection. Recently, available techniques in the context of NEMS were reviewed [54]. In the following subsections, we summarize these techniques with an emphasis on their accuracy, bandwidth, and limitations.

24.2.3.1
Inducing Motion

Similar to MEMS, electrostatic actuation of nanostructures by an applied electrical field is commonly used for the actuation of NEMS, e.g., nanotweezers [9, 12]. The

Lorenz force has been used to move small conducting beams [8, 26, 55], with alternating currents passing through them in a strong transverse magnetic field. The induced electromotive force (emf), or voltage, can be detected as a measure of the motion. This method requires a fully conducting path and works well with a beam clamped at both ends [56]. Other actuation methods include piezoelectric actuation, thermal actuation using bilayers of materials with different thermal expansion, thermal in-plane actuation due to a specially designed topography [57], and a scanning tunneling microscope (STM) [58].

24.2.3.1.1
Magnetomotive Technique

In magnetomotive induced motion, the actuation force is the Lorenz force produced by current flow through a conductor immersed in a static magnetic field. For example, a doubly clamped beam can be vibrated by flowing a varying current. The beam motion can be either in plane or out of plane depending on the orientation of the beam element with respect to the magnetic field. The Lorenz force acting on the beam is given by

$$F = lBI \,, \tag{24.1}$$

where l is the length of the beam, B is the magnetic field strength, and I is the current going through the beam. The magnetomotive actuation technique is broadband, even in the presence of parasitic capacitance. Huang et al. [59] have recently achieved actuation of a NEMS device in the gigahertz range using this technique.

24.2.3.1.2
Electrostatic Actuation

Electrostatic actuation is widely used in the actuation of MEMS and can also be extended to NEMS. The electrostatic force is calculated from the system capacitance. For instance, the force between two biased parallel plates is given by

$$F = \frac{1}{2}V^2\frac{\varepsilon A}{d^2} \,, \tag{24.2}$$

where V is the biased voltage, ε is the permittivity (for vacuum, $\varepsilon = 8.854 \times 10^{-12}$ $C^2\,N^{-1}\,m^{-2}$), A is the area of the plate, and d is the distance between the plates. It is noted that, owing to the presence of parasitic capacitance in MEMS/NEMS, the efficiency of this actuation technique in the high-frequency domain is significantly reduced.

24.2.3.2
Detecting Motion

The most straightforward method is by direct observation of the motion under electron microscopes [9, 12, 60, 61]. This visualization method, typically having resolution on the nanometer scale, projects the motion in the direction perpendicular to the electron beam. Limitations in depth of focus require that the nano-object motion be primarily in a plane, which normally is coaxial with the electron beam. Electron tunneling is a very sensitive method that can detect subnanometer motion by the exponential dependence of the electron tunneling current on the separation between tunneling electrodes; therefore, this technique is widely used in NEMS motion detection [3, 7]. Magnetomotive detection is a method based on the presence of a magnetic field, either uniform or spatially inhomogeneous, through which a conductor is moved. The time-varying flux generates an induced emf in the loop, which is proportional to the motion [26, 55, 62–64]. The displacement-detection sensitivity of this technique is less than 1 Å [65]. It is known that CNTs can act as transistors, and as such they can be utilized to sense their own motion [3, 66]. Capacitance sensors have been widely used in MEMS. They can also be used in NEMS motion sensing with a resolution of a few nanometers [57], and the resolution can potentially be increased to the angstrom range provided that the capacitance measurement can be improved by 1 order of magnitude. Piezoresistive and piezoelectric detection are techniques capable of sensing minute motion in NEMS by utilizing the material's sensitivity to strain. A material is piezoresistive if its resistance changes with strain. A material is piezoelectric if the electrical polarization changes with deformation. Optical interferometry, in particular, path-stabilized Michelson interferometry and Fabry–Perot interferometry, has been recently developed to detect motion in NEMS [67–71]. For the path-stabilized Michelson interferometric technique, a tightly focused laser beam reflects from the surface of a NEMS device in motion and interferes with a stable reference beam. For the Fabry–Perot interferometric technique, the optical cavity formed within the sacrificial gap between the NEMS surface and the substrate modulates the optical signal on a photodetector as the NEMS device moves in the out-of-plane direction. In the following subsections, we will discuss, in detail, some of the aforementioned displacement-detection techniques and their sensitivities due to the noise level in the system.

24.2.3.2.1
Magnetomotive Technique

The magnetomotive displacement-detection technique is based on the presence of a static magnetic field, through which a conducting nanomechanical element is moved. The time-varying flux induces an emf in the loop, which can be detected by the electrical circuit. For a doubly clamped beam, the generated emf is given by

$$v = \xi B l \dot{x}(t) \,, \tag{24.3}$$

where l is the length of the beam, B is the magnetic field strength. ξ is a geometric factor (for a doubly clamped beam, $\xi \approx 0.885$), and $x(t)$ is the motion of the beam element. A low-noise amplifier is typically employed to enlarge the small emf resulting from minute displacements. The coupling of a magnetomotive transducer to an amplifier is illustrated in Fig. 24.6 [54], which employs a standard amplifier model with two uncorrelated noise sources: a voltage and a current noise source with respective power spectral densities $S_V(\omega)$ and $S_I(\omega)$. Therefore,

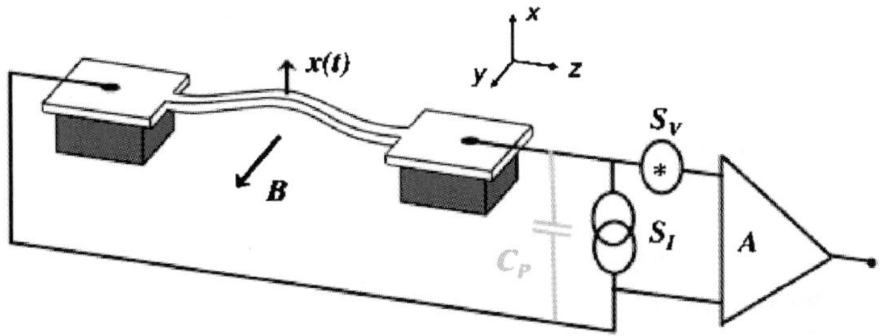

Fig. 24.6. Magnetomotive displacement detection. (Reprinted with permission from Ekinci [54]. Copyright 2005, Wiley-VCH)

the noise power generated in the amplifier determines the displacement sensitivity as [54]

$$|S_X(\omega)| = \frac{S_V(\omega)}{(\xi l B \omega)^2} + \frac{S_I(\omega) l^2 B^2}{m_{\mathrm{eff}}^2 \left(\omega_0^2 - \omega^2\right)^2 + \frac{(\omega \omega_0)^2}{Q^2}} ,$$
(24.4)

where m_{eff} is the effective mass, ω_0 is the resonant frequency, and Q is the quality factor. The first term on the right-hand side in (24.4) is related to the voltage noise in the amplifier, while the second term is related to the current noise in the amplifier.

For displacement measurement using the magnetomotive technique, the bandwidth of motion detection is primarily limited by the parasitic capacitance (C_p) in the detection circuit; thus, the cut-off frequency. If the output impedance of a NEMS device is R_e, the motion-detection bandwidth of this technique is about $R_\mathrm{e} C_\mathrm{p}$.

24.2.3.2.2
Electron Tunneling

Electron tunneling is a quantum phenomenon whereby electrons tunnel through a junction in the presence of an electric field. For example, tunneling happens between a very sharp metallic tip and a conductive surface when a subnanometer gap separates the surfaces. This is the principle of scanning electron microscopy. The tunneling current has a very important characteristic: it exhibits an exponential decay as the gap increases, namely,

$$i \approx \rho_\mathrm{s}(E_\mathrm{F}) V e^{-2kd} ,$$
(24.5)

where i is the current, V is the applied voltage, d is the gap, $\rho_\mathrm{s}(E_\mathrm{F})$ is the local density of electronic states in the test mass, and k is the decay constant for the electron wave function within the gap given by $k = \sqrt{2 m_\mathrm{e} \phi}/\hbar$; here m_e is the mass of the electron, ϕ is the work function of the metal ($\phi \approx 3-5\,\mathrm{eV}$), and $\hbar$ is Planck's constant divided by 2π. Typically, $k \approx 0.1\,\mathrm{nm}^{-1}$, meaning the current will change an order of magnitude when the gap changes by 1 Å. Therefore, motion information in the element can be converted into an electrical current signal through the tunneling transducer with extremely high sensitivity. Figure 24.7 illustrates a tunneling transducer for displacement detection in NEMS [54]. The motion-detection sensitivity

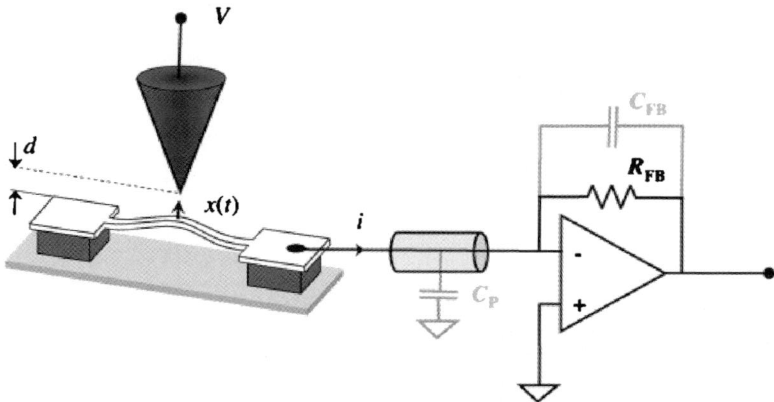

Fig. 24.7. Tunneling transducer for displacement detection in nanoelectromechanical systems (NEMS). (Reprinted with permission from Ekinci [54]. Copyright 2005, Wiley-VCH)

of this technique due to the various sources of noise in the measurement is given by [72, 73]

$$|S_\mathrm{X}(\omega)| = \frac{S_\mathrm{I}^{(\mathrm{A})}(\omega)}{(2kI)^2} + \frac{e}{2k^2 I} + \frac{S_\mathrm{F}^{(\mathrm{BA})}}{m_\mathrm{eff}^2\left(\omega_0^2 - \omega^2\right)^2 + \frac{(\omega\omega_0)^2}{Q}} \, . \tag{24.6}$$

The first term on the right-hand side is due to the equivalent current noise of the transimpedance amplifier. The second term arises from the granular nature of the electrical charge.

Tunneling in an atomic-scale junction is inherently a fast phenomenon with a speed greater than 1 GHz [74]; however, owing to the high impedance of the tunneling junction and the presence of unavoidable parasitic capacitance, the bandwidth of this detector is typically much lower than the speeds at which NEMS operate.

24.2.3.2.3
Piezoresistive Detection

Piezoresistive detection utilizes materials that are sensitive to strain from which structure displacements can be deconvoluted. Thin films made of piezoresisitive materials have been integrated into microscale cantilevers to detect their motion. High strain sensitivity has been achieved using semiconductor-based piezoresistors, primarily doped Si [75] or AlGaAs [76]. The shortcoming of this sensing technique is its low bandwidth owing to the high impedance of the sensor beam and the presence of parasitic capacitance. Recently, a down-mixing scheme was developed to detect the motion of NEMS in the high-frequency regime [77]. Figure 24.8a illustrates the difficulties in applying low-frequency techniques to high-frequency piezoresistive NEMS. Figure 24.8b illustrates the scheme used to down-mix the displacement signal to a low-frequency signal. In Fig. 24.8a, the piezoresistor R_c is placed in a bridge

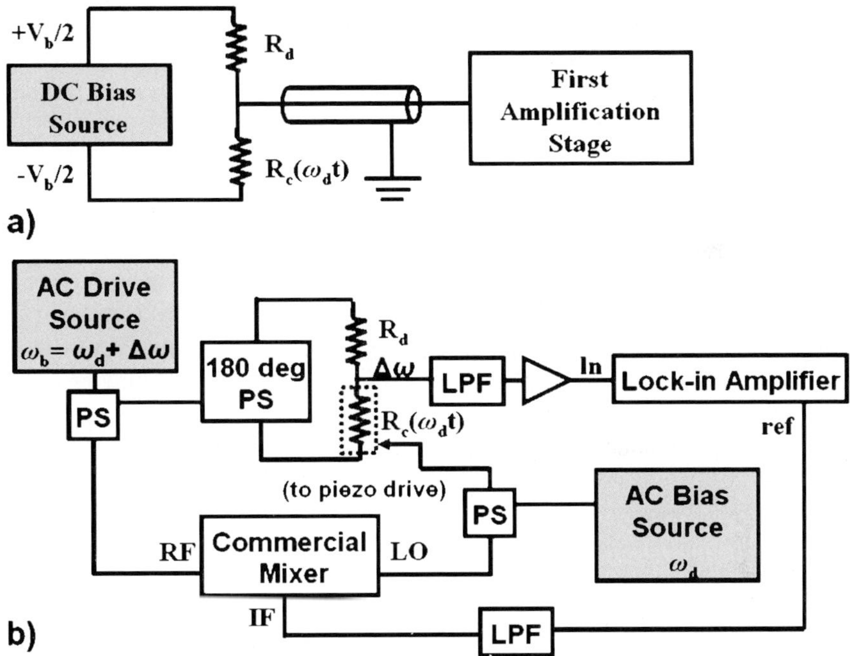

Fig. 24.8. Direct-current bias (**a**) and down-mixing (**b**) circuits. (Reprinted with permission from Bargatin et al. [77] Copyright 2005, American Institute of Physics)

configuration with a fixed dummy resistor R_d. Assuming $R_c = R_d = R$, the voltage output from the bridge is zero when the ends of the resistors are oppositely DC-biased at $+V_b/2$ and $-V_b/2$. When the NEMS is driven at frequency ω_d, the resistance R_c also has a time-dependent component: $R_c = R + \Delta R \cos(\omega_d t + \phi)$; therefore, the voltage output from the bridge is $V_{out} \approx V_b(\Delta R/4R) \cos(\omega_d t + \phi)$. To measure V_{out} the bridge output has to be connected in some way to a measuring circuit, such as the input of a preamplifier. At moderately high frequencies (below 30 MHz), this introduces a capacitance C_{par} in parallel with the cantilever and the dummy resistor, effectively forming a low-pass filter with a cutoff frequency of $(\pi R C_{par})^{-1}$. For a typical amplifier input resistance $R > 10\,\mathrm{k\Omega}$ and a cable capacitance $C_{par} > 10\,\mathrm{pF}$, the output voltage V_{out} is strongly attenuated at frequencies of $\omega_d/2\pi > 2\,\mathrm{MHz}$. For the circuit shown in Fig. 24.8b, an AC voltage $V_b(t) = V_{b0} \cos(\omega_b t)$ is applied across the resistors at a frequency ω_b, which offsets the drive frequency by an amount $\Delta\omega \equiv \omega_b - \omega_d$. Using a 180° power splitter, we apply the voltage oppositely to the ends of the resistors to null the biased voltage at the bridge point. The bias produces an AC current $I(t) = V_{b0} \cos(\omega_b t)/(R_d + R_c)$. V_{out} then becomes

$$V_{out}(t) \approx \frac{V_{b0} \cos(\omega_b t)}{4R}(\Delta R \cos(\omega_d t + \phi))$$

$$= V_{b0}\frac{\Delta R}{8R}[\cos(\Delta\omega t - \phi) + \cos((2\omega_d + \Delta\omega)t + \phi)] \ . \tag{24.7}$$

From (24.7), the output signal V_{out} has two frequency components given by the sum and the difference of the biased and the drive frequency. With $\Delta\omega$ sufficiently small, the down-mixed frequency is attenuated minimally by the capacitance C_{par}. Using this piezoresistive signal down-mixing scheme, Bargatin et al. [77] demonstrated the motion detection of NEMS cantilevers with fundamental mode frequencies in the 20-MHz range.

24.2.4
Functional NEMS Devices

In this section, we review the CNT- or nanowire-based NEMS devices reported in the literature with a special emphasis on fabrication methods, working principles, and applications.

24.2.4.1
CNT-Based NEMS Devices

24.2.4.1.1
Nonvolatile Random Access Memory

A CNT-based nonvolatile random access memory (NRAM) reported by Rueckes et al. [1] is illustrated in Fig. 24.9a. The device is a suspended SWNT crossbar array for both input/output and switchable, bistable device elements with well-defined off and on states. This crossbar consists of a set of parallel SWNTs or nanowires (lower) on a substrate composed of a conducting layer [e.g., highly doped silicon (dark gray)] that terminates in a thin dielectric layer [e.g., SiO_2 (light gray)], and a set of perpendicular SWNTs (upper) suspended on a periodic array of inorganic

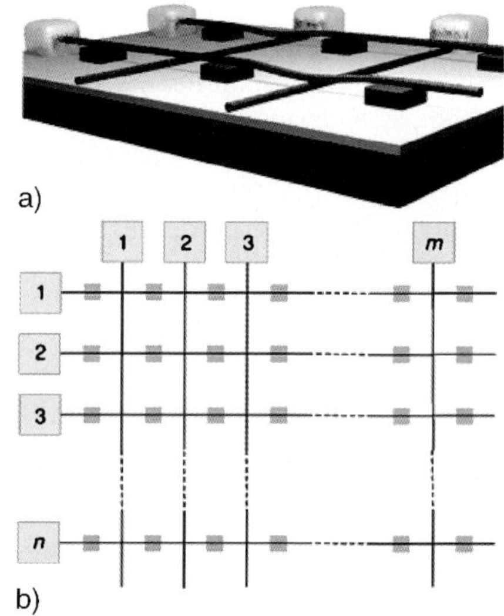

Fig. 24.9. Freestanding nanotube device architecture with multiplex addressing. (**a**) Three-dimensional view of a suspended crossbar array showing four junctions with two elements in the on (contact) state and two elements in the off (separated) state. (**b**) Top view of an n by m device array. (Reprinted with permission from Rueckes et al. [1] Copyright 2000, American Association for the Advancement of Science)

or organic supports. Each nanotube is contacted by a metal electrode. Each cross point in this structure corresponds to a device element with a SWNT suspended above a perpendicular nanoscale wire. Qualitatively, bistability can be envisioned as arising from the interplay of the elastic energy and the van der Waals energy when the upper nanotube is freestanding or the suspended SWNT is deflected and in contact with the lower nanotube. Because the nanotube junction resistance depends exponentially on the separation gap, the separated upper-to-lower nanotube junction resistance will be orders of magnitude higher than that of the contact junction; herefore, two states, "off" and "on," are well defined. For a device element, these two states can be read simply by measuring the resistance of the junction and, moreover, can be switched between off and on states by applying voltage pulses to nanotubes through the corresponding electrodes to produce attractive or repulsive electrostatic forces. A key aspect of this device is that the separation between top and bottom conductors must be on the order of 10 nm. In such a case, the van der Waals energy overcomes the elastic energy when the junction is actuated (on state) and remains on this state even if the electrical field is turned off (nonvolatile feature).

In the integrated system, electrical contacts are made only at one end of each of the lower and upper sets of nanoscale wires in the crossbar array, and thus many device elements can be addressed from a limited number of contacts (Fig. 24.9b). This approach suggests a highly integrated, fast, and macroscopically addressable NRAM structure that could overcome the fundamental limitations of semiconductor RAM in size, speed, and cost. Integration levels as high as 1×10^{12} elements per square centimeter and switching time down to approximately 5 ps (200-GHz operation frequency) using 5-nm device elements and 5-nm supports are envisioned while maintaining the addressability of many devices through the long (approximately 10-μm) SWNT wires. However, such small dimensions, in particular, the junction gap size, impose significant challenges in the nanofabrication of parallel device arrays.

24.2.4.1.2
Nanotweezers

There are two types of CNT-based nanotweezers: those reported by Kim and Lieber [12] in 1999 and Akita et al. [9] in 2001. Both nanotweezers employ MWNTs as the tweezers' arms which are actuated by electrostatic forces. Applications of these nanotweezers include the manipulation of nanostructures and two-tip STM or atomic force microscope (AFM) probes [12].

The fabrication process of the CNT-based nanotweezers reported by Kim and Lieber [12] is illustrated in Fig. 24.10a. Freestanding electrically independent electrodes were deposited onto tapered glass micropipettes with end diameters of 100 nm (Fig. 24.10b). MWNT or SWNT bundles with diameters of 20–50 nm were attached to the two gold electrodes under the direct view of an optical microscope operated in dark-field mode using an adhesive [78, 79]. A scanning electron microscope image of fabricated nanotube tweezers is shown in Fig. 24.10c. The gap between the two nanotweezers' arms was controlled by the applied bias voltage between the electrodes. The nanotube nanotweezers have been demonstrated successfully to

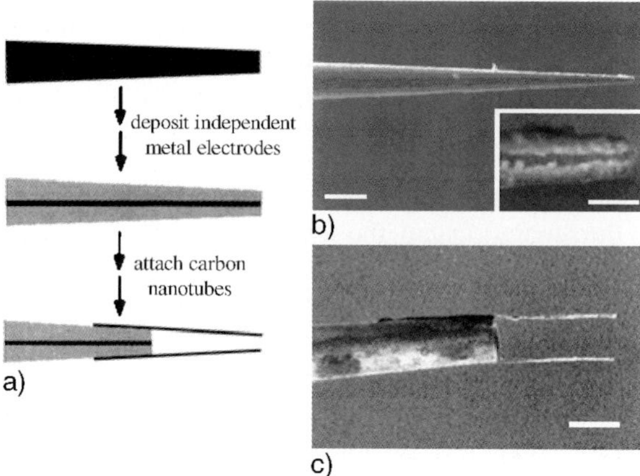

Fig. 24.10. Overview of the fabrication of CNT nanotweezers. (**a**) The deposition of two independent metal electrodes and the subsequent attachment of CNTs to these electrodes. (**b**) SEM image of the end of a tapered glass structure after the two deposition steps. *Scale bar* 1 μm. The higher-resolution *inset* shows clearly that the electrodes are separated. *Scale bar* 200 nm. (**c**) SEM image of nanotweezers after mounting two MWNT bundles on each electrode. *Scale bar* 2 μm. (Reprinted with permission from Kim et al. [12]. Copyright 1999, American Association for the Advancement of Science)

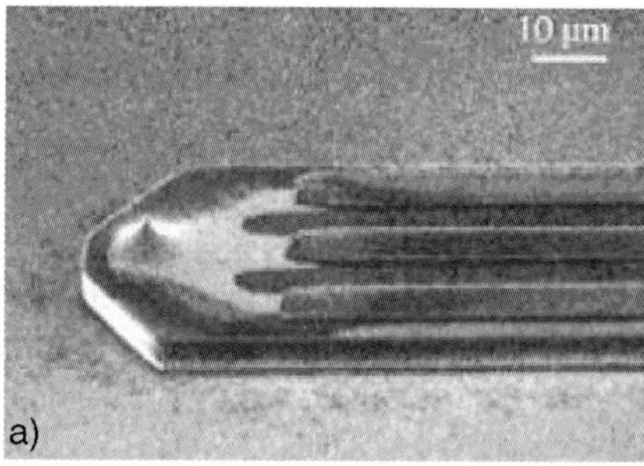

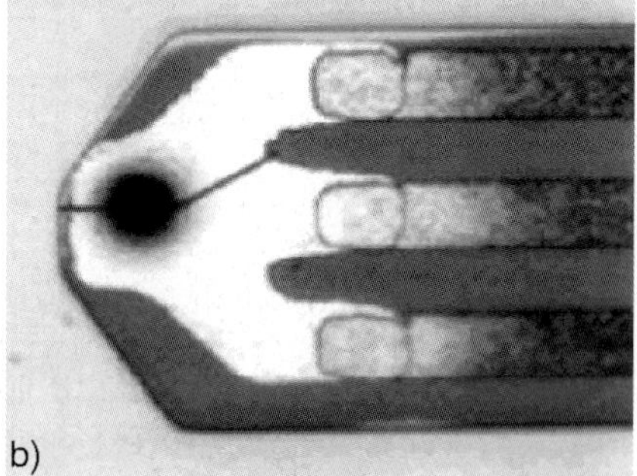

Fig. 24.11. SEM images of a Si cantilever as a base for nanotube nanotweezers. (**a**) A Ti/Pt film was coated on the tip and connected to three Al lines patterned on the cantilever. (**b**) The Ti/Pt film was separated into two by a focused ion beam and these two were connected to one and two Al lines, respectively. (Reprinted with permission from Akita et al. [9]. Copyright 2001, American Institute of Physics)

manipulate nanostructures, such as fluorescently labeled polystyrene spheres and a β-SiC nanocluster, and GaAs nanowires [12].

The CNT-based nanotweezers reported by Akita et al. [9], are shown in Fig. 24.11. Commercially available Si AFM cantilevers were employed as the device body. A Ti/Pt film was coated on the tip of the cantilever and connected to three Al interconnects that were patterned on the cantilever by a conventional lithographic technique as shown in Fig. 24.11a. The Ti/Pt film was separated into two by focused ion beam (FIB) etching. These two parts were independently connected to Al interconnects as shown in Fig. 24.11b. DC voltage was applied between the separated Ti/Pt tips through the Al interconnect, to operate the tweezers after attaching two arms of nanotubes to them using a three-stage manipulator. Figure 24.12a shows an SEM image of a typical pair of nanotube nanotweezers with a cantilevers length of about 2.5 μm and the separation between their tips of about 780 nm.

The operation of the nanotube nanotweezers was examined by in situ SEM. Various voltages were applied between the two arms to get them to close owing to the electrostatic attraction force. Figure 24.12b–d shows the motion of the nanotube arms as a function of the applied voltage V. It is clearly seen that the arms bent and the separation between the tips decreased with increasing applied voltage. The separation became 500 nm at $V = 4$ V and zero at $V > 4.5$ V. It is noted that the motion in Fig. 24.12 could be repeated many times without permanent deformation, suggesting that CNTs are ideal materials for building NEMS.

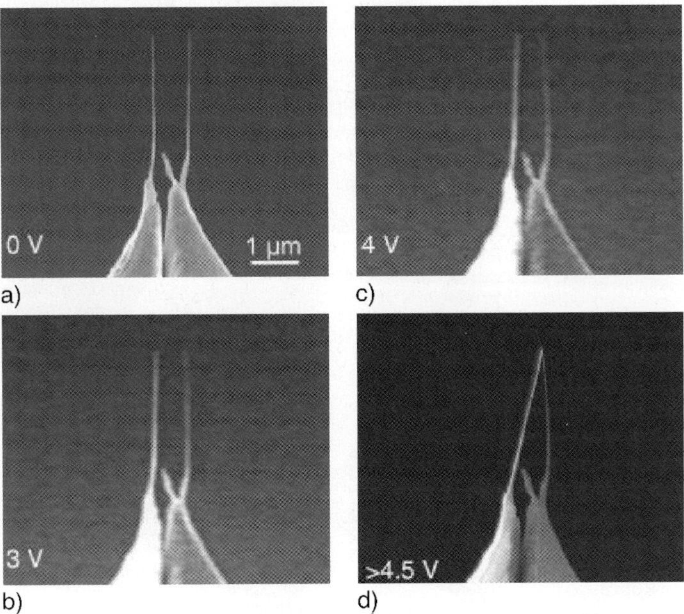

Fig. 24.12. SEM images of the motion of nanotube arms in a pair of nanotweezers as a function of the applied voltage. (Reprinted with permission from Akita et al. [9]. Copyright 2001, American Institute of Physics)

24.2.4.1.3
Rotational Motors

A CNT-based rotational motor reported by Fennimore et al. [10] in 2003 is conceptually illustrated in Fig. 24.13a. The rotational element (R), a solid rectangular metal plate serving as a rotor, is attached transversely to a suspended support shaft. The ends of the support shaft are embedded in electrically conducting anchors (A1, A2) that rest on the oxidized surface of a silicon chip. The rotor plate assembly is surrounded by three fixed stator electrodes: two "in-plane" stators (S1, S2) are horizontally opposed and rest on the silicon oxide surface, while the third "gate" stator (S3) is buried beneath the surface. Four independent (DC and/or appropriately phased AC) voltage signals, one applied to the rotor plate and three to the stators, are applied to control the position, speed, and direction of rotation of the rotor plate. The key component in the assembly is a single MWNT, which serves simultaneously as the rotor plate support shaft and the electrical feedthrough to the rotor plate; most importantly it is also the source of rotational freedom.

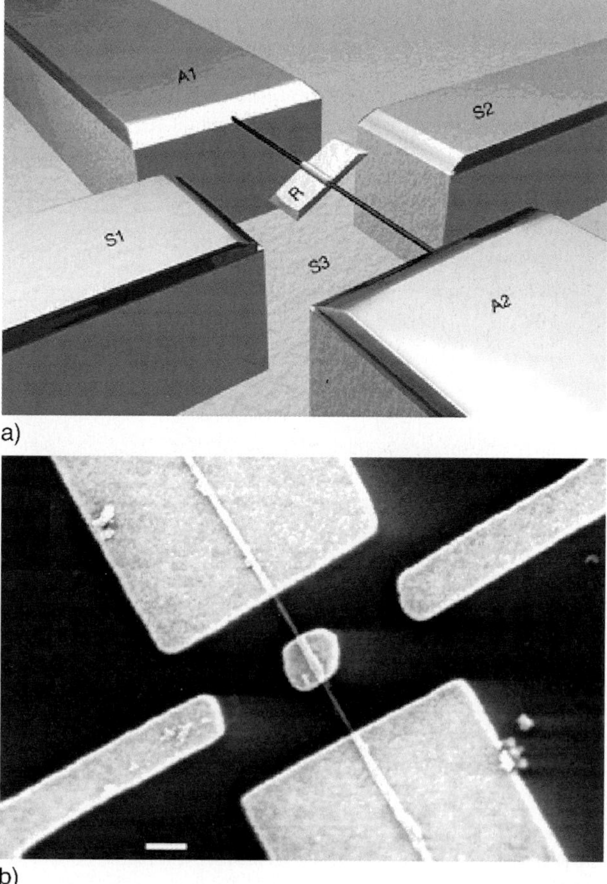

Fig. 24.13. Integrated synthetic NEMS actuator. (**a**) Conceptual drawing of nanoactuator. (**b**) SEM image of nanoactuator just prior to HF etching. *Scale bar* 300 nm. (Reprinted with permission from Fennimore et al. [10]. Copyright 2003, Nature Publishing Group)

The nanoactuator was constructed using lithographic methods. MWNTs in suspension were deposited on a doped silicon substrate covered with 1 μm of SiO_2. The nanotubes were located using an AFM or a scanning electron microscope. The remaining actuator components (in-plane rotor plate, in-plane stators, anchors, and electrical leads) were then patterned using electron-beam lithography. A HF etch was used to remove roughly 500 nm of the SiO_2 surface to provide clearance for the rotor plate. The conducting Si substrate here serves as the gate stator. Figure 24.13b shows an actuator device prior to etching. Typical rotor plate dimensions were 250–500 nm on a side.

The performance of the nanoactuator was examined in situ inside the scanning electron microscope chamber. Visible rotation could be obtained by applying DC voltages up to 50 V between the rotor plate and the gate stator. When the applied voltage was removed, the rotor plate would rapidly return to its original horizontal position. To exploit the intrinsic low-friction-bearing behavior afforded by the perfectly nested shells of MWNTs, the MWNT supporting shaft was modified in situ by successive application of very large stator voltages. The processes resulted in fatigue and eventually shear failure of the outer nanotube shells. In the "free" state, the rotor plate was still held in position axially by the intact nanotube core shells, but could be azimuthally positioned, using an appropriate combination of stator signals,

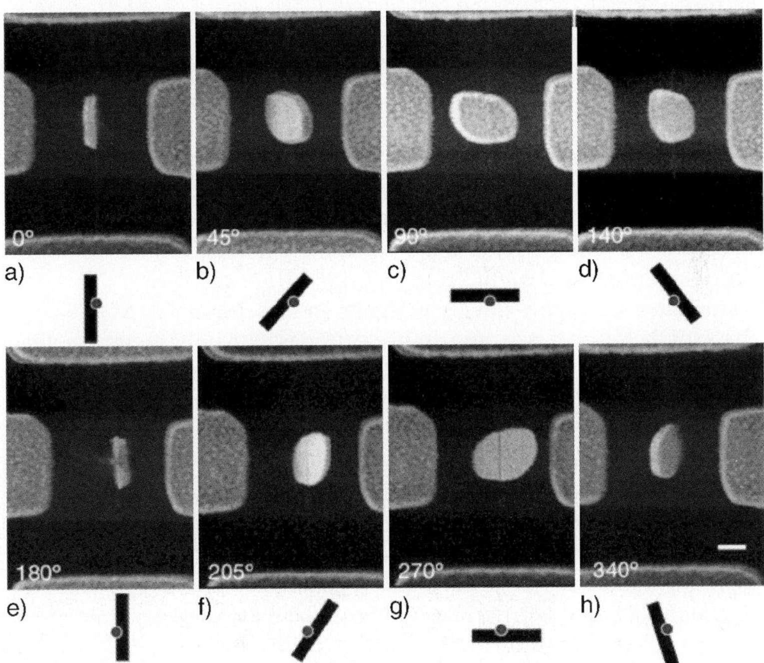

Fig. 24.14. Series of SEM images showing the actuator rotor plate at different angular displacements. The schematic diagrams located *beneath* each SEM image illustrate a cross-sectional view of the position of the nanotube/rotor-plate assembly. *Scale bar* 300 nm. (Reprinted with permission from Fennimore et al. [10]. Copyright 2003, Nature Publishing Group)

to any arbitrary angle between 0 and 360°. Figure 24.14 shows a series of still SEM images, recorded from an actuated device in the free state, being "walked" through one complete rotor plate revolution using quasi-static DC stator voltages. The stator voltages were adjusted sequentially to attract the rotor plate edge to successive stators. Reversal of the stator voltage sequence allowed the rotor plate rotation to be reversed in an equally controlled fashion. The experiments show that the MWNT clearly serves as a reliable, presumably wear free, NEMS element providing rotational freedom. No apparent wear or degradation in performance was observed after many thousands of cycles of rotations.

Potential applications of the MWNT-based actuators include ultra-high-density optical sweeping and switching elements, paddles for inducing and/or detecting fluid motion in microfluidics systems, gated catalysts in wet chemistry reactions, biomechanical elements in biological systems, or general (potentially chemically functionalized) sensor elements.

24.2.4.1.4
Nanorelays

CNT-based nanorelays were first reported by Kinaret et al. [13] in 2003 and were later demonstrated experimentally by Lee et al. [42] in 2004. The nanorelay is a three-terminal device including a conducting CNT placed on a terrace in a silicon substrate and connected to a fixed source electrode (S), as shown in Fig. 24.15a. A gate electrode (G) is positioned underneath the nanotube so that charge can be induced in the nanotube by applying a gate voltage. The resulting capacitance force q between the nanotube and the gate bends the tube and brings the tube end into contact with a drain electrode (D) on the lower terrace, thereby closing an electric circuit. Theoretical modeling of the device shows that there is a sharp transition from a nonconducting (off) to a conducting (on) state when the gate voltage is varied at a fixed source-drain voltage. The sharp switching curve allows for amplification of weak signals superimposed on the gate voltage [13].

One fabricated nanorelay device is shown in Fig. 24.15b. A MWNT was positioned on top of the source, gate, and drain electrodes with poly(methyl methacrylate) (PMMA) as a sacrificial layer using AC-electrophoresis techniques [36]. Then, a top electrode was placed over the nanotube at the source to ensure good contact. The underlying PMMA layer was then carefully removed to produce a nanotube suspended over the gate and drain electrodes. The separation between gate and drain was approximately 250 nm and the source–drain distance was 1.5 μm.

The electromechanical properties of nanotube relays were investigated by measuring the current–gate voltage ($I-V_{sg}$) characteristics, while applying a source–drain voltage of 0.5 V. Figure 24.16 shows the $I-V_{sg}$ characteristics of one of the nanotube relays with an initial height difference between the nanotube and drain electrode of approximately 80 nm. The drain current started to increase nonlinearly when the gate voltage reached 3 V (at this gate voltage the current is in the order of 10 nA). The nonlinear current increase was a signature of electron tunneling as the distance between the nanotube and the drain electrode was decreased. Beyond $V_{sg} = 20$ V there was a change in the rate of current increase. With the current increase rate becoming more linear, strong fluctuations could be detected. The deflection of the nanotube

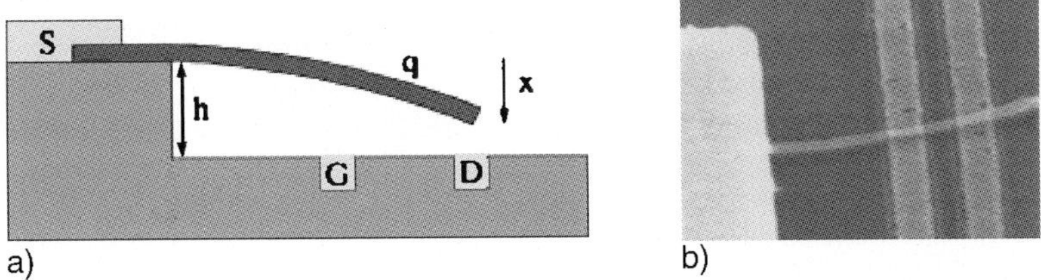

Fig. 24.15. A CNT nanorelay device (**a**) and a SEM image of a fabricated nanorelay device (**b**) ((**a**) Reprinted with permission from Kinaret et al. [13]. Copyright 2002, American Institute of Physics. (**b**) Reprinted with permission from Lee et al. [42]. Copyright 2004, American Chemical Society)

Fig. 24.16. I–V_{sg} characteristics of a nanotube relay initially suspended approximately 80 nm above the gate and drain electrodes. (Reprinted with permission from Lee et al. [42] Copyright 2004, American Chemical Society)

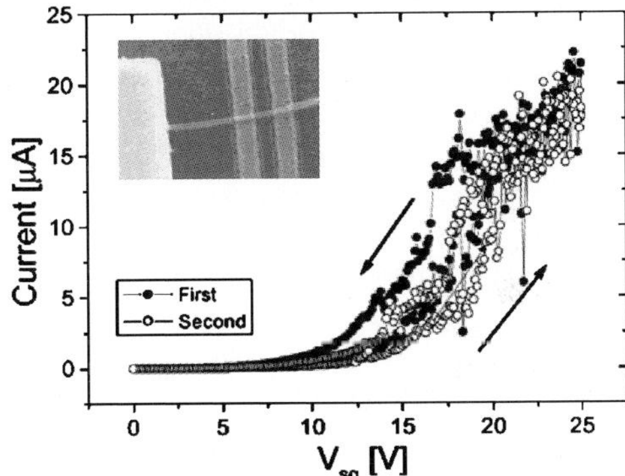

was found to be reversible. The current decreased with the reduction of gate voltage, showing some hysteresis, until it reached zero for a gate voltage below 3 V. The current measured during the increasing V_{sg} part of the second scan closely followed that of the first scan, especially in the region below $V_{sg} = 12$ V. The dynamics of the nanorelays was recently investigated by Jonsson et al. [80]. The results show that the intrinsic mechanical frequencies of nanorelays are in the gigahertz regime and the resonance frequency can be tuned by the biased voltage.

The potential applications of nanorelays include memory elements, pulse generators, signal amplifiers, and logic devices.

24.2.4.1.5
Feedback-Controlled Nanocantilevers

A feedback-controlled CNT-based NEMS device reported by Ke and Espinosa [11] in 2004, schematically shown in Fig. 24.17, is made of a MWNT placed as a cantilever over a microfabricated step. A bottom electrode, a resistor, and a power supply are parts of the device circuit. When a voltage $U < V_{PI}$ (*pull-in* voltage) is applied, the resulting electrostatic force is balanced by the elastic force from the deflection of the nanotube cantilever. The nanotube cantilever remains in the "upper" equilibrium position. When the applied voltage exceeds the pull-in voltage, the electrostatic force becomes larger than the elastic force and the nanotube accelerates towards the

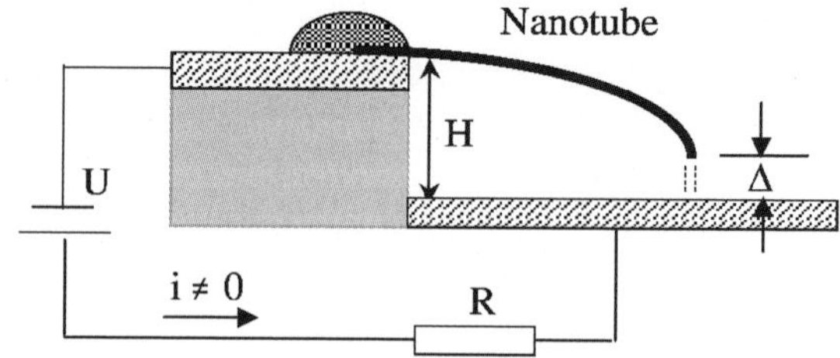

Fig. 24.17. Nanotube-based device with tunneling contact. (Reprinted with permission from Ke et al. [11]. Copyright 2004, American Institute of Physics)

bottom electrode. When the tip of the nanotube is very close to the electrode (i.e., gap $\Delta \approx 0.7$ nm), as shown in Fig. 24.17, a substantial tunneling current passes between the tip of the nanotube and the bottom electrode. Owing to the existence of the resistor in the circuit, the voltage applied to the nanotube drops, weakening the electric field. Because of the kinetic energy of the nanotube, it continues to deflect downward and the tunneling current increases, weakening the electric field further. In this case, the elastic force is larger than the electrostatic force and the nanotube decelerates and eventually changes the direction of motion. This decreases the tunneling current and the electrical field recovers. If there is damping in the system, the kinetic energy of the nanotube is dissipated and the nanotube stays at the position where the electrostatic force is equal to the elastic force, and a stable tunneling current is established in the device. This is the "lower" equilibrium position for the nanotube cantilever. At this point, if the applied voltage U decreases, the cantilever starts retracting. When U decreases to a certain value, called the *pull-out* voltage V_{PO}, the cantilever is released from its lower equilibrium position and returns back to its upper equilibrium position. At the same time, the current in the device diminishes substantially. Basically the pull-in and pull-out processes follow a hysteretic loop for the applied voltage and the current in the device. The upper and lower equilibrium positions correspond to "on" and "off" states of a switch, respectively. The existence of the tunneling current and feedback resistor make the "lower" equilibrium state very robust, which is key to some applications of interest. The representative characteristic curve of the device is shown in Fig. 24.18: Fig. 24.18a shows the relation between the gap Δ and the applied voltage U; Fig. 24.18b shows the relation between the current i in the circuit and the applied voltage U. The potential applications of the device include ultrasonic wave detection for monitoring the health of materials and structures, gap sensing, NEMS switches, memory elements, and logic devices.

In comparison with nanorelays [13, 42], the device reported in [11] is a two-terminal device, providing more flexibility in terms of device realization and control than the nanorelay. In comparison with the NRAM described in [1], the feedback-controlled device employs an electrical circuit incorporated with a resistor to adjust the electrostatic field to achieve the second stable equilibrium position. This feature reduces the constraints in fabricating devices with nanometer-gap control between the free standing CNTs or nanowires and the substrate, providing more reliability and tolerance to variability in fabrication parameters. However, the drawback of the device in memory applications is that the memory becomes volatile. The working principle and the potential applications for these two devices are somewhat complementary.

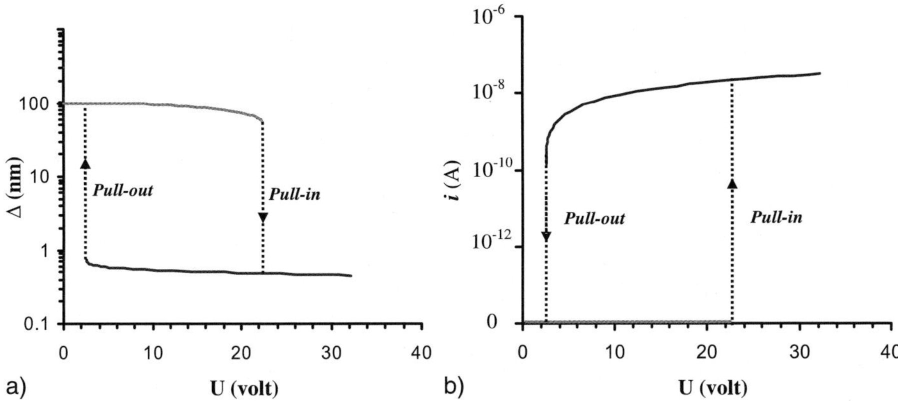

Fig. 24.18. Representative characteristic of pull-in and pull-out processes for the feedback-controlled nanocantilever device. (**a**) Relationship between the gap Δ and the applied voltage U. (**b**) Relationship between the current i in the circuit and the applied voltage U. (Reprinted with permission from Ke et al. [11]. Copyright 2004, American Institute of Physics)

Recently the electromechanical behavior of the feedback-controlled NEMS device was demonstrated experimentally by in situ SEM testing [81]. The test configuration employed is schematically shown in Fig. 24.19a [81, 82]. By employing a three-axes nanomanipulator (Klocke Nanotechnik) possessing nanometer positioning accuracy, a MWNT was welded to a tungsten probe tip by EBID of platinum [31]. A second electrode employed in the configuration (Fig. 24.19a) consisted of a silicon chip coated with a 5-nm-thick Cr adhesion layer and a 50-nm Au film. This chip was glued onto the side of a Teflon block and mounted vertically in the scanning electron microscope x–y–z stage. With use of an electric feedthrough, the two electrodes were connected to a resistor, $R = 1\,\text{G}\Omega$, and to a current–voltage electronic measurement unit (Keithley 4200 SCS). The nanotube cantilever welded to the manipulator probe was displaced until a desired distance (typically 0.5–$3\,\mu\text{m}$, depending on the length and the diameter of nanotubes) between the freestanding CNT and Au electrode was reached. The configuration shown in Fig. 24.19a is considered electromechanically equivalent to the proposed device shown in Fig. 24.17. The pull-in behavior of the device (gap–voltage curve), in particular, the pull-in voltage, and the pull-in/pull-out behavior of the device (current–voltage curve) were examined systematically using this testing configuration.

The measured pull-in behavior of a MWNT cantilever 6.8-μm long and with outer diameter of $47\,\text{nm}$ placed parallel to the electrode with a gap size of $3\,\mu\text{m}$ and the comparison with the theoretical prediction will be detailed in Sect. 24.3.2.3. Figure 24.19b shows an experimentally measured current–voltage (I–U) curve during the pull-in/pull-out processes and the theoretical prediction for a nanotube 9-μm long [81, 82]. The measured I–U curve exhibited the theoretically predicted bistability and hysteretic loop. The arrows show the direction in which the hysteretic loop traveled during the increase and decrease of the driving voltage U. The measurement exhibits a background noise of about $0.1\,\text{pA}$, which is typical in these measurements.

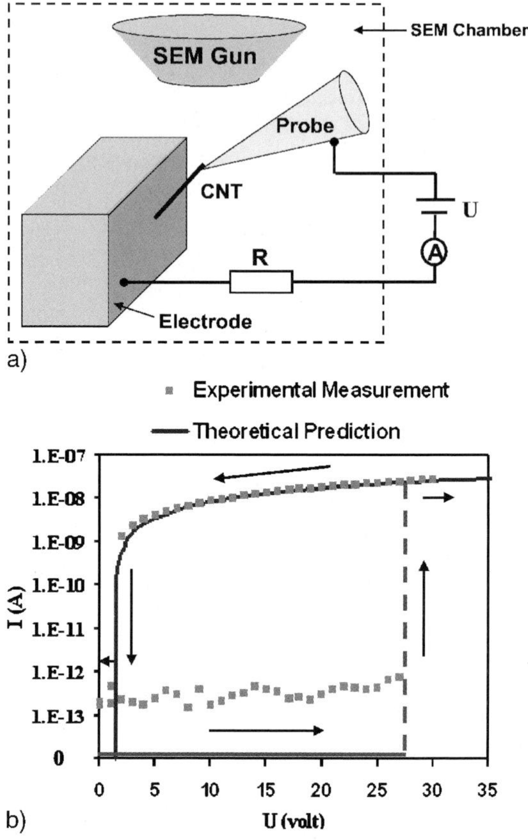

Fig. 24.19. In situ pull-in/pull-out experiments. (**a**) The setup for in situ testing of nanotube cantilever devices. (**b**) Measured I–U characteristic curve during the pull-in/pull-out processes and comparison with analytical predictions. The *arrows* show the direction in which the hysteretic loop is described during the increase and decrease of the driving voltage U

Depending on the device characteristic dimensions, failure modes were observed in the in situ SEM studies [81, 82]. These dimensions were selected to produce accelerated failure, i.e., the operational conditions were extreme. Figure 24.20a–d illustrates one of the observed failure modes consisting of partial loss of the nanotube cantilever around the tip area after each pull-in/pull-out cycle, i.e., a reduction of nanotube length. This phenomenon has been consistently observed. Detailed examination of Fig. 24.20a–d clearly reveals that the length of the nanotube becomes approximately 1 μm shorter after each pull-in/pull-out event. The gaps between the tip of the nanotube and the electrode for the sequence of pull-in/pull-out tests were controlled to be in the range 0.5–2 μm and the pull-in voltages were found to be in the range 15–30 V. From the various experiments, two possible failure sources are envisioned. One involves CNT fracture as a result of its impact with the bottom electrode during the unstable pull-in event. Another possible source is the sublimation of the CNT as a result of high instantaneous current densities and associated Joule heating. It is important to note that these two damage mechanisms may operate simultaneously. A study on the stability of charged SWNTs using the density functional based tight-binding method (DFTB) provides some insight into these failure modes [83, 84]. The study showed that the ends of nanotubes become unstable as the

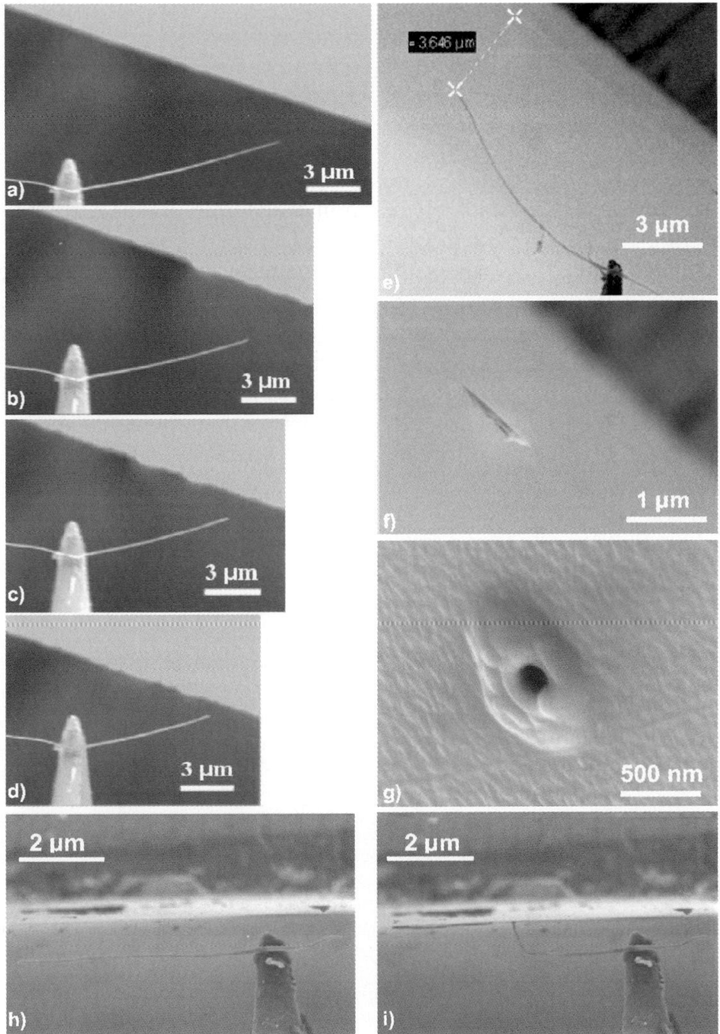

Fig. 24.20. Failure modes of the NEMS device. (**a**)–(**d**) Example of device failure due to partial loss of the nanotube cantilever after each pull-in/pull-out sequence. (**e**)–(**g**) SEM images showing electrode damage and melting at high pull-in voltages. (**h**), (**i**) Fracture of the nanotube cantilever during the pull-in event

electrical charge density is increased. This is the case because charges are concentrated at the ends, eliciting strong repulsive Coulomb electrostatic interactions that tend to eject end atoms. The ejection mechanism may be exacerbated by local mechanical deformation and high temperatures arising from spikes in current densities. Further evidence of the CNT impact and high current densities was obtained from another experiment in which the gap was set to a value of 3.646 μm. Figure 24.20e,f shows SEM images of the device before and after pull-in. It is observed that part of the electrode surface was damaged when pull-in happened at an applied voltage U of 70 V. Figure 24.20g is a zoom-in view of the damaged electrode area showing a hole formed in the center of the damaged surface area, likely the result of local gold vaporization. The diameter of the hole was measured to be approximately 150 nm, which is a little larger than the diameter of the nanotube (approximately 120 nm). The area around the hole seems "melted" and blistered, probably owing to the rapid thermal expansion and delamination of the gold film as a result of sudden energy release. Figure 24.20h,i illustrates SEM images showing evidence of nanotube fracture. Figure 24.20h shows a nanotube 6.6-μm long and with a diameter of approximately 100 nm placed parallel to the electrode with a gap of 1 μm prior to actuation. When pull-in occurred at a voltage of 19.2 V, part of the nanotube cantilever was broken and remained attached to the Au electrode. The remaining part of the nanotube pulled in again and remained in contact with the electrode, as shown in Fig. 24.20i. All these observations confirm that in order to design reliable NEMS devices it is imperative to gain a basic understanding of failure modes as a function of configuration parameters such that maps with envelops defining reliable operational conditions can be obtained.

24.2.4.1.6
Tunable Oscillators

The fabrication and testing of a tunable CNT oscillator was reported by Sazonova et al. [3]. It consists of a doubly clamped nanotube, as shown in Fig. 24.21. They demonstrated that the resonance frequency of the oscillators can be widely tuned and that the devices can be used to transduce very small forces.

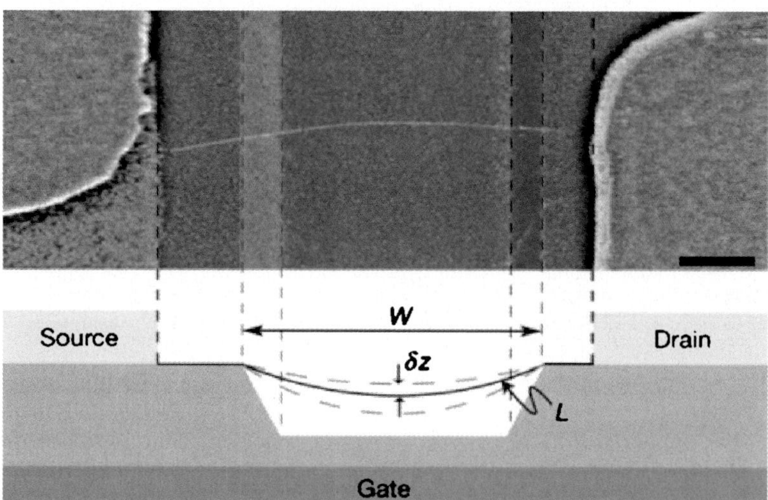

Fig. 24.21. SEM image of a suspended device (*top*) and the device geometry (*bottom*). *Scale bar* 300 nm. The sides of the trench, typically 1.2–1.5-μm wide and 500-nm deep, are marked with *dashed lines*. A suspended nanotube can be seen bridging the trench. (Reprinted with permission from Sazonova et al. [3]. Copyright 2004, Nature Publishing Group)

Single or few-walled nanotubes with diameters in the range 1–4 nm, grown by CVD, were suspended over a trench (typically 1.2–1.5-μm wide, 500-nm deep) between two metal (Au/Cr) electrodes. A small section of the tube resided on the oxide on both sides of the trench; the adhesion of the nanotube to the oxide provided clamping at the end points. The nanotube motion was induced and detected using the electrostatic interaction with the gate electrode underneath the tube. In this device, the gate voltage has both a static (DC) component and a small time-varying (AC) component. The DC voltage at the gate produces a static force on the nanotube that can be used to control its tension. The AC voltage produces a periodic electric force, which sets the nanotube into motion. As the driving frequency approaches the resonance frequency of the tube, the displacement becomes large.

The transistor properties of semiconducting [85] and small-bandgap semiconducting [86,87] CNTs were employed to detect the vibrational motion. Figure 24.22a shows the measured current through the nanotube as a function of driving frequency at room temperature. A distinctive feature in the current on top of a slowly changing background can be seen. This feature is due to the resonant motion of the nanotube, which modulates the capacitance, while the background is due to the modulating gate voltage.

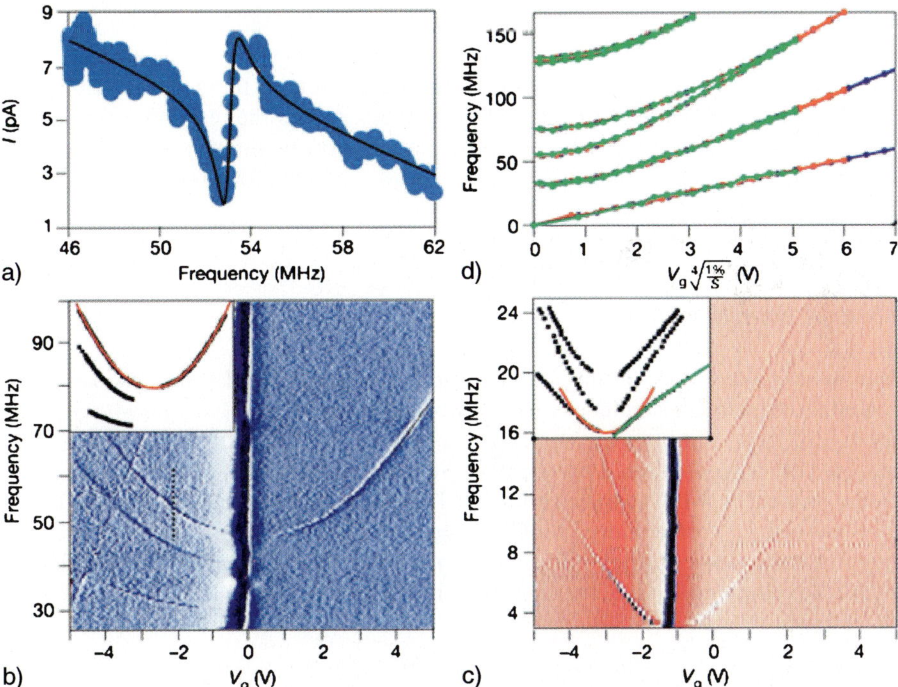

Fig. 24.22. Measurements of resonant response. (**a**) Detected current as a function of driving frequency. (**b**),(**c**) Detected current as a function of gate voltage V_g and frequency for devices 1 and 2. (**d**) Theoretical predictions for the dependence of vibration frequency on gate voltage for a representative device. (Reprinted with permission from Sazonova et al. [3]. Copyright 2004, Nature Publishing Group)

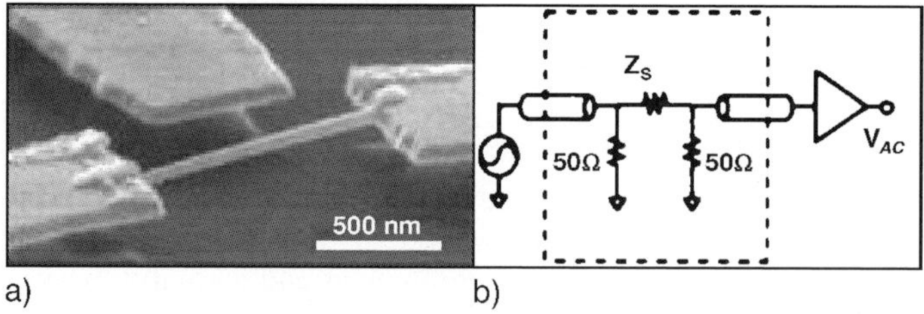

Fig. 24.23. (a) SEM image of the suspended nanowire device, 1.3-μm long and 43 nm in diameter. (b) Measurement circuit used for magnetomotive drive and detection. (Reprinted with permission from Husain et al. [26]. Copyright 2003, American Institute of Physics)

The DC voltage on the gate can be used to tune the tension in the nanotube and therefore the oscillation frequency. Figure 24.22b,c shows the measured response as a function of the driving frequency and the static gate voltage. The resonant frequency shifts upward as the magnitude of the DC gate voltage is increased. Several distinct resonances are observed, corresponding to different vibrational modes of the nanotube. Figure 24.22d shows the theoretical predictions for the dependence of the vibration frequency on gate voltage for a representative device. The predictions are based on finite-element (FE) analysis with the nanotube modeled as a long beam suspended over a trench. With the increase of the gap voltage, the deflection of the nanotube becomes larger and the stretching dominates the bending; therefore, the stiffness of the nanotube beam increases, and so does the resonance frequency. The theoretical predictions (Fig. 24.22d) show good qualitative agreement with experiments (Fig. 24.22b,c). The device showed a high force sensitivity (below 5 aN), which made it a small force transducer.

24.2.4.2
Nanowire Based NEMS Devices

Nanowires, like CNTs, are high-aspect-ratio one-dimensional nanostructures. The materials of nanowires include silicon [55,88–91], gold [92,93], silver [94–96], platinum [26], germanium [90,97–100], zinc oxide [101, 102], and so on. Besides their size, the advantages offered by nanowires when employed in NEMS are their electronic properties, which can be controlled in a predictable manner during synthesis. This has not yet been achieved for CNTs. In contrast to CNTs, nanowires do not exhibit the same degree of flexibility, which may affect device fabrication and reliability. In the following section, two nanowire-based NEMS device are briefly reviewed.

24.2.4.2.1
Resonators

Figure 24.23 shows a suspended platinum nanowire resonator (Fig. 24.23a), reported by Husain et al. [26] in 2003, and the circuit used for magnetomotive drive and detection of its motion (Fig. 24.23b).

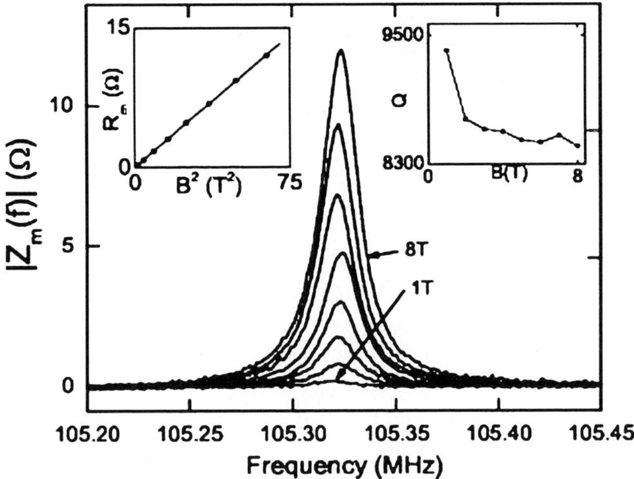

Fig. 24.24. Measured mechanical impedance of a Pt nanowire device as a function of frequency, at a series of magnetic fields from 1 to 8 T. *The left inset* shows the characteristic B^2 dependence typical of magnetomotive detection. The *right inset* shows the quality factor Q as a function of magnetic field. (Reprinted with permission from Husain et al. [26] Copyright 2003, American Institute of Physics)

Synthesized platinum nanowires were deposited on a Si substrate capped by a 300-nm-thick layer of thermally grown silicon dioxide and prepatterned with Au alignment marks. The location of the deposited wires was mapped, by means of optical microscopy, using their strong light scattering properties [95,103]. Metallic leads (5 nm Cr, 50 nm Au) to individual wires were subsequently patterned by electron-beam lithography, evaporation, and lift-off. Finally, the SiO_2 was removed by wet etching (HF) to form suspended nanowire structures. The suspended Pt nanowire shown in Fig. 24.23a has a diameter of 43 nm and a length of 1.3 μm. A magneto-motive detection scheme (Fig. 24.23b), in which an AC drives a beam in a transverse magnetic field, was used to drive and read out the resonators. Figure 24.24 shows the measured motion-induced impedance of the nanowire device, $|Z_m(f)|$, versus frequency. The measured quality factor Q was approximately 8500 and decreased slightly with the increase in magnetic field. It was noted that the characteristic curve shown in Fig. 24.24 corresponds to a linear response of the beam. Badzey et al. [55] reported a doubly clamped nanomechanical Si beam working in the nonlinear response region. The nonlinear response of the beam displays notable hysteresis and bistability in the amplitude–frequency space when the frequency sweeps upward and downward. This particular behavior shows that the device can be used as mechanical memory elements.

24.2.4.2.2
Nanoelectromechanical Programmable Read-Only Memories

A nanowire-based nanoelectromechanical programmable read-only memory (NEM-PROM) reported by Ziegler et al. [99] in 2004 is shown in Fig. 24.25a. The germanium nanowire was synthesized directly onto a macroscopic gold wire (diameter 0.25 mm). The combination of a transmission electron microscope (TEM) and a STM was used to control and visualize the nanowire under investigation. Figure 24.25b–g illustrates how the device can work as a NEMPROM. In equilibrium, the attractive van der Waals force and electrostatic interactions between the nanowire and the gold electrode are countered by the elastic force from the deflection of the nanowire. Figure 24.25b shows the position of the nanowire with relatively low applied voltage. With the increase in voltage, the nanowire moves closer to the electrode (Fig. 24.25c). When the applied voltage exceeds a certain value, a jump-to-contact happens, i.e., the nanowire makes physical contact with the electrode (Fig. 24.25d). The nanowire remains in contact with the electrode even when the electrostatic field is removed because the van der Waals force is larger than the elastic force (Fig. 24.25e). This is the "on" state of the NEMPROM. The NEMPROM device can be switched off by mechanical motion or by heating the device above the stability limit to overcome the van der Waals attractive forces. Figure 24.25f and g shows the separation of the nanowire and the electrode after imposing a slight mechanical motion, resulting in a jump-off-contact event. This is the "off" state of the NEMPROM. The working principle of a NEMPROM is similar to that of a NRAM [1] since both of them employ van der Waals energy to achieve the bistability behavior, although the usage of germanium may provide better control of size and electrical behaviors of the device than that of CNT.

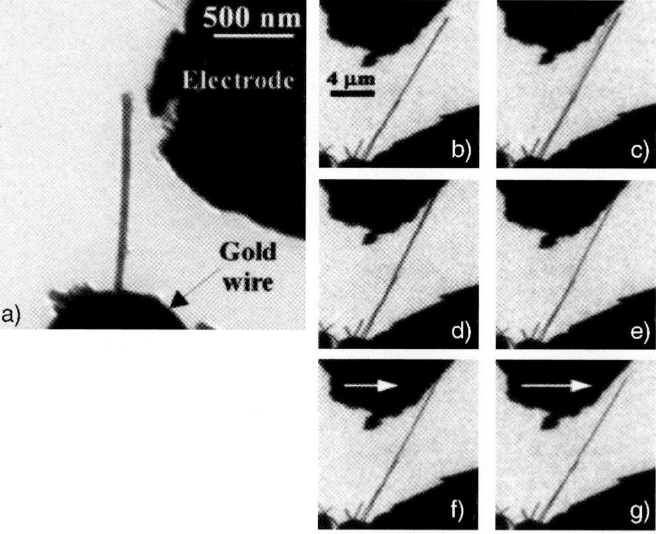

Fig. 24.25. (a) TEM image of a Ge nanowire device. (b)–(d) TEM sequence showing the jump-to-contact of a Ge nanowire as the voltage is increased. (e) TEM image demonstrating the stability of the device after removal of the electrostatic potential. (f),(g) TEM sequence demonstrating the resetting behavior of the device. (Reprinted with permission from Ziegler et al. [99] Copyright 2004, American Institute of Physics)

24.2.5
Future Challenges

NEMS offer unprecedented and intriguing properties in the fields of sensing and electronic computing. Although significant advancement has been achieved, there are many challenges that will need to be overcome before NEMS can replace and revolutionize current technologies. Among the issues that need further research and development are:

1. *Extremely high integration level*: For applications such as RAM and data storage, the density of the active components is definitely a key parameter. Direct growth and directed self-assembly are the two most promising methods to make NEMS devices with levels of integration orders of magnitudes higher than that of current microelectronics. A process for nanofabrication of the NEMS device developed by Ke and Espinosa [11], based on the directed self-assembly, is schematically shown in Fig. 24.26.

    a) A 1-μm-thick Si_3N_4 dielectric film is deposited on a Si wafer by low-pressure CVD. Then, a 50-nm-thick gold film (with 5-nm Cr film as the adhesion layer) is deposited by electron-beam evaporation and patterned by lithography to form the bottom electrodes. A 1-μm-thick SiO_2 layer is deposited by plasma-enhanced CVD (PECVD).

    b) The fountain-pen nanolithography technique [104] is then employed to functionalize specific areas, with widths down to 40 nm, either with polar chemical groups [such as the amino groups ($-NH_2/-NH_3^+$) of cysteamine] or carboxyl ($-COOH/-COO^-$) or with nonpolar groups [such as methyl ($-CH_3$)] from molecules like 1-octadecanethiol.

    c) The substrate is dipped into a solution containing prefunctionalized (with polar chemical groups) CNTs or nanowires, to adhere and self-assemble to the functionalized sites.

    d) The chip is patterned with electron-beam lithography and electron-beam evaporation of 100-nm gold film (lift-off, with 5-nm Cr film as the adhesion layer) to form the top electrodes.

    e) Removal of the SiO_2 layer using wet etching (HF) to free one end of the CNT cantilever completes the process.

The final product, a 2D array of NEMS devices with multiplexing capabilities, is schematically shown in Fig. 24.27. The top and bottom electrodes are interconnected to the pads, correspondingly. By applying voltage between the corresponding pads, the individual NEMS devices can be independently actuated.

2. *Better understanding the quality factor*: One of the keys to realizing the potential applications of NEMS is to achieve ultrahigh quality factors; however, it has been consistently observed that the quality factor of resonators decreases significantly with size scaling [7]. Defects in the bulk materials and interfaces, fabrication-induced surface damages, adsorbates on the surface, thermoelastic damping arising from inharmonic coupling between mechanical models and the phonon reservoir [105], and clamping losses [106] are a few commonly listed factors that can dampen the motion of resonators. Unfortunately, the dominant energy-dissipation mechanism in nanoscale mechanical resonators is still unclear.

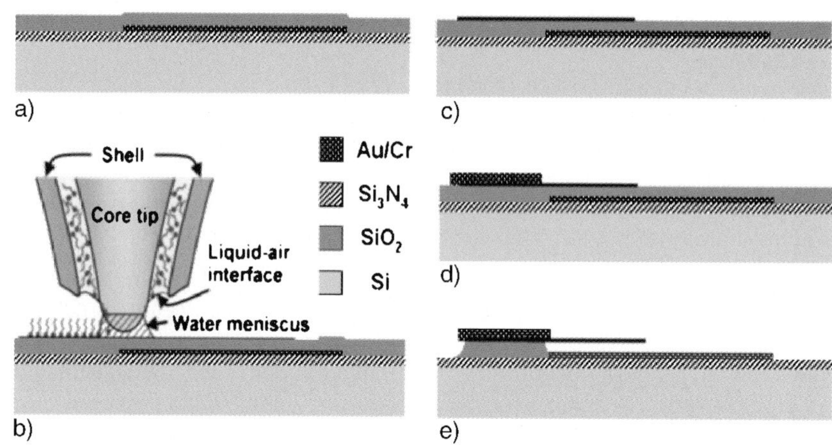

Fig. 24.26. The fabrication steps involving nanofountain probe functionalization

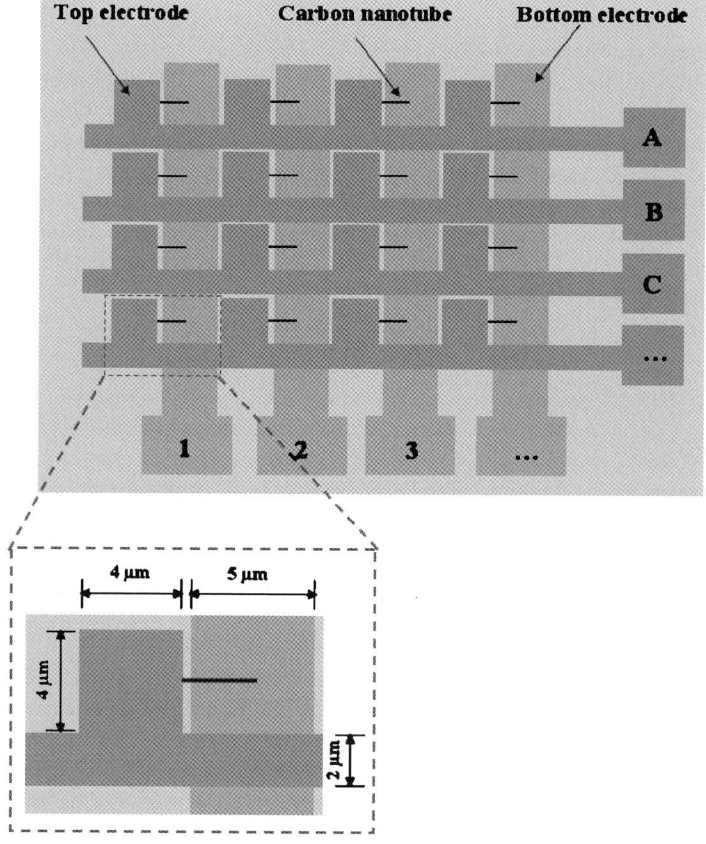

Fig. 24.27. Two-dimensional array of the NEMS device with multiplexing

3. *Reproducible and routine nanomanufacturing*: Fabrication reproducibility is key in applications such as mass sensors. Since the NEMS can respond to mass at the level of single atom or molecules, it places an extremely stringent requirement on the cleanness and precision of nanofabrication techniques. Likewise, devices that rely on van der Waals energy require dimensional control (e.g., gap dimension) on the order of a few nanometers.

4. *Quantum limit for mechanical devices*: The ultimate limit for NEMS is its operation at, or even beyond, the quantum limit [7]. In the quantum regime, the individual mechanical quanta are of the same order of magnitude as or greater than the thermal energy. Quantum theory should be utilized to understand and optimize force and displacement measurements. Recently, position resolution with a factor of 4.3 above the quantum limit was achieved for a single-electron transistor with high quality factor at millikelvin temperature [107]. The pursuit of NEMS devices operating at the quantum limit will potentially open new fields in science at the molecular level.

24.3
Modeling of NEMS

The design of NEMS depends on a thorough understanding of the mechanics of the devices themselves and the interactions between the devices and the external forces/fields. With the critical dimension shrinking from micron to nanometer scale, new physics emerges, so the theory typically applied to MEMS does not immediately translate to NEMS. For example, van der Waals forces from atomic interactions play an important role in NEMS, while they can be generally neglected in MEMS. The behavior of materials at the nanometer scale begins to be atomistic rather than continuous, giving rise to anomalous and often nonlinear effects, for example:

– The roles of surfaces and defects become more dominant.
– The devices become more compliant than continuum models predict.
– Molecular interactions and quantum effects become key issues, to the point that thermal fluctuation could make a major difference in the operation of NEMS.

For instance, the nanoresonators reported by the research groups of Roukes and Craighead are operated in the gigahertz range and usually have sizes within $200 \times 20 \times 10 \, nm^3$ [108]. Devices of this size and smaller are so miniscule that material defects and surface effects have a large impact on their performance.

In principle, atomic-scale simulations should reliably predict the behavior of NEMS devices; however, atomic simulations of the entire NEMS involve pro-hibitively expensive computational resources or exceed current computational power. Alternatively, multiscale modeling, which simulates the key region of a device with an atomistic model and other regions with a continuum model, can well serve the purpose under the circumstance of limited computational resources. Besides, it has been demonstrated that the behavior of some nanostructures, like CNTs, can be approximated by continuum mechanics models, based on the same potentials governing molecular dynamics (MD) simulation [109], if the surface nonideality of the nanostructures is neglected. Thus, continuum mechanics models are still adequate for the design of NEMS, in particular, in the initial stages.

24.3.1
Multiscale Modeling

Multiscale modeling is a technique to bridge the atomic simulations and continuum modeling. It generally makes use of coarse–fine decomposition in order to make computation more tractable. Atomistic simulation methods are typically used for the regions where individual MD information is crucial, where continuum modeling approaches are selected for all the other regions in which the deformation is considered to be homogeneous and smooth. In the atomistic domain, MD and quantum mechanics (QM) are typically employed, while in the continuum domain, the FE method (FEM) is often used. MD deals with the interaction of many thousands of atoms or more according to an interaction law. The "constitutive" behavior of each atom is governed by QM. QM involves the electronic structure, which in turn determines the interatomic force law – the "constitutive" behavior of each atom. However, in practice, the interatomic force laws have been determined empirically based on both QM and experiments. To model the response of NEMS devices, MD and continuum mechanics are generally adequate; hence, here we restrict our discussion to the basic ideas behind these two models. In some cases, QM modeling is required so the reader should consult the information on QM.

Multiscale modeling can be pursued sequentially (hierarchically) or concurrently. In the sequential method, information from each model at a given scale is passed to the next modeling level. In this fashion, "informed" or physically motivated models are developed at larger scales. Sequential techniques are based on the assumption of homogeneous lattice deformation; therefore, they are more effective for elastic single-phase problems. Challenges may arise from modeling defects in atomic lattices, dislocations, and failure phenomena [110]. In concurrent multiscale modeling, the system is split into primarily two domains: the atomistic domain and the continuum domain. These two scales are strongly dependent on each other through a smooth interscale coupling. Concurrent approaches are more relevant for studying complicated problems, such as inhomogeneous lattice deformation and fracture in multiphase macroscopic materials. Separation of the scales/domains and interscale coupling mechanism are two key issues for the concurrent techniques.

Recently, a new approach for the multiscale modeling, *multiscale boundary conditions* for MD simulations, was proposed by Karpov et al. [111] and Wagner et al. [112]. This technique does not involve the explicit continuum modeling; therefore, the issues of separating the scale and interscale coupling do not exist. The coarse grain behavior is taken into account on the fine/coarse grain interface at the atomistic level through the lattice impedance techniques. Multiscale boundary conditions are employed within concurrent coupling methods to represent atomistic behaviors in the continuum domain. That results in a smooth atomistic-continuum scale coupling, without involving a costly artificial handshake region which is typically mandatory for concurrent approaches. This approach can apply to multidimensional problems through the use of the Fourier analysis of periodic structures [113, 114].

As an example of sequential multiscale modeling, we discuss how the mechanical properties of bulk tantalum were calculated using a multiscale modeling strategy. Moriarty et al. [115] started with fundamental atomic properties and used rigorous quantum-mechanical principles calculations to develop accurate interatomic force laws that were then applied to atomistic simulations involving many thousands of atoms. From these simulations, they derived the properties of individual dislocations in a perfect crystal and then, with a new microscale simulation technique, namely dislocation dynamics, examined the behavior of large collections of interacting dislocations at the microscale in a grain-sized crystal. They modeled the grain interactions in detail with FE simulation, and from those simulations, they finally constructed appropriate models of properties such as yield strength in a macroscopic volume of tantalum. At each length scale, the models were experimentally tested and validated with available data. The concept of information passing between models, from quantum modeling to atomic to continuum scale, is quite general and can be applied in a variety of problems, including NEMS.

24.3.1.1
Concurrent Multiscale Modeling

In a concurrent method, simulations at different length scales and time scales are performed simultaneously. The behavior at each scale depends strongly on the other. Atomistic approaches, such as MD, are used to describe behaviors of the materials in the atomistic regions, where characteristics of individual atoms are crucial, while continuum approaches, such as the FEM, are selected for the continuum domain, where the deformation is uniform and smooth.

MD computes the classical trajectories of atoms by integrating Newton's law, $F = ma$, for the system. In the MD domain, the interaction force follows an empirical potential. Consider a set of n_M molecules with the initial coordinates X_1, $I = 1$ to n_M. Let the displacements be denoted by $d_I(t)$. The potential energy is then given by $W_M(d)$. For a given potential function $W_M(d)$, an equilibrium state is given by

$$dW_M(d) = 0 \; . \tag{24.8}$$

From the continuum viewpoint, the governing equations arise from conservation of mass, momentum, and energy. Using a so-called total Lagrangian description [116], we can write the linear momentum equations as

$$\frac{\partial P_{ij}}{\partial X_j} + \rho_0 b_i = \rho_0 \ddot{u}_i \; , \tag{24.9}$$

where ρ_0 is the initial density, P the nominal stress tensor, b the body force per unit mass, u the displacement, and the superposed dots denote material time derivatives.

In the following section, we review some of the concurrent methods reported in the literature, including macroscopic, atomistic, and ab initio dynamics (MAAD), quasi-continuum method, scale bridging method, and coupling method, with the focus on the scale coupling mechanism and its application to the modeling of CNTs and microsystems/nanosystems. Reviews of the applications of these methods on modeling of other nanomaterials/nanostructures can be found in [110, 117–119, 151].

24.3.1.1.1
Macroscopic, Atomistic, and ab Initio Dynamics

MAAD is a length-scale concurrent coupling method developed by Abraham et al. [120, 121] and is named after the computation at three different length scales: macroscopic, atomistic, and ab initio dynamics. It is one of the early efforts to build a concurrent multiscale model. In this method, three different computational methods, namely tight-binding (TB), MD and FE are concurrently linked together. The dynamics of the entire targeted domain is governed by a total Hamiltonian function that combines the separate Hamiltonians of the three different scales, namely [121],

$$
\begin{aligned}
H_{\text{Tot}} = {} & H_{\text{FE}}(\{u, \dot{u}\} \in \text{FE}) + H_{\text{FE/MD}}(\{u, \dot{u}, r, \dot{r}\} \in \text{FE/MD}) \\
& + H_{\text{MD}}(\{r, \dot{r}\} \in \text{MD}) + H_{\text{MD/TB}}(\{r, \dot{r}\} \in \text{MD/TB}) \\
& + H_{\text{TB}}(\{r, \dot{r}\} \in \text{TB}) \,,
\end{aligned}
\tag{24.10}
$$

where, r and $\dot{r}$ are the positions and the velocities of atoms in the TB and MD regions; u and $\dot{u}$ are the displacements and time rate of change of elements in the FE region; MD/TB and FE/MD refer to the handshake regions. The equations of motion of all the variables are obtained by taking the appropriate derivative of H_{Tot} and, for a given set of initial conditions, the system evolves in a manner that conserves the total energy. More detailed information about the implementation of this technique can be found in [122].

This approach has been successfully applied to the investigation of crack propagation in a brittle solid, such as silicon [120], and microsystems/nanosystems, such as microresonators and microgears [108], as shown in Fig. 24.28. Figure 24.28a shows the schematic diagrams of domain decomposition for a microresonator. MD is used in the regions of the device with moderate strain oscillation while the FEM is used in the region where the change of strain is small. TB is used in regions of very large strain and bond breaking, such as at defects. These three length scales are smoothly coupled through the MD/FE and TB/MD handshaking. In brief, the MD/FE handshaking is accomplished through a mean force boundary condition. That is, the FE and MD regions overlap at the interface, and FE mesh nodes are positioned at the equilibrium positions of the corresponding MD atoms. The handshaking between the MD region and the TB region is accomplished through the termination of the chemical bonds that extend from the TB cluster. Various effects such as bond breaking, defects, internal strain, surface relaxation, statistical noise, and dissipation due to internal friction are included in the simulation.

The application of MAAD on modeling of microgears is illustrated in Fig. 24.28b. Such devices are presently made at the 100-μm scale and rotate at speeds of 150,000 rpm. The next generation of devices (nanogears), based on nanofabrication, are expected to be below the 1-μm level. The effects of wear, lubrication, and friction at the nanoscale are expected to have a significant consequence for the performance of the system; hence, the need for detailed modeling of these effects. The process of nanogear teeth grinding against each other cannot be simulated accurately with the FEM because deformation and bond breaking at the points of contact can only be treated empirically or with phenomenological models. Alter-

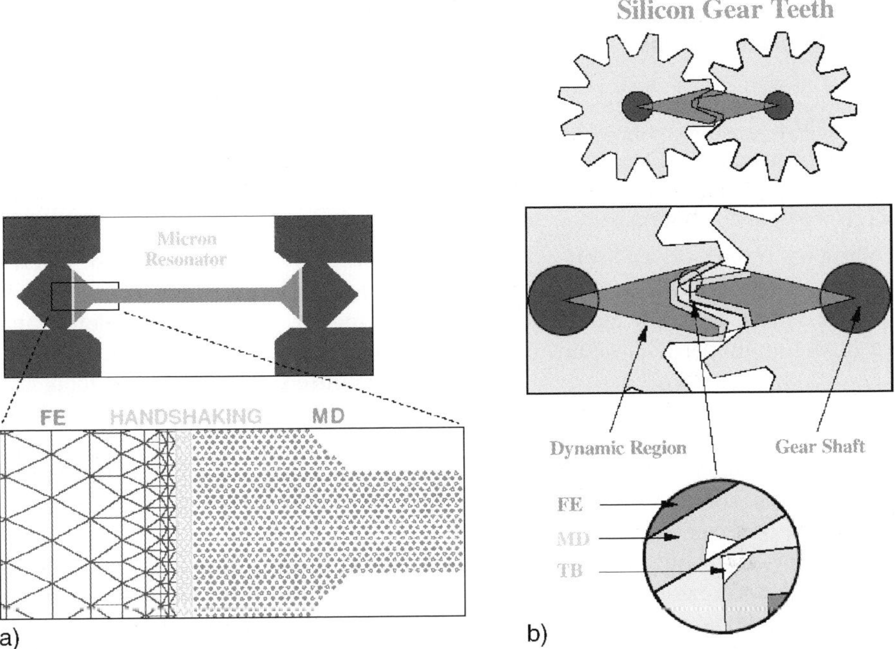

Fig. 24.28. (**a**) Scale decomposition for a microresontaor. Molecular dynamics (MD) is used in the regions of the device with moderate strain oscillators, while a finite-element (FE) method is used in the peripheral regions, where the change in strain is small. The two are joined through a consistent boundary condition in the handshake region, and both are run concurrently in lock-step. (**b**) Dynamic simulation zone and domain decomposition for coupling of length scales in microgears: from continuum (FE), to atomistic (MD) to electronic structure (tight-binding, TB). (Reprinted with permission from Rudd et al. [108] Copyright 1999, Applied Computational Research Society)

natively, multiscale modeling provides a good tool to predict chemomechanical-related issues for these devices, as shown in Fig. 24.28b. An inner region including the shaft is discretized by FEs. The handshaking between the FE and the MD region is accomplished by a self-consistent overlap region. In the gear–gear contact regions, in the nonlubricated case, a TB description is used as part of a QM simulation [108].

Two critical issues in this technique are the time step used in the simulation and the coupling of the different simulation domains. The time step used in the integration of the equations of motion (TB, MD, and FE) is determined by the time- step in the TB domain, which is the smallest and typically in the order of 10^{-15} s; therefore, many time steps are wasted in updating the MD and FE equations of motion, as the timescales governing those solutions are orders of magnitude larger. In MAAD, the coupling of the domains is accomplished by assuming in the transition region that each simulation contributes an equal amount of energy to the total energy. However, no rigorous studies have been performed to quantify the effectiveness of this method in eliminating spurious wave reflection at the simulation boundaries [110].

24.3.1.1.2
Quasi-continuum Method

The quasi-continuum method, originally developed by Tadmor et al. [123, 124], is a systematic computational approach that seeks a unified and efficient treatment of systems with a large number of atoms. It has been applied to the simulation of dislocation [123–126], grain boundary interactions [127, 128], nanoindentation [124, 128, 129], and fracture [130, 131]. In the quasi-continuum method, the continuum framework and continuum particle concept are retained, while the macroscopic constitutive equations are derived from atomistic interactions. The continuum particle is refined to the atomic level in critical regions such as near a defect and each continuum particle is considered as a small crystallite surrounding a representative atom. The quasi-continuum method has two different formulations: local and nonlocal. The local formulation of the quasi-continuum method is essentially the application of crystal elasticity by introducing a homogeneous deformation assumption. The advantage of this approach is that a large number of atoms can be lumped into a small set of representative atoms. The strain energy associated with the representative atoms can be computed by summing up the interatomic potential following the Cauchy–Born rule. A FE mesh is then imposed on top of the representative atoms. The deformation of the field variables can be approximated by the FE shape function and nodal values. The energy of the system can then be expressed in terms of the representative atoms, instead of all the atoms, with the approximate weights, resulting in significant computational savings. Hence, usage of the crystal elasticity approach, in the local formulation, allows the atoms in the system to be considered as a hyperelastic continuum material. The nonlocal formulation is introduced to allow atomic defects, e.g., dislocations, to occur when the FE approximation is insufficient to describe the deformation field. Some of the complications of this method include the following: (1) the mesh must be prescribed such that the nodes conform to the representative atoms at the interface; (2) the continuum mesh is required to be graded down to the scale of the atomic lattice in the region of the localized regions. A detailed review of the quasi-continuum and related methods can be found in [119].

24.3.1.1.3
Bridging Scale Method

The bridging scale method is a concurrent coupling method developed by Liu et al. [110–112, 132]. It assumes FE and MD solutions exist simultaneously in the entire computational domain and MD calculations are performed only in the region that is necessary; therefore, the issue of grading FE mesh down to lattice size as in quasi-continuum and MAAD methods does not arise here. The basic idea of the bridging scale method is to decompose the total displacement field $u(x)$ into coarse and fine scales:

$$u(x, t) = \bar{u}(x, t) + u'(x, t) , \qquad (24.11)$$

where $\bar{u}(x, t)$ is the coarse-scale solution and $u'(x, t)$ is the fine scale solution, corresponding to the part that has a vanishing projection onto the coarse-scale basic function. The coarse-scale solution can be integrated by a basic FE shape function as $\bar{u} = Nd$, in which d is the FE solution and N is the shape function evaluated at atomic locations. The fine-scale solution can be obtained by $u'(x, t) = Qq$ [133], where q is the MD solution and $Q = I - P$, in which P is a projection operator that depends on both shape functions and the properties of the atomic lattice, and I is the identity operator.

One example in which the bridging scale method was employed successfully was the modeling of the buckling of a MWNT [134]. In the simulation, a 15-walled CNT with length of 90 nm was considered. The original MD system contained about 3×10^6 atoms. This was replaced with a system of 27,450 particles. In addition to

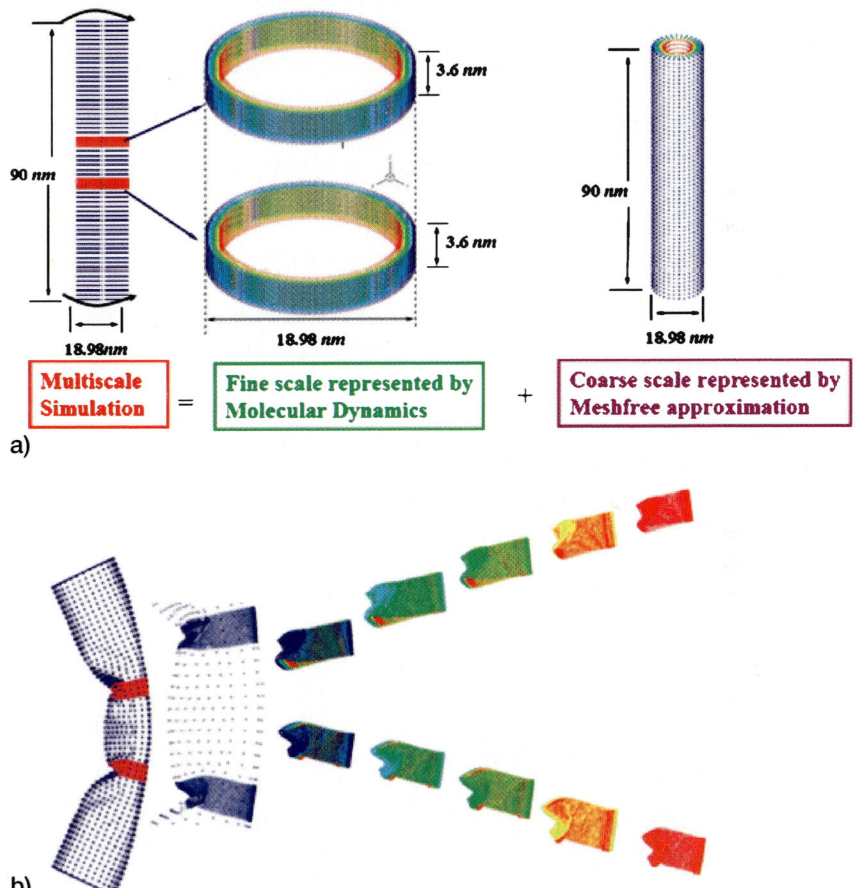

Fig. 24.29. Multiscale analysis of a 15-walled CNT by a bridging scale method. (**a**) The multiscale simulation model consists of ten rings of carbon atoms (with 49,400 atoms each) and a mesh-free continuum approximation of the 15-walled CNT by 27,450 nodes. (**b**) The global buckling pattern is captured by a mesh-free method, whereas the detailed local buckling of the ten rings of atoms is captured by a concurrent bridging scale MD simulation. (Reprinted with permission from Qian et al. [134] Copyright 2004, Elsevier)

the particles, two sections along the tube were enriched with molecular structures of MWNTs. The position of the enrichment was determined by a multiresolution analysis of the coarse-scale simulation. The multiscale configuration employed is illustrated in Fig. 24.29a. Figure 24.29b shows the buckling pattern at the final stage of loading and the energy density contour plot for each layer of the MWNT. Two distinctive buckling patterns can be seen, while the contour plot shows clearly the strain energy concentration at the buckling point. A unique feature revealed by the multiscale method is the details of the molecular structure at the kinks, which cannot be resolved by the coarse-scale representation alone. The atomic structures of the buckling region for each layer of the MWNT are plotted on the right-hand side of Fig. 24.29b.

The advantages of the bridging scale method over other concurrent methods are as follows: (1) no mesh gradation is required; (2) it does not involve calculation of any high-order tensors, such as the Piola–Kirchoff stress tensor, with associated gains in computational efficiency; (3) it can be extended to the dynamic regime [132, 133]. For these reasons, the bridging scale method has become a popular approach in the modeling of nanosystems.

24.3.1.1.4
Coupling Methods

Coupling methods are based on the coupling of continuum and molecular models through the definition of various domain decompositions [135]: (1) an edge-to-edge decomposition method, which has an interface between the two models, and (2) an overlapping-domain decomposition. A significant feature of the overlapping-domain decomposition method is that in the overlapping region, handshake region, the total potential energy is a linear combination of the continuum and atomistic potential energies.

An example of the overlapping-domain decomposition model is shown in Fig. 24.30. The complete domain in the initial configuration is denoted by Ω_0 and its boundaries by Γ_0. The domain is subdivided into the subdomains treated by

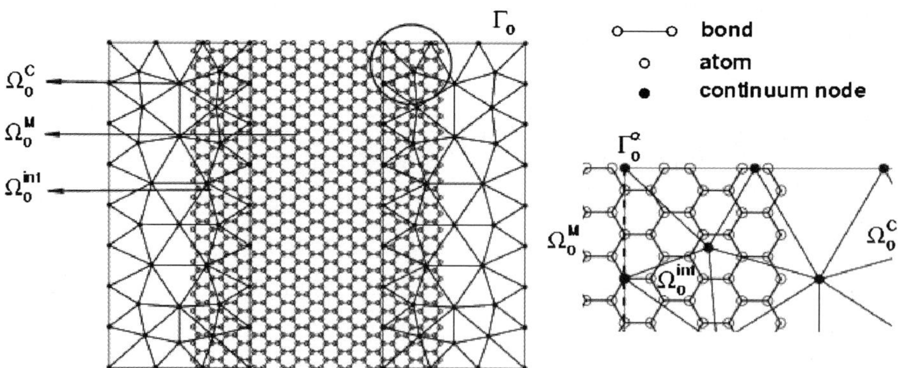

Fig. 24.30. Overlapping-domain decomposition method. (Reprinted with permission from Belytschko and Xiao [135]. Copyright 2003, Begell House)

continuum mechanics, Ω_0^C, and that treated by molecular mechanics, Ω_0^M, which is the domain encompassed by the atoms of the model. The intersection of these two domains is donated by Ω_0^{int} in the initial configuration and Ω^{int} in the current configuration (Ω^{int} is often called handshake domain). Γ_0^α denotes the edges of the continuum domain.

In the overlapping region Ω_0^{int}, the constraint imposed on the continuum model and molecular mechanics model is that the continuum displacements conform to the atomic displacements at the discrete position of the atoms, i.e.,

$$g_1 = ||\boldsymbol{u}(X_I) - \boldsymbol{d}_I|| \ . \tag{24.12}$$

In the Lagrange multiplier method, the problem consists of finding the stationary condition of

$$W_L = W^{int} - W^{ext} + \lambda^T g \ , \tag{24.13}$$

where W^{int} is the total internal potential of the system, W^{ext} is the total external potential energy, and $\lambda = \{\lambda_I\}$ is the vector of Lagrange multipliers for the above constraint for each of the atoms.

The augmented Lagrangian method can be obtained by adding a penalty to (24.13), namely,

$$W_{AL} = W^{int} - W^{ext} + \lambda^T g + \frac{1}{2} p\boldsymbol{g}^T\boldsymbol{g} \ , \tag{24.14}$$

where p is the penalty parameter. If $p = 0$, (24.14) is identical to (24.13).

It is noted that the purpose of the handshake region is to assure a smoother coupling between the atomistic and continuum regions; therefore, an extremely fine FE mesh in the handshake region is required in order to provide adequate space resolution to match the positions of interface atoms and FEM nodal positions. At the front end of the continuum interface, the FEs have to be scaled down the chemical or ion bond lengths that may call forth costly inversions of large stiffness matrices [110]. As mentioned before, the recently proposed multiscale boundary conditions approach [112] may prove advantageous because it computes positions of actual next-to-interface atoms at the intrinsic atomistic level by means of a functional operator over the interface atomic displacements. This eliminates the need to have a costly handshake region at the atomic/continuum interface and a dense FEM mesh scale down to the chemical bond lengths.

In the edge-to-edge decomposition coupling method (Fig. 24.31), there are three types of particles. Besides the nodes of the continuum domain and the atoms of the molecular domain, virtual atoms are defined to model the bond angle-bending for bonds between the continuum and the molecular domains. The virtual atoms are connected with the molecular domain by virtual bonds.

The internal potential energy for the entire domain is given by the sum of continuum and molecular energies, viz.,

$$W^{int} = W^C + W^M \ , \tag{24.15}$$

where W^C includes the stretching energy of virtual bonds and W^M includes the bond angle-bending potential resulting from the bond angle change between the

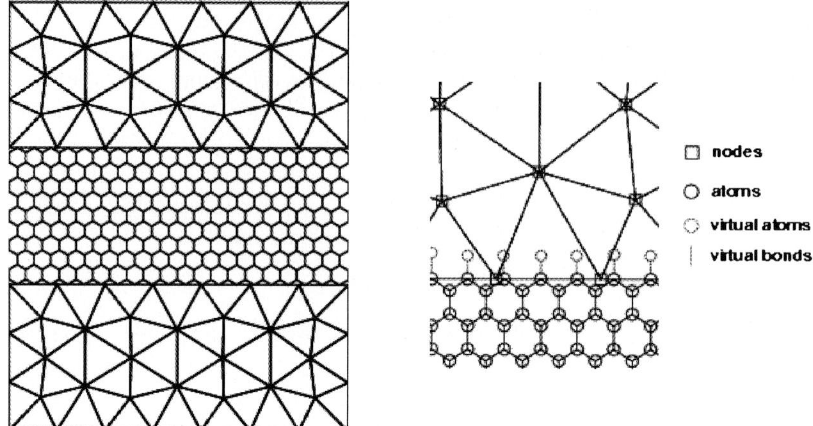

Fig. 24.31. Edge-to-edge coupling (*left*) and definition of particles and bonds in edge-to-edge coupling (*right*). (Reprinted with permission from Belytschko and Xiao [135]. Copyright 2003, Begell House)

virtual bonds and adjacent bonds in the molecular domain. The motion of each virtual atom will depend on the modes of the element that contain this virtual atom. The constraints given by (24.12) can then be imposed by means of the Lagrange multiplier method (24.13) or the augmented Lagrangian method (24.14).

An example illustrating the performance of these methods is shown in Fig. 24.32. The deformed configuration of a bent graphite sheet, with bending angle of 25°, was obtained using three different modeling techniques: (1) pure atomistic method, (2) overlapping-domain decomposition method, and (3) edge-to-edge decomposition method. A comparison of the predictions is shown in Fig. 24.32, which reveals that the deformation configurations are nearly identical. Figure 24.33 shows a comparison of predictions for the bending of a SWNT, with bending angle of 30°, using molecular mechanics (Fig. 24.33a) and overlapping coupling (Fig. 24.33b) methods. A comparison of the potential energies obtained by these two approaches is shown in Fig. 24.33c. The differences are small and likely due to the fact that the continuum model did not capture the tiny kinks predicted by the molecular mechanics model. Figure 24.34 shows another example of modeling the fracture of CNTs using the

Fig. 24.32. Bending of graphite sheets. Comparison of deformed configurations obtained by means of (**a**) molecular mechanics, (**b**) overlapping coupling, and (**c**) edge-to-edge coupling. (Reprinted with permission from Belytschko and Xiao [135]. Copyright 2003, Begell House)

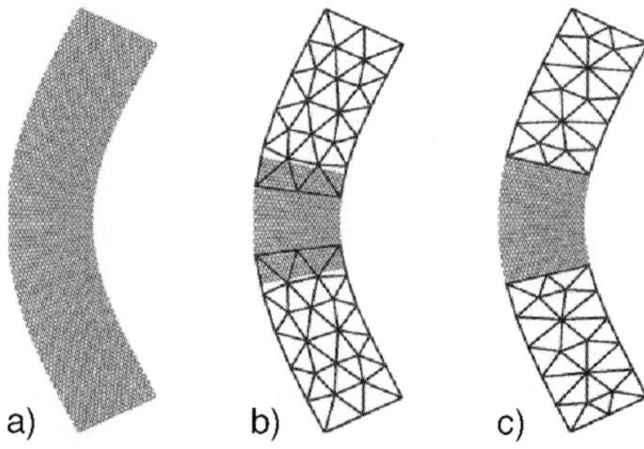

overlapping-coupling method. In the simulation, two nested nanotubes with van der Waals interactions between the shells were considered. The molecular model was used only in a small subdomain surrounding a defect, while the FE model was employed outside of the molecular model (Fig. 24.34). A modified Tersoff–Brenner

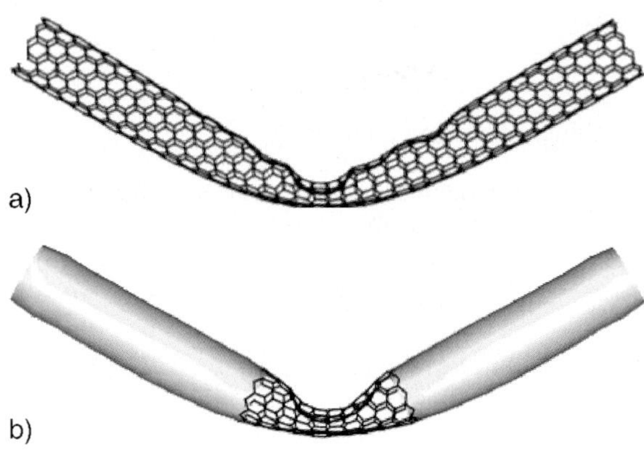

Fig. 24.33. Bending of a SWNT. Comparison of deformed configurations obtained by means of (**a**) molecular mechanics and (**b**) the overlapping-coupling method. (**c**) Comparison of potential energy as a function of bending angle. (Reprinted with permission from Belytschko and Xiao [135]. Copyright 2003, Begell House)

Fig. 24.34. CNT model for fracture study by the overlapping-coupling method. (Reprinted with permission from Belytschko and Xiao [135]. Copyright 2003, Begell House)

potential was employed to model the covalent C–C bond and the Lennard-Jones potential was employed to capture the intershell nonbonded (van der Waals) interactions. The load was only applied to the outer shell. The numerical results revealed that large vacancies are needed to explain the failure stresses experimentally measured in MWNTs. Furthermore, the simulations demonstrated that the van der Waals intershell interactions are negligible.

24.3.2
Continuum Mechanics Modeling

Many NEMS devices can be modeled either as biased cantilever beams or as fixed–fixed beams freestanding over a ground substrate, as shown in Fig. 24.35. The beams can be CNTs, nanowires, or small nanofabricated parts. The electromechanical characterization of NEMS involves the calculation of the elastic energy (E_{elas}), from the deformation of active components, the electrostatic energy (E_{elec}), and van der Waals energy (E_{vdW}) from atomic interactions. In the following we overview the continuum theory for each of these energy domains and derive the governing equations for both small and finite deformation regimes. We follow the work reported in [31, 109, 136–139].

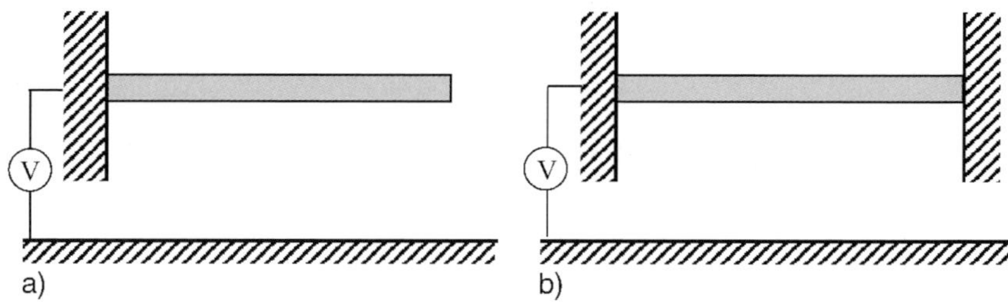

Fig. 24.35. NEMS devices: (**a**) Cantilever beam configuration; (**b**) doubly clamped beam configuration

24.3.2.1
Continuum Theory

24.3.2.1.1
Van der Waals Interactions

The van der Waals energy originates from the interaction between atoms. The Lennard-Jones potential is a suitable model to describe van der Waals interaction [140]. In the Lennard-Jones potential, there are two terms: one is repulsive and the other is attractive. The Lennard-Jones potential between two atoms i and j is given by

$$\phi_{ij} = \frac{C_{12}}{r_{ij}^{12}} - \frac{C_{6}}{r_{ij}^{6}} , \tag{24.16}$$

where r_{ij} is the distance between atoms i and j and C_6 and C_{12} are attractive and repulsive constants, respectively. For the carbon–carbon interaction, $C_6 = 15.2\,\text{eV}\,\text{Å}^6$ and $C_{12} = 24.1\,\text{keV}\,\text{Å}^{12}$ and the equilibrium spacing $r_0 = 3.414\,\text{Å}$ [141]. From (24.16), we can see that the repulsive components of the potential decay extremely fast and play an important role only when the distance is close to or smaller than r_0. The total van der Waals energy can be computed by a pairwise summation over all the atoms. The computational cost (number of operations) is proportional to the square of the number n of atoms in the system. For a NEMS device with millions of atoms, this technique is prohibitively expensive. Instead, a continuum model was established to compute the van der Waals energy by the double volume integral of the Lennard-Jones potential [142], i.e.,

$$E_{\text{vdW}} = \int_{v_1} \int_{v_2} n_1 n_2 \left(\frac{C_{12}}{r^{12}(v_1, v_2)} - \frac{C_6}{r^6(v_1, v_2)} \right) dv_1\, dv_2 \,, \tag{24.17}$$

where v_1 and v_2 represent the two domains of integration, and n_1 and n_2 are the densities of atoms for the domains v_1 and v_2, respectively. $r(v_1, v_2)$ is the distance between any point on v_1 and v_2.

Let us consider a SWNT freestanding above a ground plane consisting of layers of graphite sheets, with interlayer distance $d = 3.35\,\text{Å}$, as illustrated in Fig. 24.36a. The energy per unit length of the nanotube is given by

$$\frac{E_{\text{vdw}}}{L} = 2\pi\sigma^2 R \sum_{n=1}^{N} \int_{-\pi}^{\pi} \left(\frac{C_{12}}{10[(n-1)d + r_{\text{init}} + R + R\sin\theta]^{10}} \right.$$

$$\left. - \frac{C_6}{4[(n-1)d + r_{\text{init}} + R + R\sin\theta]^4} \right) d\theta \,, \tag{24.18}$$

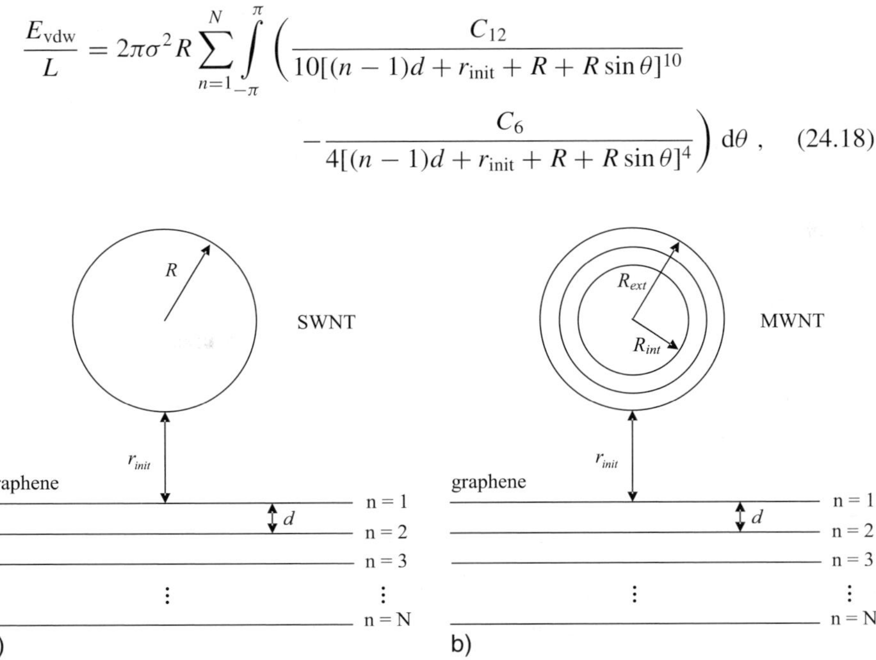

Fig. 24.36. Van der Waals integration of a SWNT (**a**) and a MWNT (**b**) over a graphite ground plane. (Reprinted with permission from Dequesnes et al. [109]. Copyright 2002, Institute of Physics)

where L is the length of the nanotube, R is the radius of the nanotube, r_{init} is the distance between the bottom of the nanotube and the top graphene sheet, N is the number of graphene sheets, and $\sigma \cong 38\,\text{nm}^{-2}$ is the graphene surface density. When r_{init} is much larger than the equilibrium spacing r_0, the repulsive component can be ignored and (24.18) can be simplified as [109]

$$\frac{E_{\text{vdW}}}{L} = C_6\sigma^2\pi^2 R \sum_{r=r_{\text{init}}}^{(N-1)d+r_{\text{init}}} \frac{(R+r)[3R^2+2(r+R)^2]}{2[(r+R)^2-R^2]^{7/2}} . \tag{24.19}$$

The accuracy of (24.19) in approximating the continuum van der Waals energy of a SWNT placed over a graphite plane is verified by the comparison with the direct pairwise summation of the Lennard-Jones potential given by (24.16) for a (16,0) tube, which is shown in Fig. 24.37 [109].

For a MWNT, as illustrated in Fig. 24.36b, the energy per unit length can be obtained by summing up the interaction between all separate shells and layers:

$$\frac{E_{\text{vdW}}}{L} = \sum_{R=R_{\text{int}}}^{R_{\text{ext}}} \sum_{r=r_{\text{init}}}^{(N-1)d+r_{\text{init}}} \frac{C_6\sigma^2\pi^2 R(R+r)[3R^2+2(r+R)^2]}{2[(r+R)^2-R^2]^{7/2}} , \tag{24.20}$$

where R_{int} and R_{ext} are the inner and outer radii of the nanotube, respectively.

The van der Waals force per unit length can be obtained as

$$q_{\text{vdw}} = \frac{\mathrm{d}\left(\frac{E_{\text{vdw}}}{L}\right)}{\mathrm{d}r} . \tag{24.21}$$

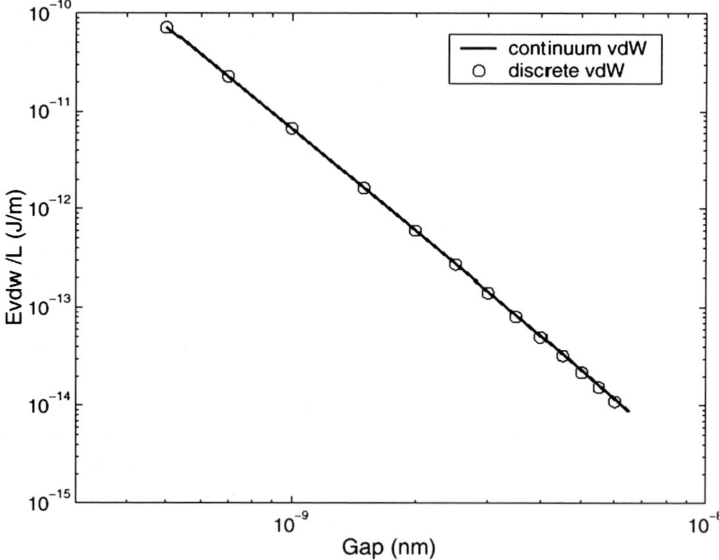

Fig. 24.37. Comparison of the continuum van der Waals energy given by (24.19) with the discrete pairwise summation given by (24.16). (Reprinted with permission from Dequesnes et al. [109]. Copyright 2002, Institute of Physics)

Thus, inserting (24.7) into (24.8) and taking the derivative with respect to r, one obtains [109]

$$q_{\text{vdw}} = \sum_{R=R_{\text{int}}}^{R_{\text{ext}}} \sum_{r=r_{\text{init}}}^{(N-1)d+r_{\text{init}}}$$
$$- \frac{C_6 \sigma^2 \pi^2 R \sqrt{r(r+2R)}(8r^4 + 32r^3 R + 72r^2 R^2 + 80rR^3 + 35R^4)}{2[2r^5(r+2R)^5]^5}$$

(24.22)

24.3.2.1.2
Electrostatic Force

When a biased conductive nanotube is placed above a conductive substrate, there are induced electrostatic charges both on the tube and on the substrate. The electrostatic force acting on the tube can be calculated using a capacitance model [143].

Let us look at the electrostatic force for a conductive nanotube with finite length and round cross section, above an infinite ground plane. Although nanotubes have hollow structures, CNTs with capped ends are more electrochemically stable than those with open ends [144]; thus, nanotubes with finite length, as well as nanowires, can be geometrically approximated by conductive nanocylinders. For small-scale nanocylinders, the density of states on the surface is finite. The screening length, the distance that the "surface charge" actually penetrates into the cylinder interior, is found to be a nanometer-scale quantity [145]. For nanocylinders with transverse dimension, i.e., diameter, approaching the screening length, such as SWNTs, the finite size and density of states (quantum effects) have to be considered thoroughly when calculating the surface–volume charge distribution [146, 147]. For nanocylinders with transverse dimension much larger than the screening length, such as MWNTs or nanowires with large outer diameter, e.g., 20 nm, this quantum effect can be considered negligible. Thus, the charge distribution can be approximated by the charge distribution on a metallic, perfectly conductive cylinder with the same geometry, to which classical electrostatic analysis can be applied.

For infinitely long metallic cylinders, the capacitance per unit length is given by [143]

$$C_{\text{d}}(r) = \frac{\pi \varepsilon}{a \cosh \left(1 + \frac{r}{R}\right)}, \tag{24.23}$$

where r is the distance between the lower fiber of the nanocylinder and the substrate, R is the radius of the nanocylinder, and ε is the permittivity of the medium. For vacuum, $\varepsilon_0 = 8.854 \times 10^{-12}\,\text{C}^2\,\text{N}^{-1}\,\text{m}^{-2}$. Equation (24.23) can be applied for infinitely long MWNTs with large diameters ($R = R_{\text{ext}}$).

For the charge distribution on infinitely long SWNTs, Bulashevich and Rotkin [147] proposed a quantum correction, rendering the capacitance per unit length as

$$C = \frac{C_{\text{d}}}{1 + \frac{C_{\text{d}}}{C_{\text{Q}}}} \approx C_{\text{d}} \left(1 - \frac{C_{\text{d}}}{C_{\text{Q}}}\right), \tag{24.24}$$

where $C_Q = e^2 \nu_m$, ν_M is the constant density of the states near the electroneutral level measured from the Fermi level.

For nanocylinders with finite length, there are two types of boundary surfaces – the cylindrical side surface and the planar end surface. Essentially classical distribution of charge density with a significant charge concentration at the cylinder end has been observed [146, 148, 149]. Here we discuss a model to calculate the electrostatic charge distribution on metallic cylindrical cantilevers based on a boundary element method, considering both the concentrated charge at the free end and the finite rotation due to the deflection of the cantilever [136].

Figure 24.38 shows the charge distribution along the length L of a freestanding nanotube, subjected to a bias voltage of 1 V. The contour plot shows the charge density (side view), while the curve shows the charge per unit length along the nanotube. The calculation was performed using the CFD-ACE+ software (a commercial code from CFD Research Corporation based on finite volume and boundary element methods). There is significant charge concentration on the free ends and uniform charge distribution in the center of the cantilever, which is found to follow (24.23). The charge distribution along a deflected cantilever nanotube is shown in Fig. 24.39. The parameters are $R_{ext} = 20$ nm, $H = 500$ nm, $L = 3$ μm, and the gap between the free end and the substrate $r(L)$ is 236 nm. From Fig. 24.39, it is seen that, besides the concentrated charge on the free end, the clamped end imposes a significant effect on the charge distribution in the region close to it [145]; however, this effect can be considered negligible because its contribution to the deflection of the nanotube is quite limited. The charge distribution in regions other than the two ends closely follows (24.23). A formula for the charge distribution including end charge effects and the deflection of the cantilever is derived from a parametric analysis, as follows [136]:

$$C(r(x)) = C_d(r(x)) \left\{ 1 + 0.85[(H + R)^2 R]^{\frac{1}{3}} \delta(x - x_{\text{tip}}) \right\} = C_d(r(x))\{1 + f_c\} , \quad (24.25)$$

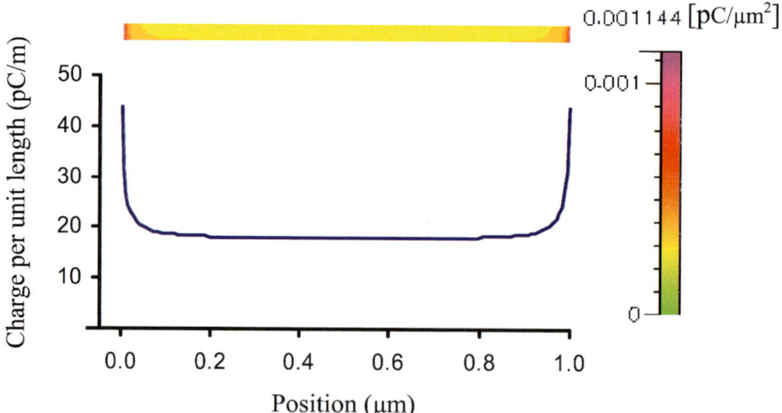

Fig. 24.38. Charge distribution for a biased nanotube. The device parameters are $R_{ext} = 9$ nm, $H = 100$ nm, and $L = 1$ μm. (Reprinted with permission from Ke et al. [137]. Copyright 2005, The American Society of Mechanical Engineers)

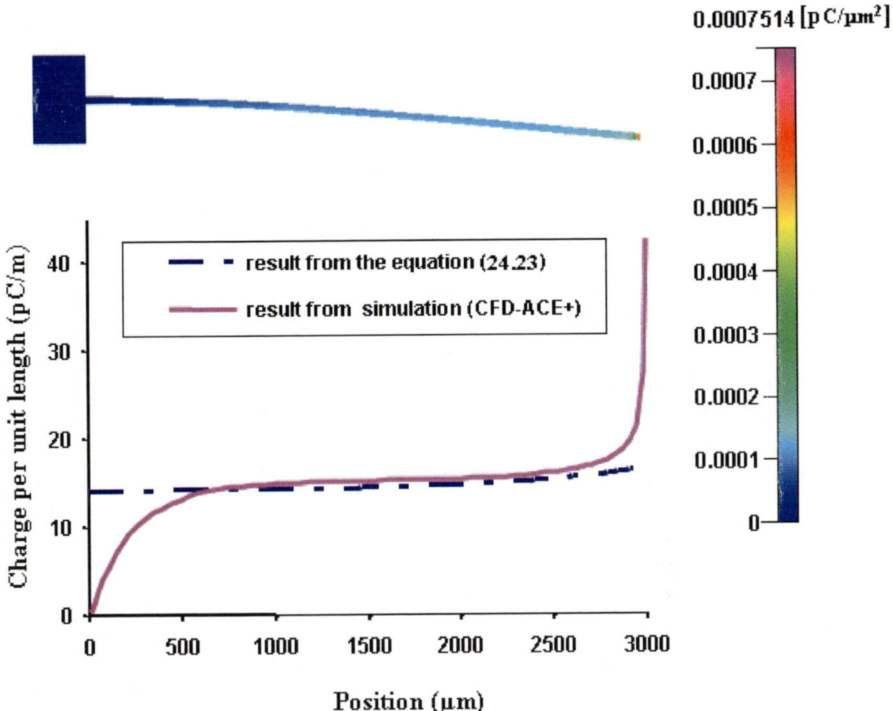

Position (μm)

Fig. 24.39. *Top*: Two-dimensional side view of the charge distribution in a deflected nanotube cantilever. *Bottom*: Charge distribution per unit length along a deflected nanotube cantilever. The *solid line* is the simulation result performed with CFD-ACE+; the *dotted line* is plotted from (24.23). (Reprinted with permission from Ke and Espinosa [136]. Copyright 2005, The American Society of Mechanical Engineers)

where the first term in the bracket accounts for the uniform charge along the side surface of the tube and the second term, f_c, accounts for the concentrated charge at the end of the tube (for a doubly clamped tube, $f_c = 0$). H is the distance between the cantilever and the substrate when the cantilever is in a horizontal position, R is the radius of the tube (for a MWNT $R = R_{ext}$), $x = x_{tip} = L$ for small deflection (when considering the finite kinematics, i.e., large displacement, $x = x_{tip} \neq L$), $\delta(x)$ is the Dirac function, and $r(x) = H - w(x)$, with w being the tube deflection.

Thus, the electrostatic force per unit length of the nanotube is given by differentiation of the energy as follows [137]:

$$q_{elec} = \frac{1}{2}V^2\frac{dC}{dr} = \frac{1}{2}V^2\left(\frac{dC_d}{dr}\right)\{1 + f_c\}$$

$$= \frac{-\pi\varepsilon_0 V^2}{\sqrt{r(r+2R)}\,a\cosh^2\left(1+\frac{r}{R}\right)}(1 + f_c)\,, \tag{24.26}$$

where V is the bias voltage.

24.3.2.1.3
Elasticity

Continuum beam theory has been widely used to model the mechanics of nanotubes [28, 31, 109, 136, 138, 139]. The applicability and accuracy of the continuum theory have been evaluated by comparison with MD simulations [109]. Figure 24.40 shows the comparison of the deflection of a 20-nm-long, doubly clamped, double-walled nanotube with a diameter of 1.96 nm, calculated by MD simulation and by the beam equation, respectively. The solid black curve – the deflection predicated by the beam equation – follows closely the shape predicted by MD calculations.

Because nanotubes have high flexibility with strain at tensile failure on the order of 30% [150], nonlinear effects such as finite kinematics accounting for large displacement need to be considered in the modeling. This is particularly important for doubly clamped nanotube beams because the stretching from the finite kinematics stiffens the beam, resulting in a significant increase of the pull-in voltage, a key parameter in NEMS devices.

Fig. 24.40. Comparison between MD and beam theory for the deflection of a 20-nm-long fixed–fixed double-walled nanotube (diameter 1.96 nm). The *solid black curve* is the deflection predicated by beam theory. (Reprinted with permission from Dequesnes et al. [109]. Copyright 2002, Institute of Physics)

24.3.2.1.4
Governing Equations

The electromechanical characteristic of nanotube cantilevers or doubly clamped nanotube beams can be determined by coupling the van der Waals, electrostatic, and elastic forces. The governing equation under the small-deformation assumption (considering only bending) is given by [109]

$$EI\frac{\mathrm{d}^4 r}{\mathrm{d}x^4} = q_{\text{elec}} + q_{\text{vdw}} \,, \tag{24.27}$$

where r is the gap between the nanotube and the ground plane, x is the position along the tube, E is Young's modulus (for a CNT $E = 1$–1.2 TPa), I is the moment of inertia (for nanotubes $I = \frac{\pi}{4}(R_{\text{ext}}^4 - R_{\text{int}}^4)$, where R_{ext} and R_{int} are the outer and inner radii of the nanotubes, respectively), and q_{elec} and q_{vdw} are given by (24.26) and (24.22), respectively.

For cantilevers exhibiting large displacements, as shown in Fig. 24.41, the curvature of the deflection should be considered and the governing equation changes into [137]

$$EI \frac{\mathrm{d}^2}{\mathrm{d}x^2} \left(\frac{\frac{\mathrm{d}^2 r}{\mathrm{d}x^2}}{\left(1 + \left(\frac{\mathrm{d}r}{\mathrm{d}x}\right)^2\right)^{\frac{3}{2}}} \right) = (q_{\mathrm{vdw}} + q_{\mathrm{elec}}) \sqrt{1 + \left(\frac{\mathrm{d}r}{\mathrm{d}x}\right)^2} \, . \tag{24.28}$$

For doubly clamped structures exhibiting finite kinematics, as shown in Fig. 24.42, stretching becomes significant as a consequence of the ropelike behavior of a doubly clamped nanotube. The corresponding governing equation is expressed as [137–139]

$$EI \frac{\mathrm{d}^4 r}{\mathrm{d}x^4} - \frac{EA}{2L} \int_0^L \left(\frac{\mathrm{d}r}{\mathrm{d}x}\right)^2 \mathrm{d}x \frac{\mathrm{d}^2 r}{\mathrm{d}x^2} = q_{\mathrm{elec}} + q_{\mathrm{vdw}} \, , \tag{24.29}$$

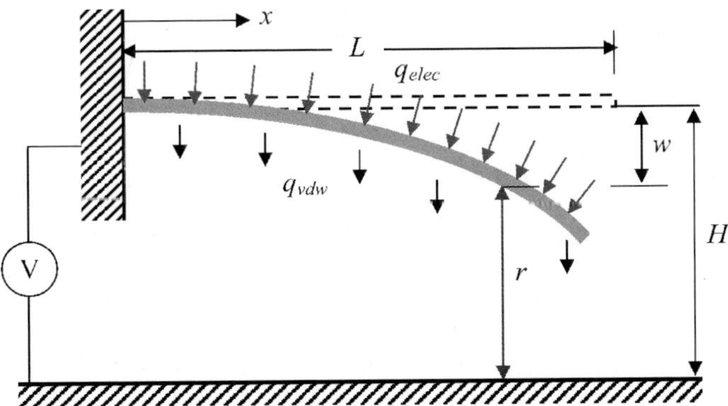

Fig. 24.41. Finite-kinematics configuration of a cantilever nanotube device subjected to electrostatic and van der Waals forces. (Reprinted with permission from Ke et al. [137]. Copyright 2005, The American Society of Mechanical Engineers)

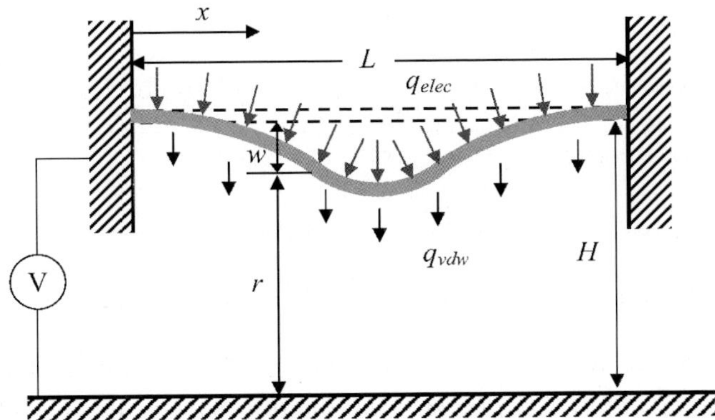

Fig. 24.42. Finite-kinematics configuration of a doubly clamped nanotube device subjected to electrostatic and van der Waals forces. (Reprinted with permission from Ke et al. [137]. Copyright 2005, The American Society of Mechanical Engineers)

where the term $\frac{EA}{2L}\int_0^L \left(\frac{dr}{dx}\right)^2 dx$ is the tension along the axis of the tube due to stretching.

The aforementioned governing equations can be numerically solved by either direct integration or the finite-difference method.

24.3.2.2
Analytical Solutions

In this section, we discuss the analytical solutions of the electromechanical characteristic of the NEMS devices consisting of both cantilevered and doubly clamped nanotubes. In particular, the pull-in voltage calculations based on the energy method are reported [31, 138].

For nanotube cantilevers (singly clamped), the deflection of the cantilevered nanotube can be approximated by the following quadratic function [31]:

$$w(x) \approx \frac{x^2}{L^2} c ,$$
(24.30)

where L is the length of the nanotube, c is a constant that represents the displacement of the end of the cantilever, and x is the coordinate along the nanotube.

The total energy of the system E_{total}, is expressed as

$$E_{total}(c) = E_{elas}(c) + E_{elec}(c) + E_{vdW}(c) ,$$
(24.31)

where the elastic energy $E_{elast}(c)$, the electrostatic energy $E_{elec}(c)$, and van der Waals energy $E_{vdW}(c)$ can be obtained by integration as

$$E_{elast}(c) = \frac{EI}{2} \int_0^L \left(\frac{d^2 w}{dx^2}\right)^2 dx$$
(24.32a)

and

$$E_{elec,vdW}(c) \approx \int_0^L \frac{dE_{elec,vdW}(r(w(x)))}{dx} dx .$$
(24.32b)

The equilibrium condition is reached when the total energy reaches a minimum value, i.e.,

$$\frac{dE_{total}}{dc} = 0 .$$
(24.33a)

Similarly, the instability of the devices, i.e., pull-in, happens when the second-order derivative of the total energy equals zero, namely,

$$\frac{d^2 E_{total}}{dc} = 0 .$$
(24.33b)

The van der Waals interaction plays an important role only for a small gap between the nanotubes and substrate, i.e., a few nanometers; thus, it can be neglected in the analysis of NEMS with large gaps. We consider $E_{vdw} \approx 0$ in this analysis.

Under the assumption that the nanotube's (external) radius R_{ext} is much smaller than the distance r between the nanotube and ground plane, i.e., $R_{ext}/r \ll 1$, the pull-in voltage, considering the nonlinear finite kinematics and the concentrated charges at the free end, is given by [31]

$$V_{\substack{S\\PI}} \approx k_S \sqrt{\frac{1 + K_S^{FK}}{1 + K_S^{TIP}}} \frac{H}{L^2} \ln\left(\frac{2H}{R_{ext}}\right) \sqrt{\frac{EI}{\varepsilon_0}} , \tag{24.34a}$$

where

$$k_S \approx 0.85 , \quad K_S^{FK} \approx \frac{8H^2}{9L^2} , \quad K_S^{TIP} \approx \frac{2.55 \left[R_{ext}(H + R_{ext})^2\right]^{\frac{1}{3}}}{L} , \tag{24.34b}$$

where subscripts S refer to singly clamped boundary conditions for cantilevers, superscript FK refers to finite kinematics, and superscript TIP refers to the charge concentration.

For doubly clamped nanotubes, the deflection is assumed to satisfy the boundary conditions $w(x = 0, L) = w'(x - 0, L) = 0$, namely, [138]

$$w(z) \approx 16 \left[\left(\frac{x}{L}\right)^2 - 2\left(\frac{x}{L}\right)^3 + \left(\frac{x}{L}\right)^4\right] c , \tag{24.35}$$

where $w(x = L/2) = c$ is here an unknown constant that represents the displacement of the central point. The pull-in voltage can be expressed as [138]

$$V_{\substack{D\\PI}} = k_D \sqrt{1 + k_D^{FK}} \frac{H + R}{L^2} \ln\left(\frac{2(H + R)}{R}\right) \sqrt{\frac{EI}{\varepsilon_0}} , \tag{24.36a}$$

where

$$k_D = \sqrt{\frac{1024}{5\pi S'(c_{PI})}\left(\frac{c_{PI}}{H + R}\right)} , \quad k_D^{FK} = \frac{128}{3003}\left(\frac{c_{PI}}{\rho}\right)^2 , \tag{24.36b}$$

where

$$\rho^2 = \frac{I}{A} = \frac{R_{ext}^2 + R_{int}^2}{4} , \quad S(c) = \sum_{i=1}^{\infty}\left(\frac{1}{\left(\ln\left(\frac{2(H+R)}{R}\right)\right)^i}\sum_{j=1}^{\infty} a_{ij}\left(\frac{c}{H + R}\right)^j\right) . \tag{24.36c}$$

Subscripts D refer to doubly clamped boundary conditions, c_{PI} is the central deflection of the nanotube at the pull-in, and the $\{a_{ij}\}$ in (24.36c) are known constants [138].

The accuracy of the analytical solutions is verified by the comparison with both numerical integration of the governing equations [137, 138] and experimental measurements (Sect. 24.3.2.3) [31]. The comparison between pull-in voltages evaluated

Table 24.1. Comparison between pull-in voltages evaluated numerically (num.) and theoretically (theo.) for doubly (D) and singly (S) clamped nanotube devices, respectively. $E = 1$ TPa, $R_{int} = 0$. For a cantilever nanotube device w denotes that the effect of charge concentration has been included. (Reprinted with permission from Ke et al. [137]. Copyright 2005, The American Society of Mechanical Engineers)

Case	BC	H (nm)	L (nm)	$R = R_{ext}$ (nm)	V_{PI} (V) (theo. linear)	V_{PI} (V) (num. linear)	V_{PI} (V) (theo. nonlinear)	V_{PI} (V) (num. nonlinear)
1	D	100	4000	10	3.20	3.18	9.06	9.54
2	D	100	3000	10	5.69	5.66	16.14	16.95
3	D	100	2000	10	12.81	12.73	36.31	38.14
4	D	150	3000	10	9.45	9.43	38.93	40.92
5	D	200	3000	10	13.53	13.52	73.50	77.09
6	D	100	3000	20	19.21	18.74	31.57	32.16
7	D	100	3000	30	38.57	37.72	51.96	50.63
8	S	100	500	10	27.28(w)	27.05(w)	27.52(w)	27.41(w)
9	S	100	500	10	27.28(w)	27.05(w)	30.87	31.66

numerically and theoretically for doubly and singly clamped nanotube devices is listed in Table 24.1 [137]. Columns six and seven in Table 24.1 compare analytical and numerical pull-in voltage predictions under the assumption of small deformations. Columns eight and nine in Table 24.1 compare analytical and numerical pull-in voltage predictions under the assumption of finite kinematics. The agreement is good (with a maximum discrepancy of 5%).

24.3.2.3
Comparison Between Analytical Predictions and Experiments

In this section, a comparison between analytical predictions and experimental data, for both small-deformation and finite-kinematics regimes, is presented.

24.3.2.3.1
Small-Deformation Regime

The nanotweezers experimental data reported by Akita et al. [9] in 2001, plotted in Fig. 24.43, is used to asses the model accuracy under small deformation. In this case, the nanotweezers are equivalent to a nanotube cantilever with length of 2.5 µm freestanding above an electrode with a gap of 390 nm. Symmetry is here exploited. In the same figure, a comparison between the analytically predicted nanotube cantilever deflection and the experimentally measured data is shown [31]. The analytical model includes the van der Waals force and charge concentration at the free end of the nanotube cantilever. Model parameters include Young's modulus, $E = 1$ TPa, external radius $R = R_{ext} = 5.8$ nm, and $R_{int} = 0$. The pull-in voltage from the analytical model is 2.34 V, while the experimentally measured pull-in voltage was 2.33 V. It is clear that the analytical prediction and experimental data for the deflection of the nanotube cantilever, as a function of applied voltage, are in very good agreement.

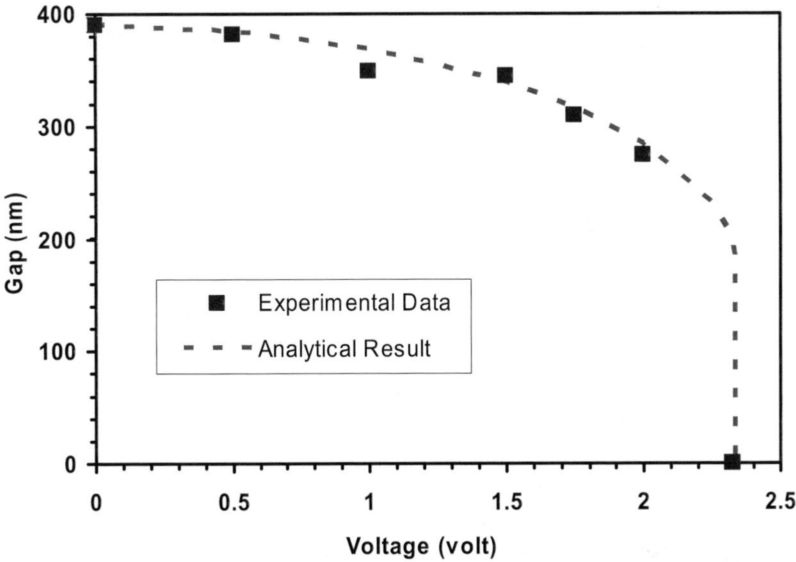

Fig. 24.43. Comparison between experimental data and theoretical prediction in the small-deformation regime. (Reprinted with permission from Ke at al. [31]. Copyright 2005, Elsevier)

24.3.2.3.2
Finite-Kinematics Regime

Experimental data corresponding to the deflection of CNT cantilevers in the finite-kinematics regime were recently obtained by in situ SEM measurements [31].

The configuration of the in situ measurement is shown in Fig. 24.44. The electrode was made of a silicon substrate coated with 50-nm Au film by electron-beam evaporation. This Si chip was attached onto the side of a Teflon block and mounted in the scanning electron microscope sample holder at an angle of 93° with respect to the holder plane. The nanotube cantilever fabricated by the method shown in Fig. 24.4 was placed horizontally and parallel to the electrode surface as schematically shown in Fig. 24.44. The distance between the top surface and the electron beam gun was 5 mm, while the distance between the nanotube and the electron-beam gun was measured to be 6.8 mm. Focusing on the electrode surface and adjusting the working distance to be 6.8 mm allowed a feature on the electrode, which was on the same horizontal plane as the nanotube, to be located. Such a feature is schematically marked as a line in Fig. 24.44. The horizontal distance between the nanotube and the line was controlled by the nanomanipulator and set to 3 μm. In the circuit, a resistor, $R_0 = 1.7\,\mathrm{M\Omega}$, was employed to limit the current. Because the ratio between the length of the nanotube and the gap between the nanotube and electrode is 2.3, the deflection of the nanotube can be considered to be in the finite-kinematics regime.

Figure 24.45 shows the SEM images of the deflection of the CNT as it is subject to increasing applied voltages. The feature on the electrode, which is in the same horizontal plane containing the cantilevered nanotube, is schematically marked as a solid black line in Fig. 24.45. These images clearly reveal changes in nanotube

Fig. 24.44. The experimental configuration employed in the electrostatic actuation of MWNT. Reprinted with permission from Ke et al. [31]. (Copyright 2005, Elsevier, Ltd)

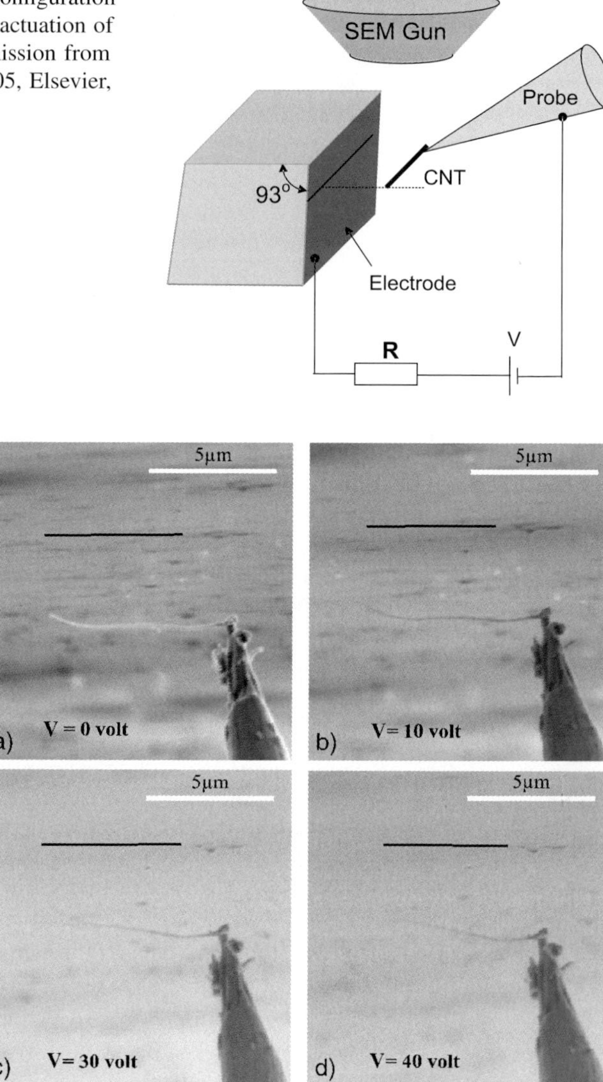

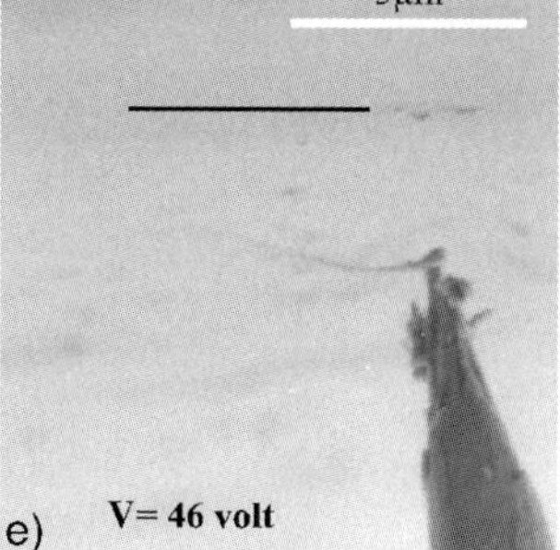

Fig. 24.45. SEM images of the deformed CNT at various bias voltages. (Reprinted with permission from Ke et al. [31]. Copyright 2005, Elsevier)

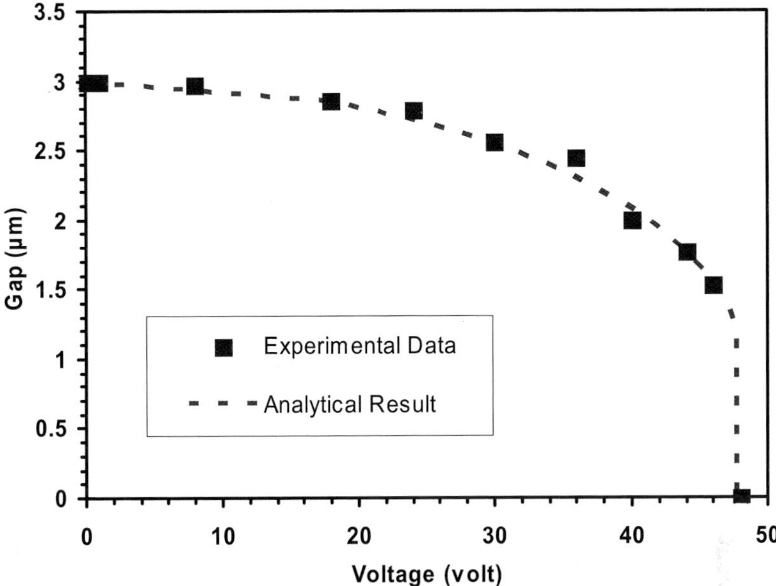

Fig. 24.46. Comparison between experimental data and theoretical prediction in the finite-kinematics regime. (Reprinted with permission from Ke et al. [31]. Copyright 2005, Elsevier)

deflection and local curvature as a function of applied voltage. A very noticeable effect, although difficult to quantify accurately, is the change in local curvature. The pull-in voltage, V_{PI}, was measured to be 48 V. Through digital image processing, the tip deflection as a function of voltage was measured.

The experimentally measured nanotube cantilever deflections, in the finite kinematics regime, are plotted in Fig. 24.46 [31]. The figure also shows a comparison between analytical prediction and experimental data. The analytical model includes finite kinematics, the van der Waals force, and charge concentration at the free end of the nanotube cantilever. For these predictions, the following parameters were employed: length of the nanotube, $L = 6.8\,\mu$m, initial gap between nanotube and electrode, $H = 3\,\mu$m, $R = R_{ext} = 23.5\,$nm, $R_{int} = 0$, $E = 1\,$TPa. The pull-in voltage given by the analytical analysis is 47.8 V, while the pull-in voltage measured experimentally was 48 V.

Acknowledgements. The authors acknowledge the support from the FAA through award no. DTFA03-01-C-00031 and the NSF through award no. CMS-0120866. We would like to express our appreciation to J. Newcomb and J. Larsen-Base for supporting this work. Work was also supported in part by the Nanoscale Science and Engineering Initiative of the NSF under NSF award no. EEC-0118025. Special thanks is due to E.G. Karpov, N. Pugno, and N. Moldovan for many useful discussions in relation to nanoscale modeling.

References

1. Rueckes T, Kim K, Joselevich E, Tseng GY, Cheung CL, Lieber CM (2000) Carbon nanotube-based nonvolatile random access memory for molecular computing, Science 289:94–97
2. Ekinci KL, Roukes ML (2005) Nanoelectromechanical systems. Review of Scientific Instruments 76:061101
3. Sazonova V, Yaish Y, Ustunel H, Roundy D, Arias TA, Mc Euen PL (2004) A tunable carbon nanotube electromechanical oscillator. Nature 431:284–287
4. Ilic B, Craighead HG, Krylov S, Senaratne W, Ober C, Neuzil P (2004) Attogram detection using nanoelectromechanical oscillators. Journal of Applied Physics 95:3694–3703
5. Davis ZJ, Abadal G, Kuhn O, Hansen O, Grey F, Boisen A (2000) Fabrication and characterization of nanoresonating devices for mass detection. Journal of Vacuum Science and Technology B 18:612–616
6. Roukes ML (1999) Yoctocalorimetry: phonon counting in nanostructures. Physica B 263:1–15
7. Roukes ML (2000) Nanoelectromechanical systems, presented at Technical Digest of the 2000 Solid-State Sensor and Actuator Workshop
8. Cleland AN, Roukes ML (1996) Fabrication of high frequency nanometer scale mechanical resonators from bulk Si crystals. Applied Physics Letters 69:2653–2655
9. Akita S, Nakayama Y, Mizooka S, Takano Y, Okawa T, Miyatake Y, Yamanaka S, Tsuji M, Nosaka T (2001) Nanotweezers consisting of carbon nanotubes operating in an atomic force microscope. Applied Physics Letters 79:1691–1693
10. Fennimore AM, Yuzvinsky TD, Han WQ, Fuhrer MS, Cumings J, Zettl A (2003) Rotational actuators based on carbon nanotubes. Nature 424:408–410
11. Ke CH, Espinosa HD (2004) Feedback controlled nanocantilever device. Applied Physics Letters 85:681–683
12. Kim P, Lieber CM (1999) Nanotube nanotweezers. Science 286:2148–2150
13. Kinaret JM, Nord T, Viefers S (2003) A carbon-nanotube-based nanorelay. Applied Physics Letters 82:1287–1289
14. Jang JE, Cha SN, Choi Y, Amaratunga GAJ, Kang DJ, Hasko DG, Jung JE, Kim JM (2005) Nanoelectromechanical switches with vertically aligned carbon nanotubes. Applied Physics Letters 87:163114
15. Iijima S (1991) Helical microtubules of graphitic carbon. Nature 354:56–58
16. Ajayan PM (1999) Nanotubes from carbon, Chemical Reviews 99:1787–1799
17. Ebbesen TW, Ajayan PM (1992) Large-scale synthesis of carbon nanotubes. Nature 358:220–222
18. Journet C, Maser WK, Bernier P, Loiseau A, delaChapelle ML, Lefrant S, Deniard P, Lee R, Fischer JE (1997) Large-scale production of single-walled carbon nanotubes by the electric-arc technique. Nature 388:756–758
19. Thess A, Lee R, Nikolaev P, Dai HJ, Petit P, Robert J, Xu CH, Lee YH, Kim SG, Rinzler AG, Colbert DT, Scuseria GE, Tomanek D, Fischer JE, Smalley RE (1996) Crystalline ropes of metallic carbon nanotubes. Science 273:483–487
20. Li WZ, Xie SS, Qian LX, Chang BH, Zou BS, Zhou WY, Zhao RA, Wang G (1996) Large-scale synthesis of aligned carbon nanotubes. Science 274:1701–1703
21. Qian D, Wagner GJ, Liu WK, Yu MF, Ruoff RS (2002) Mechanics of carbon nanotubes. Applied Mechanics Review 55:495–533
22. Mc Euen PL, Fuhrer MS, Park HK (2002) Single-walled carbon nanotube electronics. IEEE Transactions on Nanotechnology 1:78–85

23. Kuzumaki T, Mitsuda Y (2004) Dynamic measurement of electrical conductivity of carbon nanotubes during mechanical deformation by nanoprobe manipulation in transmission electron microscopy. Applied Physics Letters 85:1250–1252
24. Liu B, Jiang H, Johnson HT, Huang Y (2004) The influence of mechanical deformation on the electrical properties of single wall carbon nanotubes. Journal of the Mechanics and Physics of Solids 52:1–26
25. Tombler TW, Zhou CW, Alexseyev L, Kong J, Dai HJ, Lei L, Jayanthi CS, Tang MJ, Wu SY (2000) Reversible electromechanical characteristics of carbon nanotubes under local-probe manipulation. Nature 405:769–772
26. Husain A, Hone J, Postma HWC, Huang XMH, Drake T, Barbic M, Scherer A, Roukes ML (2003) Nanowire-based very-high-frequency electromechanical resonator Applied Physics Letters 83:1240–1242
27. Williams PA, Papadakis SJ, Patel AM, Falvo MR, Washburn S, Superfine R (2003) Fabrication of nanometer-scale mechanical devices incorporating individual multiwalled carbon nanotubes as torsional springs. Applied Physics Letters 82:805–807
28. Wong EW, Sheehan PE, Lieber CM (1997) Nanobeam mechanics: Elasticity, strength, and toughness of nanorods and nanotubes. Science 277:1971–1975
29. Falvo MR, Clary GJ, Taylor RM, Chi V, Brooks FP, Washburn S, Superfine R (1997) Bending and buckling of carbon nanotubes under large strain. Nature 389:582–584
30. Falvo MR, Taylor RM, Helser A, Chi V, Brooks FP, Washburn S, Superfine R (1999) Nanometre-scale rolling and sliding of carbon nanotubes. Nature 397:236–238
31. Ke CH, Pugno N, Peng B, Espinosa HD (2005) Experiments and modeling of carbon nanotube-based NEMS devices. Journal of the Mechanics and Physics of Solids 53:1314–1333
32. Taylor RMII, Superfine R (1999) Advanced Interfaces to Scanning Probe Microscopes, In: Nalwa HS (ed) Handbook of Nanostructured Materials and Nanotechnology 2. New York, Academic
33. Yu MF, Dyer MJ, Skidmore GD, Rohrs HW, Lu XK, Ausman KD, Von Ehr JR, Ruoff RS (1999) Three-dimensional manipulation of carbon nanotubes under a scanning electron microscope. Nanotechnology 10:244–252
34. Yu MF, Lourie O, Dyer MJ, Moloni K, Kelly TF, Ruoff RS (2000) Strength and breaking mechanism of multiwalled carbon nanotubes under tensile load. Science 287:637–640
35. Smith PA, Nordquist CD, Jackson TN, Mayer TS, Martin BR, Mbindyo J, Mallouk TE (2000) Electric-field assisted assembly and alignment of metallic nanowires. Applied Physics Letters 77:1399–1401
36. Chen XQ, Saito T, Yamada H, Matsushige K (2001) Aligning single-wall carbon nanotubes with an alternating-current electric field. Applied Physics Letters 78:3714–3716
37. Chung J, Lee J (2003) Nanoscale gap fabrication and integration of carbon nanotubes by micromachining. Sensors and Actuators A 104:229–235
38. Yamamoto K, Akita S, Nakayama Y (1998) Orientation and purification of carbon nanotubes using ac electrophoresis. Journal of Physics D: Applied Physics 31:L34–L36
39. Hughes MP, Morgan H (1998) Dielectrophoretic trapping of single sub-micrometre scale bioparticles. Journal of Physics D: Applied Physics 31:2205–2210
40. Ramos A, Morgan H, Green NG, Castellanos A (1998) Ac electrokinetics: a review of forces in microelectrode structures. Journal of Physics D: Applied Physics 31:2338–2353
41. Krupke R, Hennrich F, von Lohneysen H, Kappes MM (2003) Separation of metallic from semiconducting single-walled carbon nanotubes. Science 301:344–347
42. Lee SW, Lee DS, Morjan RE, Jhang SH, Sveningsson M, Nerushev OA, Park YW, Campbell EEB (2004) A three-terminal carbon nanorelay. Nano Letters 4:2027–2030
43. Huang Y, Duan XF, Wei QQ, Lieber CM (2001) Directed assembly of one-dimensional nanostructures into functional networks. Science 291:630–633

44. Fujiwara M, Oki E, Hamada M, Tanimoto Y, Mukouda I, Shimomura Y (2001) Magnetic orientation and magnetic properties of a single carbon nanotube. Journal of Physical Chemistry A 105:4383–4386

45. Huang SM, Dai LM, Mau AWH (1999) Patterned growth and contact transfer of well-aligned carbon nanotube films. Journal of Physical Chemistry B 103:4223–4227

46. Dai HJ (2000) Physics World 13:43

47. Kong J, Soh HT, Cassell AM, Quate CF, Dai HJ (1998) Synthesis of individual single-walled carbon nanotubes on patterned silicon wafers Nature 395:878–881

48. Zhang YG, Chang AL, Cao J, Wang Q, Kim W, Li YM, Morris N, Yenilmez E, Kong J, Dai HJ (2001) Electric-field-directed growth of aligned single-walled carbon nanotubes. Applied Physics Letters 79:3155–3157

49. Dai HJ (2002) Accounts of Chemical Research 35:1035

50. Nojeh A, Ural A, Pease RF, Dai HJ (2004) Electric-field-directed growth of carbon nanotubes in two dimensions. Journal of Vacuum Science and Technology B 22:3421–3425

51. Rao SG, Huang L, Setyawan W, Hong SH (2003) Large-scale assembly of carbon nanotubes. Nature 425:36–37

52. Piner RD, Zhu J, Xu F, Hong SH, Mirkin CA (1999) Dip-pen nanolithography. Science 283:661–663

53. Roukes ML (2001) Nanoelectromechanical systems face the future, Physics World 14:25–31

54. Ekinci KL (2005) Electromechanical transducers at the nanoscale: Actuation and sensing of motion in nanoelectromechanical systems (NEMS). Small 1:786–797

55. Badzey RL, Zolfagharkhani G, Gaidarzhy A, Mohanty P (2004) A controllable nanomechanical memory element. Applied Physics Letters 85:3587–3589

56. Craighead HG (2000) Nanoelectromechanical systems. Science 290:1532–1535

57. Zhu Y, Moldovan N, Espinosa HD (2005) A microelectromechanical load sensor for in situ electron and X-ray microscopy tensile testing of nanostructures. Applied Physics Letters 86:013506

58. Zalalutdinov M, Ilic B, Czaplewski D, Zehnder A, Craighead HG, Parpia JM (2000) Frequency-tunable micromechanical oscillator. Applied Physics Letters 77:3287–3289

59. Huang XMH, Zorman CA, Mehregany M, Roukes ML (2003) Nanodevice motion at microwave frequencies. Nature 421:496–496

60. Poncharal P, Wang ZL, Ugarte D, De Heer WA (1999) Electrostatic deflections and electromechanical resonances of carbon nanotubes. Science 283:1513–1516

61. Treacy MMJ, Ebbesen TW, Gibson JM (1996) Exceptionally high Young's modulus observed for individual carbon nanotubes. Nature 381:678–680

62. Greywall DS, Yurke B, Busch PA, Pargellis AN, Willett RL (1994) Evading amplifier noise in nonlinear oscillators. Physical Review Letters 72:2992–2995

63. Cleland AN, Roukes ML (1999) External control of dissipation in a nanometer-scale radiofrequency mechanical resonator. Sensors and Actuators A 72:256–261

64. Ekinci KL, Yang YT, Roukes ML (2004) Ultimate limits to inertial mass sensing based upon nanoelectromechanical systems. Journal of Applied Physics 95:2682–2689

65. Mohanty P, Harrington DA, Ekinci KL, Yang YT, Murphy MJ, Roukes ML (2002) Intrinsic dissipation in high-frequency micromechanical resonators. Physical Review B 66:085416

66. Dresselhaus MS, Dresselaus G, Avouris P (2001) Carbon Nanotubes. Berlin, Heidelberg, New York, Springer

67. Carr DW, Evoy S, Sekaric L, Craighead HG, Parpia JM (2000) Parametric amplification in a torsional microresonator. Applied Physics Letters 77:1545–1547

68. Carr DW, Sekaric L, Craighead HG (1998) Measurement of nanomechanical resonant structures in single-crystal silicon. Journal of Vacuum Science and Technology B 16:3821–3824

69. Meyer C, Lorenz H, Karrai K (2003) Optical detection of quasi-static actuation of nano-electromechanical systems. Applied Physics Letters 83:2420–2422

70. Keeler BEN, Carr DW, Sullivan JP, Friedmann TA, Wendt JR (2004) Experimental demonstration of a laterally deformable optical nanoelectromechanical system grating transducer. Optics Letters 29:1182–1184

71. Kouh T, Karabacak D, Kim DH, Ekinci KL (2005) Diffraction effects in optical interferometric displacement detection in nanoelectromechanical systems. Applied Physics Letters 86:013106

72. Bocko MF, Stephenson KA, Koch RH (1988) Vacuum tunneling probe – a nonreciprocal, reduced-back-action transducer. Physical Review Letters 61:726–729

73. Presilla C, Onofrio R, Bocko MF (1992) Uncertainty-principle noise in vacuum-tunneling transducers. Physical Review B 45:3735–3743

74. Nunes G, Freeman MR (1993) Picosecond resolution in scanning-tunneling-microscopy. Science 262:1029–1032

75. Kanda Y (1982) A graphical representation of the piezoresistive coefficients in silicon. IEEE Transactions on Electron Devices 29:64–70

76. Hjort K, Soderkvist J, Schweitz JA (1994) Gallium-Arsenide as a Mechanical Material. Journal of Micromechanics and Microengineering 4:1–13

77. Bargatin I, Myers EB, Arlett J, Gudlewski B, Roukes ML (2005) Sensitive detection of nanomechanical motion using piezoresistive signal downmixing. Applied Physics Letters 86:133109

78. Dai HJ, Hafner JH, Rinzler AG, Colbert DT, Smalley RE (1996) Nanotubes as nanoprobes in scanning probe microscopy. Nature 384:147–150

79. Wong SS, Joselevich E, Woolley AT, Cheung CL, Lieber CM (1998) Covalently functionalized nanotubes as nanometre-sized probes in chemistry and biology. Nature 394:52–55

80. Jonsson LM, Axelsson S, Nord T, Viefers S, Kinaret JM (2004) High frequency properties of a CNT-based nanorelay. Nanotechnology 15:1497–1502

81. Ke CH, Espinosa HD (2006) In-situ electron microscopy electro-mechanical characterization of a NEMS bistable device, Small, (in press)

82. Ke CH (2006) Development of a Feedback Controlled Carbon Nanotube-Based Nanoelectromechanical Device. PhD dissertation, Northwestern University

83. Zapol P, Sternberg M, Curtiss LA, Frauenheim T, Gruen DM (2002) Tight-binding molecular-dynamics simulation of impurities in ultrananocrystalline diamond grain boundaries. Physical Review B 65:045403

84. Sternberg M, Curtiss LA, Gruen DM, Kedziora G, Horner DA, Redfern PC, Zapol P (2006) Carbon ad-dimer defects in carbon nanotubes. Physical Review Letters 96:075506

85. Tans SJ, Verschueren ARM, Dekker C (1998) Room-temperature transistor based on a single carbon nanotube. Nature 393:49–52

86. Zhou CW, Kong J, Dai HJ (2000) Intrinsic electrical properties of individual single-walled carbon nanotubes with small band gaps. Physical Review Letters 84:5604–5607

87. Minot ED, Yaish Y, Sazonova V, Mc Euen PL (2004) Determination of electron orbital magnetic moments in carbon nanotubes. Nature 428:536–539

88. Hu JT, Min OY, Yang PD, Lieber CM (1999) Controlled growth and electrical properties of heterojunctions of carbon nanotubes and silicon nanowires. Nature 399:48–51

89. Hu JT, Odom TW, Lieber CM (1999) Chemistry and physics in one dimension: Synthesis and properties of nanowires and nanotubes. Accounts of Chemical Research 32:435–445

90. Morales AM, Lieber CM (1998) A laser ablation method for the synthesis of crystalline semiconductor nanowires. Science 279:208–211

91. Yu DP, Lee CS, Bello I, Sun XS, Tang YH, Zhou GW, Bai ZG, Zhang Z, Feng SQ (1998) Synthesis of nano-scale silicon wires by excimer laser ablation at high temperature. Solid State Communications 105:403–407

92. Ji CX, Searson PC (2002) Fabrication of nanoporous gold nanowires. Applied Physics Letters 81:4437–4439

93. Wong TC, Li CP, Zhang RQ, Lee ST (2004) Gold nanowires from silicon nanowire templates. Applied Physics Letters 84:407–409

94. Bhattacharyya S, Saha SK, Chakravorty D (2000) Silver nanowires grown in the pores of a silica gel. Applied Physics Letters 77:3770–3772

95. Barbic M, Mock JJ, Smith DR, Schultz S (2002) Single crystal silver nanowires prepared by the metal amplification method. Journal of Applied Physics 91:9341–9345

96. Malandrino G, Finocchiaro ST, Fragala IL (2004) Silver nanowires by a sonoself-reduction template process. Journal of Materials Chemistry 14:2726–2728

97. Heath JR, Legoues FK (1993) A liquid solution synthesis of single-crystal germanium quantum wires. Chemical Physics Letters 208:263–268

98. Greytak AB, Lauhon LJ, Gudiksen MS, Lieber CM (2004) Growth and transport properties of complementary germanium nanowire field-effect transistors. Applied Physics Letters 84:4176–4178

99. Ziegler KJ, Lyons DM, Holmes JD, Erts D, Polyakov B, Olin H, Svensson K, Olsson E (2004) Bistable nanoelectromechanical devices. Applied Physics Letters 84:4074–4076

100. Wu YY, Yang PD (2000) Germanium nanowire growth via simple vapor transport. Chemistry of Materials 12:605–607

101. Banerjee D, Lao JY, Wang DZ, Huang JY, Ren ZF, Steeves D, Kimball B, Sennett M (2003) Large-quantity free-standing ZnO nanowires. Applied Physics Letters 83:2061–2063

102. Dai Y, Zhang Y, Bai YQ, Wang ZL (2003) Bicrystalline zinc oxide nanowires. Chemical Physics Letters 375:96–101

103. Mock JJ, Oldenburg SJ, Smith DR, Schultz DA, Schultz S (2002) Composite plasmon resonant nanowires. Nano Letters 2:465–469

104. Kim KH, Moldovan N, Espinosa HD (2005) A nanofountain probe with sub-100 nm molecular writing resolution. Small 1:632–635

105. Lifshitz R, Roukes ML (2000) Thermoelastic damping in micro- and nanomechanical systems. Physical Review B 61:5600–5609

106. Huang XMH, Prakash MK, Zorman CA, Mehregany M, Roukes ML (2003) Free-free beam silicon carbide nanomechanical resonators, presented at TRANSDUCERS '03 Proceedings of the 12th International Conference on Solid State Sensors. Actuators and Microsystems, Boston

107. La Haye MD, Buu O, Camarota B, Schwab KC (2004) Approaching the quantum limit of a nanomechanical resonator. Science 304:74–77

108. Rudd RE, Broughton JQ (1999) Journal of Modeling and Simulation of Microsystems 1:29

109. Dequesnes M, Rotkin SV, Aluru NR (2002) Calculation of pull-in voltages for carbon-nanotube-based nanoelectromechanical switches. Nanotechnology 13:120–131

110. Liu WK, Karpov EG, Zhang S, Park HS (2004) An introduction to computational nanomechanics and materials. Computer Methods in Applied Mechanics and Engineering 193:1529–1578

111. Karpov EG, Wagner GJ, Liu WK (2005) A Green's function approach to deriving non-reflecting boundary conditions in molecular dynamics simulations. International Journal for Numerical Methods in Engineering 62:1250–1262

112. Wagner GJ, Karpov EG, Liu WK (2004) Molecular dynamics boundary conditions for regular crystal lattices. Computer Methods in Applied Mechanics and Engineering 193:1579–1601

113. Karpov EG, Stephen NG, Dorofeev DL (2002) On static analysis of finite repetitive structures by discrete Fourier transform. International Journal of Solids and Structures 39:4291–4310

114. Karpov EG, Stephen NG, Liu WK (2003) Initial tension in randomly disordered periodic lattices. International Journal of Solids and Structures 40:5371–5388

115. Moriarty JA, Belak JF, Rudd RE, Soderlind P, Streitz FH, Yang LH (2002) Quantum-based atomistic simulation of materials properties in transition metals. Journal of Physics: Condensed Matter 14:2825–2857

116. Belytschko T, Liu WK, Moran B (2000) Nonlinear Finite Elements for Continua and Structures. New York, Wiley

117. Curtin WA, Miller RE (2003) Atomistic/continuum coupling in computational materials science. Modelling and Simulation in Materials Science and Engineering 11:R33–R68

118. Vvedensky DD (2004) Multiscale modeling of nanostructures. Journal of Physics: Condensed Matter 16:1537–1576

119. Miller RE, Tadmor EB (2002) The quasicontinuum method: Overview, applications and current directions. Journal of Computer-Aided Materials Design 9:203–239

120. Abraham FF, Broughton JQ, Bernstein N, Kaxiras E (1998) Spanning the continuum to quantum length scales in a dynamic simulation of brittle fracture. Europhysics Letters 44:783–787

121. Broughton JQ, Abraham FF, Bernstein N, Kaxiras E (1999) Concurrent coupling of length scales: Methodology and application. Physical Review B 60:2391–2403

122. Rudd RE, Broughton JQ (2000) Concurrent coupling of length scales in solid state systems. Physica Status Solidi B 217:251–291

123. Tadmor EB, Ortiz M, Phillips R (1996) Quasicontinuum analysis of defects in solids. Philosophical Magazine A 73:1529–1563

124. Tadmor EB, Phillips R, Ortiz M (1996) Mixed atomistic and continuum models of deformation in solids. Langmuir 12:4529–4534

125. Rodney D, Phillips R (1999) Structure and strength of dislocation junctions: An atomic level analysis, Physical Review Letters 82:1704–1707

126. Shin CS, Fivel MC, Rodney D, Phillips R, Shenoy VB, Dupuy L (2001) Formation and strength of dislocation junctions in FCC metals: A study by dislocation dynamics and atomistic simulations. Journal De Physique IV 11:19–26

127. Shenoy VB, Miller R, Tadmor EB, Phillips R, Ortiz M (1998) Quasicontinuum models of interfacial structure and deformation. Physical Review Letters 80:742–745

128. Shenoy VB, Miller R, Tadmor EB, Rodney D, Phillips R, Ortiz M (1999) An adaptive finite element approach to atomic-scale mechanics – the quasicontinuum method. Journal of the Mechanics and Physics of Solids 47:611–642

129. Tadmor EB, Miller R, Phillips R, Ortiz M (1999) Nanoindentation and incipient plasticity. Journal of Materials Research 14:2233–2250

130. Miller R, Ortiz M, Phillips R, Shenoy V, Tadmor EB (1998) Quasicontinuum models of fracture and plasticity. Engineering Fracture Mechanics 61:427–444

131. Miller R, Tadmor EB, Phillips R, Ortiz M (1998) Quasicontinuum simulation of fracture at the atomic scale. Modelling and Simulation in Materials Science and Engineering 6:607–638

132. Park HS, Karpov EG, Liu WK, Klein PA (2005) The bridging scale for two-dimensional atomistic/continuum coupling. Philosophical Magazine 85:79–113

133. Wagner GJ, Liu WK (2003) Coupling of atomistic and continuum simulations using a bridging scale decomposition. Journal of Computational Physics 190:249–274

134. Qian D, Wagner GJ, Liu WK (2004) A multiscale projection method for the analysis of carbon nanotubes, Computer Methods in Applied Mechanics and Engineering 193:1603–1632

135. Belytschko T, Xiao SP (2003) Coupling methods for continuum model with molecular model. International Journal for Multiscale Computational Engineering 1:115–126

136. Ke CH, Espinosa HD (2005) Numerical analysis of nanotube-based NEMS devices – Part I: Electrostatic charge distribution on multiwalled nanotubes. Journal of Applied Mechanics–Transactions of the ASME 72:721–725

137. Ke CH, Espinosa HD, Pugno N (2005) Numerical analysis of nanotube based NEMS devices – Part II: Role of finite kinematics, stretching and charge concentrations. Journal of Applied Mechanics – Transactions of the ASME 72:726–731

138. Pugno N, Ke CH, Espinosa HD (2005) Analysis of doubly clamped nanotube devices in the finite deformation regime. Journal of Applied Mechanics – Transactions of the ASME 72:445–449

139. Dequesnes M, Tang Z, Aluru NR (2004) Static and dynamic analysis of carbon nanotube-based switches. Journal of Engineering Materials and Technology – Transactions of the ASME 126:230–237

140. Lennard-Jones JE (1930) Perturbation problems in quantum mechanics. Proceeding of the Royal Society of London Series A 129:598–615

141. Girifalco LA, Hodak M, Lee RS (2000) Carbon nanotubes, buckyballs, ropes, and a universal graphitic potential. Physical Review B 62:13104–13110

142. Girifalco LA (1992) Molecular-Properties of C-60 in the Gas and Solid-Phases Journal of Physical Chemistry 96:858–861

143. Hayt WABJ (2001) Engineering Electromagnetics, 6th edn. New York, McGraw-Hill

144. Lou L, Nordlander P, Smalley RE (1995) Fullerene nanotubes in electric-fields. Physical Review B 52:1429–1432

145. Krcmar M, Saslow WM, Zangwill A (2003) Electrostatics of conducting nanocylinders. Journal of Applied Physics 93:3495–3500

146. Rotkin SV, Bulashevich KA, Aluru NR (2002) Atomistic capacitance of a nanotube electromechanical device. International Journal of Nanoscience 1:337–346

147. Bulashevich KA, Rotkin SV (2002) Nanotube devices: A microscopic model. JEPT Letters 75:205–209

148. Keblinski P, Nayak SK, Zapol P, Ajayan PM (2002) Charge distribution and stability of charged carbon nanotubes. Physical Review Letters 89:255503

149. Smythe WR (1956) Charged Right Circular Cylinder. Journal of Applied Physics 27:917–920

150. Yakobson BI, Campbell MP, Brabec CJ, Bernholc J (1997) High strain rate fracture and C-chain unraveling in carbon nanotubes. Computational Materials Science 8:341–348

151. Liu WK, Karpov EG, Park HS (2006) Nano Mechanics and Materials: Theory, Multiscale Methods and Applications. New York, Wiley

25 Application of Atom-resolved Scanning Tunneling Microscopy in Catalysis Research

Jeppe Vang Lauritsen · Ronny T. Vang · Flemming Besenbacher

25.1
Introduction

Heterogeneous catalysis is of immense importance to society. Most world-wide chemical production is performed via catalysis, and it is estimated that the product value enters through 15–20% of the GNP in industrialized countries. Catalysis also provides the extremely important platform for new vital environmental protection technologies for the sustainable energy production in for example fuel cells, and for the production of hydrogen as an alternative to fossil fuel. Many of the current challenges related to the protection of our environment and the sustainable energy production rely on the development of new and better catalysts. In spite of the great importance of catalysis in society and the fact that its application has expanded dramatically throughout the preceding century, the detailed microscopic principles governing catalyzed chemical reactions are in general still far from being understood and established. This is in part a consequence of the immense complexity of catalyst structures and their dependence on external parameters (temperature, pressure, etc.).

The activity of a solid catalyst scales with the number of exposed active sites on the surface, and industrially the activity is optimized by dispersing the active material as nanometer-sized particles onto highly porous supports with surface areas often in excess of $500 \, \text{m}^2/\text{g}$. The catalytically active sites are the molecular docking sites, which temporarily bind reactants during reaction, thereby offering an alternative and energetically favorable reaction pathway. Fundamental understanding of catalytic processes requires the ability to characterize the structure and location of catalytically active sites at the single-atom level. However, the structural complexity of the materials, combined with the high temperatures and pressures of catalysis, often limit the possibilities for detailed structural characterization of real catalysts. Furthermore, when the dimensions of the catalytic material become sufficiently small, the properties become size dependent, and it is often insufficient to model a catalytically active material from its macroscopic properties. The level of insight needed to understand catalysis requires characterization of the catalyst and individual reactants with the highest spatial and temporal resolution. The ultimate goal of catalysis research is to achieve full understanding and control of the constituents at the molecular and atomic levels. With this ability, it may be possible to construct tailor-made, high-performance catalysts for even highly specialized reactions, with obvious benefits for the chemical industry, the environment, and consumers of the countless products made in catalytic processes.

One approach to a detailed atomic-scale understanding of catalysis is to investigate well-characterized and simpler model systems under well-controlled conditions [1–8]. Within this *surface-science approach*, the investigation of a catalyst is broken down into simplified components that can be dealt with in detail. The active catalyst particles have typically been represented by idealized flat single-crystal surfaces kept under well-controlled ultrahigh vacuum (UHV) conditions at pressures of 10^{-10} mbar or less. This approach allows for a detailed characterization of the surface structures by applying a wide range of surface science tools. Such fundamental and idealized investigations have provided valuable atomic-scale insight for a number of catalytic reactions.

However, in such studies one has to take into account the large phenomenological gaps that separate the idealized surfaces under vacuum and complex catalysts operating at high temperatures and pressures referred to as a pressure gap and a materials gap, respectively [9]. The materials gap refers to the discrepancy between catalytically active clusters on supports and the corresponding model systems. A description based on investigations of single-crystal surfaces has often been shown to work very well, but only if the nanoclusters in the real catalyst expose the same facet as in the model, and only if this facet is the catalytically active part of the cluster. In some cases (as in the hydrodesulfurization catalysts referred to below), the properties of the catalytically active material in the form of nanoclusters are quite different from those of the bulk material, and geometric or electronic features in the microscopic clusters control the reactivity on the macroscale [10, 11]. A fully functional oxidation catalyst consisting of TiO_2-supported gold nanoclusters is an excellent example of how the properties of the materials may change as the dimensions are reduced, inasmuch as bulk gold is known to be noble and notoriously inert [12]. To bridge the materials gap, efforts have recently been focused on the investigation of new and more realistic model systems consisting of particles on flat supports [11, 13–19].

A number of techniques have been developed providing detailed atomic-scale structural and chemical insight into both real and model catalysts both in vacuum or in realistic environments [3, 20]. These traditional scattering and diffraction techniques have provided much valuable new insight, but they are in general averaging techniques, and information about individual particles and defect sites is a priori not obtained. Consequently, these techniques have somewhat limited applicability when analyzing complex model systems containing, for example, nanoclusters. This limitation makes the attempts to characterize and understand the active structures in catalysts a great challenge to researchers. Currently, only transmission electron microscopy (TEM) provides the means to resolve individual nanoclusters in real catalysts, even with atomic resolution and with samples in reactive gaseous atmospheres [21–23]. It is, however, generally impossible with these techniques to image adsorbates and single defects, which often turn out to be very important active sites in catalysis. Here we focus on an alternative approach involving investigations of model catalysts with the scanning tunneling microscope (STM). The STM is the technique of choice for exploring the direct space atomic structure of matter. In the preceding two decades following the development of the STM microscope [24–26], the technique has provided much information about the structures and electronic properties of surfaces, and since it also resolves individual signatures of adsorbed species, it has become a very valuable tool for monitoring chemical reactions on surfaces [27–30]. The ability of the STM to achieve atom-resolved real-space images of localized regions of a surface and to directly resolve the local atomic-scale structure has provided essential insight into the active sites on catalysts and emphasized the importance of edges, kinks, atom vacancies, and other defects, which often are difficult to detect with other techniques [31–34]. It is evident, however, that STM cannot be used to image real catalysts supported on high surface-area, porous oxide carriers, because STM requires flat and conductive samples. Nonetheless, it is fair to say that STM has provided unprecedented new insight into surface chemistry in general, and STM studies of catalyst model systems can, as shown below, in some cases even trigger new ideas that have led to the development of new catalysts.

25.2
Scanning Tunneling Microscopy

Although the fundamental principle of STM is conceptually quite simple, it is well known that the interpretation of atomic-scale structures in STM images is not straightforward. Especially when it comes to identification of adsorbates, a careful analysis of STM images is required, and before we proceed a small comment on this issue is warranted.

Since STM is dealing with tunneling between the tip and the surface by electrons located energetically near the Fermi level, the atomically resolved STM images are associated with the local density of electronic states at the Fermi level, rather than the total electron charge density. Indeed, the theoretical description due to Tersoff and Hamann [35, 36] shows that constant-current STM images represent contours of constant local density of states (LDOS) at the Fermi level projected to the tip apex. Consequently, STM images may often reflect a rather complicated convolution of electronic and geometric features of the surface. Features of the LDOS of a metal surface measured at some distance from the surface generally coincide with the total electron charge density, and the contours in STM images of metal are therefore mostly interpreted as simple topographic maps of the surface. When adsorbates are present on the surface, they alter the LDOS in the surrounding area and are in general found to be imaged either as protrusions or, counter-intuitively, as depressions with respect to the bare surface. The contrast is decided by the way the adsorbates change the LDOS at the Fermi level; that is, if the LDOS is enhanced (or depleted), the tip moves away from (or closer to) the surface when the affected area is scanned [37]. Consequently, one cannot in general assume that maxima expected from the surface topography coincide with maxima observed in STM images. This effect becomes especially evident on oxide- and sulfide-semiconductor surfaces, e.g., MoS_2 or TiO_2 which are both used extensively in catalysis either as catalysts or as support materials for catalysts. In contrast to metals, semiconductors show a very strong variation of the LDOS with bias voltage. In particular, the LDOS changes discontinuously at the band edges. Adding to the complexity of the contrast in STM images, the electronic structure may also be dominated by spatially localized states or dangling bonds, which generally change the picture relative to what would be expected on geometric grounds. On the other hand, the electronic surface states can be determined by opening the feedback loop and measuring atom-resolved I–V curves, generally referred to as scanning tunneling spectroscopy (STS) [38]. Such STS data provide additional, valuable information about the surface electronic structure, i.e. STM is sensitive to low-dimensional electron localization in the surface.

STM images provide no direct insight into the chemical identities of the observed structures. This lack of chemical specificity makes it difficult a priori to relate the observed structures of complex clusters, molecular adsorbates, or reaction intermediates to their chemical nature and conformation on the surface. However, a common and successful approach is to combine the STM experiment with theoretical electronic-structure calculations (see, e.g., [39]) to assist in the interpretation of STM results [40–42]. The theoretical calculations provide complementary information about the possible ground-state configurations of surface structures and it is indeed possible from such theoretically calculated surface structures to calculate theoretical simulated STM images.

25.3
STM Studies of a Hydrotreating Model Catalyst

Hydrotreating processes are catalytic treatments of compounds in petroleum, with the primary goal of removing the heteroatoms; sulfur, nitrogen, and oxygen. Sulfur and nitrogen are removed in so-called hydrodesulfurization (HDS) and hydrodenitrogenation (HDN) reactions with gaseous hydrogen during oil refining to reduce the environmentally harmful SO_2 and NO_X, resulting from combustion of the fuel. Furthermore, sulfur and nitrogen are removed because compounds of these elements generally inhibit the performance of transition-metal containing catalysts like, e.g., Pt, Ni used in downstream treatment of the petroleum fractions. In the past, most catalysts have been discovered and optimized using the so-called trial-and-error approach, but the area of hydrotreating catalysis is an excellent example of the improvement of a technological catalyst through application of new and better catalyst characterization tools [43]. However, the traditional catalyst characterization techniques [1,3], which have made these advances possible, seem to have reached a limit in the details they can provide.

Today, the more severe legislation is currently forcing refiners to further upgrade and purify petroleum fuels, thus requiring further improved catalysts [44–46]. To meet these requirements, there is considerable interest to address a number of fundamental questions concerning the operation of hydrotreating catalysts on the atomic scale which has remained unanswered. The challenge is great because the hydrotreating catalyst is very complex in composition and structure. Previously, the MoS_2-based hydrodesulfurization (HDS) catalyst has been the subject of extensive studies using a large variety of different tools, and several extensive reviews on the subject exist [44,47–50]. A considerable effort has been aimed at relating fundamental characteristics such as catalyst activity and selectivity to microscopic properties, e.g. catalyst composition, electronic structure and geometric structure. In particular, in situ EXAFS studies and Mössbauer spectroscopy studies have provided information on the structure, and today the "CoMoS" model [51–54] is widely accepted to describe the active structures in catalysts. In these studies, it is concluded that the active phase is present as supported single-layer MoS_2-like nanoclusters with a size of 10–20 Å under operating conditions, and that the promoting effect of Co in the CoMoS structures is caused by Co promoter atoms replacing Mo at positions near the edges of the MoS_2 clusters [51–54]. Despite the large number of studies of HDS catalysts, some very fundamental questions remain, however, unanswered. Much of the controversy is directly related to the fact that the traditional spectroscopy-based techniques have not been able to unequivocally map the real-space atomic structure of the few nanometer wide CoMoS nanoclusters. It was discovered early that only the edges of the S–Mo–S layers in MoS_2 are catalytically active [55], but in order to pinpoint and understand more precisely the active sites, it is necessary to obtain more detailed information on the atomic-scale structure of MoS_2 nanoclusters. For instance, what is the preferential morphology of the MoS_2 nanoclusters, which are the active edge structures, and what is the role of the Co promoters in CoMoS?

The ability of the STM to achieve atom-resolved real-space images of the nanoclusters has made it possible, for the first time, to directly address the MoS_2 morphology and the structure of the catalytically important edges experimentally. By synthesizing ensembles of single-layer MoS_2 or CoMoS nanoclusters on a flat Au(111) single-crystal substrate, it is shown in this section how it is possible to form a realistic model system of the catalyst and study the atomic-scale structure of the catalytically relevant phases.

Normally, oxide carriers are used in the industrial catalysts, but these are generally nonconducting and therefore unsuitable as supports for STM studies. We therefore use gold supports since the close-packed Au(111) is rather inert, and it is therefore possible to study the intrinsic properties of MoS_2, i.e. the MoS_2 nanoclusters with a weak support interaction. Furthermore, the Au(111) reconstructs in the characteristic herringbone pattern, which is used as a template for the synthesis of a highly dispersed ensemble of nanoclusters [56]. MoS_2 nanoclusters are synthesized on gold, by deposition of pure Mo from an e-beam evaporator onto the gold substrate kept in a background of H_2S gas, followed by a postannealing step at 673 K while maintaining the sulfiding environment. The detailed preparation method for the synthesis of MoS_2 and CoMoS nanoclusters supported on Au(111) is reported elsewhere in [57, 58]. As illustrated in Fig. 25.1, in the case of MoS_2, the clusters synthesized by this method are characterized by a high degree of dispersion and a fairly narrow size distribution. The average size of the clusters is approximately 500 $Å^2$. This corresponds to a side length of $\sim$ 30 Å, which matches well the spatial extension of the active particles in typical HDS catalysts. A detailed analysis of STM images reveals that the morphology of the nanoclusters is also remarkably uniform with respect to shape, and as shown in Fig. 25.1, we find that nearly all of the MoS_2 nanoclusters are triangular in shape under the conditions of the experiment. As a model system for the HDS catalyst, the synthesized MoS_2 clusters therefore form a well-characterized reference for experiments elucidating details on the atomic structure of the catalytically active MoS_2 edge structures and their reactivity with adsorbed molecules.

An atomically resolved STM image of a triangular MoS_2 nanocluster consisting of a single S–Mo–S layer is depicted in Fig. 25.2. The nanocluster is seen to exhibit a hexagonally close-packed plane of protrusions arranged with the 3.15 Å interatomic distance of MoS_2. The height of the cluster is around 2.1 Å, and the cluster is thus concluded to reflect a single S–Mo–S layer oriented with the (0001) basal plane

Fig. 25.1. STM image of MoS$_2$ nanoclusters synthesized on Au(111) surface at 673 K in a sulfiding atmosphere. The size is 744 Å × 721 Å. Reproduced from [57]

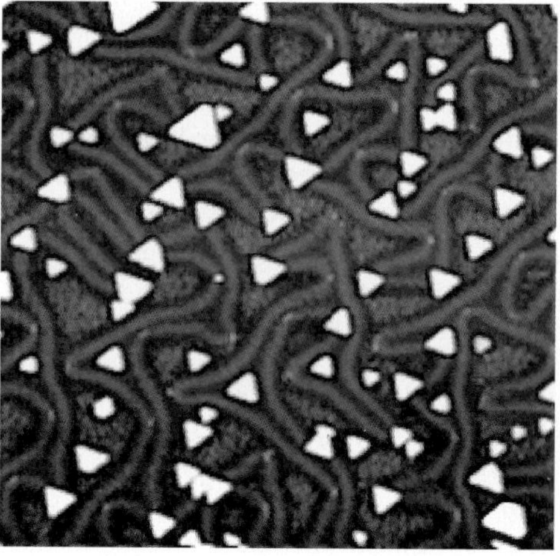

in parallel to the Au substrate and with the protrusions reflecting the hexagonally arranged S atoms in the topmost layer. At the edges the protrusions are, however, clearly seen to be imaged out of registry with the basal plane S atoms. In fact, the protrusions are shifted exactly half a lattice constant along the edge, but retain their interatomic distance of 3.15 Å.

The shifted registry along the edge might indicate that the catalytically active edges are severely reconstructed compared to a model based on bulk properties of MoS$_2$, but this is not the case. As discussed briefly in the introduction, it is important to stress that STM images obtained in the constant-current mode reflect contours of constant local density of states (LDOS) at the Fermi level measured at the position of the STM tip [35]. This implies that the contrast in STM images, in general, reflects a rather complicated convolution of geometric and electronic features of the surface/nanoclusters. This is particularly the case for materials that exhibit a bandgap, e.g. some metal oxides or sulfides. MoS$_2$ is a semiconductor with a bandgap $\sim$ 1.2 eV in the bulk form, and for the MoS$_2$ nanoclusters in the present study we indeed find that a purely geometrical model is not adequate to account for the STM images. In addition to the observed shifted registry of protrusion at the edge, a pronounced bright brim of high electron state density is also observed to extend all the way around the cluster edge adjacent to the edge protrusions in Fig. 25.2. This bright brim is attributed to an electronic effect probed by the STM, reflecting the existence of localized electron states at the cluster perimeter, so-called edge states. From an interplay with density functional theory (DFT) calculations [59], it was possible to unambiguously identify these signatures in STM simulations. It was shown that the MoS$_2$ nanoclusters exclusively expose one specific type of edge, the ($10\bar{1}0$ Mo) edges, and it was concluded that these are fully saturated with sulfur under the conditions of the experiment. Full coordination was achieved by placing two sulfurs per Mo atom on the edge in a dimer configuration (Fig. 25.2b). The DFT calculations also directly show that the electronic structure near the Mo edges of the triangular MoS$_2$ nanoclusters is indeed significantly perturbed relative to the bulk, and they reveal that the edges are in fact metallic due to the existence of two distinct

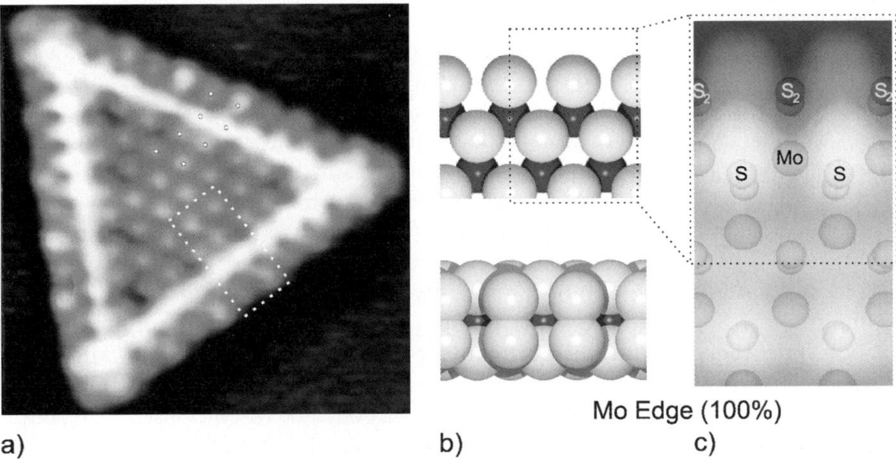

a)   b)   c)

Fig. 25.2. (**a**) Atomically resolved STM image ($V_t = 5.2\,\mathrm{mV}$, $I_t = 1.28\,\mathrm{nA}$) of triangular, single-layer MoS_2 nanoclusters on Au(111). The size of the image is 41 Å × 42 Å. Adapted from [57]. (**b**) *Side view* and *front view* of the Mo edge fully saturated with S dimers. (**c**) Simulated STM image of a two-atom-wide section of the Mo edge with S dimers. The simulation is based on DFT calculations including the effect of the Au substrate (from [59]). The positions of the individual atoms in the simulated image are represented by shadowed atom balls (small: S, big: Mo). For clarity, the Au atoms of the support are omitted

electronic edge states [59, 60]. An STM simulation based on the Tersoff–Hamann model [35] (i.e. a contour map of the constant surface LDOS) shows that both the bright brim and the apparent shifted registry of the edge protrusions (Fig. 25.2c) can be traced back to the existence of the two edge states on the fully sulfided Mo edges, and the theory thus fully reproduces the experimentally observed features. Under the sulfiding conditions in the experiment, it is therefore concluded that triangular single-layer MoS_2 nanoclusters are terminated with the Mo edge fully covered with S dimers, and that the electronic structure of these edges is dominated by one-dimensional electronic "brim" states, which give the edges a metallic character.

From a coordination chemistry point-of-view, fully sulfur-saturated edges, like the one in Fig. 25.2b, are not normally considered particularly reactive, but due to the metallic character of the edges we find evidence of a rather different chemistry from what has ordinarily been assumed. We investigate this by selectively adsorbing a sulfur-containing molecule, thiophene (C_4H_4S), under ultrahigh vacuum conditions on the MoS_2 clusters at different temperatures. In this way we can reveal the adsorption sites and map out the interaction strength. Specifically, we find that thiophene adsorbs nondissociatively onto sites near the bright brims associated with a metallic one-dimensional edge state of MoS_2 at temperatures below 200 K, whereas we do not see evidence of thiophene adsorption on the inert (0001) basal plane of the MoS_2 clusters [61]. The thiophene molecules thus seem to bind considerably more strongly near the metallic edge state of the fully sulfided Mo edges than to internal regions of the MoS_2 basal plane, although the edges are still fully sulfur-saturated.

However, when the MoS_2 nanoclusters are treated with predissociated hydrogen atoms, we observe an even stronger chemisorbed state of thiophene [62]. Figure 25.3

shows an atom-resolved STM image of a triangular MoS_2 nanoclusters first exposed to atomic hydrogen and subsequently to thiophene. Adjacent to the edges, bean-like features that mark the position of thiophene related species are now visible. STM movies show that these molecular species are mobile and diffuse along the cluster edges, and that they therefore do not reflect molecules bound rigidly to a vacancy [62]. The molecular species are identified as reaction intermediates resulting from a partial hydrogenation reaction occurring on the metallic brim states. The hydrogen that drives this reaction derives from H atoms adsorbed on the terminal S atoms on the edges from S–H groups [63], which are also observed under reaction conditions in the real catalyst. The combination of hydrogen atoms adsorbed on the edges in the form of S–H groups and the unusual sites for thiophene adsorption on the metallic brim presents a favorable situation for a hydrogenation reaction. Comparing the experimental observed STM images with simulated STM images from extensive density functional theory (DFT) calculations, we conclude that the observed reaction intermediates are cis-but-2-ene-thiolates (C_4H_7S-) coordinated through the terminal sulfur atom to sites near the metallic brim. These species are formed by a sequential hydrogenation of one of the double bonds in thiophene by hydrogen adsorbed on the edges (from the S–H groups) followed by C–S bond cleavage. DFT calculations show that the reaction barrier associated with the rate limiting step, the C–S bond

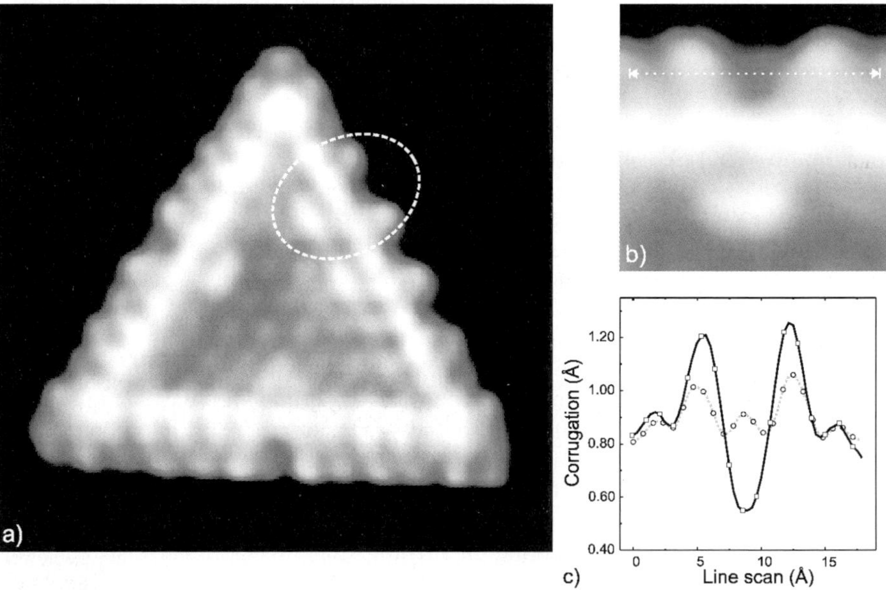

Fig. 25.3. (a) Atom-resolved STM image ($V_t = -331$ mV, $I_t = 0.50$ nA) of an atomic hydrogen pretreated MoS_2 cluster subsequently exposed to thiophene. Image dimensions are 50 Å × 54 Å. The *dashed circle* indicates the features that are associated with individual, adsorbed molecules. (b) A close-up that illustrates in detail the features associated with the adsorption of individual thiolate species at the edge of a MoS_2 nanocluster. (c) STM line scans along the edge protrusions of a cluster (*black*) corresponding to the line in (b). Line scan of an equal section of an unreacted, fully sulfided edge (*gray dashed*) is also shown. From [62]

breaking, is actually quite modest $\sim 100\,kJ/mol$. The configuration observed in the STM image associated with a ring-opened structure is simply an ordinary thiol in which the S is much more reactive. The final extrusion of this S from the hydrocarbon remainder may proceed on sulfur vacancies, which are traditionally considered to be the main active sites in hydrodesulfurization.

Sulfur vacancies are believed to be formed continuously under HDS reaction conditions by reaction with H_2 provided from the gas phase [44]. They are thus expected to be abundant under reaction conditions at elevated temperatures, but they have not previously been observed directly and their exact location remained thus uncertain. Under vacuum conditions it was found that vacancies could be formed only by exposing the clusters to predissociated (or atomic) hydrogen at 673 K. Figure 25.4 shows an STM image of a triangular single-layer MoS_2 nanocluster subsequent to its exposure to predissociated hydrogen. The dosing of hydrogen leaves the internal region of the cluster unaltered, but at the edge vacancies are clearly identified as dark spots. The atom-resolved STM images show that sulfur vacancies are indeed formed at the edge region of MoS_2. It is not possible to observe directly a sulfur-extrusion process of the ring-opened thiolate species observed in Fig. 25.3 on one of the observed vacancy sites, due to the fact that such a reaction occurs on a timescale much faster than the time resolution of STM. Nevertheless, the findings show that it is possible with STM to identify a complex route for an initial activation of a relatively inert S-bearing molecule like thiophene on MoS_2 nanoclusters, and the reaction intermediates observed in the STM image may therefore be the result of an important first step of hydrodesulfurization. Interestingly, the observed process takes place on the metallic brim states of the fully saturated Mo edges of the MoS_2 nanoclusters, which have the ability to accept or donate electrons and thus act as catalytic sites just like ordinary metal surfaces.

It is well known that cobalt acts as a promoter for the hydrotreating catalyst, and the increased activity has been correlated with the formation of so-called CoMoS clusters. The CoMoS clusters represent a nonstoichiometric phase of MoS_2 with Co atoms substituted at sites on the edge of the MoS_2 layer. Figure 25.5 shows an atom-

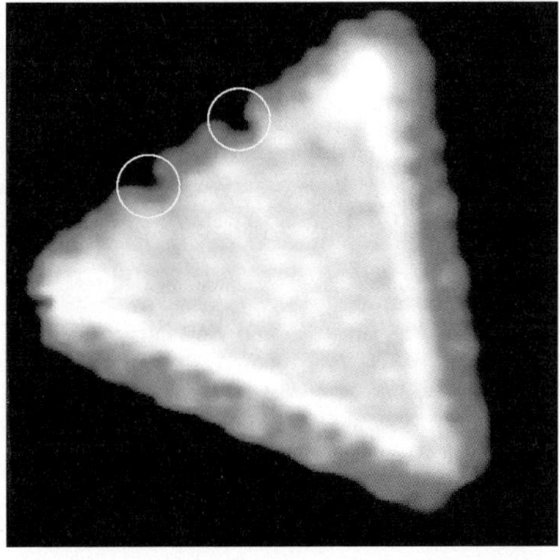

Fig. 25.4. Atom-resolved STM image ($V_t = -8.5\,mV$, $I_t = 1.12\,nA$) of a triangular single-layer MoS_2 nanocluster after reduction with atomic hydrogen atoms at 673 K in 5 min. Image size $46\,\text{Å} \times 47\,\text{Å}$. The location of two sulfur vacancies is illustrated with *circles*. Reproduced from [57]

ically resolved STM image of a single-layer CoMoS cluster formed by codeposition of Mo and Co onto the Au substrate during exposure to an H_2S atmosphere and subsequent annealing. We found that the CoMoS nanoclusters adopt a hexagonal shape as opposed to the triangular morphology of unpromoted MoS_2. This change in the equilibrium shape is attributed to the incorporation of cobalt into the MoS_2 structure, i.e. the formation of the CoMoS phase. The predominant hexagonal morphology implies that both fundamental types of low-indexed edge terminations of MoS_2 must be present, i.e. the Mo edge and the S edge. One edge type in the CoMoS is found to be similar to that observed for the MoS_2 triangles, with the edge protrusions clearly imaged out of registry with the lattice of S atoms on the basal plane and a bright brim along the edge. These edges are therefore identified as Mo edges, fully sulfided with two S dimers per Mo edge atom like in the triangles in Fig. 25.2a. From the symmetry of MoS_2, the other shorter edges are consequently attributed to S-type edges. On the basis of the detailed atomic-scale information provided by the STM images, a structural model of the CoMoS nanoclusters is obtained, in which Co atoms have substituted Mo atoms along the S edges of hexagonally truncated nanoclusters. As depicted in the ball model in Fig. 25.5, a tetrahedral environment of the Co atoms is produced if the outermost protrusions are assumed to be S monomers, which agrees well with previously published EXAFS results on supported CoMoS catalysts [52,64–66] and previous DFT studies [67,68]. Interestingly, the promoted edges are seen to exhibit an even brighter brim in the STM images. This finding suggests that metallic brim states also exist in the promoted CoMoS structures, and in view of the result of the unpromoted clusters it is tentatively proposed that 1D-metallic brim states located at the edges may play a role also for the catalytic properties of the promoted phase.

In summary, the STM studies of model systems consisting of single-layer MoS_2 and CoMoS nanoclusters are thus shown to offer unprecedented atomic-scale insight

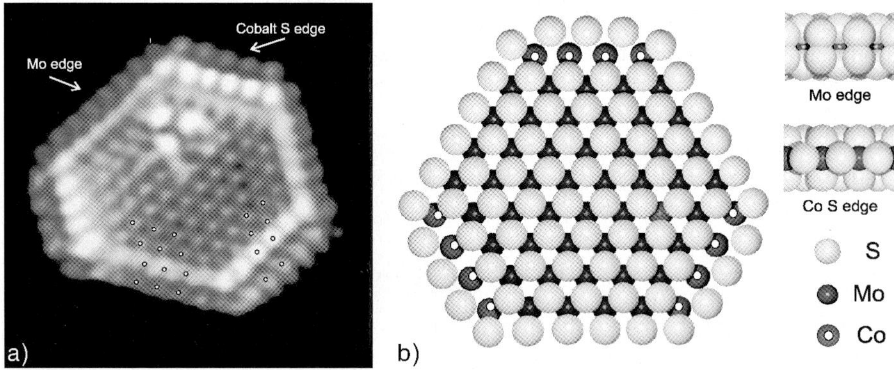

Fig. 25.5. (**a**) Atom-resolved STM image of a CoMoS nanocluster. Size is 51 Å × 52 Å and $V_t = -95.2$ mV, $I_t = 0.81$ nA. Notice the very intense brim associated with the Co-substituted S edge (shorter edges). (**b**) Ball model of the proposed CoMoS structure. The CoMoS cluster is shown in *top view* exposing the unpromoted Mo edge and a Co-promoted S edge (Mo: *dark*, S: *bright*, Co: *dark with spot*). Also shown on the basal plane is a single Co inclusion. The Mo edge appears to be unaffected by Co and is shown in a side-view ball model. The Co-substituted S edge with a tetrahedral coordination of each edge Co is also shown

into a very complex technical catalyst, for which very little microscopic knowledge was available before. However, due to the diversity of reactants in crude oil engaging in hydrotreating reactions and the complexity of the technical catalyst, many variables still need to be addressed in order to provide a full understanding of the catalytic role of active sites in the MoS_2 and promoted MoS_2 clusters. Hopefully, such information will increase our fundamental understanding of hydrotreating catalysis further and spark the development of better catalysts.

25.4
Selective Blocking of Active Sites on Ni(111)

The concept of active sites on a catalytic surface is fundamental to heterogeneous catalysis and the identification and characterization of active sites can potentially reveal important information on a given catalytic system. Very often, the catalytic activity is associated only with a minority of the surface atoms of a nanocluster, and in particular defects such as step edges, kink sites and/or corner atoms often provide the active sites. The concentration of defect sites is particularly low on low-index single-crystal surfaces, which constitute the standard model systems in surface-science studies of catalytic systems, and consequently these catalytically important sites are difficult to probe with the averaging and integrating techniques most widely used in surface-science studies. For this reason the role of defect sites in catalysis has most often in the past been addressed in an indirect manner by studying, e.g., the structure sensitivity, that is, the variation in reactivity over different crystal facets including surfaces with high step and kink density [69, 70]. The situation has, however, improved tremendously following the development of the STM, which provides the unique capability of imaging single defects with atomic resolution and the STM and the related family of scanning probe techniques are indeed ideally suited for the study of active sites in catalysis. The power of this technique was first demonstrated by Zambelli et al. [27] who show that the step edges on the Ru(0001) surface are the active sites for the dissociation of NO. Indeed it has been showed from both experiment and theory that the rate of dissociation of a number of diatomic molecules is orders of magnitude higher at the step edges as compared to the regular crystal facets [27, 71–76]. Since these studies are all concerned with simple diatomic molecules, the relevant catalytic parameter is the activity measured at the turnover frequency.

For more complicated molecules where several different reaction pathways are possible, the *selectivity* between end products is often equally or even more important than the activity. One of the simplest molecules displaying multiple reaction pathways for the decomposition process is ethylene (C_2H_4). In particular there are two possible reactions for the initial step of decomposition, either dehydrogenation (breaking of one of the C–H bonds) or dissociation of the C=C bond. Ethylene is thus an ideal test molecule to study the influence of defect sites on the selectivity of a catalytic reaction. Recently, we studied the decomposition of ethylene on the Ni(111) surface by STM in combination with ab initio computer simulations.

A series of experiments were performed in which a Ni(111) surface was exposed to ethylene at different temperatures. The lowest temperature at which decomposition is observed was room temperature (RT), and at this temperature we thus probe only the most reactive sites. We found that exposure to ethylene at RT led to the formation of a narrow region of reaction products along the upper step edges on Ni(111) as can be seen by comparing the two STM images of Fig. 25.6. The figure shows a Ni(111) surface before and after exposure of 1 L (10^{-8} torr × 100 s) ethylene at RT. The reaction products appear in the STM images as areas of different corrugation as compared to the (111) facets. These areas were found all along the step edges and with a width of typically no more than 40 Å. The area covered by the reaction products along the step edges was apparently independent of the size of the terraces. The cleanliness of the sample, and in particular that of the step edges was checked by STM prior to the ethylene exposure, and the formation of the reaction products could therefore unambiguously be associated with the adsorption of ethylene. No indication of nucleation of islands on the terraces of the surface was observed upon exposure to ethylene. When the exposure time was increased by a factor of 20 while the ethylene pressure was kept at 10^{-8} torr, we observed a similar formation of reaction products along the upper step edges, and, furthermore, we found that the coverage was identical to the one observed for the lower exposure time. This observation shows that the process of forming the reaction products is self-poisoning, i.e. the reaction products prevent further reaction in taking place. Since the only adsorbates observed on the surface are the ones at the step edges, we conclude that the mechanism behind the formation of the reaction products is decomposition of ethylene at the step-edge atoms. We cannot say whether or not decomposition is complete into adsorbed C and H, but we can rule out that we simply observe molecularly adsorbed ethylene. The latter is based on the fact that the islands extend up to 40 Å into the terrace, and so most of the reaction products are adsorbed on regular terrace sites. If these adsorbates were molecular ethylene, then one would expect an increasing coverage

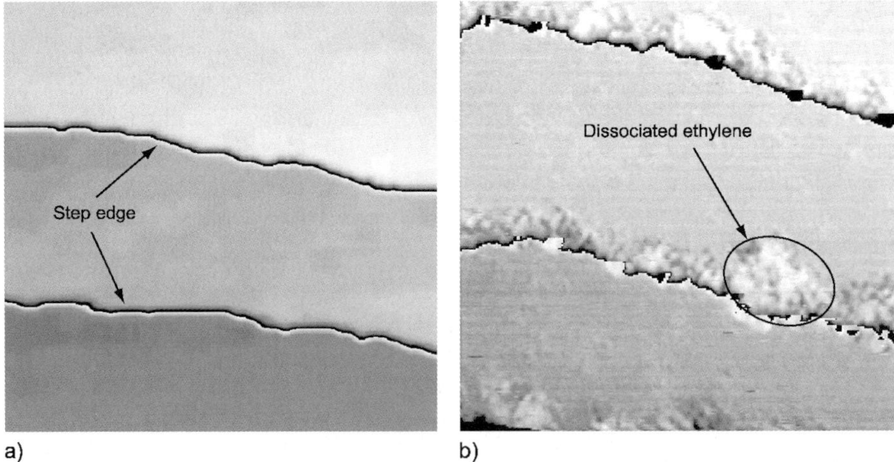

Fig. 25.6. STM images (200 Å × 200 Å and 150 Å × 150 Å of a Ni(111) surface before (**a**) and after (**b**) exposure to 1 L ethylene at room temperature. Reaction products are adsorbed along the step edges upon exposure. A cyclic color scaling has been used to enhance the features of each terrace

with increasing exposure time, in contrast to the experimental findings. A further proof was found when we exposed the Ni(111) surface to 1 L ethylene at 200 K. At these conditions no reaction products were formed, strongly indicating that the process is thermally activated. From the STM measurements at RT we therefore conclude that the step edges of the Ni(111) surface have a higher reactivity of ethylene decomposition than the regular atoms of the (111) facets.

At higher temperatures we found a change in the structure of the adsorbed reaction products after ethylene exposure. At 500 K we observed ordered islands growing from the upper step edges of the Ni(111) surface as illustrated in Fig. 25.7a. A zoom-in on these islands (Fig. 25.7b) reveals a quadratic unit cell with a size of ca. $5 \times 5\,\text{Å}^2$. The structure can be identified as the carbidic phase of carbon atoms adsorbed on Ni(111), which is known to form for ethylene exposure at this temperature [77,78]. In the carbidic phase the C atoms induce a reconstruction of the topmost layer of Ni(111), and the structure is illustrated in a ball model in Fig. 25.7c. A detailed discussion of this carbidic phase can be found in [77]. The existence of the carbidic phase shows that ethylene decomposes fully into atomic C and H at 500 K and leaves adsorbed C on the surface. Just as for the decomposed ethylene formed after RT exposure, the carbidic islands, formed by dissociated ethylene at 500 K, always nucleated at the upper step edges and no isolated islands were observed on the terraces. In contrast to the RT experiments the growth of the carbidic islands was, however, not found to be a self-poisoning process. This was seen by investigating the dependency of the island coverage on the ethylene dose. This investigation was done by measuring the carbon coverage for different dosing times at a constant ethylene pressure of 1.0×10^{-9} torr. The analysis of this series of experiments is presented in Fig. 25.7d, where the coverage (measured as the width of the carbidic islands averaged over the length of the step edges) is plotted as a function of the dosing time. From the graph it is seen that an increased dose of ethylene led to an increased coverage of carbon on the Ni(111) surface, which indicates that decomposition of ethylene is no longer restricted to the step sites, as was the case for the RT exposure. This is seen even more clearly, when the data is fitted to a Langmuir adsorption curve, in which the rate of dissociation is assumed to be proportional to the coverage of free Ni(111) terrace atoms. We plot the average width of carbon islands, $\varphi = l\theta$, where l is the average terrace width, and θ is the coverage measured in monolayers. Then the Langmuir adsorption curve changes from its usual expression, $\theta(t) = 1 - e^{-kt}$, to

$$\varphi(t) = l(1 - e^{-kt}) \, .$$

This expression fits nicely with the data, and furthermore the deduced average terrace width of approx. 600 Å agrees well with that of the used crystal, when the average terrace width is only calculated for the terraces large enough to enter into the analysis. On this basis we conclude that at 500 K the process of ethylene decomposition on Ni(111) is completely dominated by the terrace atoms. This finding does not conflict with the observation at RT, where only the step edges were found to be active for the decomposition of ethylene. This is due to the fact that the step edges are poisoned by the reaction products after which all the activity is related to the regular atoms in the (111) facets.

From the STM experiments it is not possible to identify the intermediate reaction products, and therefore we cannot determine through which reaction pathway

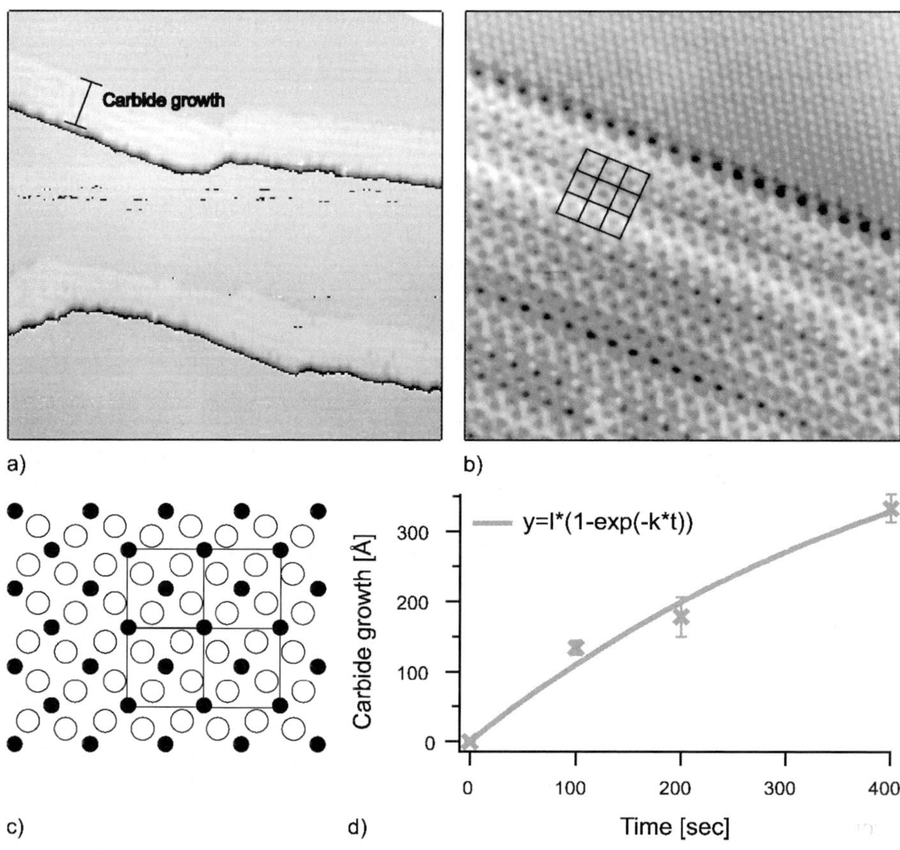

Fig. 25.7. (**a**) STM image (500 Å × 500 Å) of carbidic islands at the step edges of Ni(111) formed upon exposure to 0.1 L ethylene at 500 K. (**b**) Zoom-in (100 Å × 100 Å) on a carbidic island. The square unit cell of the carbidic structure is indicated. (**c**) Ball model of the carbon atoms (*filled circles*) and the topmost layer of Ni atoms (*open circles*) in the carbidic structure of C adsorbed on Ni(111). The c(2 × 2) grid of the C atoms is indicated. (**d**) Plot of the coverage of carbidic islands vs. the exposure time

decomposition proceeds, or what the final products are for the RT decomposition. In other words we have only discussed the activity of the step edges relative to the regular (111) facets and so far left out the question of how the step edges influence the selectivity in this reaction. In order to address these issues the STM studies were complimented by DFT calculations.

The potential energy diagram computed by DFT is presented in Fig. 25.8. Step edges have been introduced by performing DFT calculations on a Ni(211) surface that displays narrow (111)-oriented facets separated by monoatomic steps. When focusing on the initial step leading either to two adsorbed CH_2 (dissociation) or to adsorbed $CHCH_2$ plus H (dehydrogenation), it is seen that both of the transition state energies at the step edge are lower by 0.5 eV than the lowest transition state energy (dehydrogenation) on the flat surface. This finding supports the experimental finding

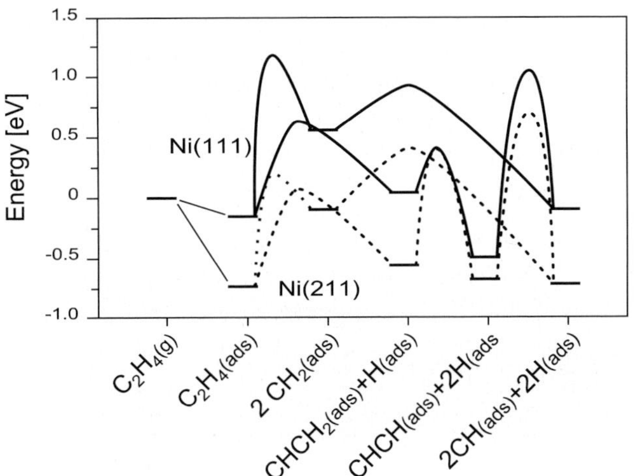

Fig. 25.8. Potential energy diagram for C=C bond breaking and C–H bond breaking on Ni(111) (*full lines*) and Ni(211) (*dotted lines*)

that the reactivity at the step edges on the Ni(111) surface is much higher than that of the (111) facets. The potential energy diagram also shows that a markedly change in the difference between the dehydrogenation and the dissociation barriers occurs when comparing the stepped Ni(211) surface to the flat Ni(111) surface. On Ni(111) the barrier for dissociation is more than 0.5 eV higher than that for dehydrogenation, on the Ni(211) surface the two barriers differ by less than 0.2 eV. The consequence is that on the stepped Ni(211) surface the initial step in decomposition is likely to be a competition between dehydrogenation and dissociation, whereas the flat Ni(111) surface will show a clear favoring of going through initial dehydrogenation. In other words, the bond-breaking selectivity is greatly improved in changing from a stepped to a flat Ni(111) surface. This bond-breaking selectivity will eventually show up in the overall selectivity of catalytic reaction involving ethylene. In particular it is worth noticing that dehydrogenation of ethylene leads to intermediates with stronger C=C bonds as illustrated by the large barrier for the dissociation of CHCH. This implies that any dissociation should take place at the level of molecular ethylene and so one would expect a high selectivity for dehydrogenation products on a perfectly flat Ni(111) surface.

As just pointed out, a perfectly flat Ni(111) surface would be expected to show a better selectivity towards dehydrogenation reactions. But since such a perfectly flat surface can only exist in theory, another approach is needed to prevent reaction from taking place at the highly active step edges. One such approach is to selectively block or poison the step edges by introducing some impurities that bind preferentially at the step edge sites. From earlier work within our group it is known that Ag nucleates preferentially at the step edges of Ni(111) [79], and also it is in general a less reactive atom than Ni [80]. For these reasons we considered Ag to be a good candidate for blocking the step edge sites on the Ni(111) surface.

Ag evaporated onto a Ni(111) surface at RT nucleates via step flow into large islands as shown in Fig. 25.9a. Because of the nucleation into very large islands, the majority of the step-edge atoms are unaffected by this RT evaporation of Ag. Postannealing of the Ni(111) surface after Ag evaporation, however, causes the Ag

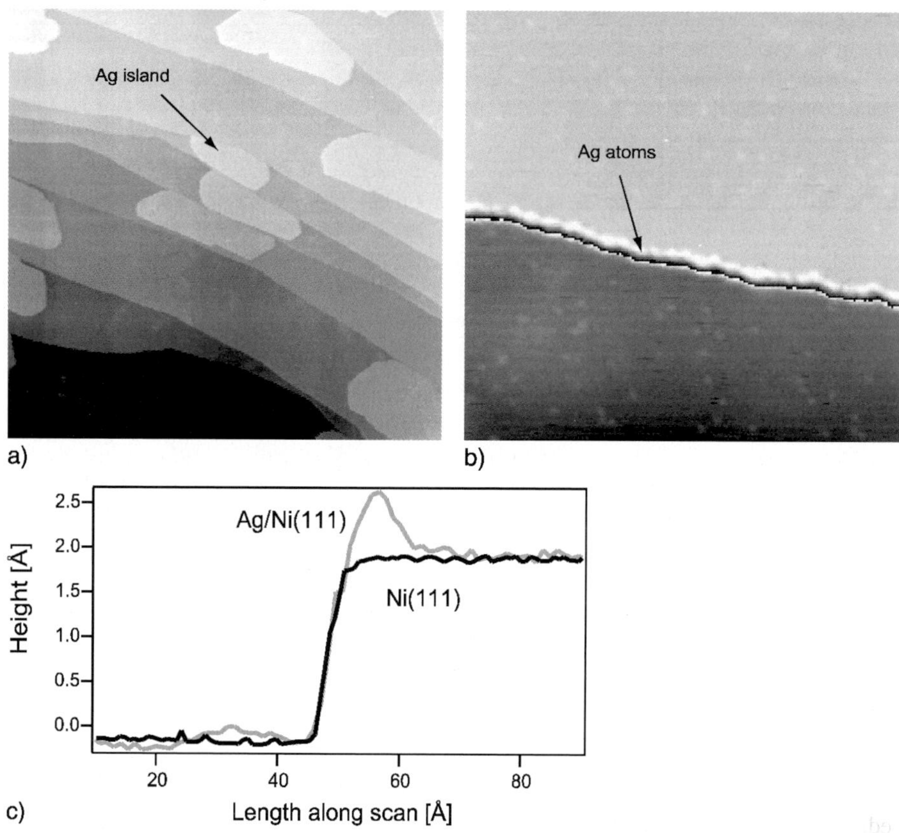

Fig. 25.9. (**a**) STM image (7000 Å × 7000 Å) of Ag islands nucleated along the step edges of Ni(111) after evaporation of Ag at room temperature. (**b**) STM image (400 Å × 400 Å) of a Ag/Ni(111) surface after postannealing at 800 K. A bright row of Ag atoms is seen along the step edges. (**c**) Line scans across the step edges of clean Ni(111) and the postannealed Ag/Ni(111) surface

atoms to completely wet all the step edges. An STM image of a Ni(111) surface annealed to 800 K after evaporation of Ag at RT is seen in Fig. 25.9b. When comparing this to the clean Ni(111) (see Fig. 25.6a), one finds a row of protrusions along the step edges. These protrusions appear bright in the STM image, and they are also seen when comparing a line scan across a step edge on the Ag/Ni(111) surface with a similar scan on the clean Ni(111) surface, as illustrated in Fig. 25.9c. These protrusions are ascribed to Ag atoms wetting the step edges, and for this surface there are no free Ni step-edge atoms.

When the postannealed Ag/Ni(111) surface was exposed to 1 L ethylene at RT, we observed no formation of reaction products along the step edges (or anywhere else) as seen from the STM image in Fig. 25.10a. This finding shows that the Ag atoms are effectively blocking the step-edge atoms of the Ni(111) surface, so that these can no longer act as the active sites for ethylene decomposition at RT. In this way we both provide further evidence that the step-edge atoms are indeed the active sites for

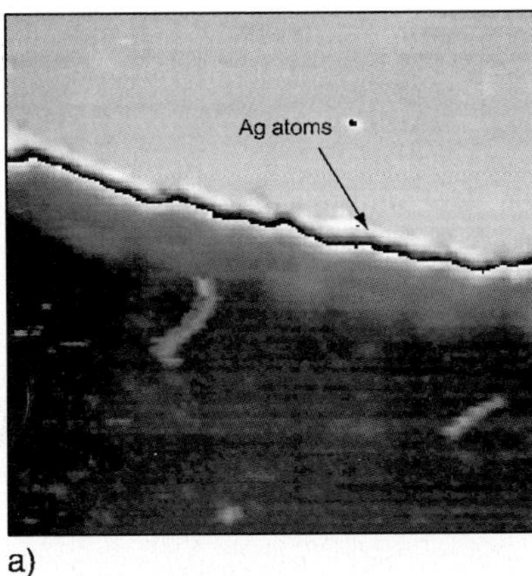

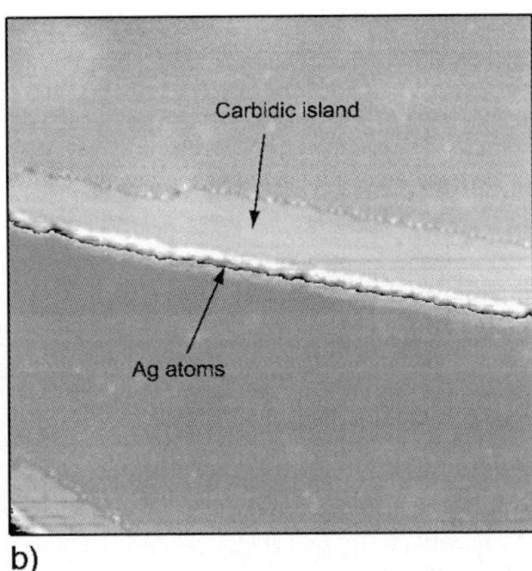

Fig. 25.10. STM images (300 Å × 300 Å and 500 Å × 500 Å) of a Ni(111) surface with the step edges decorated by Ag atoms, after exposure to ethylene at (**a**) room temperature and (**b**) 500 K. No adsorption is observed upon room-temperature exposure, whereas the exposure at 500 K leads to a nucleation of carbidic islands along the Ag-decorated step edges

ethylene decomposition at RT and more importantly we also demonstrate how the reactivity of the Ni(111) surface can be altered by blocking these very reactive sites.

At 500 K the exposure of ethylene to the Ag/Ni(111) leads to the growth of carbidic islands at the step edges (Fig. 25.10b) exactly as it was found for the clean Ni(111) surface. This correlates nicely with the observation that the atoms of the (111) facets are the active sites for ethylene decomposition at 500 K, and hence it also demonstrates that the addition of Ag to the Ni(111) surface only blocks the step edges and not the terraces. In this way the Ag/Ni(111) surface would be expected to show a high selectivity for dehydrogenation reactions.

The results obtained in the UHV-STM and DFT study were put to the test by performing reactivity measurements on a high surface area oxide-supported AgNi bimetallic catalyst. This catalyst was tested for the hydrogenolysis of ethane, which is the simplest reaction used to probe the activity of C–C bond-breaking (see, e.g., [81]). Figure 25.11 summarizes the results of the integral flow reactor measurements performed on both a pure and a Ag-modified Ni high-surface-area catalyst. From the figure, it is evident that the addition of Ag to the Ni catalyst led to a decrease in the rate constant for ethane hydrogenolysis by approximately an order of magnitude. It was, however, also seen that the apparent activation energy (the slope of the curves) was the same for the two catalysts, which strongly indicates that the active sites on the two catalysts are the same. From this observation together with the STM and DFT results, we conclude that the decrease in the rate constant was caused by Ag atoms partly covering the Ni step edges. It seems highly plausible that not all step sites are blocked, when taking into account the high degree of complexity of the supported high surface area catalyst as compared to the well-defined Ni(111) single-crystal surface on which a complete blocking of all the step sites could be achieved with a relatively low amount of Ag. The AgNi bimetallic catalyst is thus an example that fundamental insight may be exploited in the development of new heterogeneous high surface area catalysts.

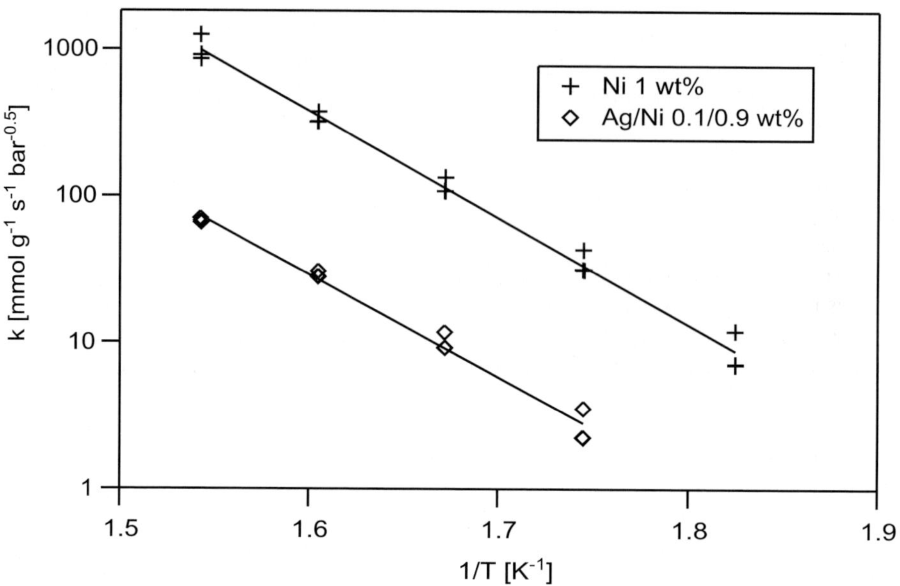

Fig. 25.11. Arrhenius plots of the rate constant for ethane hydrogenolysis over $Ni/MgAl_2O_4$ and $Ag/Ni/MgAl_2O_4$. The *slopes* of the *straight line fits* give the activation energies for the hydrogenolysis process

25.5
High-Pressure STM: Bridging the Pressure Gap in Catalysis

The standard surface-science approach to the study of catalysts, in which the catalytic process is separated into elementary surface processes that can be studied under ideal UHV conditions, has brought about a tremendous amount of new insight. It is, however, very important always to recall that the surface-science approach deals with highly idealized model systems, and care has to be taken when extrapolating the results to the conditions of a real working catalyst. One of the gaps between typical surface-science experiments and applied catalysis is the *pressure gap* caused by the ~ 13 orders of magnitude pressure span from a standard surface-science experiment $\sim 10^{-10} - 10^{-6}$ mbar) to an operating catalyst ($\sim 1 - 100$ bar).

One key issue is thus to characterize the adsorption structures of the gases present during the catalytic process. The high coverage structures formed at high gas pressures and high temperatures are usually studied by performing experiments at low pressures and low temperatures, i.e. at a similar chemical potential of the gas-phase molecules. This, however, only holds if entropy effects can be neglected and the adsorbed gas phase is in thermodynamical equilibrium with the gas phase. Another very important issue is whether the surface studied in surface-science experiments is in fact identical to the active phase of the working catalyst. A very clear example that the thermodynamically stable surface in UHV may differ from the surface of the active catalyst is found in the oxidation of CO on transition metals. In UHV experiments Ru is found to be the least active of the late transition metals for CO oxidation, whereas at high pressures of oxygen a RuO_2 structure is formed, which is superior in activity to all the other metals [31, 82]. To better understand the atomic mechanism of working catalysts, it is therefore essential to extend experimental methods to a pressure range closer to the catalytic relevant regime.

Over the past two decades much progress has been achieved in the development of in situ surface-characterization tools, and several groups are currently exploring catalysts at or close to operating conditions. Most tools for in situ characterization are based on optical principles, since electron-based techniques suffer from the low mean-free path of the electrons at high gas pressures. Modern X-ray synchrotron sources provide beams that enable the use of X-ray diffraction spectroscopy for the structure determination at high pressures [83]. This technique has been developed with success in the group of Ferrer, who used it to study the high-pressure-induced faceting of CO/Ni(110) [84], the compression of the CO adsorption structure on Ni(111) [85], and also the hydrogenation of CO on Ni(111) [86]. Synchrotron radiation can also be used for in situ X-ray absorption spectroscopy as, e.g., done by Moggridge et al. [87], and Ogletree et al. [88] have developed an in situ system for photoemission spectroscopy. Laser light may also be used to obtain spectroscopic information such as vibrational frequencies of adsorbates in, e.g., sum frequency generation. This method was employed by Rupprechter et al. and Su et al. to study the high-pressure structures of CO on Pt(111) [89, 90], and in a very recent work Lu et al. [91] studied the electrochemical oxidation of CO on Pt electrodes. Polarization modulation infrared reflection absorption spectroscopy is another laser-based technique which has been used, e.g., in the group of Goodman to study NO adsorption on Pd(111) [92].

Despite the fact that electron spectroscopy, as mentioned, is not the optimal choice for in situ studies, a transmission electron microscope, which allows for a local gas pressure at the sample of up to ~ 20 mbar by means of several stages of differential pumping along the electron beam, has been developed at the Haldor Topsøe research laboratories [22, 23, 93]. Although the STM uses electrons to probe the electric and morphological structure of a surface, it is not limited to low pressures since the electrons only have to tunnel through the narrow gap between the tip and the surface, and the STM can be operated over a wide pressure range ranging from UHV pressure to high-pressure (~ 1 atm) conditions. The STM thus facilitates studies of surface-morphological changes at the atomic scale over 13 orders of magnitude in pressure, and as such it is a powerful tool for the study of pressure gaps. Today, a number of groups have taken up the challenge of building an STM capable of operating in a high-pressure cell [94–99]. A major challenge in imaging surfaces with the STM over a wide pressure range is the sensitivity of the tunneling current to minute changes at the tunneling junction due to tip and/or sample instabilities due to the ambient gas. Furthermore, the gas atmosphere and gas flow transmits vibrations from the surroundings to the STM in contrast to the situation under vacuum conditions, and thus special care must be taken to reduce vibrational noise. Gas purification is another very important issue as the STM gives a priori no direct clues on the chemical nature of adsorbates, and even much diluted impurities may entirely pollute the surface when exposing the surface to high pressures.

The challenges involved in the design and operation of STM under high gas pressure have been overcome, and it has been possible to image adsorbate structures on single-crystal surfaces, even with atomic resolution at pressures ranging from the idealized UHV conditions to pressures of 1 bar [100, 101]. Hendriksen and Frenken [102] made a further development and implemented a reactor cell in connection with the STM, which enabled them to perform *in situ* STM investigations in a flow of reactant gases. Using this setup, the authors were able to monitor the partial pressures of the reactants and CO_2 during CO oxidation on a Pt(110) surface. From their data it was possible to correlate an observed surface roughening of the platinum(110) in an O_2-rich atmosphere directly to an increase in oxidation activity as shown in Fig. 25.12. The surface roughening was attributed to the formation of a platinum oxide in the topmost layer. According to the Mars–Van Krevelen mechanism, the platinum oxide serves as an oxygen reservoir, which is continuously reduced by reaction with CO and reoxidized under the O_2-rich conditions to produce a steady-state reaction. Such a Mars–van Krevelen mechanism was also found by Over and coworkers in their study of CO oxidation over a ruthenium surface [31, 82, 103–105]. These two systems are examples where the catalytically active surface is best characterized as a metal oxide rather that the pure metallic phase.

Several other studies on other metal-adsorbate systems have shown that the traditional surface-science approach is applicable, i.e. that raising the pressure is equivalent to lowering the temperature as long as the thermodynamical equilibrium structures remain kinetically accessible. This implies that it is legitimate to extrapolate results obtained under the well-characterized and well-controlled UHV conditions to high pressures. These systems include H adsorption on Cu(110) [106], CO adsorption on Pt(110) [100] and Pt(111) [107, 108], and NO on Pd(111) [109] all of which were found to form adsorption structures at high pressures (1 bar) that could also be formed by lowering the temperature at low-pressure conditions. In contrast to these studies Rider et al. [110] identified a new adsorption structure for NO on Rh(111) that only formed in equilibrium with the gas phase at high NO pressures, and could not be formed at low temperature.

Another issue that may be addressed by in situ studies is that of catalyst stability. In particular the surface composition in two- (or multi-) component catalysts may change in a high-pressure environment due to strong metal–adsorbate interactions [111, 112]. Using a dedicated high-pressure system with the home-built Aarhus STM incorporated [97], we have investigated the stability of a Au/Ni(111) surface alloy under high pressures of CO [113]. The Au/Ni(111) surface alloy is formed by the evaporation of Au onto a Ni(111) surface followed by an annealing to 800 K. This leads to a surface as shown in Fig. 25.13, where the dark depressions represent Au atoms alloyed into the topmost layer of the Ni(111) surface. The fact that the Au atoms counter intuitively are imaged as depressions although the Au atoms have a larger radius than the Ni atoms cannot be ascribed to an inwards relaxation but is rather an electronic effect, i.e. the local density of states at the Fermi level is depleted at the positions of the Au atoms. It is also observed that the Ni atoms with a Au nearest neighbor appear brighter in the STM image, indicating that the electronic state and hence reactivity of these atoms are altered as compared to Ni atoms with pure Ni nearest neighbors. This interpretation of the STM images is confirmed by DFT calculations, and in particular it was shown that the presence of Au atoms significantly destabilized the binding of C atoms to the Ni(111) surface [114]. The

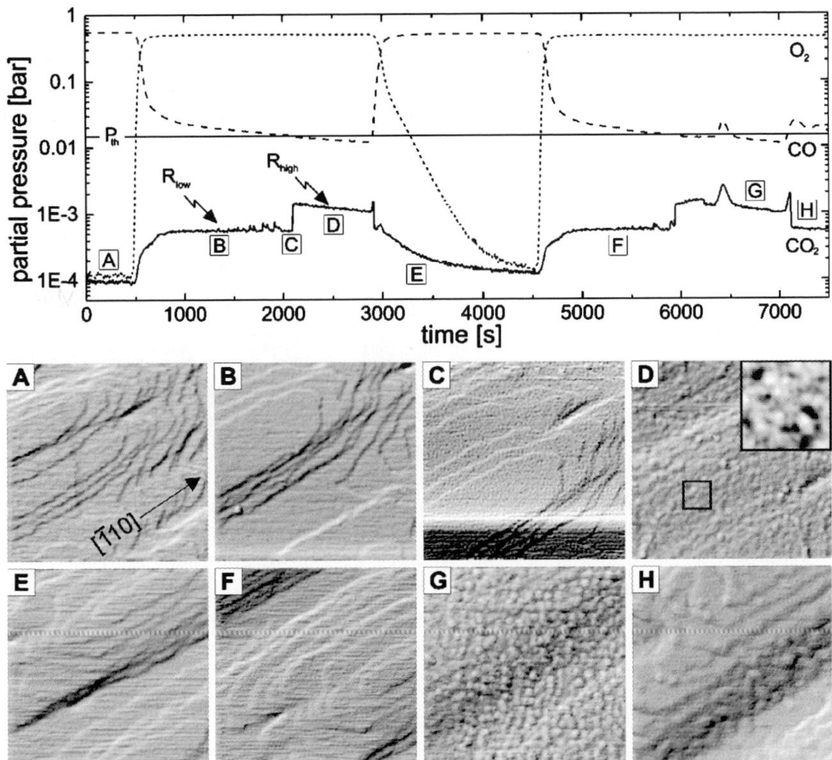

Fig. 25.12. Data recorded for CO oxidation on platinum(110) with an STM in a flow reactor. The *upper panel* shows mass spectrometer signals recorded directly from the reactor cell. The STM images show the surface morphology at different stages, corresponding to the curves in the mass spectrometer signal. High CO oxidation activity was correlated with the observation of a rough, oxidic platinum surface. Reprinted with permission from [102]. Copyright (2002) by the American Physical Society

formation of graphite on Ni-based catalysts during the steam reforming of natural gas is a severe problem that leads to whisker growth and eventually a total catalyst breakdown. The finding that Au destabilizes C atoms on the Ni(111) surface therefore led to the idea that using Au/Ni alloy nanoparticles as the active material could improve the lifetime of the steam reforming catalyst. This is another example of how a new catalyst was synthesized based on fundamental STM and DFT studies.

Besides being an improved catalyst for the steam-reforming process, the Au/Ni surface alloy could potentially be of interest to many other catalytic reactions. This circumstance is due to the fact that the electronic and chemical properties of the alloy surface can be "tuned" by varying the amount of Au as has been demonstrated, e.g., in a series of CO temperature-programmed desorption experiments performed by Holmblad et al. [115]. Here it was shown that both the CO binding energy and saturation coverages dropped with increasing Au concentration in the surface layer of Au/Ni(111). The possibility of controlling the reactivity of the Ni surface by the alloying of Au is particularly interesting because Ni is close to Pt in the periodic

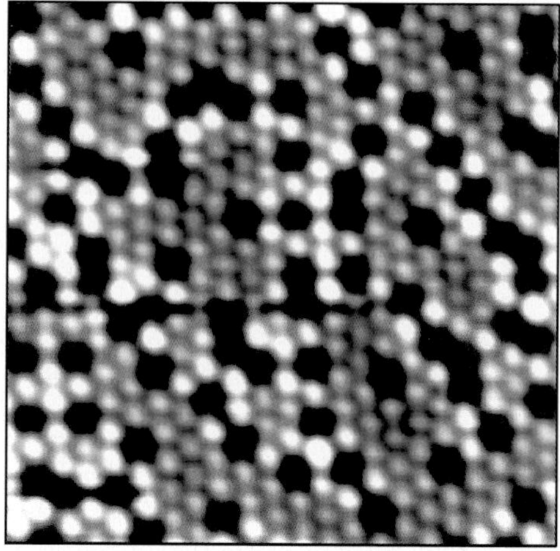

Fig. 25.13. STM image (50 Å × 50 Å) of the Au/Ni(111) surface alloy formed after evaporation of Au and postannealing at 800 K. The *dark depressions* are single Au atoms substituted into the topmost layer of the Ni(111) surface

system (same main group). Pt is one of the most versatile catalysts, but it suffers from high costs and a low abundance, and cheaper and more abundant alternatives are therefore sought. Returning to the issue of the pressure gap, it is of course relevant to examine the stability of the Au/Ni(111) surface alloy under more realistic conditions. CO is present in numerous processes involving both Ni- and Pt-based catalysts, and CO in general interacts relatively strong with these surfaces, and CO was thus chosen as the test gas for the first stability studies of the Au/Ni(111) surface alloy.

From STM movies obtained in a high CO background pressure we observed that CO indeed induces severe morphological changes on the Au/Ni(111) surface. This can be seen in the series of images taken from an STM movie recorded at room temperature and a CO pressure of 10 torr shown in Fig. 25.14a–f. (The full-length movie can be seen at http://www.phys.au.dk/spm/movies/carbonyl.mpg.) The movie shows a $1000 \times 1000\,\text{Å}^2$ area of a Ni(111) surface with 0.3 monolayer Au. Due to the large scanning area, the single atoms are not seen, but the surface before CO exposure is similar to that presented in Fig. 25.13. Once CO is introduced into the chamber, holes start to develop at the step edges, and the step edges begin to move across the surface while small clusters are nucleated behind them. The movement of the step edges is apparently induced by the removal of atoms from the topmost layer, and we suggest that the mechanism is the removal of the Ni atoms through the formation of Ni carbonyl:

$$Ni(s) + 4CO(g) \rightarrow Ni(CO)_4(g) \ .$$

Nickel carbonyl is known to form on Ni surfaces at temperatures below ~ 525 K and CO pressures above ~ 0.1 mbar [116]. Furthermore, it has been shown that the rate of carbonyl formation is greatly enhanced on Ni thin films with a high step density [117], which corresponds well with the fact that we observe the removal of surface atoms only at the step edges. The proposed model is therefore that Ni surface atoms react with gas-phase CO molecules to form carbonyl, which is volatile at room temperature and thus leaves the surface. The Au atoms are left on the surface and are dragged along the moving step edge until a critical size is achieved and a Au cluster is nucleated.

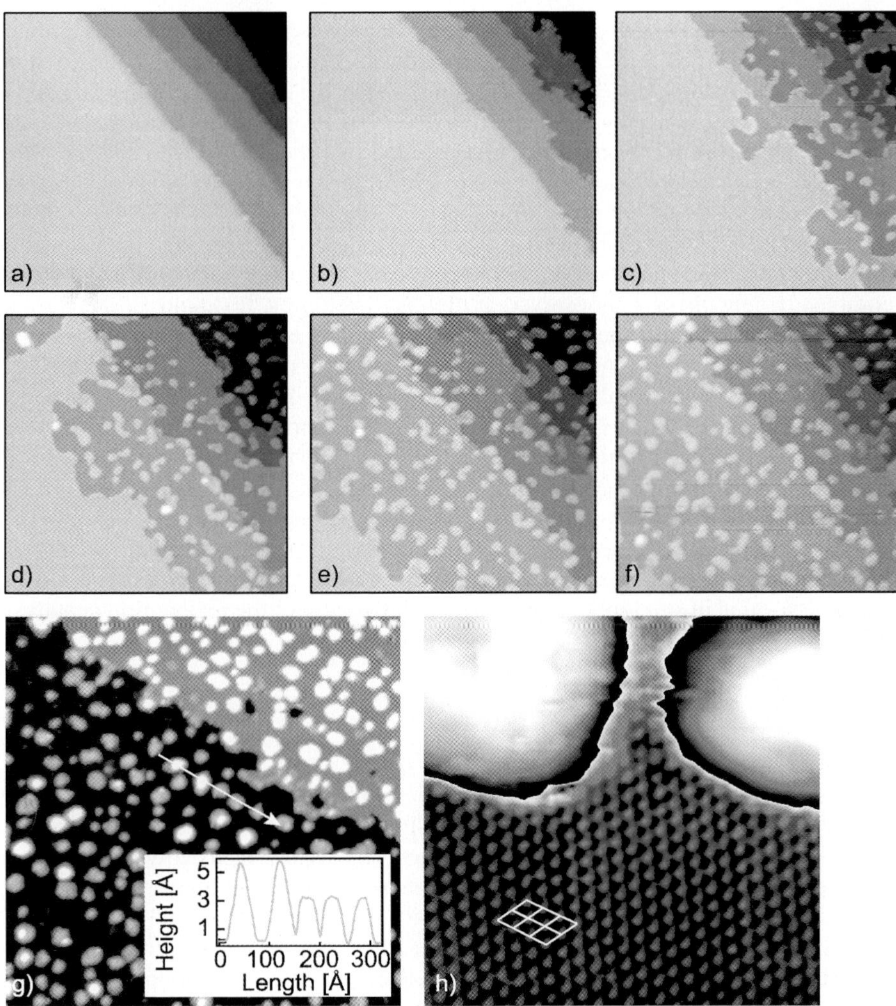

Fig. 25.14. (**a**)–(**f**) STM images (1000 Å × 1000 Å) of a Au/Ni(111) surface alloy after exposure to 10 torr CO for (**a**) 0 min, (**b**) 25 min, (**c**) 50 min, (**d**) 75 min, (**e**) 100 min, and (**f**) 125 min. (**g**) The Au/Ni(111) surface after exposure to 1 bar of CO at room temperature. The *arrow* indicates the direction of the inserted line scan. (**h**) A zoom-in (60 Å × 60 Å) on the area between the clusters reveals a clean Ni(111) substrate

The proposed model was further supported by STM measurements carried out after a prolonged CO exposure at high pressure. Figure 25.14g shows a Au/Ni(111) surface after exposure to 1 bar CO at room temperature, and as in the end of the movie obtained at 10 torr (Fig. 25.14f) the surface is seen to be fully covered with clusters, and a zoom-in between the clusters reveals a clean Ni(111) surface (Fig. 25.14h). A height analysis of the clusters shows that most islands have a height of 3.2 Å, and $\sim 10\%$ have a height of 5.6 Å (see the line scan in Fig. 25.14g). These heights are identical to the heights of single- and double-layer Au islands on Ni(111) as measured by STM. Furthermore, there is a good agreement between the coverage of the amount of Au evaporated and the coverage as calculated by assuming that the clusters consist of Au. Finally, annealing of the sample after the exposure to CO leads to the formation of a surface alloy similar to that observed prior to the CO exposure. Based on these observations, it can thus be concluded that the Au/Ni(111) surface alloy undergoes a phase separation into Au clusters dispersed on a pure Ni(111) substrate at high pressures of CO at room temperature, and the mechanism behind the phase separation is the removal of the Ni atoms in the topmost layer through the formation of Ni carbonyl. The study on Au/Ni(111) is a clear example of a pressure gap where completely new physics is observed at high gas pressures. It should, however, also be stressed that these experiments, although done at a high CO pressure, do not represent the operating conditions of any real catalytic reaction. In relation to, e.g., the use of Au/Ni alloy catalysts in steam reforming, we have neglected the presence of coadsorbates (mainly H_2O, CH_4, H_2, and CO_2) and the elevated temperatures at which the steam-reforming process is operated. The formation of Ni carbonyl is exothermic, and thus the equilibrium is shifted towards solid Ni and gaseous CO at high temperatures, and the reaction is found not to take place above ~ 525 K [116], which is well below typical operating temperatures in steam-reforming reactors. In general it is always important to bear in mind that surface-science experiments, even when performed in situ, only constitute a model system of the real working catalyst.

25.6
Summary and Outlook

As an analytical technique to study the atomic-scale realm of surfaces, the scanning tunneling microscope has matured enormously since the development of the first microscope 25 years ago [24, 26]. STM instruments have since then evolved from being relatively complicated home-built instruments sensitive to vibrations into compact, rigid, stable variable-temperature microscopes, available commercially at fairly low cost. The availability of the instrumentation for STM has led to a tremendous diversification of the application of the technique, and new areas of applications are constantly being developed. One area in which STM has turned out to be a particularly powerful and versatile tool is for the study of surface chemistry and catalysis. According to the so-called *surface-science approach to catalysis*, detailed investigations are made by applying the arsenal of modern analytical methods to well-defined single-crystal surfaces as catalyst model systems, often kept under well-controlled high-vacuum conditions. This approach is capable of providing fundamental microscopic insight into the principles underlying the elementary steps of heterogeneous catalysis. The development of the STM to study such model systems has in particular enabled a direct atomic-scale view of active sites such as edges, kinks, or atomic defects, which turn out to be present in only low densities, but nonetheless to dominate the catalytic properties in many cases. The results of these investigations reveal that, in general, one cannot neglect the dispersed nature of the active materials in catalysts and model them from the usual well-known bulk properties. Atomic-scale insight is essential, and the STM is an excellent tool to resolve structures on this scale on nanoclusters deposited on conducting substrates. In this chapter, we have presented a few good examples that illustrate how information at the atomic scale obtained by STM can increase our general understanding of surface chemistry and catalysis. This progress seems to indicate that a new era in heterogeneous catalysis is within reach in the sense that new high surface area catalysts may be designed from detailed fundamental science input. However, one should be aware of the fact that catalyst model systems studied under well-controlled conditions may differ significantly from the real high surface area catalysts, both with respect to the nature of the surface morphology and the applied pressure range. Today, different routes are being developed to overcome both the material and pressure gaps, and it is very likely that the STM will play a central part in this progress. For instance, STM investigations of models systems more complex than single crystals have also enabled us to bridge the materials gap. In particular, systems consisting of nanoclusters supported on flat conducting surfaces (e.g., thin oxide films) have moved STM studies closer to real catalysis. The STM is, however, limited to the study of conducting surfaces and can, e.g., not be used for the study of poorly conducting or insulating materials such as the oxide supports widely used in catalysis. For this purpose the closely related atomic force microscope (AFM) is a well-suited instrument [118, 119], capable of probing, conducting as well as the many insulating surfaces relevant to catalysis, with a high degree of detail similar to STM [120–123]. In addition, the new developments of the STM instruments operating at high pressures [100, 102, 106] or with high scanning speeds [124–126] have already shown promising results with clear significance for the fundamental studies of catalysis.

Acknowledgements. We gratefully acknowledge our many colleagues who have contributed in various ways to the work presented in this article, including T. An, M. Bollinger, I. Chorkendorff, B. S. Clausen, S. Dahl, B. Hammer, K. Honkala, S. Helveg, J. Knudsen, E. Lægsgaard, M. Nyberg, J. K. Nørskov, J. Schnadt, I. Stensgaard, H. Topsøe, and E. K. Vestergaard. We also thank B. L. M. Hendriksen and J. W. M. Frenken for permission to use their figures.

References

1. Somorjai GA (1994) Surface Chemistry and Catalysis. Wiley, New York
2. Ertl G, Freund H-J (1999) Phys Today, p 32
3. Niemantsverdriet JW (2000) Spectroscopy in Catalysis. Wiley-VCH Verlag GmbH, Weinheim
4. Goodman DW (1996) J Phys Chem 100:13090
5. King DA (1994) Surf Sci 299/300:678
6. King DA (1997) In: Froment GF, Waugh KC (eds) Dynamics of Surfaces and Reaction Kinetics in Heterogeneous Catalysis (Studies in Surface Science and Catalysis), Vol 109. Elsevier Science, Amsterdam p 79
7. Gunter PLJ, Niemantsverdriet JW, Ribeiro FH, Somorjai GA (1997) Catal Rev-Sci Eng 39:77
8. Sinfelt JH (2002) Surf Sci 500:923
9. Freund H-J, Kuhlenbeck H, Libuda J, Rupprechter G, Bäumer M, Hamann H (2001) Top Catal 15:201
10. Valden M, Lai X, Goodman DW (1998) Science 281:1647
11. Henry CR (1998) Surf Sci Rep 31:231
12. Haruta M (1997) Catal Today 36:153

13. Campbell CT (1997) Surf Sci Rep 27:3
14. Hansen KH, Worren T, Stempel S, Lægsgaard E, Bäumer M, Freund H-J, Besenbacher F, Stensgaard I (1999) Phys Rev Lett 83:4120
15. Bäumer M, Freund H-J (1999) Prog Surf Sci 61:127
16. Freund H-J (2002) Surf Sci 500:271
17. Campbell CT, Parker SC, Starr DE (2002) Science 298:811
18. Becker C, von Bergmann K, Rosenhahn A, Schneider J, Wandelt K (2001) Surf Sci 486:L443
19. Perrot E, Humbert A, Piednoir A, Chapon C, Henry CR (2000) Surf Sci 445:407
20. Topsøe H (2000) Stud Surf Sci Cat 130:1
21. Boyes ED, Gai PL (1997) Ultramicroscopy 67:219
22. Hansen PL, Wagner JB, Helveg S, Rostrup-Nielsen JR, Clausen BS, Topsøe H (2002) Science 295:2053
23. Hansen TW, Wagner JB, Hansen PL, Dahl S, Topsøe H, Jacobsen CJH (2001) Science 294:1508
24. Binnig G, Rohrer H, Gerber C, Weibel E (1982) Phys Rev Lett 49:57
25. Binnig G, Rohrer H, Gerber C, Weibel E (1983) Surf Sci 131:L379
26. Binnig G, Rohrer H (1987) Rev Mod Phys 59:615
27. Zambelli T, Wintterlin J, Trost J, Ertl G (1996) Science 273:1688
28. Wintterlin J, Völkening S, Janssens TVW, Zambelli T, Ertl G (1997) Science 278:1931
29. Renisch S, Schuster R, Wintterlin J, Ertl G (1999) Phys Rev Lett 82:3839
30. Mitsui T, Rose MK, Fomin E, Ogletree DF, Salmeron M (2003) Nature 422:705
31. Over H, Kim YD, Seitsonen AP, Wendt S, Lundgren E, Schmid M, Varga P, Morgante A, Ertl G (2000) Science 287:1474
32. Zambelli T, Barth JV, Wintterlin J, Ertl G (1997) Nature 390:495
33. Schaub R, Wahlström E, Rønnau A, Lægsgaard E, Stensgaard I, Besenbacher F (2003) Science 299:377
34. Schaub R, Thostrup R, Lopez N, Lægsgaard E, Stensgaard I, Nørskov JK, Besenbacher F (2001) Phys Rev Lett 87:6104
35. Tersoff J, Hamann DR (1985) Phys Rev B 31:805
36. Tersoff J, Hamann DR (1983) Phys Rev Lett 50:1998
37. Lang ND (1985) Phys Rev Lett 55:230
38. Stroscio JA, Pierce DT, Davies A, Celotta RJ (1995) Phys Rev Lett 75:2960
39. Hammer B, Nørskov JK (2000) Adv Catal 45:71
40. Hofer WA, Foster AS, Shluger AL (2003) Rev Mod Phys 75:1287
41. Sautet P (1997) Chem Rev 97:1097
42. Magonov SN, Whangbo M-H (1994) Adv Mater 5:355
43. Wivel C, Candia R, Clausen BS, Mørup S, Topsøe H (1981) J Catal 68:453
44. Topsøe H, Clausen BS, Massoth FE (1996) Hydrotreating Catalysis. Springer Verlag, Berlin-Heidelberg
45. Landau MV (1997) Catal Today 36:393
46. Grisham JL (1999) Chem Eng News 23:21
47. Prins R (1997) In: Ertl G, Knözinger H, Weitkamp J (eds) Handbook of Heterogeneous Catalysis, Vol 4. VHC, Wienheim, p 1908
48. Whitehurst DD, Isoda T, Mochida I (1998) Adv Catal 42:345
49. Kabe T, Ishihara A, Qian W (1999) Hydrodesulfurization and Hydrogenation – Chemistry and Engineering. Wiley-VCH, Kodansha
50. Gosselink JW (1998) Cattech 4:127
51. Parham TG, Merrill RP (1984) J Catal 85:295
52. Clausen BS, Lengeler B, Candia R, Als-Nielsen J, Topsøe H (1981) Bull Soc Chim Belg 90:1249
53. Boudart M, Dalla Betta RA, Foger K, Loffler DG, Samant MG (1985) Science 228:717

54. Topsøe H, Clausen BS (1984) Catal Rev-Sci Eng 26:395
55. Salmeron M, Somorjai GA, Wold A, Chianelli R, Liang KS (1982) Chem Phys Lett 90:105
56. Barth JV, Brune H, Ertl G, Behm R (1990) Phys Rev B 42:9307
57. Helveg S, Lauritsen JV, Lægsgaard E, Stensgaard I, Nørskov JK, Clausen BS, Topsøe H, Besenbacher F (2000) Phys Rev Lett 84:951
58. Lauritsen JV, Helveg S, Lægsgaard E, Stensgaard I, Clausen BS, Topsøe H, Besenbacher F (2001) J Catal 197:1
59. Bollinger MV, Lauritsen JV, Jacobsen KW, Nørskov JK, Helveg S, Besenbacher F (2001) Phys Rev Lett 87:196803
60. Bollinger MV, Jacobsen KW, Nørskov JK (2003) Phys Rev B 67:085410
61. Lauritsen JV, Nyberg M, Nørskov JK, Clausen BS, Topsøe H, Lægsgaard E, Besenbacher F (2004) J Catal 224:94
62. Lauritsen JV, Nyberg M, Vang RT, Bollinger MV, Clausen BS, Topsøe H, Jacobsen KW, Besenbacher F, Lægsgaard E, Nørskov JK, Besenbacher F (2003) Nanotech 14:385
63. Lauritsen JV, Bollinger MV, Lægsgaard E, Jacobsen KW, Nørskov JK, Clausen BS, Topsøe H, Besenbacher F (2004) J Catal 221:510
64. Niemann W, Clausen BS, Topsøe H (1990) Catal Lett 4:355
65. Louwers SPA, Prins R (1992) J Catal 133:94
66. Bouwens SMAM, van Veen JAR, Koningsberger DC, de Beer VHJ, Prins R (1991) J Phys Chem 95:123
67. Byskov LS, Nørskov JK, Clausen BS, Topsøe H (1999) J Catal 187:109
68. Raybaud P, Hafner J, Kresse G, Kasztelan S, Toulhoat H (2000) J Catal 190:128
69. Gwathmey AT, Cunningham RE (1958) Adv Catal 10:57
70. Yates JT (1995) J Vac Sci Technol A 13:1359
71. Ciobicã IM, Santen RAv (2003) J Phys Chem B 107:3808
72. Dahl S, Logadottir A, Egeberg RC, Larsen JH, Chorkendorff I, Törnquist E, Nørskov JK (1999) Phys Rev Lett 83:1814
73. Gambardella P, Sljivancanin Z, Hammer B, Blanc M, Kuhnke K, Kern K (2001) Phys Rev Lett 87:056103
74. Hammer B (1999) Phys Rev Lett 83:3681
75. Liu ZP, Hu P (2003) J Am Chem Soc 125:1958
76. Zubkov T, Morgan GA Jr, Yates JT Jr, Kühlert O, Lisowski M, Schillinger R, Fick D, Jänsch HJ (2003) Surf Sci 526:57
77. Klink C, Stensgaard I, Besenbacher F, Lægsgaard E (1995) Surf Sci 342:250
78. Nakano H, Ogawa J, Nakamura J (2002) Surf Sci 514:256
79. Besenbacher F, Nielsen LP, Sprunger PT (1997) In: King DA, Woodruff DP (eds) The Chemical Physics of Solid Surfaces and Heterogeneous Catalysis. Surface alloying in Heteroepitaxial metal-on-metal growth. Elsevier Science Publishers 8, Chapter 10
80. Hammer B, Nørskov JK (1995) Nature 376:238
81. Sinfelt JH (1983) Bimetallic catalysts, Discoveries, Concepts, and Applications. Wiley, New York
82. Over H, Muhler M (2003) Prog Surf Sci 72:3
83. Bernard P, Peters K, Alvarez J, Ferrer S (1999) Rev Sci Instrum 70:1478
84. Peters KF, Walker CJ, Steadman P, Robach O, Isern H, Ferrer S (2001) Phys Rev Lett 86:5325
85. Quiros C, Robach O, Isern H, Ordejon P, Ferrer S (2003) Surf Sci 522:161
86. Ackermann M, Robach O, Walker C, Quiros C, Isern H, Ferrer S (2004) Surf Sci 557:21
87. Moggridge GD, Rayment T, Ormerod RM, Morris MA, Lambert RM (1992) Nature 358:658
88. Ogletree DF, Bluhm H, Lebedev G, Fadley CS, Hussain Z, Salmeron M (2002) Rev Sci Instrum 73:3872

89. Rupprechter G, Dellwig T, Unterhalt H, Freund HJ (2001) J Phys Chem B 105:3797
90. Su XC, Cremer PS, Shen YR, Somorjai GA (1996) Phys Rev Lett 77:3858
91. Lu GQ, Lagutchev A, Dlott DD, Wieckowski A (2005) Surf Sci 585:3
92. Ozensoy E, Hess C, Loffreda D, Sautet P, Goodman DW (2005) J Phys Chem B 109:5414
93. Helveg S, Lopez-Cartes C, Sehested J, Hansen PL, Clausen BS, Rostrup-Nielsen JR, Abild-Pedersen F, Nørskov JK (2004) Nature 427:426
94. McIntyre BJ, Salmeron M, Somorjai GA (1993) J Vac Sci Technol A 11:1964
95. Rasmussen PB, Hendriksen BLM, Zeijlemaker H, Ficke HG, Frenken JWM (1998) Rev Sci Instrum 69:3879
96. Jensen JA, Rider KB, Chen Y, Salmeron M, Somorjai GA (1999) J Vac Sci Technol B 17:1080
97. Lægsgaard E, Österlund L, Thostrup P, Rasmussen PB, Stensgaard I, Besenbacher F (2001) Rev Sci Instrum 72:3537
98. Kolmakov A, Goodman DW (2003) Rev Sci Instrum 74:2444
99. Rössler M, Geng P, Wintterlin J (2005) Rev Sci Instrum 76:023705
100. Thostrup P, Vestergaard EK, An T, Laegsgaard E, Besenbacher F (2003) J Chem Phys 118:3724
101. Rider KB, Hwang KS, Salmeron M, Somorjai GA (2002) J Am Chem Soc 124:5588
102. Hendriksen BLM, Frenken JWM (2002) Phys Rev Lett 89:046101
103. Over H, Seitsonen AP, Lundgren E, Schmid M, Varga P (2001) J Am Chem Soc 123:11807
104. Over H, Knapp M, Lundgren E, Seitsonen AP, Schmid M, Varga P (2004) Chem Phys Chem 5:167
105. Over H, Seitsonen AP, Lundgren E, Schmid M, Varga P (2002) Surf Sci 515:143
106. Österlund L, Rasmussen PB, Thostrup P, Lægsgaard E, Stensgaard I, Besenbacher F (2001) Phys Rev Lett 86:460
107. Longwitz SR, Schnadt J, Vestergaard EK, Vang RT, Laegsgaard E, Stensgaard I, Brune H, Besenbacher F (2004) J Phys Chem B 108:14497
108. Vestergaard EK, Thostrup P, An T, Laegsgaard E, Stensgaard I, Hammer B, Besenbacher F (2002) Phys Rev Lett, p 88
109. Vang RT, Wang JG, Knudsen J, Schnadt J, Laegsgaard E, Stensgaard I, Besenbacher F (2005) J Phys Chem B 109:14262
110. Rider KB, Hwang KS, Salmeron M, Somorjai GA (2001) Phys Rev Lett 86:4330
111. Christoffersen E, Stoltze P, Nørskov JK (2002) Surf Sci 505:200
112. Greeley J, Mavrikakis M (2004) Nature Mater 3:810
113. Vestergaard EK, Vang RT, Knudsen J, Pedersen TM, An T, Lægsgaard E, Stensgaard I, Hammer B, Besenbacher F (2005) Phys Rev Lett 95:126191
114. Besenbacher F, Chorkendorff I, Clausen BS, Hammer B, Molenbroek AM, Nørskov JK, Stensgaard I (1998) Science 279:1913
115. Holmblad PM, Larsen JH, Chorkendorff I (1996) J Chem Phys 104:7289
116. de Groot P, Coulon M, Dransfeld K (1980) Surf Sci 94:204
117. Medvedev VK, Borner R, Kruse N (1998) Surf Sci 401:L371
118. Binnig G, Quate CF, Gerber C (1986) Phys Rev Lett 56:930
119. Giessibl FJ (2003) Rev Mod Phys 75:949
120. Fukui K, Onishi H, Iwasawa Y (1997) Phys Rev Lett 79:4202
121. Namai Y, Fukui KI, Iwasawa Y (2003) Catal Today 85:79
122. Barth C, Henry CR (2003) Phys Rev Lett 91(19):196102
123. Barth C, Reichling M (2001) Nature 414:54
124. Rost MJ, Crama L, Schakel P, van Tol E, van Velzen-Williams G, Overgauw CF, ter Horst H, Dekker H, Okhuijsen B, Seynen M, Vijftigschild A, Han P, Katan AJ, Schoots K, Schumm R, van Loo W, Oosterkamp TH, Frenken JWM (2005) Rev Sci Instrum 76:053710
125. Besenbacher F, Lægsgaard E, Stensgaard I (2005) Materials Today 5:26
126. Wintterlin J (2000) Adv Catal 45:131

26 Nanostructuration and Nanoimaging of Biomolecules for Biosensors

Claude Martelet · Nicole Jaffrezic-Renault ·
Yanxia Hou · Abdelhamid Errachid · François Bessueille

Abbreviations

AFM   atomic force microscopy
LB   Langmuir–Blodgett
SAM   self-assembled monolayer
ODT   octadecanethiol
OBP   odorant-binding protein
MHDA 16-mercaptohexadecanoic acid
EDC N-ethyl-N-[dimethylaminopropyl] carbodiimide
NHS N-hydroxysuccinimide

26.1
Introduction and Definition of Biosensors

Recognition between a selective molecule and an analyte constitutes the primary event of mechanisms involved in biosensing. Thus the engineering of the biospecies in the recognizing layer is critical for the design of biosensors. For a few years now, such devices tended to be a reliable alternative to classical bioanalytical methods in numerous fields such as diagnostics, health care, pharmaceuticals, food and process control, environmental monitoring, defense and security, etc.

26.1.1
Definition

Biosensors are analytical devices containing immobilized biological sensitive materials in contact with or integrated within a transducer, which ultimately converts a biological signal into a quantitatively measurable electrical signal. From the concept we can see that a biosensor is composed of two essential components: a bioreceptor and a transducer. Figure 26.1 shows a schematic view of a biosensor.

26.1.2
Biosensor Components

A bioreceptor is the most critical component of a biosensor. It is responsible for the selective recognition of the analytes, generating the physicochemical signal monitored by the transducer. Ultimately, it determines the sensitivity of the device.

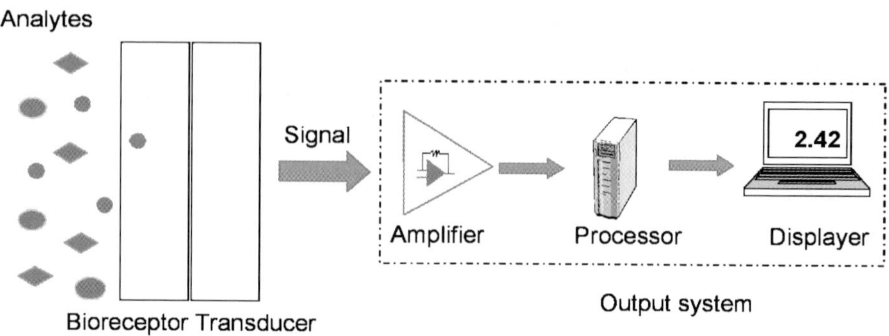

Fig. 26.1. Schematic view of a biosensor

Table 26.1. Fundamental components used for biosensor construction

Receptors	Transducers
Enzymes	Optical
Antibodies	Electrochemical
Organelles	Thermometric
Cell receptors	Piezoelectric
Tissues	Magnetic
Microorganisms	
Nucleic acids	
Synthetic receptors	

There is a wide range of naturally produced molecules from plants, animals, and microorganisms that can be used as the receptor in biosensor elaboration.

A transducer is the another crucial component for the biosensor, which converts a specific biological response to a useful electronic output, and must also be suitable for the receptor immobilization on, or close to, its surface. The output system involves amplifying, processing and displaying the signal in an appropriate format. In Table 26.1, the most frequently used sensing elements and transducers in the construction of biosensors are listed.

26.1.3
Immobilization of the Bioreceptor

For designing a biosensor, the biological component has to be properly attached to the transducer. This process is known as immobilization. One key point in biosensor fabrication is the development of immobilization technologies for stabilizing biomolecules and tethering them to surfaces. During the immobilization of the bioreceptor the recognizing properties have to be kept. Two immobilization techniques LB and SAMs, allowing the build up of the sensing layer at a molecular level, have been proven to be quite efficient, used alone or in combination. In this chapter these strategies will be detailed and illustrated with specific cases, AFM techniques appearing as a valuable tool to understand the mechanisms involved at a molecular level.

26.2
Langmuir–Blodgett and Self-Assembled Monolayers as Immobilization Techniques

26.2.1
Langmuir–Blodgett Technique

26.2.1.1
History

When describing the history of Langmuir or Langmuir–Blodgett films one is bound to start with an American statesman, Benjamin Franklin, who showed the first interest in monomolecular layers and reported the following to the British Royal Society in 1774 [1]:

> *At length at Clapman where there is, on the common, a large pond, which I observed to be one day very rough with the wind, I fetched out a cruet of oil, and dropped a little of it on the water. I saw it spread itself with surprising swiftness upon the surface; but the effect of smoothing the waves was not produced; or I had applied it first on the leeward fide of the pond, where the waves were largest, and the wind drove my oil back upon the shore. I then went to the windward fide, where they began to form; and there the oil, though not more than a teaspoonful, produced an instant calm over a space several yards square, which spread amazingly and extended itself gradually until it reached the leeside, making all that quarter of the pond, perhaps half an acre, as smooth as a looking glass.*

Over one hundred years later Lord Rayleigh suspected that the maximum extension of an oil film on water represents a layer with the thickness of one molecule. At the same time the foundation for our ability to characterize monolayers at an air/water interface was set by Agnes Pockels. She developed a rudimentary surface balance in her kitchen sink, which she used to determine (water) surface contamination as a function of area of the surface for different oils. Publication of Pockels's work in 1891 in Nature set the stage for Langmuir's quantitative work on fatty acid, ester and alcohol monolayers [2, 3].

Irwing Langmuir was the first one to perform systematic studies on floating monolayers on water in the late 1910s and early 1920s. He investigated the pressure-area relationship of molecules on an aqueous surface. The areas occupied by molecules such as acids, alcohols and esters were found to be independent of the hydrocarbon chain length, thus showing that only the hydrophilic head groups were immersed in the subphase. Langmuir was awarded the Nobel Prize for Chemistry in 1932 for his studies of surface chemistry, using floating monolayers to learn about the nature of intermolecular forces.

Katherine Blodgett, who worked with Langmuir on the properties of floating monolayers, developed the technique of transferring the films onto solid substrates and hence building up multilayer films. These built-up monolayer assemblies are therefore referred to as Langmuir–Blodgett films. The term "Langmuir film" is normally reserved for a floating monolayer at the air/water interface.

Interest in Langmuir–Blodgett films subsided with the outbreak of the Second World War, and remained low until the 1960s when Kuhn and Möbius showed how monolayers could be used to construct precise supermolecular structures. They used the Langmuir–Blodgett technique to demonstrate the fluorescence and quenching of dye molecules attached to fully saturated fatty acids under ultraviolet light [4]. This work and the publication of Gaines' "Insoluble Monolayers at Liquid Gas Interfaces" in 1966 initiated a revival of interest in the field. The first international conference on LB films was held in 1979 and since then the use of this technique has been increasing widely among scientists working on various different fields of research.

26.2.1.2
Langmuir Films

The formation of Langmuir films consists of two fundamental steps:

(i) Spread

To form a Langmuir film, the surface-active molecule, normally amphiphile, is dissolved in a volatile organic solvent (frequently chloroform or hexane) that will not react with or dissolve into the subphase. A very small quantity of this solution is spread on the surface of the subphase with a micropipette, and as the solvent evaporates, the surfactant molecules spread.

(ii) Compress

Sweeping the barrier over the subphase surface causes the molecules to come closer together and eventually to form a compressed, ordered monolayer.

26.2.1.2.1
Amphiphilic Molecules

An amphiphile is a molecule that is insoluble in water, with one hydrophilic end, which is preferentially immersed in the water; and the other hydrophobic part, which preferentially resides in the air (or in the nonpolar solvent) [5]. With correct "amphiphatic balance", namely the balance between the hydrophilic and hydrophobic constituents within the same molecules, such amphiphilic molecules can orient themselves at an air/water interface with their hydrophobic chains upwards in the air and their hydrophilic head groups downwards into the aqueous phase.

The hydrophobic part of amphiphile molecule usually consists of hydrocarbon or fluorocarbon chains, while the hydrophilic part consists of a polar group ($-OH$, $-COOH$, $-NH_3^+$, $-SO_3^-$, etc.). The hydrocarbon chain of the substance used for monolayer studies has to be long enough for formation of an insoluble monolayer. A rule of thumb is that there should be more than 12 hydrocarbons or groups in the chain ($(CH_2)_n$, $n > 12$). If the chain is shorter, though still insoluble in water, the amphiphile on the water surface tends to form micelles. On the other hand, if the chain is too long the amphiphile is prone to crystallize on the water surface, not to form a monolayer at the air/water interface. It should be noted that the solubility of an amphiphile in water depends on the balance between the alkyl chain length

Table 26.2. Effect of different functional groups on Langmuir films formation with C16-compounds

No film	Unstable film	Stable film	Dissolve in water
Hydrocarbon	$-CH_2OCH_3$	$-CH_2OH$	$-SO_3^-$
$-CH_2I$	$-C_6H_4OCH_3$	$-COOOH$	$-OSO_3^-$
$-CH_2Br$	$-COOCH_3$	$-CN$	$-C_6H_4SO_4^-$
$-CH_2Cl$		$-CONH_2$	$-NR_4^+$
$-NO_2$		$-CH = NOH$	
		$-C_6H_4OH$	
		$-CH_2COCH_3$	
		$-NHCONH_2$	
		$-NHCONH_3$	

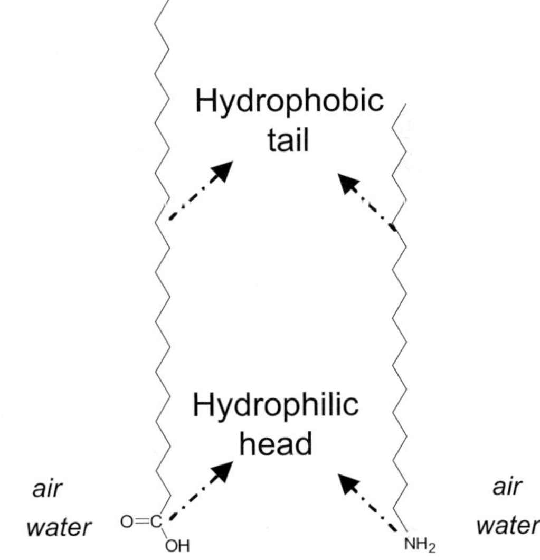

Fig. 26.2. Chemical structures of BA and ODA, and their spreading at the air/water interface

and the strength of its hydrophilic part. Abraham Ulman presented the effect of different functional groups on monolayer formation of C16-compounds as shown in Table 26.2.

The amphiphilic molecules used in our study are behenic acid (BA) ($C_{21}H_{43}$ COOH) and octadecylamine (ODA) ($C_{18}H_{37}NH_2$). Their chemical structures are shown in Fig. 26.2.

26.2.1.2.2
LB Instrument

The formation of Langmuir and Langmuir–Blodgett films is performed in a Langmuir trough, which is equipped with a Wilhelmy balance in order to measure surface pressure of the monolayer, one or two moveable barriers to compress the monolayer at the air/water interface, and a dipper mechanism permitting transfer of the monolayer

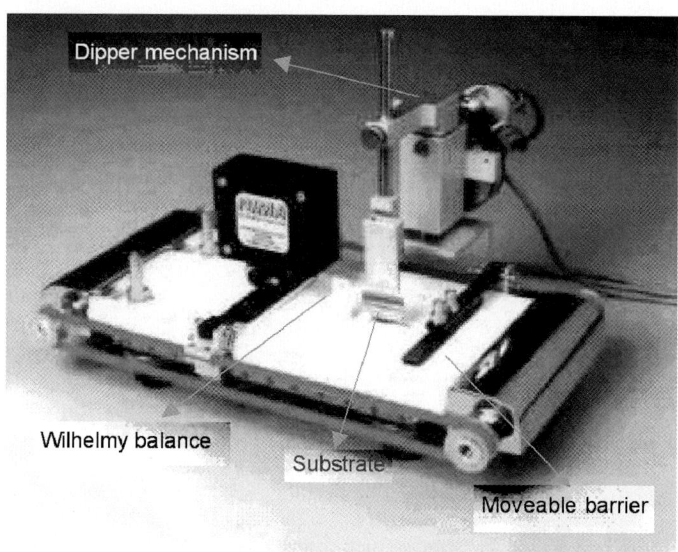

Fig. 26.3. Langmuir trough produced by NIMA

onto solid substrate, as shown in Fig. 26.3. Usually the trough is placed in a chamber in order to prevent external contamination.

26.2.1.2.3
Surface Pressure–Molecular Area Isotherm

The most important indicator of the monolayer properties is given by measuring the surface pressure as a function of the area of each molecule. This is carried out at a constant temperature and is known as the surface pressure–molecular area isotherm (referred to as Π-A). Usually an isotherm is recorded by compressing the film at a constant rate while continuously monitoring the surface pressure. A schematic Π-A isotherm is shown in Fig. 26.4.

The surface pressure–molecular area isotherm is rich in information on stability of the monolayer at the air/water interface, the reorientation of molecules in the two-dimensional system, phase transitions, and conformational transformations.

A simple terminology used to classify different monolayer phases of fatty acids has been proposed by Harkins as early as 1952 [6]. As the monolayer is compressed, it shows three phases: gas, liquid or a solid one.

In the gas state, the area per molecule is large, and ideally there should be no interaction (lateral adhesion) between the molecules, and therefore, the surface pressure is low.

As the barriers move, the monolayers are compressed, accompanyied by an intermolecular distance decrease and a surface pressure increase. A phase transition (from the gas to the liquid state) takes place. In the liquid phase, the monolayer is coherent, and the molecules have more degrees of freedom and gauche conformations can be found in the alkyl chains.

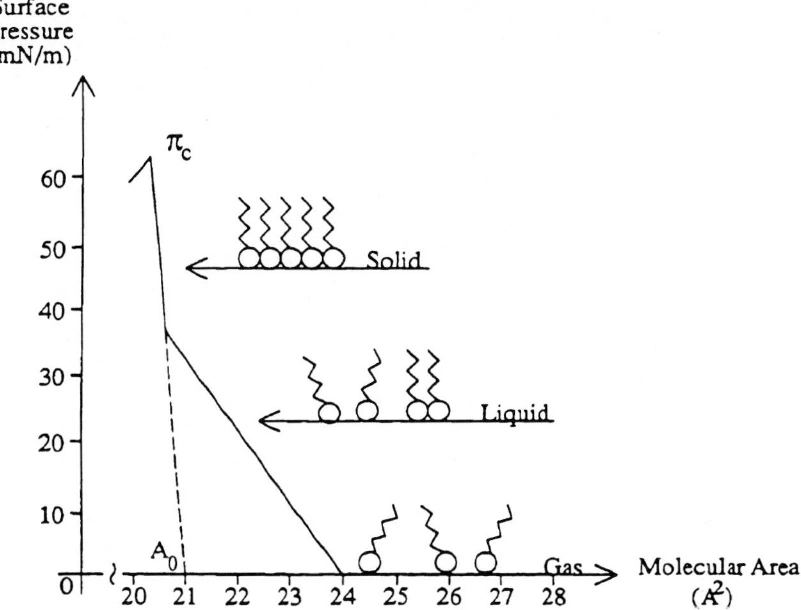

Fig. 26.4. Schematic Π-A isotherm

When the barriers compress the film further, a second phase transition can be observed, from the liquid to the solid state. In the condensed solid phase, the molecules are closely packed and uniformly oriented. If further pressure is applied on the monolayer, it collapses due to mechanical instability, and consequently a sharp decrease in the surface pressure is observed.

26.2.1.3
Langmuir–Blodgett Films

26.2.1.3.1
Principle

The term "Langmuir–Blodgett film" traditionally refers to monolayers that have been vertically transferred off the water subphase and onto a solid support. Vertical deposition is the most common method of LB film transfer, called the Langmuir–Blodgett technique. An alternative way to deposit the monolayer is the Langmuir–Schaeffer (LS) technique, which differs from the LB technique only in the sense that the solid substrate is horizontally lowered in contact with the monolayer for transferring the monolayer.

The LB deposition is traditionally carried out in the "solid" phase with surface pressure kept constant. In this case the surface pressure is high enough to ensure sufficient cohesion in the monolayer, so that the monolayer will not fall apart during transfer to the solid support. This also ensures building up of homogeneous multilayers.

When the solid surface is hydrophobic the first layer is deposited during the first down pass. When the solid support is hydrophilic, the first layer will be deposited during the first up-stroke pass, because the hydrophobic tails are repelled by the hydrophilic support as it is immersed in the water. Once the first layer is deposited, more layers will be deposited on each subsequent pass of the support through the air/water interface. Multilayers can therefore be deposited onto the support.

There are several parameters that can affect quality of LB films:

- Quality of the monolayer at the air/water interface, particularly the stability;
- Properties of the subphase (temperature, ingredient, pH);
- Surface pressure for transferring LB films, namely the target pressure;
- Transfer rate;
- Surface properties of support (cleanliness, surface free energy, chemical composition, etc).

In order successfully to deposit LB films onto the support, usually the substrate is previously functionalized to provide a favorable platform for the deposition.

26.2.1.3.2
Types of Deposition

Different film architectures can be obtained upon deposition by the Langmuir–Blodgett technique (Fig. 26.5). Y-type multilayers are most common and can be prepared on either hydrophilic or hydrophobic substrates with the reversed head–tail arrangement. They are the most stable architectures because of strong head–head and tail–tail interactions. X-type and Z-type films are reported less often. In fact, with some amphiphiles, even when films have been deliberately transferred by an X- or Z-method, the spacing between the hydrophilic headgroups shows a packing similar to that of Y-type films, implying that some molecular reorganization could occur during or shortly after deposition [7].

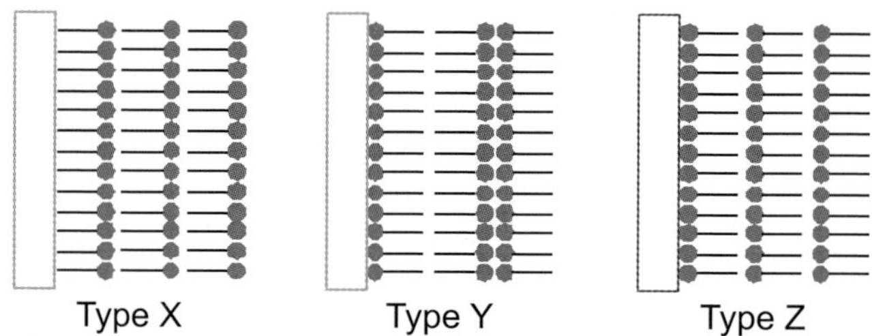

Fig. 26.5. Possible architectures for Langmuir–Blodgett films

26.2.1.3.3
Transfer Ratio

The quantity and the quality of the deposited monolayer on the solid support can be measured by transfer ratio (TR), given by the equation:

$$TR = A_l/A_s \,,$$

where A_l is the area of monolayer removed from subphase at a constant pressure, A_s is the area of substrate immersed in water. Such a ratio is usually measured for each substrate pass through the air/water interface. For an ideal transfer the TR is equal to 1.

26.2.1.4
Mixed Amphiphile/Protein LB Films

Langmuir–Blodgett films have been studied for application in a great variety of domains, such as nonlinear optics, molecular electronics, sensors, etc. Since the 1980s, extensive studies have been carried out to immobilize biomolecules by the LB technique for elaboration of biosensors. Proteins were introduced in the amphiphile monolayer to form mixed amphiphile/protein LB films. There are several different ways to prepare mixed amphiphile/protein LB films.

26.2.1.4.1
Methods for Preparation of the Mixed Amphiphile/Protein LB Films

Adsorption of Proteins Under the Compressed Amphiphile Monolayer In 1971, Fromherz used this method to make the mixed layer with a multicompartment Langmuir trough [8]. As shown in Fig. 26.6, in the first compartment, the amphiphile was injected on the subphase of pure water and the monolayer was compressed till the surface pressure in the range of 5 to 30 mN m^{-1}. After that, under a constant surface pressure, the layer was moved to the second compartment, in which the proteins were present in the subphase. The proteins were adsorbed onto the head groups of amphiphile molecules by electrostatic interaction and the mixed monolayer formed. Finally, the mixed layer was moved to the third compartment with pure water as the subphase, where the mixed layer was transferred onto solid supports. Arisawa and Yamamoto [9] developed a urea sensor by immobilizing urease onto an ISFET by using this LB technique.

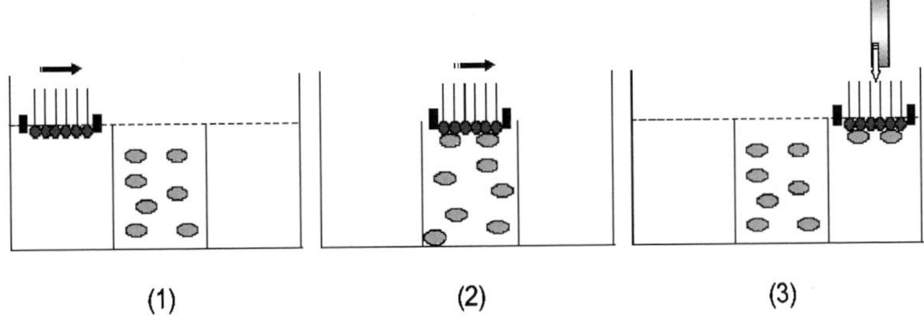

Fig. 26.6. Formation of the mixed LB films by adsorption of proteins under the compressed amphiphile monolayer

Injection of Proteins Under the Compressed Amphiphile Monolayer This method was first developed by Zaitsev et al. [10, 11], they injected a solution of protein in

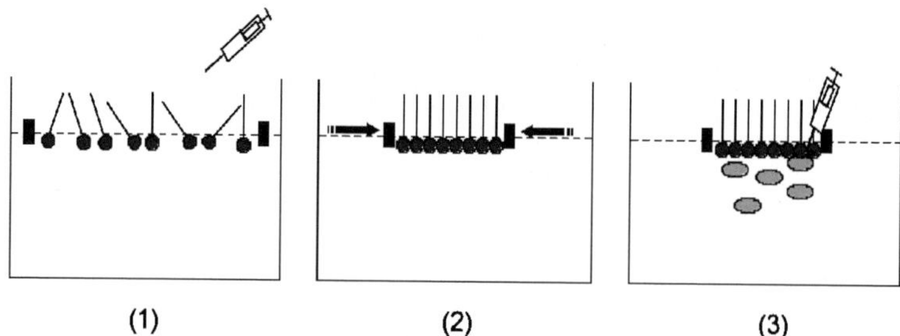

Fig. 26.7. Formation of the mixed LB films by injection of proteins under the compressed amphiphile monolayer

the water subphase under the amphiphile layer that was previously formed at the air/water interface, as shown in Fig. 26.7. However, these two methods show poor reproducibility for the formation of mixed LB films.

Adsorption of Proteins Under Noncompressed Amphiphile Monolayer The most commonly used method is based on adsorption of proteins under a noncompressed amphiphile monolayer. As shown in Fig. 26.8, first, the amphiphile solution is injected on the subphase of protein solution, proteins are gradually adsorbed onto the amphiphile layer by the electrostatic interaction. When the adsorption reaches equilibrium, the mixed monolayer is compressed to the target pressure. And finally at this surface pressure the stable mixed layer is transferred onto solid substrates.

Such mixed LB films have been investigated by the group of Valleton [12–20] and in our laboratory [21–23]. Compared to the first two methods mentioned above, this method has several advantages: simplicity for obtaining well-ordered multilayers

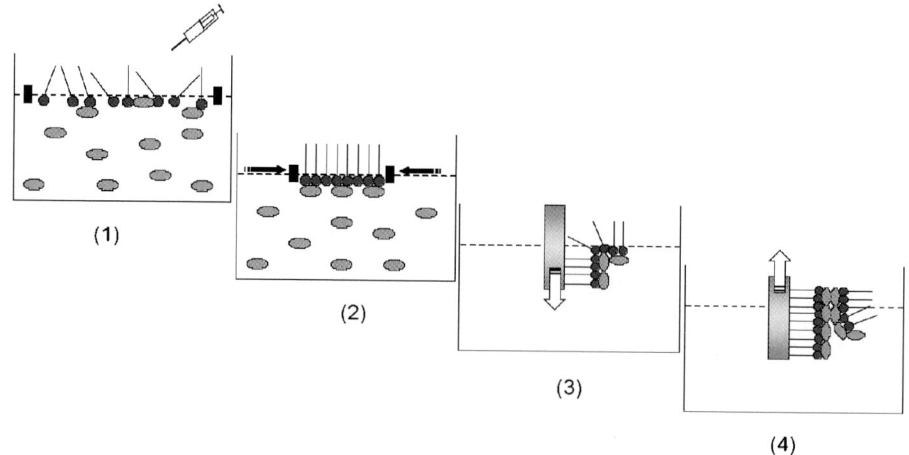

Fig. 26.8. Formation of the mixed LB films by adsorption of protein under the noncompressed amphiphile monolayer

of LB films, high reproducibility for depositing of LB films, and highly homogenous structures of LB films. There are still some other LB methods, which are not commonly used, to elaborate the mixed amphiphile/protein LB films. For example, Girard-Egrot et al. [24,25] first deposited amphiphile LB films on the substrates, and then immersed the coated substrates in a solution of enzyme glutamate dehydrogenase for adsorption onto an amphiphile multilayer. Okahata et al. [26] prepared an enzyme–lipid complex in the first step, and then transferred the monolayer of the complex onto a Pt substrate as the sensor membrane.

26.2.1.4.2
Structure of the Mixed Monolayer at the Air/Water Interface

Protein can be adsorbed onto amphiphile monolayers spread at the air/water interface by electrostatic interaction between the protein and the polar heads of amphiphile molecules. Zaitsev [27] investigated the influence of the electrostatic effect on the adsorption of protein by studying the mixed monolayer of glucose oxidase with different kinds of amphiphiles: phosphatidylcholine (neutrally charged), cetyltrimethylammonium bromide (positively charged) and stearic acid (negatively charged). The results showed that the enzyme GOD, which is negatively charged in buffer solution at pH 7.0, is more strongly adsorbed on positively charged lipid monolayers than on zwitterionic ones, and only slightly adsorbed on negatively charged monolayers of stearic acid.

The dynamic formation and structure of mixed monolayer with BA/GOD at the air/water interface proposed by the group of Valleton is shown in Fig. 26.9 [18].

The results showed that before spreading the amphiphile behenic acid, the surface pressure was about $2\,\mathrm{mN\,m^{-1}}$ showing that some of enzyme molecules were adsorbed at the air/water interface. Enzyme GOD was allowed to adsorb under the

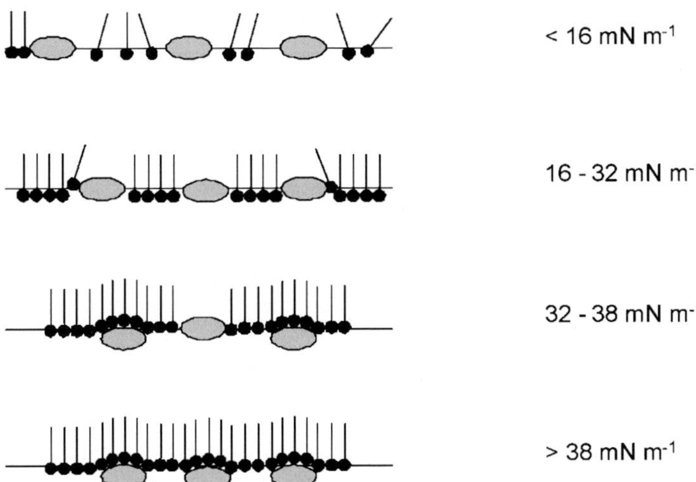

Fig. 26.9. Dynamic form ation and structure of mixed monolayer with BA/GOD at the air/water interface

behenic acid film at the air/water interface. It took about 45 min to reach adsorption equilibrium and the surface pressure reached $16\,\mathrm{mN\,m^{-1}}$. Then the monolayer was compressed at a constant rate, during the compression the surface pressure increased. In the range of surface pressure from $16\,\mathrm{mN\,m^{-1}}$ to $32\,\mathrm{mN\,m^{-1}}$, the mixed monolayer was thought to be in a liquid phase. With further compression, the enzyme molecules were gradually expelled from the air/water interface and pushed down under the polar heads of BA molecules, the monolayer is considered to reach gradually the solid phase. At a high surface pressure (above $38\,\mathrm{mN\,m^{-1}}$), the mixed film was constituted of a behenic acid monolayer in a solid phase, under which a GOD sublayer was adsorbed by the electrostatic interaction as shown in Fig. 26.9.

However, the phase change above $40\,\mathrm{mN\,m^{-1}}$ was not mentioned. If further compression was applied on the mixed monolayer, there would be a risk of overlapping of the layer and even a complete collapse of the monolayer.

26.2.1.4.3
Application of LB Technique for Biosensor Elaboration

The quest for biosensors has brought the need for highly selective and sensitive organic layers, with tailored biological properties that can be incorporated into electronic, optical, or electrochemical devices. The molecular dimension of LB films makes the use of even highly expensive molecules economically attractive, therefore, the development of sophisticated detection systems becomes possible.

Up to now, the LB technique has been widely used to immobilize enzymes [28–36], antibodies [37–44] and DNA [45], etc. on the substrate for the elaboration of various biosensors.

26.2.2
Self-Assembled Monolayers

26.2.2.1
Introduction

26.2.2.1.1
Definition of Self-Assembled Monolayers

Self-assembled monolayers are molecular assemblies that are formed spontaneously by the immersion of an appropriate substrate into a solution of a reaction-active surfactant in a suitable solvent.

From the energetics point of view, a self-assembling surfactant molecule can be divided into three parts, as shown in Fig. 26.10. The first part is the head-anchoring group that provides the most exothermic process while anchoring on the substrate surface by the chemisorption. The energy associated with the chemisorption is of the order of tens of kcal/mol (e.g., $\sim 40-45\,\mathrm{kcal\,mol^{-1}}$ for thiolate on gold). The second part is the alkyl chain, and the energy associated with its interchain van der Waals interaction is of the order of a few (< 10)

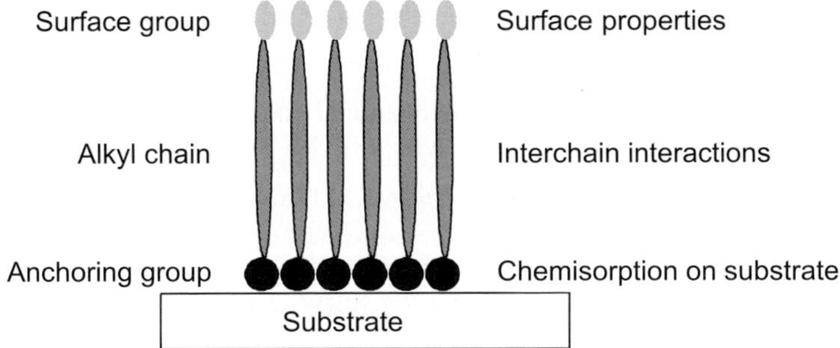

Fig. 26.10. A schematic view of the self-assembling surfactant molecule in SAMs

$kcal\,mol^{-1}$ (exothermic). The third part in the surfactant is the terminal surface group, which determines the surface properties (e.g. wetting, friction, adhesion, etc.) of the SAMs functionalized substrate surface. For example, in the case of a simple alkyl chain, the terminal methyl (CH_3) group provides a hydrophobic surface. The ordering of alkanethiol is driven by the strong affinity between the head groups and the substrates, the lateral van der Waals interactions between the tethered alkyl chains, and the dipole interactions between the polar end groups.

26.2.2.1.2
Types of Self-Assembled Monolayers on Electrode

SAMs can form on many materials that possess well-defined surface chemistries, including metals, semiconductors, oxides and so on [46]. In Table 26.3, surfactants and the related substrates, on which they can form SAMs are listed.

Table 26.3. Chemical systems of substrates and the corresponding surfactants that form SAMs

Surface	Substrates	Surfactants
Metal	Au	R–SH, R–SS–R, R–S–R, R–NH$_2$, R–NC, R–Se, R–Te
	Ag	R–COOH, R–SH
	Pt	R–SH, R–NC
	Pd	R–SH
	Cu	R–SH
	Hg	R–SH
Semiconductor	GaAs	R–SH
	InP	R–SH
	CdSe	R–SH
	ZnSe	R–SH
Oxide	Al$_2$O$_3$	R–COOH
	TiO$_2$	R–COOH, R–PO$_3$H
	ITO	R–COOH, R–SH, R–Si(x)$_3$
	SiO$_2$	R-Si(x)$_3$

26.2.2.1.3
Advantages of Self-Assembled Monolayers

The advantages of SAMs arise from the chemical binding of the anchoring group at the substrate surface and the ease with which SAMs are formed.

- SAMs are very stable as mentioned above and they are able to survive after long exposure to vacuum.
- Different surface properties of the substrates can be provided by terminal groups of the surfactants without disrupting the self-assembly process or destabilizing the SAMs.
- SAMs provide a means of attaching the biomolecules to the electrode with a diverse set of structures ranging from modified monolayers to multilayers.
- In electrochemistry, the thiol-based SAMs survive the electrochemical treatment and experiments. The anchored sulfur atoms are believed to resist oxidation, reduction, and desorption in a range of potentials.
- A monolayer is deposited on the metal in a manner of seconds to minutes.
- Usually, this method does not require anaerobic or anhydrous conditions; nor does it require vacuum.
- Weakly adsorbed impurities have a slight effect on the formation of SAMs, since high affinity interaction between the anchoring group and the substrate enables the assembling layer to displace more weakly adsorbed impurities.
- Curvature or accessibility of the substrate surface is not a serious obstacle factor; because substrates can range from the macro- to the nanoscale, from smooth to porous, as well as from plane to spherical (nanoparticles).

Another advantage is the amenability to application in controlling interfacial (physical, chemical, electrochemical) properties.

Compared to LB films SAMs are more stable. Since LB films are deposited onto the substrates by physisorption, mainly by hydrophilic and/or hydrophobic interaction, they are more sensitive to the external environment, such as temperature, pH, solvent, etc.

26.2.2.2
Preparation of Self-Assembled Monolayers

In further examples shown, SAMs of alkanethiol on gold and silver were performed. Therefore, only preparations of SAMs on gold and silver substrates will be discussed here.

Usually, the gold and silver substrates are obtained by evaporating or sputtering gold and silver films (typically 50–200 nm thick) on glass, silicon, or cleaved mica. To improve the adhesion of the gold and silver film to the oxide surface, a thin layer of chromium or titanium is often deposited first.

26.2.2.2.1
Preparation of SAMs on Gold Substrates

Pretreatment of Gold Substrates   Gold is the most popular substrate for thiol SAMs. Owing to its noble character, gold substrate can be handled in air without the formation of an oxide surface layer, and can survive harsh chemical treatments for the removal of organic contaminants. The formation of a monolayer on gold is strongly dependent on the cleanliness and structure of gold prior to modification [47], since it is well known that gold is a soft metal and is readily contaminated by organic and inorganic species during manual handling [48].

Before use, gold substrates need to be cleaned well in order to remove the contaminants on the surface that may affect the integrality of the SAMs formation. The importance of the cleanliness of the gold surface on formation and reproducibility of SAMs has been recognized by many groups working in the field of SAMs. In the literature numerous different cleaning procedures have been proposed. Most of them involve gold surface oxidation to remove organic and inorganic contaminants on the gold surface: electrochemical oxidation [49], immersion into strong oxidizing solutions [50], a combination of both [51], or exposure to reactive oxygen species formed by either UV irradiation or oxygen plasma [52, 53].

In most examples discussed hereafter, a strong oxidizing solution, H_2SO_4:H_2O_2 (7:3 v/v) was used. First, the gold substrates were cleaned with acetone in an ultrasonic bath for 10 min in order to remove the protection layer of polymer and dried under nitrogen flow, after they were immersed into the hot cleaning solution for 1 min. Subsequently, the gold substrates were rinsed thoroughly in absolute ethanol, and dried under nitrogen flow. The immersion time in acetone and piranha was optimized.

The wetting properties of gold surface before and after pretreatment were studied with a contact-angle meter. It is widely accepted that clean gold is hydrophilic. In this thesis, the contact angle of gold substrate as received is about $90°$, which is due to the presence of a carbon- and oxygen-containing contaminant layer on the gold surface [54]. After cleaning, the contact angle is $< 10°$, which proves that the gold surface is free of contaminants. The contact-angle images are shown in Table 26.4.

Table 26.4. Contact angle values and images of the gold substrates before and after different treatments

Samples	as received	after cleaning	with SAMs of ODT	with mixed SAMs
Contact angle	$90° \pm 3°$	$< 10°$	$110° \pm 3°$	$9° \pm 3°$
Contact angle image				

Formation of the SAMs on Gold Substrate   The most common approach for formation of SAMs on solid substrate is by immersing the substrate into a homogeneous solution of the self-assembling molecule at room temperature. There are a number of experimental factors that can affect the structure of the resulting SAMs: solvent, temperature, concentration of the adsorbate, immersion time, purity of the adsorbate, concentration of oxygen in solution, cleanliness of the substrate, and chain length or more generally, structure of the adsorbate.

Usually, any solvent capable of dissolving the adsorbate is suitable. Ethanol is the most popular solvent because some plausible gold oxide will be reduced in ethanol, but sometimes other solvents are preferred. The thiol concentration can be varied from micromolar levels to the net liquid thiol. However, very low concentrations of thiols are favorable for obtaining large crystalline domains of alkanethiols via slow self-assembly. Immersion time varies from several minutes to several hours for alkanethiol, while for sulfides and disulfides, immersion times of several days may be necessary.

In the examples presented, gold substrates were functionalized with two kinds of SAMS. In the first case, as soon as the cleaning procedure is finished, the gold substrates were put into a 1 mM 1-octadecanethiol (ODT) ethanol solution for 21 h at room temperature to render their surfaces hydrophobic. Since such a hydrophobic surface is favorable to the deposition of LB films. Then the substrates were thoroughly cleaned with ethanol to remove physically adsorbed thiol on the surface and dried under a N_2 flow. The contact-angle image of the gold substrate modified with SAMs of ODT is given in Table 26.4.

In the second case, after cleaning, gold substrates were immersed immediately into a mixed solution of 16-mercaptohexadecanoic acid (MHDA) and 1,2-Dioleoyl-*sn*-Glycero-3-Phosphoethanolamine-*N*-(Biotinyl) sodium salt (biotinyl-PE) in absolute ethanol at the concentration of 1 mM and 0.1 mM, respectively, for 21 h in order to obtain mixed self-assembled monolayers. Then the substrates were thoroughly cleaned with ethanol to remove physically adsorbed thiol and biotinyl-PE on the surface, and finally dried under a N_2 flow. The contact-angle image of the gold substrate with the mixed SAMs is also shown in Table 26.4. Figure 26.11 shows the procedures for the functionalization of gold substrate by thiol SAMs.

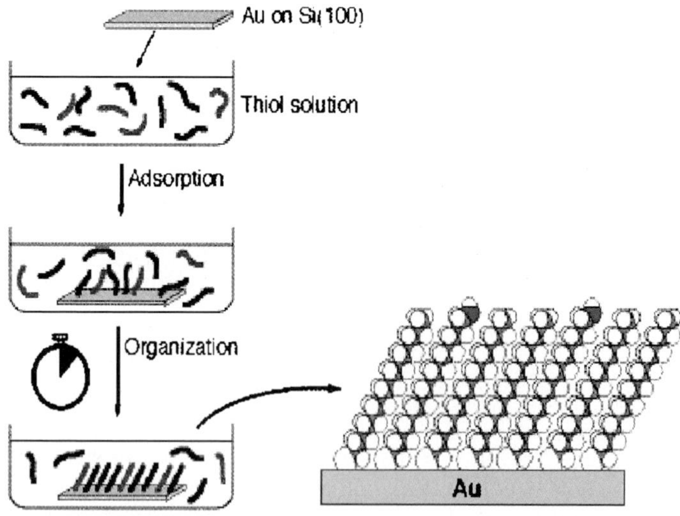

Fig. 26.11. Different steps of the functionalization of gold substrate by thiol SAMs

There are some other deposition methods for SAMs. When the thiol is sufficiently volatile, SAMs deposition can be performed in the vapor phase of the thiol, both in a vacuum [56] and at ambient pressures [57]. SAMs deposition can also be performed in a solid phase, Xia and Whitesides [58] developed the technique of microcontact printing μCP: It uses the relief pattern on the surface of a PDMS stamp to form patterns of self-assembled monolayers on the substrate surface by contact. In μCP of alkanethiols on gold, for example, the PDMS stamp is wetted with an "ink" (typically, a 2 mM solution of thiol in ethanol) and is brought into contact with the gold surface for 20 s. The thiol is transferred from the stamp to the gold surface upon contact, SAMs with predesigned patterns are stamped on the gold surface.

26.2.2.2.2
Pretreatment and Formation of SAMs on Silver Substrates

For silver substrates, mild cleaning procedures are necessary due to their high reactivity to the oxygen in air [55]. Samples were only cleaned with hot ethanol for 5 min, and then in order to minimize the effects of oxidation and contamination, they were immediately immersed into thiol solution of ODT with a concentration of 1 mM for 24 h. Since they are easily oxidized, the formation of SAMs was performed in the presence of N_2 flow, as shown in Fig. 26.12. After the substrates were thoroughly cleaned with ethanol to remove physically adsorbed thiol and dried under a N_2 flow. The contact angle of ODT-functionalized silver surface is $115 \pm 3°$.

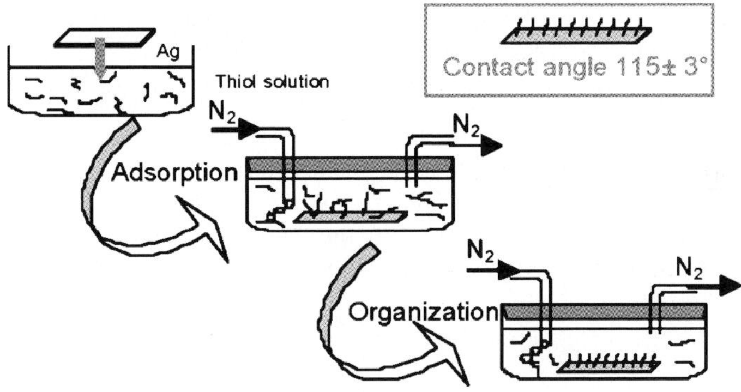

Fig. 26.12. SAMs functionalization of silver substrate with 1 mM ODT

26.2.2.3
Kinetics and Mechanism of Formation of SAMs

There are substantial differences among reports in the literature regarding the kinetics of formation of the SAMs on the solid substrates from liquid solution. The most important factors affecting the kinetics features appear to be the presence of preadsorbed contaminants, the solution phase hydrodynamics, and solvent effects [59].

The general kinetic of alkanethiol adsorption onto Au(111) in relatively dilute solution (10^{-3} M) is believed to have two distinct adsorption kinetics: a very fast step, which takes a few tens of seconds to minutes, by the end of which the contact angles are close to their desired values and the thickness are about 80–90% of their maximum; a slow step, which may last several hours, at the end the thickness and contact angles will reach their final values [60].

The initial step is a diffusion-controlled Langmuir adsorption, its kinetics is governed by the surface–head group reaction. The second step can be described as a surface reorganization process, during which the alkyl chains undergo a phase change from the disordered state to an ordered state and form a two-dimensional quasicrystal. Its kinetics is related to the chain disorder, different components of chain–chain interaction, and surface mobility of chains.

Extensive X-ray photoelectron spectroscopy (XPS) experiments suggest that chemisorption of alkanethiols on a gold surface yields the gold thiolate species. In the case of alkanethiols, the mechanism of alkanethiols binding to gold substrates is considered to be an oxidative addition of the S–H bond to gold substrate followed by reductive elimination of the hydrogen, thus resulting in the formation of a thiolate species, as shown in reaction (I):

$$R\text{–}SH + Au_n^0 \rightarrow RS^-\text{–}Au^+ + (1/2)\, H_2 + Au_{n-1}^0 \,. \tag{I}$$

In the case of disulfides, the mechanism is thought to be:

$$R\text{–}S\text{–}S\text{–}R + Au_n^0 \rightarrow 2RS^-\text{–}Au^+ + Au_{n-2}^0 \,. \tag{II}$$

Another common SAMs system is silanization of silicon oxide. For the formation of the SAMs on the silicon oxide, for example, with octadecyltricholorosilane (OTS), two possible monolayer growth modes were proposed by Dutta and coworkers: island growth and uniform growth [61, 62]. In the island mode, the molecules form dense islands of (nearly) vertical molecules. Additional molecules attach at island edges or form new islands as the coverage increases. In the latter mode, molecules deposit randomly on the substrate, falling over at low coverage. As more molecules adsorb on the substrate, the molecules begin to stand up.

26.2.2.4
Structure of SAMs on Gold and Silver Substrates

Alkanethiols on Au $\langle 111 \rangle$ form a ($\sqrt{3} \times \sqrt{3}$) R30° (R = rotated) structure. Sulfur atoms bond to the three-fold hollow sites on the Au surface and the extremely strong surface bond ($44\,\text{kcal mol}^{-1}$) contributes to the stability of the SAMs. Alkanethiols on Ag $\langle 111 \rangle$ form a ($\sqrt{7} \times \sqrt{7}$) R10.9° structure with an S–S distance of 0.441 nm. Because of this shorter S–S distance, alkanethiols on Ag are more densely packed, which results in near-perpendicular orientation of molecules on the Ag surfaces. However, in the case of Au $\langle 111 \rangle$, the alkyl chain is titled 30° from the surface normal to decrease the lattice energy. The difference in orientation is mainly due to the interchain van der Waals interaction [63].

26.2.2.5
Protein Immobilization Using SAMs

As referred to above, SAMs are well-ordered, once formed they are quite stable due to their high affinity to the substrate. Moreover, the variety of terminal functionality of SAMs can provide the needed design flexibility, both at the individual molecular and at the material levels, and offer a vehicle for investigation of specific interactions at interfaces. Such well-ordered monolayers are suitable for protein immobilization, close to the electrode surface with a high degree of control over the molecular architecture of the recognition interface. In practice, SAMs have been used for fundamental investigations of the interactions of proteins with surfaces and particularly for the fabrication of a variety of biosensors including enzyme biosensors [64–66], immunosensors [67], and DNA hybridization biosensors [68], etc.

There are two basic approaches for proteins immobilization onto gold substrates surface by SAMs. Proteins can be attached to thiol-containing moieties and then bind to the gold surfaces by SAMs, or they can be bound to previously SAMs-functionalized gold substrates.

26.2.2.5.1
Protein Modification

In this case, proteins are first tagged in the solution with an appropriate sulfur-containing molecule via reaction with specific reactive groups present on the biomolecule surface, e.g. the amino group. And then the modified proteins are self-assembled onto the gold substrate via the sulfur moiety.

It has been reported that proteins can be chemically modified by N-succinimidyl 3-(2-pyridyldithiol)propionate (SPDP) (Reaction A) and 2-iminothiolane (Reaction B). During such modification, a disulfide moiety is connected to the primary amine group of the protein via an amido linkage [69, 70]. Then modified proteins were immobilized onto the gold surface by SAMs [71], as shown in Fig. 26.13.

Such a method allows a monolayer of protein to be attached to the electrode. In addition, single-point labeling of the protein allows control over the orientation of the protein. However, Dong et al. [72] showed that the immobilization of thiol-labeled glucose oxidase produced a less stable sensing interface than enzyme immobilized onto a preformed SAM.

26.2.2.5.2
Substrate Modification

Through such a method the substrate is functionalized by a SAM monolayer of a sulfur-containing surfactant. Then proteins are bound to the SAMs via reaction using specific moieties present on the biomolecule structure.

Noncovalent Binding   One common noncovalent approach for immobilizing protein to the surface of SAMs is by electrostatic binding. This method is simple and

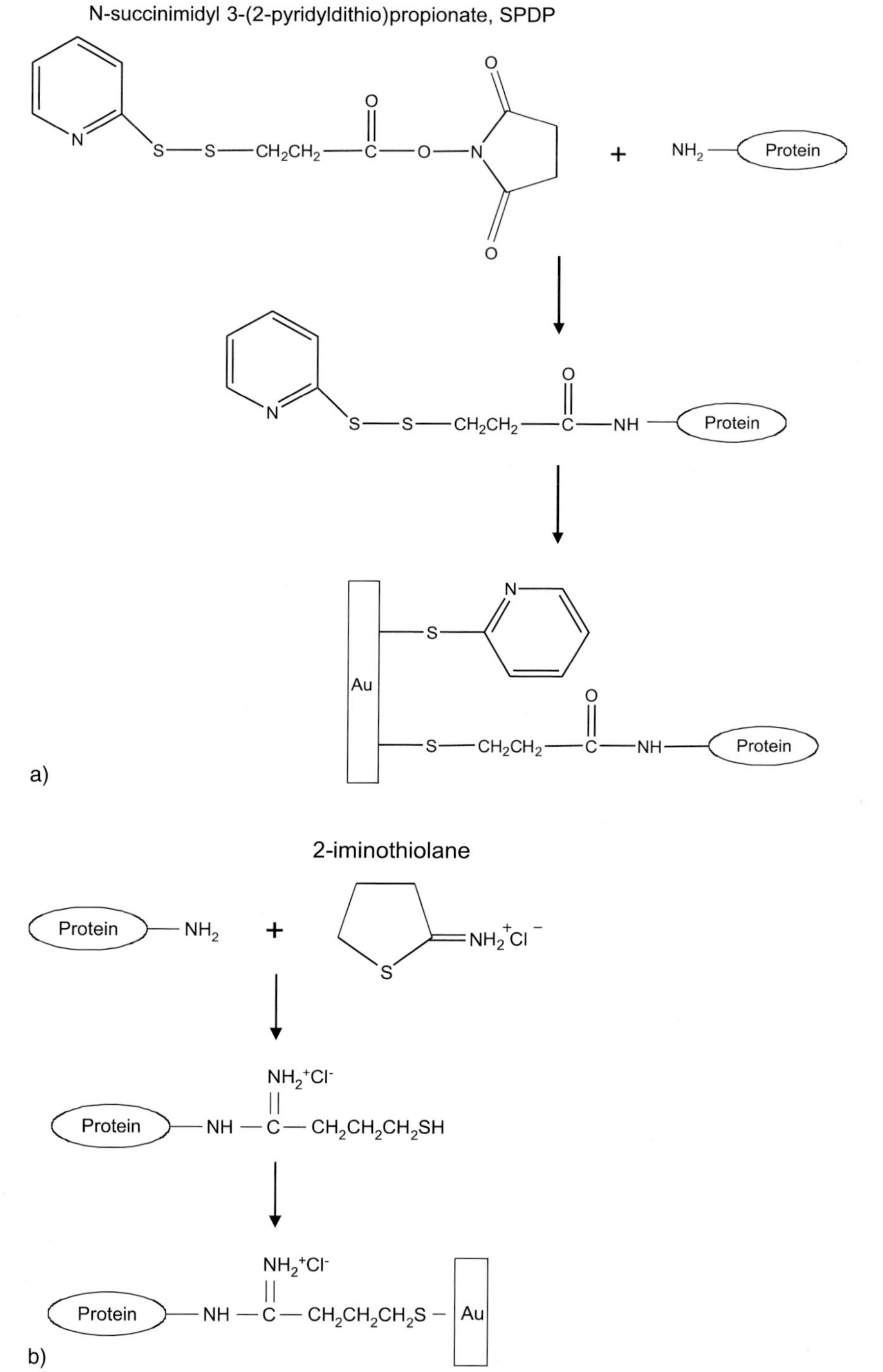

Fig. 26.13. Previously modified protein immobilization by SAMs

gentle, and it can provide a potential control over the orientation of the immobilized protein molecules depending on the charge distribution on the shell of the protein.

Mixed protein/amphiphile LB films have been transferred onto SAMs prefunctionalized gold and silver substrates by noncovalent binding and used for designing impedimetric immunosensors [85, 86].

The major drawback of the electrostatic binding is that the strength of the bond is dependent on the solution conditions. The change in ionic strength and pH can lead to removal of proteins from the surfaces of the SAMs.

Crosslinking   Crosslinking proteins to the surfaces of SAMs with bifunctional reagent glutaraldehyde is a relatively uncontrolled method. It suffers from poor reproducibility and protein leaching. Usually multilayers of protein are produced on the substrate surface.

Direct Covalent Attachment   Direct covalent attachment is the most potential method for elaboration of biosensor due to the high stability of the resulting bond. The immobilization is conducted in two stages: activation of the carrier, and coupling of the biomolecule.

Usually proteins are covalently bound to COOH-terminated SAMs by using carbodiimide coupling that allows amines to be coupled to carboxylic acids. The most often used carbodiimide coupling are N-ethyl-N-[dimethylaminopropyl] carbodiimide (EDC) and N-hydroxysuccinimide (NHS). In the reaction EDC converts the carboxylic acids into a reactive intermediate which is susceptible to be attacked by amines. In some cases, EDC and NHS or N-hydroxysulfosuccinimide (NHSS) are used together as they produce a more stable reactive intermediate that has been shown to give a greater reaction yield.

For covalently coupling proteins onto amine-terminated SAMs, a bifunctional linker glutaraldehyde is often used. GA can react with amino tails of SAMs to create –C–N=C– bond, after that the proteins with the amine group are coupled with the other aldehyde group.

Mirsky et al. [67] reviewed surface-activation methods for immobilization of proteins on the monolayers of 16-mercaptohexadecanoic and 16-mercaptohexadecylamine on gold electrodes. Here, we only list the most commonly used activation methods of carboxy and amino groups for the elaboration of biosensors, as shown in Fig. 26.14.

Direct Affinity Attachment   More recently a new strategy for more site-specific binding of biomolecules based on a streptavidin/biotin pair has been extensively investigated. Streptavidin is a tetrameric protein, approximately $4.5 \times 4.5 \times 5.8$ nm, and with a molecular weight of 60 kDa. It can bind biotin at four binding sites, two binding sites on each of two opposite faces, with an extremely high binding constant ($K_d = 10^{-15}$ M), which makes the binding process effectively irreversible [73]. Moreover, diverse biomolecules can be easily functionalized by biotin with no disturbing effect on their specific sites. All these properties make the avidin/biotin binding pair highly suitable for immobilizing proteins on the substrates and for the construction of biosensors.

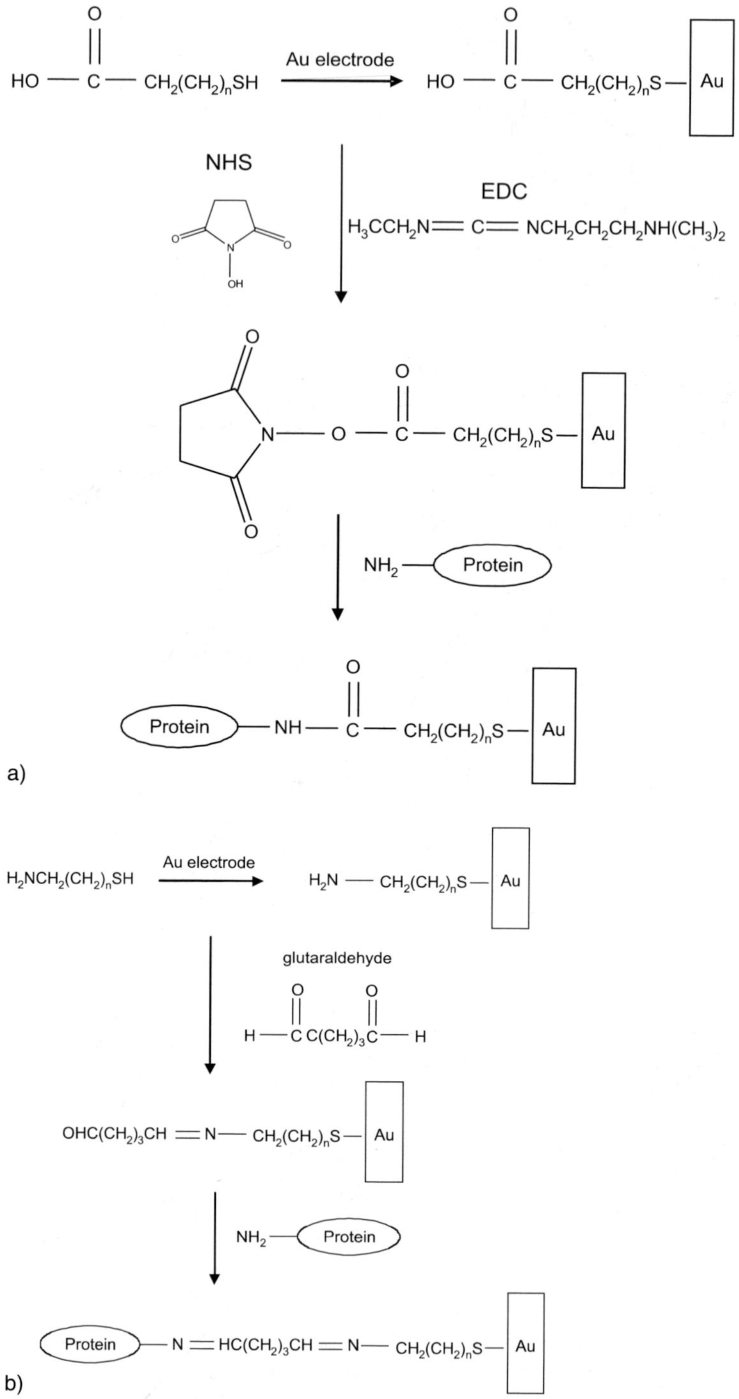

Fig. 26.14. Protein immobilization on previously SAMs-functionalized substrates by covalent attachment

The formation of SAMs with biotin molecules on gold substrate can be realized either by covalently coupling biotin to previously chemisorbed ω-functionalized thiolates [74–77], or by directly self-assembling of biotinylated thiols or disulfides [78, 79]. Usually, in order to optimize the binding properties of the monolayer, accordingly to reduce the nonspecific interaction between streptavidin and the surface, SAMs of biotinylated thiol are diluted by hydroxyalkanethiol with a spacer segment [73, 80–83].

However, these methods suffer from the complicated chemical treatment or organic synthesis of biotinylated thiol, which is almost not commercially available. As a basis of such molecular building a mixed SAMs of MHDA and biotinyl-PE, commercially available, can be chosen. The mixed SAMs, in which the biotinyl-PE is inserted and bound to MHDA by the electrostatic interaction, were simply prepared by immersing the pretreated gold substrates into the mixed solution of MHDA and Biotinyl-PE in ethanol. The ratio between two components determines the density of specific biotinyl-anchoring sites.

The biotin-containing SAMs are a desirable platform for construction of self-assembled multilayer of protein on biosensors, in particular, immunosensors. The structure of the system is shown in Fig. 26.15: the mixed SAMs containing biotinyl sites, streptavidin, biotinylated antibody and antigen in the way of a stepwise assembled system.

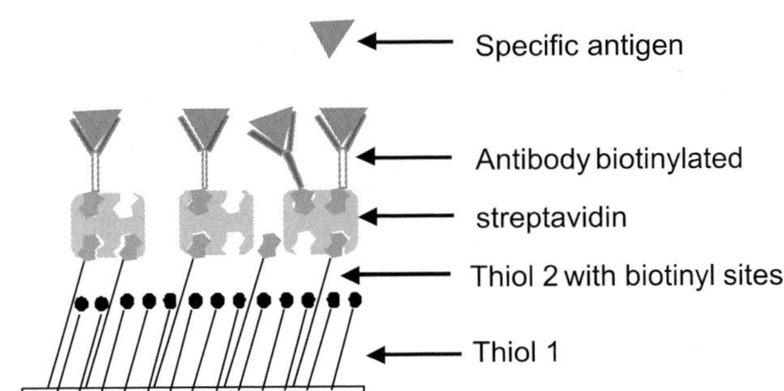

Fig. 26.15. Self-assembled multilayer structure based on biotin/avdin affinity interaction

26.2.2.6
Application of SAMs to Biosensor Elaboration

As discussed above, SAMs are effective methods for protein immobilization and they have many important advantages for the construction of biosensors, as listed in the following:

- Easy formation of ordered, pinhole-free and stable monolayers;
- Flexibility to design the head group of SAMs with various functional groups in order to obtain hydrophobic or hydrophilic surfaces as the requirement;
- Only a minimum amount of biomolecule is needed for immobilization on theSAMs;

- Reasonable stability for extended periods, allowing several reliable measurements and reproducible immobilization of the recognition molecule in a favorable microenvironment;
- Molecular level control of the interfacial chemistry can be exploited to enable ready approach of the target analyte to the interface with the limitation of the access of interfering substances.

Up to now, SAMs have been used as a major immobilization technique for elaboration of all kinds of biosensors, such as enzyme biosensors, immunosensors, DNA-based biosensors and so on. With the development of electronics and nanotechnology, the size of electrodes is gradually being minimized to the micro- and even the nanometer scale. SAMs can be expected to become an increasingly prevalent and critical tool due to their clear advantages.

26.2.3
Characterization of SAMs and LB Films

To obtain accurate information on the structure of thin organic films such as SAMs and LB films, it is essential to use a variety of characterization techniques that are able to determine certain properties of the layer. The most frequently used techniques are divided into five groups and listed in Table 26.5.

Contact angle can be used to examine the general hydrophobicity or hydrophilicity of a surface, it is a simple and effective method to obtain a first impression about the structure and composition of the layer.

A quartz crystal microbalance can detect mass change on a scale of several nanograms. It can be used to study the kinetics of layer adsorption of SAMs and LB films.

Table 26.5. Analytical techniques for characterizing the ultrathin films [84]

Analytical techniques	Structural information
General	
Contact angle	Hydrophobicity, order
QCM	Change in mass
Optical	
IR spectroscopy/FTIR	Functional groups, molecular orientation
Fluorescence spectroscopy	Density of adsorbates
Ellipsometry	Layer thickness
SPR	Layer thickness
Vacuum	
XPS	Elemental composition
Microscopy	
AFM	Molecular packing
STM	Molecular packing
Electrochemical	
Cyclic voltammetry	Thickness, order/defects, electrical properties
Impedance spectroscopy	Thickness, order/defects, electrical properties

Infrared (IR) spectroscopy is an important method for characterization of SAMs and LB films. The spectra can provide not only the usual structural data, but also information about the average orientation of the molecules in the thin layer.

Fluorescence spectroscopy can be used to observe the packing of SAMs and LB films on the substrates and it can provide information on the density of adsorbates.

Ellipsometry is one of the most common optical techniques to estimate the thickness of the thin organic films. It is based on a plane-polarized laser beam that is reflected by the substrate, which results in a change of the phase and amplitude of the light. Then, by comparing these two sets of parameters for an uncovered and covered substrate, the thickness of the layer can be calculated.

Surface plasmon resonance is based on the angle-dependent reflection of a p-polarized laser beam at thin metal layers and it can give important information on thickness of the layer. It is sensitive to the variation of the mean thickness and refractive index of the SAMs layer.

X-ray photoelectron spectroscopy is a useful technique for quantitative analysis of surface composition. It provides information on the chemical state of the constituents, but gives very little, or no, lateral resolution. XPS has been used to probe the chemical nature of the SAMs. Briefly, incident X-rays bombard the sample, and electrons are ejected from the core shells of the atoms within the SAMs. These electrons are collected in an analyzer; by measuring the energies of the electrons entering the analyzer, the binding energies are calculated. These are specific to each element and give indications of the oxidation states of the elements.

Atomic force microscopy and scanning tunneling microscopy (STM) allow characterization of the organic thin films at a molecular resolution. The generated images are the result of the forces acting between the tip and the surface or the surface electron density for AFM and STM, respectively.

The electrochemical characterization of SAMs on gold is usually performed by cyclic voltammetry (CV) and electrochemical impedance spectroscopy. For CV, large potential sweeps are applied to the gold electrode and the resulting current is recorded. However, for EIS only very small sinusoidal potential sweeps with frequencies varying from 100 mHz to 10 kHz are applied to the gold electrode, which makes it a less disturbing technique than CV. Associated with the use of a redox probe CV curves allow evaluation of the electrical blocking character of the layer. Furthermore impedance spectroscopy allows probing of the biomolecular interactions occurring on the bioengineered layers and can serve as a basis to develop impedimetric biosensors.

26.2.3.1
AFM Technique: A Tool for Characterization of Biofunctionalized Surfaces

The AFM technique has become increasingly a reference technique for the characterization of biofunctionalized surfaces at the nanometer level. In order to increase the resolution obtained with AFM measurements when imaging biological samples, the knowledge and control of the interactions between the tip and the sample is a crucial factor. The tip surface can be functionalized with chemicals in order to lead to lower adhesion force interactions. Silicon tip functionalization with a highly hydrophobic compound (organosilane: perfluorosilane) allows the imaging of cells (Fig. 26.16) and proteins (Fig. 26.17), in contact mode, in water, with quite a good resolution.

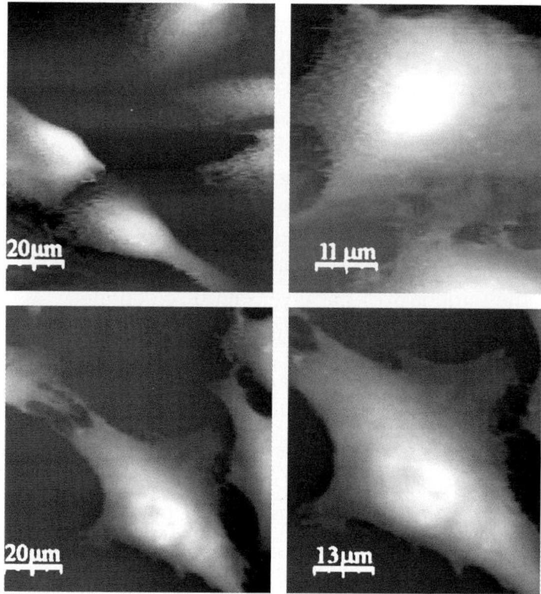

Fig. 26.16. Cell images obtained with Pico Plus II (MI) AFM system equipped with a silicon tip (0.9 N/m) in contact mode in water. At the *top*: Standard Si tip, high friction forces, unstable image. At the *bottom*: Silicon tip fucntionalized with perfluorosilane, low friction forces, stable and well-defined image

Fig. 26.17. AFM images of a gold surface functionalized with a mixed biotinylated self-assembled monolayer and an antihaemoglobin antibody, after haemoglobin recognition. Pico Plus II (MI) AFM system with a silicon tip functionalized with perfluorosilane, in contact mode (avoid any protein motion in contact mode)

26.2.3.2
Examples of Characterization of Nanostructured Biomolecules by AFM

26.2.3.2.1
Characterization of LB Films for Odorant Biosensors Applications

This example deals with the preparation of ultrathin films in which odorant-binding proteins (Rat OBP-1) were incorporated by the combination of LB films and SAMs, with the aim of drawing up a preliminary design for an artificial olfactory system to detect specific odorant molecules. Under optimized experimental conditions (phosphate buffer solution, pH 7.5, OBP-1F concentration of $4\,mg\,L^{-1}$, target pressure $35\,mN\,m^{-1}$), the mixed monolayer (OBP-1F/ODA LB film) at the air/water interface is very stable and has been efficiently transferred onto gold supports, which were previously functionalized by self-assembled monolayers with 1-octadecanethiol [87].

Topographic images (Fig. 26.18a,b) of the mixed OBP-1F/ODA LB films remain almost unchanged within the resolution of the imaging conditions of the experiments after exposure to a small, hydrophobic odorant molecule, isoamyl acetate. The white spots on the topographic images are probably due to dust or salts remaining after the rinsing process.

However, a strong change in the AFM phase images was induced by the exposure to the odorant (Fig. 26.18a',b'), the sample surface is almost uniform (variations of $\sim 1°$ maximum are observed), there is no phase contrast between the oval domains and the rest. This proves that the LB films obtained are well ordered and uniform, and OBP-1F aggregates are well covered with ODA molecules. After exposure to the odorant, a sharp contrast was observed between the aggregates and the other sample constituents, with phase variations of up to $\sim 9°$. Because phase atomic force microscopy images by recording the phase lag between the signal that drives the cantilever and the cantilever oscillation output signal, it provides information about

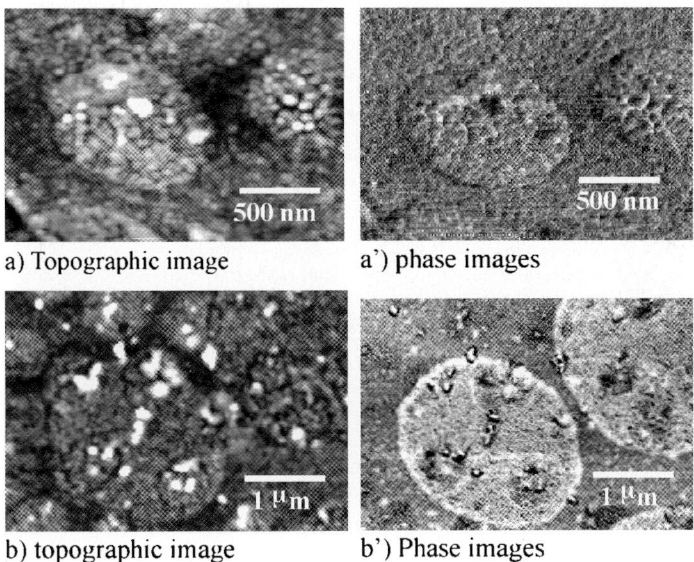

Fig. 26.18. AFM topographic (**a**) and phase (**b**) images of two layers of mixed OBP-1F/ODA LB film transferred onto functionalized gold substrate at $35\,mN\,m^{-1}$ before exposure to odorant. Scan area $2.1 \times 2.1\,\mu m^2$. Vertical scale (**a**) 10 nm, (**b**) 1.3°. AFM topographic (**c**) and phase (**d**) images of a mixed OBP-1F/ODA LB film transferred onto functionalized gold substrate at $35\,mN\,m^{-1}$ after exposure to odorant. Scan area $4 \times 4\,\mu m^2$. Vertical scale (**c**) 10 nm, (**d**) 9.4°

surface properties such as elasticity, adhesion and friction, and it can be concluded that one or all of these mechanical properties in the aggregates were modified.

Small odorant molecules can be assumed to penetrate the outer lipid layer to reach the OBP-1F aggregates. Interaction between the odorant molecules and OBP-1F can induce a conformational change of the protein, as observed by Nespoulous et al. [88] and consequently a stronger interaction between the hydrophilic part of OBP-1F and ODA polar heads, leading to a partial breakdown of LB films.

26.2.3.2.2
Characterization of Rhodopsin Immobilized on a Self-Assembled Multilayer

Rhodopsin, the G protein-coupled receptor that mediates the sense of vision, was immobilized onto gold electrode by two different techniques: Langmuir–Blodgett and a strategy based on a self-assembled multilayer. It was observed that LB films of rhodopsin are not stable on gold electrode. Thus a new protein multilayer was prepared on gold electrode by building up layer-by-layer a self-assembled multilayer. It is composed of a mixed self-assembled monolayer, as described in Sect. 26.2.5.2.4, formed by MHDA and biotinyl-PE, followed by a biotin-avidin system that allows binding of biotinylated antibody specific to rhodopsin. Figure 26.19 shows the schematic diagram for self-assembled multilayer formation on gold electrode.

In order to characterize the formation of the multilayer film and to obtain information on its architecture, AFM measurements were done in tapping mode. After the

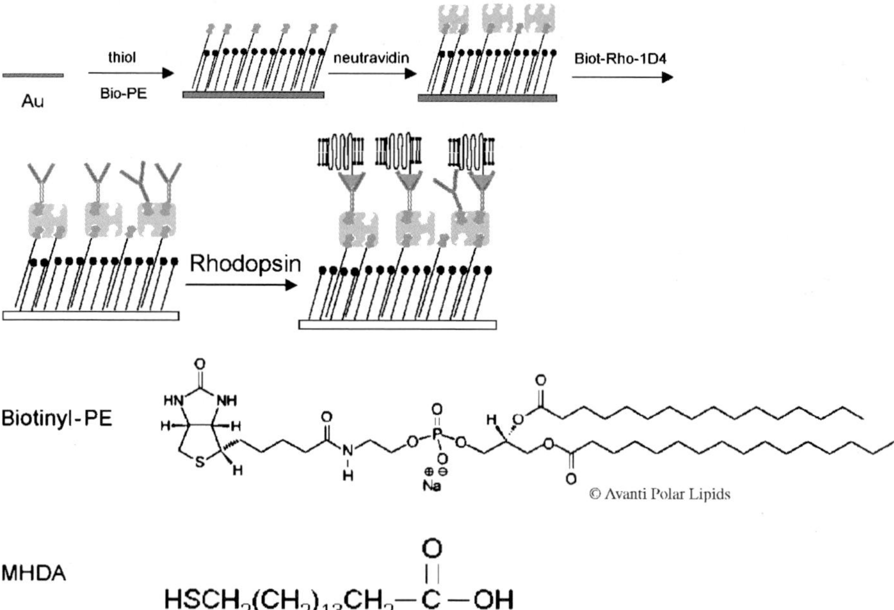

Fig. 26.19. Schematic diagram for immobilization of rhodopsin on gold, using self-assembled multilayer formation via biotin/avidin pairs

rhodopsin membrane fraction was immobilized on the self-assembled multilayers, an AFM image was taken, as presented in the image in Fig. 26.20, which reveals randomly located grainy features that range in size from 30–100 nm and are about 5 nm high. These sizes are consistent with those observed by negative staining of the rhodopsin membrane fraction preparation using electron microscopy [89]. In addition, the thickness of the rhodopsin-containing layer is approximately equivalent to the thickness of the cell membrane, indicating that membrane vesicles open upon immobilization. Further, during the scanning, we found that on some areas of fraction membrane, the tip interaction with the surface increased, which confirms that the rhodopsin is present as a membrane fraction.

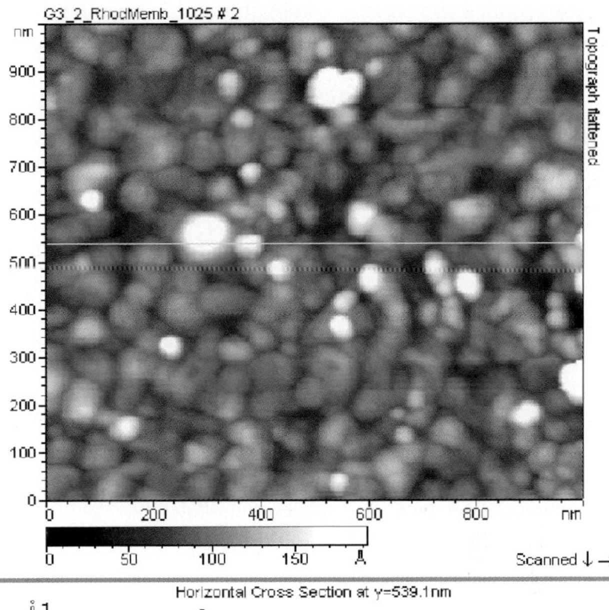

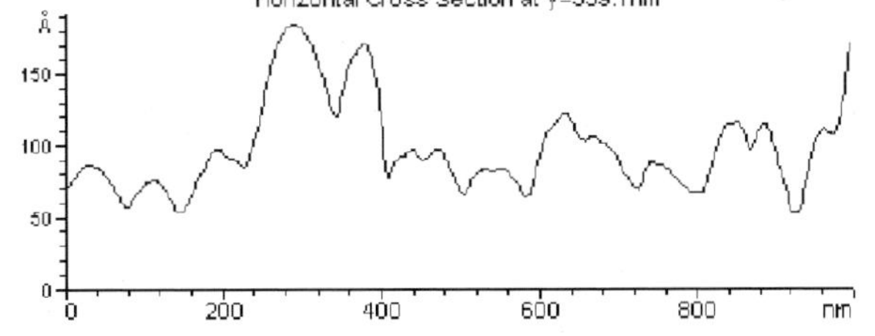

Fig. 26.20. Topographic AFM image (1 µm × 1 µm) obtained in tapping mode: after immobilization of rhodopsin membrane fraction on the self-assembled multilayers

26.3
Prospects and Conclusion

The fast development of microarrays, labs on a chip and other multimicrosensing devices need practical processes for a local molecular engineering of functional layers. Among these techniques microcontact printing appears as a promising and simple way for rapid patterning of molecules. Microcontact printing (μCP) using self-assembled monolayer (SAM) "inks" has been developed for rapid patterning of molecules over large surface areas. μCP is used to pattern gold surfaces using thiols, glass using silanes and so forth [90]. Unfortunately, conventional μCP suffers from problems limiting pattern resolution. These include limits on the structural aspect ratio (defined as the ratio of structure width vs. structure height) and the printing conditions dictated by the elastomeric stamp. The problem of structural collapse could be alleviated if the μCP is carried out in a liquid medium that supports the stamp during printing and coincidentally confines the inking molecules to the area of mutual contact. Xia and Whitesides [91, 92] have published results to show that the spreading of thiols microcontact printed under water can be used to reduce the feature dimensions of subsequently etched gold substrates. The schematic diagram of submersed microcontact printing method (SμCP) is presented in Fig. 26.21.

As shown in Fig. 26.22, it is only possible to pattern gold surface with PDMS stamp (aspect ratio 46:1) (inked using 10 mM alkylthiol solution in ethanol) if the SμCP method is used and not the classical μCP method in air. The SμCP method greatly increases the feasibility of designing microarrays of SAMs on large surface areas.

In conclusion, molecular engineering of functional layers, joined to the imaging power of AFM, has opened new research directions in the field of supramolecular engineering leading to practical applications for chemical and biosensing at the nanolevel.

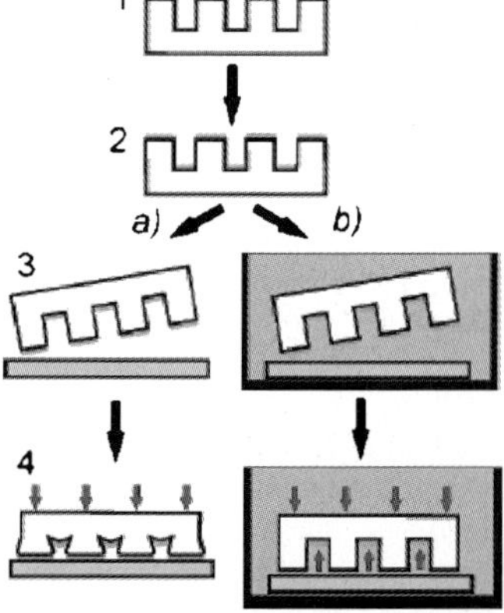

Fig. 26.21. Schematic diagram of (**a**) conventional and (**b**) submersed microcontact printing methods. The PDMS stamp is produced with the required surface structures (*1*). It is then "inked" with the thio (*2*) and placed in contact with the gold substrate (*3*). A small amount of pressure is applied manually (*4*) before the stamp is removed, leaving the patterned SAM on the substrate. In air, the applied pressure causes the stamp to deform; when submersed, the stamp is supported by the water. (From Bessueille F, Pla-Roca M, Mills C, Martinez E, Samitier J, Errachid A (2005) Langmuir 21:12060)

Fig. 26.22. Optical images μ contact printing obtained with stamps with ratio 46:1 (*Bar* = 75 μm) *1*) shows the collapse of the stamp using conventional μCP, *2*) shows the stamp under SμCP conditions. (From Bessueille F, Pla-Roca M, Mills C, Martinez E, Samitier J, Errachid A (2005) Langmuir 21:12060)

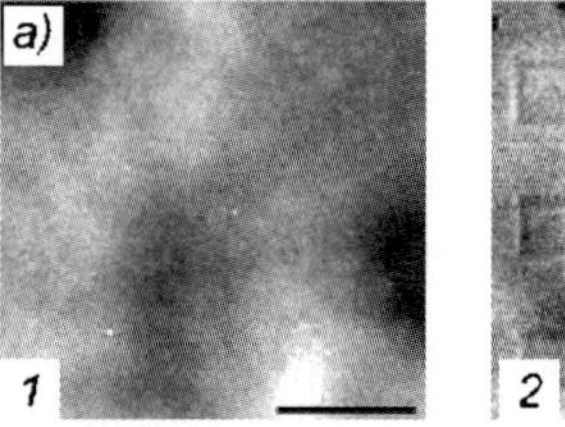
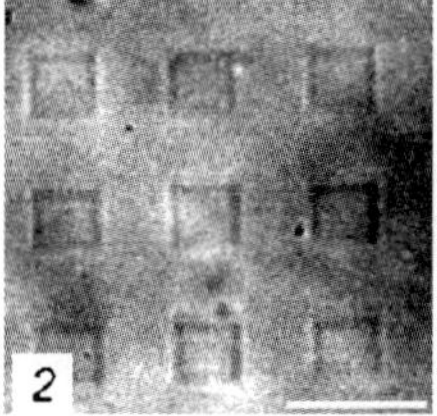

References

1. Roberts G (1990) Langmuir–Blodgett films, Plenum Press, New York
2. Pockels A (1891) Nature 43:437
3. Pockels A (1892) Nature 46:418
4. Langmuir–Blodgett Troughs-operating manual, 6th Edition, Nima Technology Ltd.:21
5. Ulman A (1991) An Introduction to Ultrathin Organic Films: From Langmuir–Blodgett to Self-Assembly, Academic Press, Boston
6. Harkins WD (1952) The Physical Chemistry of Surface Films, Reinhold, New York
7. Talham DR (2004) Chem Rev 104:5479
8. Fromherz P (1971) Biochim Biophys Acta 225:382
9. Arisawa S, Yamamoto R (1992) Thin Solid Films 210/211:443
10. Zaitsev SY (1993) Colloids Surf A 75:211
11. Zaitsev SY (1995) Sens Actuators B 24–25:177
12. Fiol C, Alexandre S, Delpire N, Valleton JM (1992) Thin Solid Films 215:88
13. Dubreuil N, Alexandre S, Fiol C, Sommer F, Valleton JM (1995) Langmuir 11:2098
14. Fiol C, Alexandre S, Dubreuil N, Valleton JM (1995) Thin Solid Films 261:287–295
15. Dubreuil N, Alexandre S, Fiol C, Valleton JM (1996) J Colloid Interface Sci 181:393
16. Alexandre S, Dubreuil N, Fiol C, Lair D, Duc TM, Valleton JM (1997) Thin Solid Films 293:295
17. Dubreuil N, Alexandre S, Lair D, Sommer F, Duc TM, Valleton JM (1998) Colloids Surf B 11:95
18. Chovelon JM, Provence M, Jafrrezic-Renault N, Derue V, Lair D, Alexandre S, Valleton JM (2001) J Biol Phys Chem 1:68
19. Alexandre S, Lafontaine C, Valleton JM (2001) J Biol Phys Chem 1:21
20. Alexandre S, Derue V, Valleton JM, Sommer F, Duc TM (2002) Colloids Surf B 23:183
21. Zhang A, Jaffrezic-Renault N, Wan K, Hou Y, Chovelon JM (2002) Mater Sci Eng C 21:91
22. Chovelon JM, Wan K, Jaffrezic-Renault N (2000) Langmuir 16:6223
23. Chovelon JM, Gaillard F, Wan K, Jaffrezic-Renault N (2000) Langmuir 16:6228
24. Girard-Egrot AP, Morelis RM, Coulet PR (1997) Thin Solid Films 292:282
25. Girard-Egrot AP, Morelis RM, Coulet PR (1997) Langmuir 13:6540
26. Okahata Y, Tsuruta T, Ijiro K, Ariga K (1998) Langmuir 4:1373
27. Zaitsev SY (1993) Colloids Surf A 75:211
28. Fiol C, Chovelon JM, Dubreuil N, Barbey G, Barraud A, Ruaudel-Teixier A (1992) Thin Solid Films 210/211:489
29. Tatsuma T, Tsuzuki H, Okawa Y, Yoshida S, Watanabe T (1991) Thin Solid Films 202:145
30. Wan K, Chovelon JM, Jaffrezic-Renault N (2000) Talanta 52:663
31. Choi JW, Kim YK, Lee IH, Min J, Lee WH (2001) Biosens Bioelectron 16:937
32. Zhang A, Hou Y, Jaffrezic-Renault N, Wan J, Soldatkin AP, Chovelon JM (2002) Bioelectrochemistry 56:157–158
33. Singhal R, Gambhir A, Pandey MK, Annapoorni S, Malhotra BD (2002) Biosens Bioelectron 17:697
34. Hou Y, Jaffrezic-Renault N, Zhang A, Wan J, Errachid A, Chovelon JM (2002) Sens Actuators B 86:143
35. Sharma SK, Singhal R, Malhotra BD, Sehgal N, Kumar A (2004) Electrochim Acta 49:2479
36. Sharma SK, Singhal R, Malhotra BD, Sehgal N, Kumar A (2004) Biosens Bioelectron 20:651
37. Vikholm I, Teleman O (1994) J Colloid Interface Sci 168:125
38. Barraud A, Perrot H, Billard V, Martelet C, Therasse J (1993) Biosens Bioelectron 8:39

39. Katsube T, Yaji T, Suzuki K, Kobayashi S, Kawaguchi T, Shiro T (1998) Appl Surf Sci 33–34:1332
40. Dubrovsky TB, Demcheva MV, Savitsky AP, Yu. Mantrova E, Yaropolov AI, Savransky VV, Belovolova LV (1993) Biosens Bioelectron 8:377
41. Chudinova GK, Chudinov AV, Savransky VV, Prokhorov AM (1997) Thin Solid Films 307:294
42. Lepesheva GI, Azeva TN, Knyukshto VN, Chashchin VL, Usanov SA (2000) Sens Actuators B 68:27
43. Oh BK, Chun BS, Park KW, Lee W, Lee WH, Choi JW (2004) Mater Sci Eng 24:65
44. Nicolini C, Adami M, Dubrovsky MT, Erokhin, Facci VP, Paschkevitsch P, Sartore M (1995) Sens Actuators B 24:121
45. Sukhorukov GB, Montrel MM, Petrov AI, Shabarchina LI, Sukhorukov BI (1996) Biosens Bioelectron 11:913
46. Smith RK, Lewis PA, Weiss PS (2004) Prog Surf Sci 75:1
47. Bain CD, Troughton EB, Tao YT, Evall J, Whitesides GM, Nuzzo RG (1989) J Am Chem Soc 111:321
48. Yang Z, Gonzalez-Cortes A, Jourquin G, Vire JC, Kauffmann JM (1005) Biosens Bioelectron 10:789
49. Strong L, Whitesides GM (1998) Langmuir 4:546
50. Evans SD, Sharma R, Ulman A (1991) Langmuir 7:156
51. Finklea HO, Snider DA, Fedyak J (1990) Langmuir 6:371
52. Samant MG, Brown CA, Gordon JG (1991) Langmuir 7:437
53. Ron H, Rubinstein I (1994) Langmuir 10:4566
54. Ron H, Matlis S, Rubinstein I (1998) Langmuir 14:1116
55. Majid A, Bensebaa F, L'Ecuyer P, Pleizier G, Deslandes Y (2003) Rev Adv Mater Sci 4:25
56. Poirier GE, Pylant ED (1996) Science 272:1145
57. Chailapakul O, Sun L, Xu C, Crooks RM (1993) J Am Chem Soc 115:12459
58. Xia Y, Whitesides GM (1998) Annu Rev Mater Sci 28:153
59. Pan W, Durning CJ, Turro NJ (1996) Langmuir 12:4469
60. Ulman A (1996) Chem Rev 96:1533
61. Richter AG, Yu CJ, Datta A, Kmetko J, Dutta P (2000) Phys Rev E 61:607
62. Richter AG, Yu CJ, Datta A, Kmetko J, Dutta P (2000) Materials Res CAT Sector 10:136–139
63. Sandhyarani N, Pradeep T (2003) Int Rev Phys Chem 22:221
64. Gooding JJ, Hibbert DB (1999) Trends Anal Chem 18:523
65. Creager SE, Olsen KG (1995) Anal Chim Acta 307:277
66. Chaki NK, Vijayamohanan K (2002) Biosens Bioelectron 17:1
67. Mirsky VM, Riepl M, Wolfbeis OS (1997) Biosens Bioelectron 12:977
68. Gooding JJ, Mearns F, Yang W, Liu J (2003) Electroanalysis 15:81
69. Kenny JW, Fanning TG, Lambert JM, Traut RR (1979) J Mol Biol 135:151
70. Carlsson J, Drevin H, Axen R (1978) Biochem J 173:23
71. Ferretti S, Paynter S, Russell DA, Sapsford KE, Richardson DJ (2000) Trends Anal Chem 19:530
72. Dong X, Lu J, Cha C (1997) Bioelectrochem Bioenerg 42:63
73. Spinke J, Liley M, Schmitt FJ, Guder HJ, Angermaier L, Knoll W (1993) J Chem Phys 99:7012
74. Yam C, Pradier CM, Salmain M, Marcus P, Jaouen G (2001) J Colloid Interface Sci 235:183
75. Yam C, Pradier CM, Salmain M, Fischer-Durand N, Jaouen G (2002) J Colloid Interface Sci 245:204
76. Lasseter TL, Cai W, Hamers RJ (2004) Analyst 129:3

77. Cui X, Pei R, Wang Z, Yang F, Ma Y, Dong S, Yang X (2003) Biosens Bioelectron 18:59
78. Häussling L, Ringsdorf LH, Schmitt FJ, Knoll W (1991) Langmuir 7:1837
79. Herron J, Müller W, Paudler M, Riegler H, Ringsdorf H, Suci PA (1992) Langmuir 8:1413
80. Pérez-Luna VH, O' Brien MJ, Opperman KA, Hampton PD, Lopez GP, Klumb LA, Stayton PS (1999) J Am Chem Soc 121:6469
81. Jung LS, Nelson KE, Stayton PS, Campbell CT (2000) Langmuir 16:9421
82. Nelson KE, Gamble L, Jung LS, Boeckl MS, Castner DG, Campbell CT, Stayton PS (2001) Langmuir 17:2807
83. Spinke J, Liley M, Guder HJ, Angermaier L, Knoll W (1993) Langmuir 9:1821
84. Flink S, Frank C, Van Veggel JM, Reinhoudt DN (2000) Adv Mater 12:1315
85. Hou Y, Tlili C, Jaffrezic-Renault N, Zhang A, Martelet C, Ponsonnet L, Errachid A, Samitier J, Baussels J (2004) Biosens Bioelectron 20:1126
86. Tlili C, Hou Y, Hafsa KY, Ponsonnet L, Martelet C, Errachid A, Jaffrezic-Renault N (2005) Sens Lett 2:1
87. Hou Y, Jaffrezic-Renault N, Martelet C, Tlili C, Zhang A, Pernollet JC, Briand L, Gomila G, Errachid A, Samitier J, Salvagnac L, Torbiero B, Temple-Boyer P (2005) Langmuir 21:4058
88. Nespoulous C, Briand L, Delage MM, Tran V, Pernollet JC (2004) Chemical Senses 29:189
89. Minic J, Grosclaude J, Aioun J, Persuy MA, Gorojankina T, Salesse R, Pajot-Augy E, Hou Y, Helali S, Jaffrezic N, Bessueille F, Errachid A, Gomila G, Ruiz O, Samitier J (2005) Biochim Biophys Acta 1724:324
90. Xia Y, Whitesides GM (1998) Angew Chem Int Ed 37:550
91. Xia Y, Whitesides GM (1997) Langmuir 13:2059
92. Xia Y, Whitesides GM (1995) J Am Chem Soc 117:3274

27 Applications of Scanning Electrochemical Microscopy (SECM)

Gunther Wittstock · Malte Burchardt · Sascha E. Pust

Abbreviations and Symbols

c	concentration (moles per volume of solution)
c^*	concentration in bulk solution
d	working distance, distance between the active area of the UME and the sample surface
D	diffusion coefficient
dsDNA	double-stranded desoxy ribonucleic acid
ECSFM	electrochemical scanning force microscope
ECSTM	electrochemical scanning tunncling microscope
$E^{\circ\prime}$	formal potential of a redox couple
E_S	sample potential applied in SECM
E_T	potential applied at the SECM probe
F	Faraday constant
$F(L, \Lambda)$	$= (11/\Lambda + 7.3)/(110 - 40L)$
Fc	shortcut for a ferrocene derivative
Fc^+	shortcut for a ferrocinium derivative
GC	generation-collection mode of SECM
GDH	glucose dehydrogenase
GOx	glucose oxidase
Γ_{enz}	Surface concentration of an immobilized enzyme
HOR	hydrogen-oxidation reaction
i_S	electrolytic sample current or its equivalent in case of an enzyme reaction
I_S	normalized sample current $= i_S/i_{T,\infty}$
ISE	ion-selective electrodes
i_T	electrolytic current at the SECM probe
I_T	normalized electrolytic current at the SECM probe $= i_T/i_{T,\infty}$
I_T^{cond}	normalized electrolytic current at the SECM probe if the reaction at the sample is diffusion-controlled
I_T^{ins}	normalized electrolytic current at the SECM probe if there is no reaction at the sample
$I_{T,k}$	normalized electrolytic current at the SECM probe if a first-order reaction proceeds at the sample with a finite rate constant
$i_{T,\infty}$	electrolytic current at the SECM probe far from any surface
k_{cat}	catalytic turnover number
k_{eff}	first-order heterogeneous rate constant of the reaction at the sample

K_M Michaelis–Menten constant

κ dimensionless normalized first-order heterogeneous rate constant at the sample $= k_{\text{eff}} r_T / D$

L normalized working distance $= d/r_T$

Λ $= \kappa L = k_{\text{eff}} d / D$

n number of transferred electrons per converted molecule

ORR   oxygen-reduction reaction

PAP p-aminophenol

PAPG p-aminophenyl-β-D-galactopyranoside

PQQ   pyrroloquinoline quinone

r_S radius of the active region on a sample

r_T radius of the active electrode area of a SECM probe

r_g radius of the total probe diameter (radius of glass sheath)

RG ratio between total probe radius and active electrode area $= r_g / r_T$

SECM   scanning electrochemical microscopy/microscope

SFM   scanning force microscopy/microscope

SG/TC   substrate generation/tip collection mode

ssDNA   single-stranded desoxy ribonucleic acid

STM   scanning tunneling microscopy/microscope

TG/SC   tip-generation/substrate collection mode

27.1
Introduction

27.1.1
Overview

Scanning electrochemical microscopy (SECM) is a technique that allows us to record spatially resolved maps of chemical reactivities, i.e. images that reflect the *rate of heterogeneous chemical reactions*. As with other scanning probe techniques, the acronym is used for both, the method and the instrument. SECM works in electrolyte solution and can be used to study reactions at insulating, semiconducting and conducting samples. The technique has found application in biotechnology (study of immobilized enzymes, development of biochips), cell biology (cell–cell communication processes), physiology (mass transport through biological tissue), corrosion science, and fuel-cell development. The SECM setup can be used to generate electrochemically precursor agents for surface modification. The combination of local surface modification and subsequent imaging of changes in chemical reactivities opens up new possibilities for the design of functional materials. Finally, the technique had and still has an enormous methodical impact on fundamental electrochemistry because the signal–distance curves (approach curves) contain a wealth of kinetic information. Besides the investigation of solid/liquid interfaces, it can also be applied to liquid/liquid and liquid/gas interfaces, where completely new insights have been obtained [1–5].

SECM instruments are commercially available and the number of groups using this technique is rapidly increasing. However, it can be noted that until now most successful groups are rooted in fundamental electrochemistry and the application potential in other fields such as cell biology and material science has not yet been fully explored. This chapter aims at providing an overview of some important applications that reach beyond the scope of fundamental electrochemistry. The experiments will be explained qualitatively and, where it seems justified, quantitative treatment will be provided with a minimum of mathematical artwork. Some directions and requirements for future developments will be discussed as a summary and outlook that are closely interconnected with the Chapt. 8 in this volume by Kranz et al. For a systematic overview on the electrochemical principles and a comprehensive coverage of the theory the reader is referred to the authoritative book edited by Bard and Mirkin [1].

27.1.2
Relation to Other Methods

SECM is historically rooted in two experimental areas. As a scanning probe technique it shares many similarities with related techniques such as scanning force microscopy in an electrochemical cell (ECSFM) or electrochemical scanning tunneling microscopy (ECSTM). However, the signal in SECM is based on an electrochemical signal specific for a certain chemical compound dissolved in an electrolyte solution. In this respect the SECM probe can also be regarded as a positionable chemical microsensor. Two different microsensors can be used: amperometric electrodes and potentiometric electrodes. An amperometric electrode operates at an electrochemical potential (vs. a reference electrode). The measured Faradaic current results from the electrolysis of a compound contained in the solution. In most cases, the current is proportional to the concentration of the dissolved compound. Potentiometric probes have been used much less frequently (vide infra). They measure the equilibrium potential difference (i.e. at zero current) between an ion-selective electrode and a reference electrode. The potential difference is proportional to the *logarithm* of the activity (i.e. the effective concentration) of the potential-determining ion. Already in the early 1980s amperometric ultramicroelectrodes (UME) gained popularity among electrochemists because of the new opportunities offered by micrometer-sized electrodes for studying fast electrochemical reactions or for the detection of substances in restricted space [6], such as the detection of neurotransmitters in living organisms [7]. SECM has found broad application in the electrochemical community. This is not surprising, because a quantitative understanding of electrochemical reactions of various substances at microelectrodes had been already accumulated in the context of chemical sensing.

Looking back at almost two decades of SECM development, it seems logical to use an UME in order to measure the concentration distribution of reactants and products close to a macroscopic specimen (sample) electrode. The first experiment with scanning electrochemical microscopy was performed by Engstrom et al. [8]. At about the same time the group of Bard reported experiments with an electrochemical scanning tunneling microscope (ECSTM) where currents were observed at unusually large sample–tip distances. The signals were considered to be Faradaic currents, i.e. currents that result from the electrolysis of a chemical species at the metal tip. Bard and coworkers [9] realized that such a tip current can be influenced by the presence of the sample in various ways. By 1989 Bard et al. had worked out an experiment in an analogous micrometer-sized system where all the effects could clearly be ascribed to well-understood physical processes such as diffusion or heterogeneous electrochemical reactions. With such an instrument quantitative agreement between the experimentally measured Faradaic currents and digital simulations was demonstrated for a few model systems [10, 11]. The simulation can provide quantitative results because the mass transport between the ultramicroelectrode and the sample under investigation is usually controlled by diffusion only, a theoretically well-understood phenomenon that can be described by continuum models with sufficient accuracy under the prevailing experimental conditions. Very soon it became evident that SECM is not just suitable to measure local solute concentrations but also, and more importantly, represents a tool to map local (electro)chemical reactivities, to induce localized electrochemical surface modifications, or to investigate heterogeneous and homogeneous kinetics. While lateral resolution in routine SECM experiments does not reach the resolution easily achieved with scanning force microscopy (SFM) or scanning tunneling microscopy (STM), the importance of SECM derives from its unique ability to analyze local fluxes of electroactive species. A brief introduction to SECM appeared in an earlier volume of this series [12].

27.1.3
Instrument and Basic Concepts

The general experimental setup is shown in Fig. 27.1. The local probe is an amperometric microdisk electrode (e.g. Pt) that is embedded in an insulating sheath, typically made from glass. Typical microelectrodes have electroactive probe di-

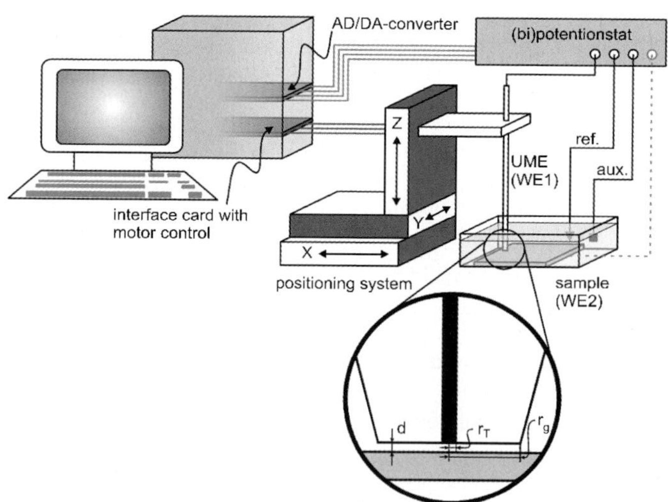

Fig. 27.1. Schematic setup of the SECM. Depending on the experiment, the sample may or may not be connected to the second channel of a bipotentiostat

ameters of 10 or 25 µm. Of course, electrodes with smaller diameters may be used and concepts of using nanoelectrodes are currently extensively explored. Prospects and limitations are discussed in Sect. 27.4. It is convenient to characterize the probe by two important radii: The radius r_T of the active electrode area and the radius r_g of the insulating glass sheath leading to the so-called RG value $RG = r_g/r_T$. The potential of the UME relative to a thermodynamically defined reference electrode is controlled by a (bi)potentiostat. For the investigation of enzymes and biological samples the specimen is usually not connected to an external voltage source. The probe is scanned in a distance d of $(0.5–3)\,r_T$ across the sample. Various positioning systems have been used. It is important that the instrument allows a vertical movement of at least $20\,r_T$ in order to record useful approach curves.

27.1.3.1
Generation-Collection Mode

The generation-collection (GC) mode can be performed with amperometric and potentiometric probes. Usually, the detected compound is initially not present in solution and therefore, the concentration far from the surface approaches zero. The chemical microsensor (the collector) is used to detect dissolved chemical species that were generated or released from the sample (the generator). Specifically, this arrangement (mode) is called the substrate-generation/tip-collection mode (SG/TC) and is shown schematically in Fig. 27.2a. By detecting dissolved chemical species in the immediate vicinity of the specimen surface, this techniques lends itself to the characterization of surfaces at which substances are locally released into the solution or the study of localized mass transport, for instance through membranes.

Recently, the so-called tip-generation/substrate collection mode gained importance for combinatorial testing of electrocatalysts (see Sect. 27.3.2.2). In TG/SC mode the UME is used to produce a chemical species (e.g. O_2) from electrolysis of the solvent with a constant rate. The UME-generated O_2 diffuses to the sample region underneath the UME, where it is reduced at a finite rate (Fig. 27.2b). The reduction current at the sample is plotted versus the location of the UME and provides reaction-rate imaging for irreversible electrochemical reactions.

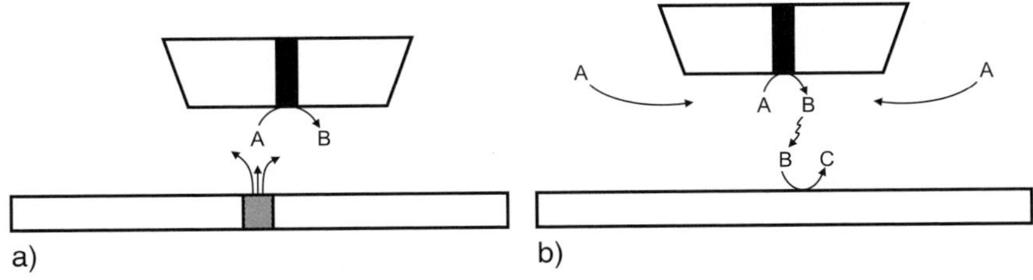

Fig. 27.2. Schematics of the GC mode. (**a**) Substrate-generation/tip-collection mode; (**b**) tip-generation-substrate-collection mode

27.1.3.2
Feedback Mode: Hindered Diffusion and Mediator Recycling

The feedback mode can be performed with amperometric probes only. Measurements are carried out in a solution containing an inert electrolyte (to provide ionic conductivity, typical $0.1-1\,\mathrm{mol\,L^{-1}}$) and the oxidized or reduced form of a quasireversible redox couple ($R \rightleftharpoons O + ne^-$, e.g. $(0.2-10) \times 10^{-3}\,\mathrm{mol\,L^{-1}}$). R and O stand as generic symbols for a reduced compound (R, can be oxidized) and its corresponding oxidized compound (O, can be reduced). In close proximity of the UME to the sample, this compound shuttles electrons between the sample and the probe and is often called a "mediator". For further discussion it is assumed that initially only the reduced form R of the redox species is present.[1] The diffusion-limited current at an UME for the reaction $R \rightarrow O + ne^-$ in bulk solution is given by

$$i_{\mathrm{T},\infty} = 4nFDr_{\mathrm{T}}c^* , \tag{27.1}$$

where n is the number of electrons transferred from each mediator molecule to the UME, F is the Faraday constant, D is the diffusion coefficient, r_{T} is the radius of the UME and c^* is the concentration of the mediator in bulk solution [13]. In analogy to other scanning probe techniques, the UME is frequently called "tip" from which the subscript "$_{\mathrm{T}}$" was taken to specify quantities of the UME. Correspondingly, the subscript "$_{\mathrm{S}}$" is used for quantities of the specimen (sample, substrate). The subscript "$_\infty$" indicates that the UME has a quasi-infinite distance to the surface. Practically $i_{\mathrm{T},\infty}$ is observed if the UME-sample distance d is larger than $20\,r_{\mathrm{T}}$.

In order to ensure comparability, the current response in SECM is usually provided in dimensionless normalized units. The dimensionless UME current I_{T} is obtained by dividing the UME current i_{T} at a given position above the sample by $i_{\mathrm{T},\infty}$ ($I_{\mathrm{T}} = i_{\mathrm{T}}/i_{\mathrm{T},\infty}$). The dimensionless working distance L is the working distance d in units of r_{T} ($L = d/r_{\mathrm{T}}$). If the UME approaches an inert and insulating surface, the diffusional flux of the electroactive species from the bulk solution to the active electrode area is hindered by the insulating glass sheath and the presence of the sample. The current, which is experimentally observed for inert insulating samples (e.g. glass) at an UME with $RG = 10$ is given by (27.2) [14] (Fig. 27.3, curves 1).

$$I_{\mathrm{T}}^{\mathrm{ins}}(L) = \frac{i_{\mathrm{T}}}{i_{\mathrm{T},\infty}} = \frac{1}{0.40472 + \frac{1.60185}{L} + 0.58819\exp\left(\frac{-2.37294}{L}\right)} . \tag{27.2}$$

The oxidized form O can be rereduced ($O + ne^- \rightarrow R$) at the sample surface, if i) the sample is electronically conducting, ii) O is consumed in a catalytic reaction at the sample surface, or iii) the sample is etched by O. After a fraction of a second, a steady-state current will be established, at which the consumption of R at the UME is balanced by the diffusional resupply of R from the bulk phase of the solution and the diffusion of R regenerated by the reaction at the sample surface. Only sample regions located under the active area of the UME will significantly contribute to the regeneration process of the mediator [15]. Plotting changes of the current versus tip location provides the basis for imaging in SECM.

---

[1] Of course all experiments can be performed by initially providing only the oxidized form of the mediator. In this case all reactions are reversed.

If the reaction at the sample and at the UME are both diffusion-controlled, the maximum current for any given UME position and solution composition is obtained. This upper limit of the current response is given by (27.3) [14].

$$I_T^{cond} = \frac{i_T}{i_{T,\infty}} = 0.72627 + \frac{0.76651}{L} + 0.26015 \exp\left(\frac{-1.41332}{L}\right). \tag{27.3}$$

The curve is given in Fig. 27.3, curve 2. Experimental approach curves to samples with finite kinetics are located in-between the two limiting cases (Fig. 27.3, curves 3–6, for details see Sect. 27.2.1.2). This situation allows two basic experiments to be carried out: Scanning the UME at constant distance provides an image that reflects the distribution of heterogeneous reaction rates of the sample. Moving the UME vertically towards the sample, allows a more detailed kinetic investigation of the reaction $O + ne^- \rightarrow R$ at a specific location of the sample. Examples for both cases will be discussed in Sects. 27.2.1.1 and 27.2.1.2.

The communication and interdependence of the reactions at the sample and at the UME lead to the term "feedback mode" for this special mode within SECM [11,16]. In the context of SECM the term "feedback" does *not* refer to an electronic circuit and actuator maintaining a constant probe-sample interaction as is common in other scanning probe microscopies.

When comparing the feedback mode and the GC mode, it has become evident that the GC mode is more sensitive to slow reaction rates and the feedback mode provides better lateral resolution. These guidelines will be illustrated and further substantiated within the following section.

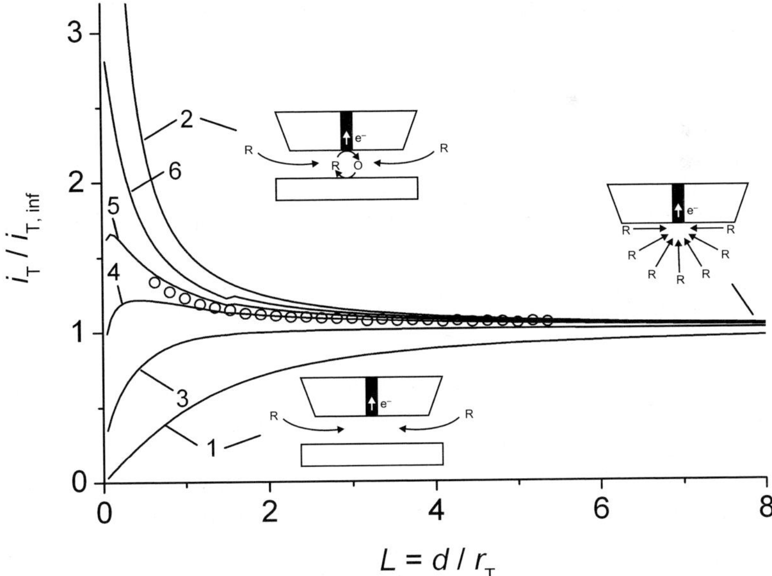

Fig. 27.3. SECM approach curve towards a GDH-modified modified surfaces (*open symbols*). For comparison calculated curves for hindered diffusion (*1*), diffusion-controlled reaction at the sample (*2*) and normalized rate constants $\kappa = k_{eff} r_T / D$ are given: 0.3 (*3*), 1.0 (*4*), 1.8 (*5*), 3.6 (*6*)

27.2
Application in Biotechnology and Cellular Biology

Since the inception of SECM, attempts have been made to apply the technique to investigations of biologically or biochemically relevant problems [17]. Measurements with the SECM are conveniently performed in buffered aqueous solutions, a preferred environment for most biological samples. The scanning electrode does not physically contact the specimen and therefore does not interfere with the sample compared to alternative scanning probe techniques, e.g., STM and SFM. The electrochemical detection of metabolites via SECM can be used to map biochemical activity thereby complementing techniques that image the topography of biological specimens or biomolecules. A systematic overview of SECM investigations at biological systems can be found in two authoritative review chapters [18, 19]. When sorting the various applications, grouping them according to the complexity of the investigated biological entity seems adequate and will be used in this chapter.

27.2.1
Investigation of Immobilized Enzymes

With respect to biological samples, isolated enzymes immobilized to a solid support were the first systems investigated with SECM. Enzymes can be divided into seven classes depending on the catalyzed reaction type. The largest group are oxidoreductases, i.e. enzymes that catalyze redox reactions. Such enzymes are technically relevant as biorecognition elements in biosensors, in enzyme-loaded columns or membranes for biotechnological conversions or in biofuel cells.

27.2.1.1
Characterization of Oxidoreductases in the SECM Feedback Mode

In order to illustrate the imaging principle, we consider the investigation of pyrroloquinoline quinone (PQQ)-dependent glucose dehydrogenase (GDH, EC 1.1.99.17) [20]. It is the classic example of PQQ-dependent quinoproteins, a large group of enzymes that can convert alcohols and amines to their corresponding aldehydes/lactones [21]. PQQ-dependent GDH catalyzes the transfer of 2 electrons and 2 protons from glucose to a soluble n-electron acceptor M_{ox} ($n = 1, 2$).

$$\text{D-glucose} + (3 - n)M_{\mathrm{ox}} \xrightarrow{\text{[GOx]}} \text{D-gluconolactone} + (3 - n)M_{\mathrm{red}} + 2\text{H}^+ .$$

$$(27.4)$$

The electron acceptors can serve as electron mediators between the active center of the enzyme and an electrode surface. GDH has a very high catalytic activity. For example, approximately 3 mmol of glucose is oxidized per minute per milligram of protein with 2,6-dichlorophenol indophenol, which is up to 20 times the activity of pure glucose oxidase (GOx) [22]. The high activity and independence of oxygen make GDH extremely interesting for developing biosensors and immunoassays technology [22, 23]. These applications require that the enzyme kinetics of the immobilized enzyme can be characterized. Furthermore, miniaturized sensors, which are increasingly demanded for high throughput testing or multianalyte sensor arrays would benefit from spatially resolved measurements.

A model sample was prepared by binding GDH to streptavidin-coated magnetic microbeads [20]. The magnetic microbeads were arranged as a mound supported on a polymer foil via an external magnetic field [24]. The resulting mound is shown in Fig. 27.4a.

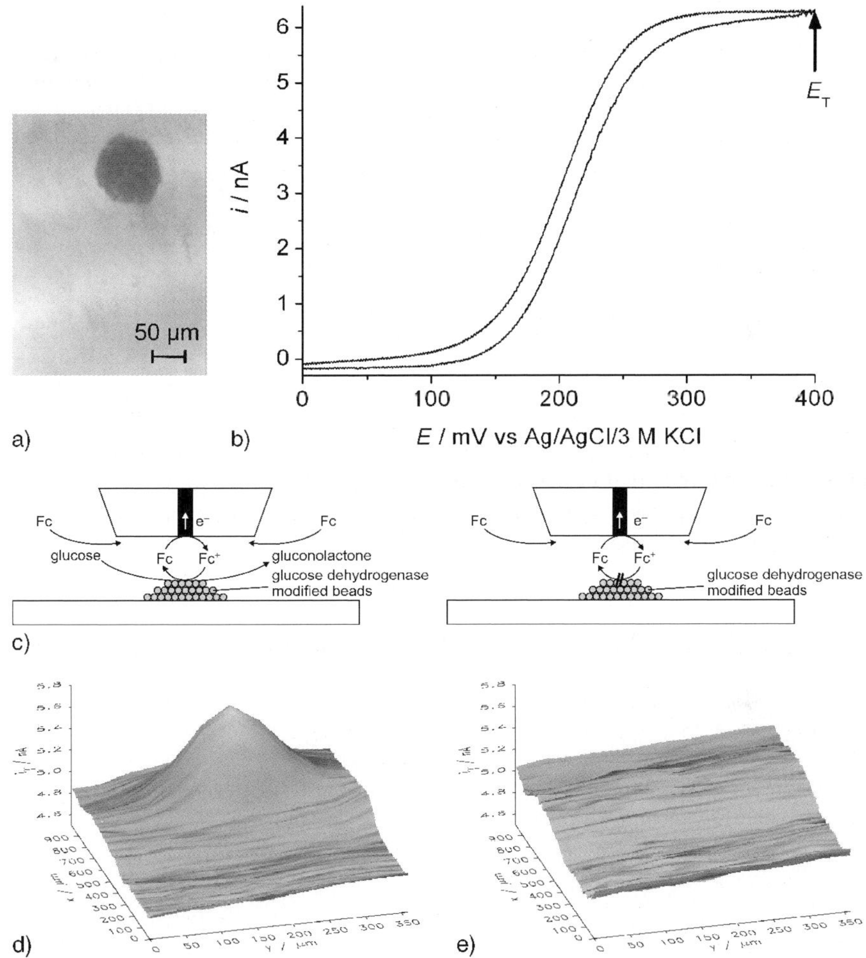

Fig. 27.4. SECM feedback imaging of GDH-modified microbeads. (**a**) optical photograph of the bead assembly. Individual beads have a diameter of 3 μm. They are not resolved in the optical image; (**b**) cyclic voltammogram for the oxidation of Fc to Fc$^+$ at the UME; (**c**) Schematic of the SECM experiment in the presence and absence of glucose; Fc, ferrocene methanol; (**d**) SECM feedback imaging in a solution containing 2 mM ferrocene methanol as mediator and 50 mM D-glucose in buffered solution. (**e**) Control experiment the same area as in (**d**) in a buffer solution containing 2 mM ferrocene methanol but *no* glucose. From [20] Copyright 2004, American Chemical Society

A set of experiments is performed in the presence and absence of the enzymatic substrate glucose. This approach is outlined in Fig. 27.4c. The buffered solution keeps the pH value close to the optimum pH value of the enzyme. In addition, it provides the ionic conductivity without being electrochemically converted at the electrode. Secondly, the solution contains a redox mediator, specifically ferrocene methanol (Fc) in Fig. 27.4. Ferrocenes are metal organic compounds in which a divalent iron ion is sandwiched between two cyclopentadienyl anions. Substitutions at the cyclopentadienyl rings allow the solubility, the overall charge and the redox potential of the compound, to be tuned. In this experiment ferrocene methanol was selected for the following reasons: i) It can be reversibly oxidized to the ferrocinium form (Fc^+) at the UME. The ferrocinium ion consists of a trivalent iron ion that is sandwiched between two cyclopentadienyl anions. Because of the absence of major structural changes during this redox reaction, the electrochemical reaction is fast and reversible. ii) Fc^+ is able to act as an electron acceptor (M_{ox} in (27.4)). iii) The solution contains glucose as a substrate for the enzyme. Glucose is not converted at the working potential E_T of the UME.

Table 27.1. SECM feedback imaging of local activity of immobilized oxidoreductases

Enzyme imaged	Substrate	Mediator/reaction at the UME	Ref.
Glucose oxidase, EC 1.1.3.4	50–100 mM glucose	0.05–2 mM ferrocene monocarboxylic acid, dimethylaminomethyl ferrocene (oxidation)	[24–26]
		0.05–2 mM $K_4[Fe(CN)_6]$ (oxidation)	[25]
		0.02–2 mM hydroquinone (oxidation)	[25, 27]
		0.5 mM [Os fpy (bpy)$_2$Cl]Cl, fpy = formylpyridine, bpy = bipyridine (oxidation)	[26]
PQQ-dependent glucose dehydrogenase, EC 1.1.99.17	50 mM glucose	0.05–2 mM ferrocene monocarboxylic acid 0.05–2 mM ferrocene methanol 0.05–2 mM p-aminophenol (oxidation)	[20]
NADH-cytochrome C reductase, EC 1.6.99.3 within mitochondria	50 mM NADH	0.5 mM N,N,N′,N′-tetramethyl-p-phenylenediamine (oxidation)	[27]
Diaphorase (NADH acceptor oxidoreductase, EC 1.6.99.-)	5.0 mM NADH	0.5 mM hydroxymethyl ferrocene (oxidation)	[28]
Horseradish peroxidase, EC 1.11.1.7	0.5 mM H_2O_2	1 mM hydroxymethyl ferrocenium (reduction)	[29]
Nitrate reductase, EC 1.7.99.4	23–65 mM NO_3^-	0.25 mM methylviologen (reduction)	[30, 31]

In order to perform an experiment, the UME is first positioned in the bulk solution and a cyclic voltammogram is recorded. In this experiment, a triangular potential profile is applied to the working electrode and the resulting current is recorded (Fig. 27.4b). If the UME potential E_T exceeds the formal potential $E^{\circ\prime}$ of the Fc/Fc$^+$ redox couple, a steady-state diffusion-limited current, $i_{T,\infty}$ is observed. The potential E_T of the UME for subsequent imaging is selected such that a diffusion-controlled current is observed.

When the UME is located above the sample regions with immobilized enzymes, Fc$^+$ formed at the UME reaches the enzyme and acts as an electron mediator enabling a conversion at the sample according to (27.4). Figure 27.4d shows an experimental image. The increased currents in the center are caused by the additional flux of the mediator when the UME is located above the enzyme-modified beads. A negative control experiment was performed by imaging the same area after exchanging the solution with a glucose-free buffered solution containing Fc. In this case the enzymatic reaction cannot proceed because the substrate glucose providing electrons to the enzyme is absent. The corresponding image in Fig. 27.4e shows a rather uniform current across the entire area. This current exclusively originates from the diffusional transport of Fc from the solution bulk to the active UME area.

The described principle has been applied to a number of enzymes within the oxidoreductase family that are important in the context of biosensors or biofuel cells (Table 27.1).

27.2.1.2
Quantification of Enzyme Kinetics Using the Feedback Mode

Soon after developing the initial SECM experiments, it was realized that this method could evolve into a new tool for the *quantitative* determination of the kinetics for a variety of interfacial reactions. The basic concept is outlined here under omission of exact deviations. Other reviews provide a broad coverage on this subject [1]. The diffusion-limited current according to (27.1) is the experimental current that is observed at sufficiently positive potentials, e.g. at $E_T = 400\,\text{mV}$ (Fig. 27.4b). The current at any given position of the UME above the sample is obtained from a partial differential equation (or a set of them). The kinetic behavior of the sample is entered as one of the boundary conditions. Such partial differential equations have been numerically solved for a variety of relative UME–sample positions (indexed k) and kinetic properties of the sample [15,32]. Results of several simulations $I_{T,k} = f(L_k)$ for discrete working distances L_k have been interpolated by analytical functions $I_T = f(L)$ facilitating further analysis. Several proposed analytical functions are available in the literature [14,33–35]. We restrict our discussion to one selected example in closest agreement with the presented experimental results [14]. In general, the differences between these functions do not exceed experimental uncertainties. The calculated function $I_T = f(L)$ is experimentally obtained by moving the UME slowly ($1\,\mu\text{m s}^{-1}$) from a large distance ($d > 10\,r_T$) towards the sample surface. Therefore, such experiments are commonly called approach curves, although the curve superimposes with the retract curve because a quasisteady-state situation is reached at each location. Such approach curves are the equivalent to force–distance curves in SFM and contain a wealth of kinetic information. The experimental approach curves (Fig. 27.3, open symbol) towards a GDH-modified surface using ferrocene methanol as mediator is located in-between the two limiting cases (Fig. 27.3, curves 1 and 2). This is the typical situation for sample surfaces with limited heterogeneous kinetics.

The experimental approach curves can be described by

$$I_T(L) = \frac{i_T}{i_{T,\infty}} = I_T^{ins}(L) + I_S^{kin}(L)\left(1 - \frac{I_T^{ins}(L)}{I_T^{cond}(L)}\right).\tag{27.5}$$

The normalized substrate current $I_S^{kin} = i_S/i_{T,\infty}$ is the current equivalent at the sample normalized by the UME current in bulk solution. It can be estimated considering the following three conditions: i) the distance range is $0.1 < L < 1.6$, ii) $RG \approx 10$ and iii) the reaction at the sample is of first order with respect to the mediator.

$$I_S^{kin}(L, k_{eff}) = \frac{0.78377}{L\left(1 + \frac{1}{\Lambda}\right)} + \frac{0.68 + 0.3315\exp\left(\frac{-1.0672}{L}\right)}{1 - F(L, \Lambda)},\tag{27.6}$$

where $\Lambda = \kappa L$; $F(L, \Lambda) = (11/\Lambda + 7.3)/(110 - 40L)$, and $\kappa = k_{eff}r_T/D$ [33]. The curves for various κ values are also given in Fig. 27.3. From the agreement of the experimental curve and the calculated curve 4 in Fig. 27.3, one can conclude $\kappa = 0.55$ for this experiment. Using the experimentally determined diffusion coefficient of the mediator $D = 7.25 \times 10^{-6}\ \mathrm{cm^2\ s^{-1}}$, and $r_T = 12.5\ \mu m$ (from an optical micrograph or from an experimental approach curve to glass and a fit to (27.2)), an effective first-order rate constant for the enzymatic reaction at the sample of $6 \times 10^{-3}\ \mathrm{cm\ s^{-1}}$ is obtained.

The general form of the expression for an enzyme that transfers two electrons to a one-electron mediator [25] is given by

$$f_{Fc^+} = -f_{Fc} = \frac{k_{cat}\Gamma_{enz}}{\dfrac{K_{M,gluc}}{c_{gluc}} + \dfrac{K_M}{c_{Fc^+}} + 1}.\tag{27.7}$$

$K_{M,gluc}$ and K_M are the Michaelis–Menten constants of the enzyme with regard to glucose and Fc^+, respectively. Γ_{enz} is the surface concentration of the enzyme and k_{cat} is the catalytic turnover number of the enzyme. Using glucose concentrations $c_{gluc} \gg K_{M,gluc}$ and mediator concentrations $c_{Fc}^* \ll K_{M,Fc}$, one obtains a pseudofirst-order rate law with an effective rate constant given by

$$k_{eff} = \frac{k_{cat}\Gamma_{enz}}{K_M}.\tag{27.8}$$

Such approach curves have been obtained for several mediator concentrations. By subtraction of I_T^{ins} (27.2) from the experimentally obtained curve using (27.5), $I_{S,K}$ is obtained and can be used in order to verify the first-order nature of the reaction and to obtain a K_M value for various mediators [20]. If Γ_{enz} is known from an independent experiment, k_{cat} of the immobilized enzyme can be determined [20].

The analysis outlined can be applied in essence to a number of different reactions at samples that can be described by first-order heterogeneous kinetics, e.g. electrochemical reactions, ion and electron transfer at liquid/liquid interfaces, surface etching, etc. Hence, SECM is an extremely versatile technique for the investigation of interfacial reactivities.

27.2.1.3
Investigation of Immobilized Enzymes in the Generation-Collection Mode

Despite the possible quantification of the immobilized enzyme activity, it should be noted that feedback-mode analysis (and imaging) of enzymes has to deal with some principal limitations: The analysis is only valid if the enzyme is immobilized at an insulating support, e.g. a membrane. If the enzyme is immobilized at a conductive surface, e.g. an electrode of an amperometric biosensor, recycling of the mediator will in most cases be superseded by the electrochemical recycling of the mediator at the sample (electron transfer from the electron conductor to the mediator "bypassing" the enzyme) [26]. If the sample is not connected to an external voltage source, it will assume an electrochemical potential controlled by the concentration of the reduced and oxidized mediator in the solution. Since the oxidized form (in our example) is only present in the vicinity of the UME and everywhere else the reduced form is in large excess, the conducting sample will assume a potential that corresponds to the reduced mediator. Between the region directly under the UME (where the oxidized form prevails) and the sample regions further away, a *concentration cell* is formed that can drive the reduction of the mediator very efficiently even in the absence of an enzyme or an externally applied potential. Most oxidoreductases already work in the regime of zero-order kinetics at millimolar concentration levels of the mediator, which are convenient for SECM measurements. This substantially decreases the sensitivity of the analysis. Furthermore, only oxidoreductases can be investigated, other enzymes cannot be imaged. The generation-collection (GC) mode, specifically the sample-generation/tip-collection mode provides a viable alternative for all these limitations since it is more sensitive and the response is independent of the nature of the immobilization substrate (conductive or insulating).

The schematic for GC imaging of GDH is shown in Fig. 27.5a. In contrast to the feedback experiment in Fig. 27.4c, the electron acceptor for GDH ($[Fe(CN)_6]^{3-}$) is contained in the bulk of the working solution and the enzymatic reaction independently proceeds of the presence of the UME. A diffusion layer develops above the active regions of the sample and the local concentration is probed by the UME. A typical experimental example is shown in Fig. 27.6 for the enzyme galactosidase [36]. Galactosidase is an important marker enzyme and is also used as a label in various biotechnological applications and in im-

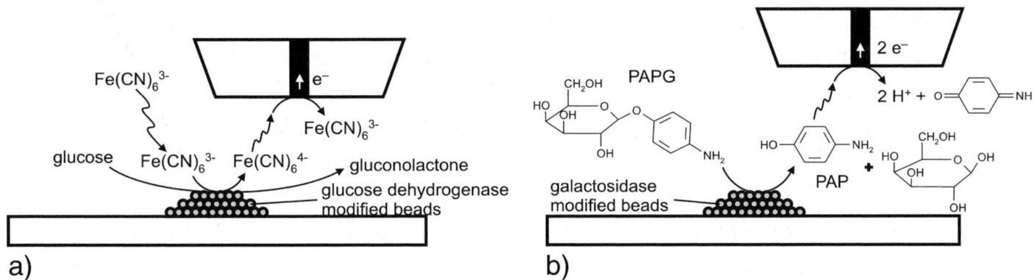

Fig. 27.5. Schematic for the GC mode for the enzyme (**a**) GDH and (**b**) galactosidase

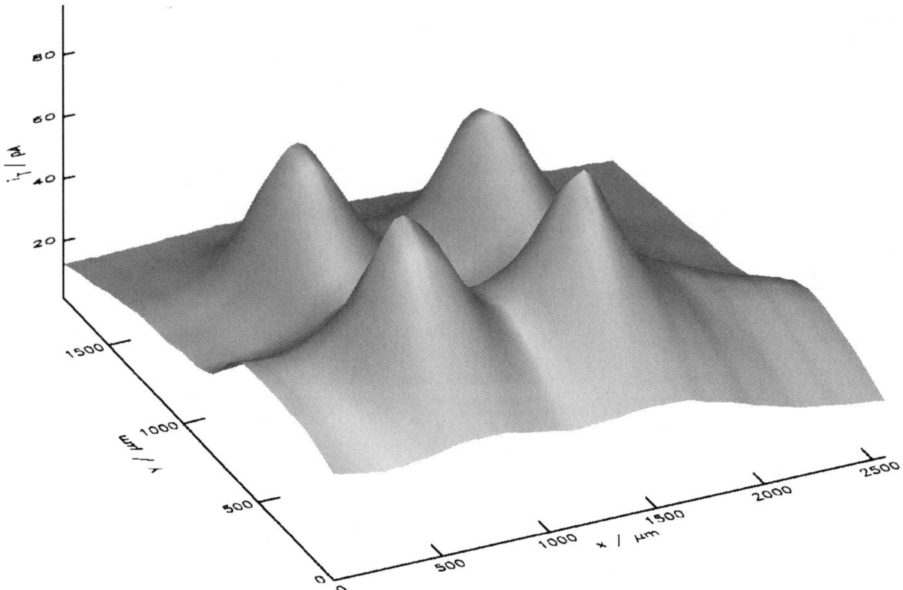

Fig. 27.6. GC-mode imaging of an array of surface-modified bead mounds labeled with galactosidase. Reprinted from J. Electroanal. Chem., 561, C. Zhao, J.K. Sinha, C.A. Wijayawardhana, G. Wittstock, Monitoring β-galactosidase activity by means of scanning electrochemical microscopy. 83–91, Copyright 2004, with permission from Elsevier Science

munoassays. In principle, this enzyme can not be imaged by the feedback mode, since it does not belong to the oxidoreductases. The "trick" to make its activity "visible" to SECM is based on the chosen substrate (p-aminophenyl-β-D-galactopyranoside, PAPG). Galactosidase catalyzes the formation of p-aminophenol (PAP) from PAPG (Fig. 27.5b). PAP can be oxidized at a lower potential than PAPG, which allows imaging of the diffusion layers formed above immobilized galactosidase. There is a wide variety of substrates available for photometric activity tests of enzymes that can be used directly or after appropriate chemical modifications.

Table 27.2 provides an overview of the enzymes that have successfully been imaged in GC mode. The following guidelines may be helpful for investigation of immobilized enzymes via a feedback experiment or a GC experiment. If the enzymes are immobilized at a conducting surface, the GC mode is more suitable. The achievable lateral resolution in GC mode is lower than in the feedback mode, however, it can be improved by coupling to further homogeneous or heterogeneous reactions [38]. The GC mode for investigating enzymes is significantly more sensitive than the feedback mode. It is less sensitive to topographic features and can operate at larger working distances, which is an advantage for corrugated samples. An experimental comparison of the two operation modes in the context of biosensor research is available in the literature [44]. The feedback mode allows a more precise quantification of the kinetics. The GC mode can also be used with ion-selective electrodes as probes, provided that there is a mecha-

Table 27.2. SECM imaging of local enzyme activities in the GC mode

Enzyme	Species detected at the UME	Probe	Ref.
Glucoseoxidase EC 1.1.3.4	H_2O_2	Amperom. Pt UME	[37,38]
		Amperom. enzyme electrode	[39]
PQQ-dependent glucose dehydrogenase, EC 1.1.99.17	$[Fe(CN)_6]^{4-}$	Amperom. Pt UME	[20]
urease EC 3.5.1.5	H^+	Potentiom. Sb UME	[40]
	NH_4^+	Liquid membrane ISE	[41]
Horse radish peroxidase EC 1.11.1.7	Ferrocene derivatives	Amperom. Pt UME	[29,42,43]
Alkaline phosphatase EC 3.1.3.1	4-aminophenol	Amperom. Pt UME	[24,44–47]
Galactosidase EC 3.2.1.23	4-aminophenol	Amperom. Pt UME	[36,48]
Alcoholdehydrogenase EC 1.1.1.1	H^+	Potentiom. Sb UME	[49]
NADPH-dependent oxidase in osteoclasts	$O_2^{\bullet-}$	Cytochrome-c-modified Au UME (amperom.)	[50]

nism to vertically position the electrode. After all, it is not surprising that many more investigations related to practical applications have been performed in GC mode, whereas feedback mode dominates in fundamental studies. This observation extends beyond imaging of enzymes. However, a successful design of SECM experiments requires fundamental understanding of both mechanisms: A complete experiment usually includes both working modes, e.g. feedback experiments are used to position the UME at a defined distance from the sample using (27.2) and then a GC image is recorded. Secondly, GC contributions may disturb feedback experiments and feedback contributions can be present in GC experiments that either disturb a quantitative analysis, or which can be exploited for enhancing the sensitivity [48]. A substantial extension of the GC-mode applications can be expected if miniaturized biosensors are used as probes, as demonstrated by Kueng et al. [51,52].

27.2.1.4
Quantification in the GC Mode

With some limitations, GC experiments can also be quantified. Most applications approximate the UME as a noninteracting probe, i.e. a probe that only senses the local concentration but does not affect the local environment by conversion of one species. While this is fulfilled for ISE, it is only an approximation for an amperometric microelectrode. The approximation is justified if the UME is considerably smaller than the active region at the sample, and if the working distance is larger than $3r_T$. It is advantageous if the RG value of the UME is small ($RG < 5$). In order to form a steady-state diffusion layer above the sample, the active regions at the sample must be microstructures and spaced with at a sufficiently large distance ensuring that the diffusion layers do not overlap. The analysis basically follows a simple diffusion model that describes how a compound diffuses through a disk-shaped pore into an acceptor compartment [53, 54]. The current at the UME can then be calculated in analogy to (27.1) where the bulk concentration c^* is replaced by the local concentration at the position of the microelectrode. This local concentration $c = c_S\theta$ can be derived from the concentration at the pore opening c_S and a geometry-dependent, dimensionless dilution factor θ. They are related to i_T by

$$i_T = 4nFDr_Tc_S\theta \ . \tag{27.9}$$

The factor θ depends on the lateral distance r and vertical distance d of the UME from the center of the active region. For a single horizontal scan (x) across the center of the spot, θ can be described as follows:

$$\theta = \frac{2}{\pi} \arctan \frac{\sqrt{2}r_S}{\sqrt{\left(\Delta x^2 + d^2 - r_S^2\right) + \sqrt{\left(\Delta x^2 + d^2 - r_S^2\right)^2 + 4d^2r_S^2}}} \ . \tag{27.10}$$

A fit of an experimental line scan to (27.9) and (27.10) provides c_S. An example of a line scan across a GDH-modified surface is given in Fig. 27.7. The current for the oxidation of $[\mathrm{Fe(CN)_6}]^{4-}$ formed during the enzymatic reaction is plotted vs. location. The experimental current can be fitted to (27.9) and (27.10). The total flux Ω from the bead spot can be calculated by [13]

$$\Omega = 4Dc_Sr_S \ . \tag{27.11}$$

From the total flux Ω, the area (πr_S^2) of the active region, the surface concentration of the enzyme Γ_{enz}, and the catalytic turnover number $k_{\mathrm{cat}} = \Omega(\pi r_S^2)/\Gamma_{\mathrm{enz}}$ can be

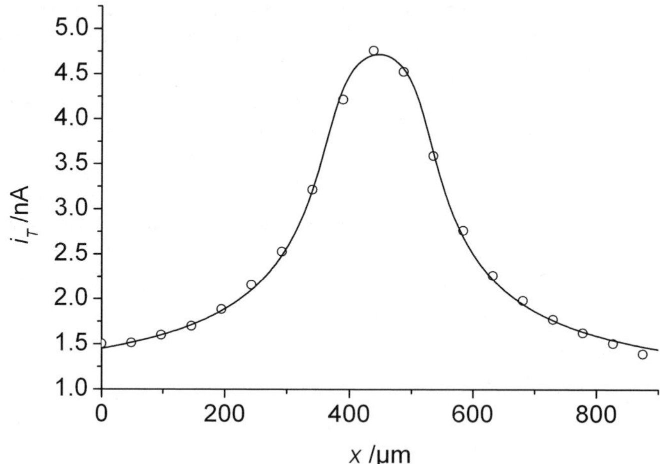

Fig. 27.7. Fitting of a GC line scan across a GDH-modified bead mound to (27.9) and (27.10). Experimental parameters $c^*_{[\mathrm{Fe(CN)_6}]^{-3}} = 10\,\mathrm{mM}$, $d = 30\,\mu\mathrm{m}$, $r_S = 75\,\mu\mathrm{m}$. The *solid line* was calculated from (27.9) and (27.10) with the following parameters $c_{S,[\mathrm{Fe(CN)_6}]^{-4}} = 1.69\,\mathrm{mM}$, $d = 30.45\,\mu\mathrm{m}$, $r_S = 74.35\,\mu\mathrm{m}$, $D_{[\mathrm{Fe(CN)_6}]^{-3}} = 6.19 \times 10^{-6}\,\mathrm{cm^2\,s^{-1}}$, and a constant current offset of 0.91 nA for all points. From [20] Copyright 2004, American Chemical Society

estimated for the immobilized enzyme, when the substrate concentrations are much higher than K_M (zero-order reaction) [20].

27.2.1.5
SECM for Read-Out and Optimization of Biochips and Biosensors

Microstructured biosensor surfaces have been investigated by SECM [17,55]. These investigations were mainly prompted by the idea to produce compartmentalized surfaces, where some regions are surface modified in order to provide an ideal environment for the biochemical component while other regions are optimized for electrochemical detection [56,57]. Furthermore, SECM can be applied to locally modify surfaces [28,43,58–67] by different defined electrochemical techniques. The combination of the imaging capabilities for specific enzymatic reactions and the possibility of modifying the surfaces in a buffered solution make SECM an ideal tool to explore the potential of such micropatterned surfaces for sensing applications.

Another set of applications relies on the opportunity of recognizing many biochemicals by a specific antibody chemically linked to an enzyme label. Herein, SECM can be used to detect the enzyme label [47]. In this application, SECM can be an alternative to fluorescence detection. However, the main disadvantage of SECM in this respect can be seen in long experimental time, lack of high throughput and the required skills of the experimentator. On the other hand, the sensitivity of the detection is comparable, the required instrumentation is less expensive and common problems in optical detection such as background fluorescence are of no concern in SECM. Given these circumstances, electrochemical techniques will be important for protein arrays, where a limited number of analytes is detected. Due to its flexibility, SECM is ideally suited for the optimization of such protocols. A readout for a mass-produced assay can then be realized in microstructured electrochemical cells using the SECM working modes but avoiding the mechanically demanding and time-consuming scanning [68,69].

Matsue et al. [29,42,70,71] have developed a number of biochips based on the described principle. Among these biochips are multianalyte assays for human placental lactogen (HPL), human chorionic gonadotropin (HCG) [42] and leukocidin, a toxic protein produced by methicillin-resistant *Staphylococcus aureus* [71]. Figure 27.8 shows an example of a dual immunoassay with SECM detection. Different antibodies were immobilized at specific regions of the chip. After exposure to a solution containing HPL and HCG, the analytes were bound by the antibody. The chip was then exposed to a mixture of horesradish-peroxidase (HRP)-labeled antibodies against HPL and HCG. After adding the substrate for the enzymatic reaction H_2O_2 and ferrocene methanol (Fc), as the electron acceptor the readout is made by recording the current for the reduction of ferrocinium methanol (Fc^+) at the UME. Fc^+ is produced locally at the chip surface by the enzyme HRP under consumption of H_2O_2. In this way the analyte is defined by the position on the chip and the amount of analyte is quantified via the collection current at the UME.

Microbeads can be used as a new platform for immunoassays and a model assay for immunoglobulin G (IgG) has been demonstrated by Wijayawardhana et al. [45,72] with alkaline phosphatase as the labeling enzyme. Detection limits down to 6.4×10^{-11} mol L^{-1} or 1.4×10^{-15} mol mouse IgG could be demonstrated.

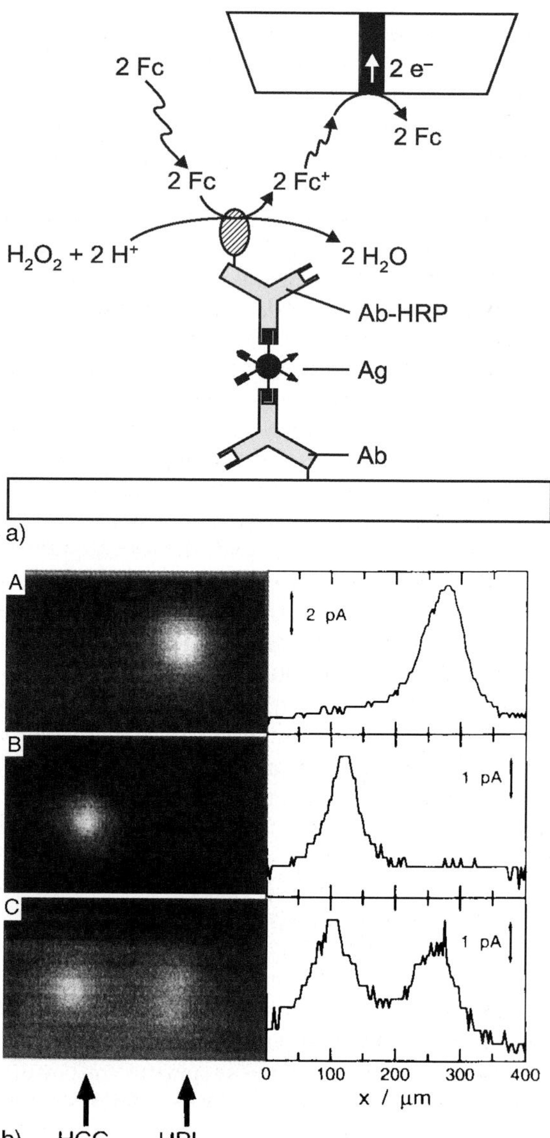

Fig. 27.8. (a) Scheme of the SECM detection in a sandwiched immunoassay. (b) SECM images and cross-sectional profiles of microstructured glass supports with (anti-HCG)-Ab and (anti-HPL)-Ab immobilized at two distinct regions, treated with **A** 56 ng mL^{-1} HPL; **B** 2.0 IU mL^{-1} HCG; **C** a mixture containing 31 ng mL^{-1} HPL + 0.63 IU mL^{-1} HCG. After rinsing, the glass support was dipped into a solution of 20 μg mL^{-1} (anti-HCG)-Ab-HRP + 7 μg mL^{-1} (anti-HPL)-Ab-HPL. Working solution: 1.0 mM Fc + 0.5 mM H$_2$O$_2$ + 0.1 M KCl + 0.1 M phosphate buffer, pH 7.0; v_t = 9.8 μm s^{-1}; E_T = +0.05 V (Ag/AgCl). Reprinted from J. Electroanal. Chem., 438, H. Shiku, Y. Hara, T. Matsue, I. Uchida, T. Yamauchi, Dual immunoassay of human chorionic gonadotropin and human placental lactogen at microfabricated substrate by scanning electrochemical microscopy. 187–190, Copyright 1997, with permission from Elsevier Science

Since the p-aminophenol produced by the enzyme-catalyzed reaction was detected close to the point of origin, an incubation step for enzymatic amplification could be avoided.

More recently, another application of SECM detection in DNA and protein chips and in electrophoresis gels has emerged with different detection principles. Wang et al. [73] labeled single-stranded DNA (ssDNA) with gold nanoparticles. After binding to their complementary strand at the chip surface, silver was electroless deposited at the metallic particles. The formed metal features were imaged in the feedback mode with $[Ru(NH_3)_6]^{3+}$ as mediator, in which the mediator is regenerated by a electron transfer to silver metal at those regions of the sample surface, where a thin silver film was formed. A similar approach was used by Carano et al. [74]. Proteins were blotted on poly(vinylidene difluoride) membranes. The model protein bovine serum albumin was then tagged with previously prepared silver nanoparticles. The position of the protein bands was detected by SECM using $[Os(bpy)_3]^{2+}$ as a mediator, which was oxidized at the UME. The oxidized mediator diffused to the sample surface and induced the oxidation and dissolution of the deposited silver nanoparticles, which were not interconnected to each other. In these applications, the exact quantitative relation between the SECM signal and the amount of analyte is not well understood and it is empirically established by calibration curves that might be sensitive to a number of interfering factors that are difficult to control.

Wang and Zhou [75] detected hybridized DNA on a chip by electrogenerating $[Ru(bpy)_3]^{3+}$ at the UME in a SECM experiment. $[Ru(bpy)_3]^{3+}$ is a strong oxidation reagent and can oxidize the base guanine, present in DNA. Where DNA was located at the chip, the reaction between guanine and the oxidized form of the mediator resulted in an enhanced current at the UME.

Turcu et al. [76, 77] immobilized ssDNA at microspots at a gold surfaces. The location of the ssDNA could be imaged using the negatively charged redox mediator $([Fe(CN)_6]^{3-/4-})$ in SECM feedback mode. The mediator regeneration proceeded by a heterogeneous electron transfer to the gold surface of the chip. At the spots of the negatively charged ssDNA, the diffusion of the negatively charged mediator to the gold surface was blocked and a reduced feedback current was observed. The observed current was further reduced, when hybridization with the complementary strand occurred and the negative charge at the surface was increased.

Fortin et al. [78] reported an alternative detection scheme based on an enzymatic amplification without imaging the enzyme activity in feedback mode. They produced the DNA microarray by depositing a polypyrrole pattern using a procedure developed earlier by Kranz et al. [79]. By using a mixture of pyrrole and functionalized monomers covalently attached to a 15mer ssDNA oligomer, polypyrrole spots were formed carrying the 15mer oligomer as single DNA strand [80]. Hybridization was performed with biotinylated complementary ssDNA strands. After reaction with streptavidin and biotinylated horseradish peroxidase, the enzymatically catalyzed oxidation of soluble 4-chloro-1-napthol to insoluble 4-chloro-1-naphton was used to create an insoluble, insulating film at the polypyrrole spots. SECM imaging with the mediator $[Ru(NH_3)_6]^{3+}$ visualized these regions by reduced currents compared to the bare gold surface [78].

27.2.2
Investigation of Metabolism of Tissues and Adherent Cells

Electroanalytical techniques such as the patch clamp technique, miniaturized ISE and surface-modified carbon microelectrodes for the detection of neurotransmitter release from nerve cells are routinely used in physiological studies, both at the cellular level and at the level of entire organs [7]. Therefore, it seems viable to use SECM to gain deeper insights into structure–function relationships important in biological research. This potential had been recognized very early in SECM development and it essentially relies on the same principles that were outlined for the investigation of immobilized enzymes. However, some additional challenges emerge when working with cellular systems and current instrumental developments are aiming to overcome these obstacles. In conventional SECM imaging, the UME is scanned parallel to a flat surface in order to keep the working distance d constant. This approach is sufficient if the roughness of the sample is negligible compared to the (electro)active UME radius r_T, which is typically 5–20 μm. In contrast, cells represent corrugated objects with height differences in the micrometer range, i.e. similar or even larger in size compared to r_T. Hence, the UME has to follow the topology of the sample in order to keep a constant d similar to the approaches used in other scanning probe techniques such as STM. Derived from Fig. 27.3 it becomes evident that this task is more challenging in conventional SECM. The current–distance relationship may vary significantly depending on the chemical nature of the sample. Therefore a constant-current imaging is not feasible. The current may decrease upon approach of the probe (inert surface) or increase (reactive surface). Therefore, it is impossible to select a setpoint that would be appropriate for an entire sample. Furthermore, using a constant-current approach would limit the possibility to quantify the SECM response with respect to chemical reactivities. In order to overcome the difficulties in positioning the UME above corrugated objects, a number of approaches have been used. The microelectrode can be positioned with the help of an optical microscope at a fixed location with respect to the biological specimen and the release of a specific substance is followed after stimulation [50, 81, 82]. Alternatively, cells can be cultivated in specifically designed conical microcavities, forming a flat surface [65, 83–85]. The UME scans in a constant distance over the openings. Alternatively, dual-electrodes have been used for positioning [86]. One electrode records the redox current of a mediator conversion that does not interact with the sample. For this compound the current is controlled exclusively by (27.2) and is used as the distance-dependent signal. The second electrode of the dual electrode assembly detects the species of biological interest that is released from the biological entity. Several attempts exist to design a setup by means of feeding back a *current*-independent signal into an actuator that keeps d constant. The most convincing results could be obtained by systems that detect mechanical shear forces between an UME that vibrates parallel to the sample surface with amplitude in the nanometer range ($\ll r_T$). Since the electrodes are immersed in a viscous solution, stable detection of the shear forces appears to be challenging. The vibrations can be excited by a tuning fork similar to the concept in scanning near-field optical microscopy [87–90]. However, since the resonance frequency is determined by the tuning fork but the UME represents usually a higher mass than the legs of the tuning fork, the mechanical resonator tends to be ill-defined and the number of published examples is rather limited, despite the high expectations that have been raised earlier. Another approach excites the microelectrode with a frequency tuned to the mechanical properties of the microelectrode itself. The vibration amplitude is detected either by projecting a diffraction pattern produced by a light-emitting diode onto a split photodiode [91] or by detecting the vibration amplitude with a second piezo attached to the UME [92]. Both approaches have been largely popularized in the SECM community by the work of Schuhmann et al. [62, 93–98]. Meanwhile, a commercially self-standing shear force system is available that can be attached to any well-designed SECM instrumentation. Nevertheless, the use of a shear-force system adds another level of complexity to the experiment because the mechanical properties of the UME, the sample and the entire setup become important and have to be optimized.

27.2.2.1
Imaging Photosynthetic Oxygen Production

SECM was used to map the topography and the rate of photosynthetic oxygen production of a leaf still attached to an intact plant, *Tradescantia fluminensis* [81]. The leaf was fixed at the base of the electrochemical cell. The entire setup was placed inside a glove bag to control of the environment (Fig. 27.9). Illumination of the leaf was achieved by shining light through an optical fiber attached to a xenon lamp from the base of the electrochemical cell. Since electrochemical redox mediators are often toxic to living organisms, oxygen was chosen to image the topography of the leaf in the dark (no O_2 production by the leaf) by exploiting (27.2) [23]. Although the electrochemical reduction of oxygen involves a more complicated scheme, it is

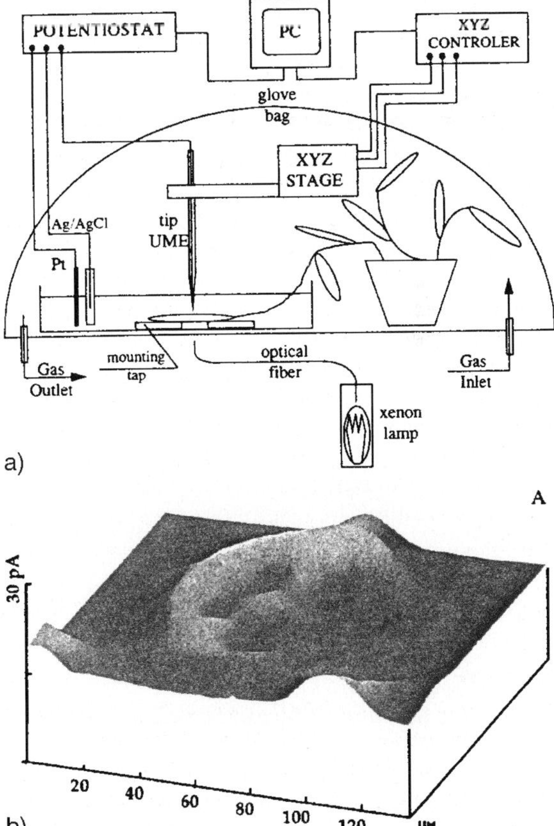

Fig. 27.9. (a) Experimental setup to study local oxygen production at leaf surfaces. (b) Oxygen release from a single stomatal complex in the white leaf region of *Tradescantia fluminensis*. (Reprinted with permission from [81] Copyright American Society of Plant Biologists)

a logical choice for studying living systems. It has to be assured that E_T is poised at a sufficiently negative value to ensure a current that is proportional to the bulk concentration of O_2 and the solution is buffered to compensate the pH shift induced by the proton transfer accompanying the O_2 reduction. Individual stomatal complexes were clearly resolved with the protruding guard cells appearing as depressions in the UME current image and the stomatal pore in the center of the complex as a current peak in agreement with (27.2) that predicts a decrease of i_T if d is reduced by protruding sample regions. *T. fluminensis* possesses white regions in which chloroplasts are only present in the guard cells around the stomata. The stomata were sufficiently widely separated that the oxygen diffusion layers did not overlap and hence, generation-collection mode imaging could be performed. The images recorded during illumination (O_2 production in the chloroplasts of the guard cells) showed a peak current roughly a factor of two higher than the O_2 background over the stomata. Tsionsky et al. [81] were able to image the photosynthetic oxygen production in a single stomatal complex following changes in illumination cycles. This work illustrates the flexibility of the SECM technique: Topography and biochemical reactivity can be imaged and a particular location at the sample can be picked to perform kinetic studies of the response to an external stimulus in essentially one experiment.

Yasukawa et al. [86, 99] studied the oxygen production of individual protoblasts from *Bryopsis plumosa* and the response to the administration of several chemicals.

27.2.2.2
Imaging Respiratory Activity via Oxygen Consumption

Oxygen consumption by individual cells or a tissue sample has been detected in a similar way. In this case the UME senses less O_2 in the vicinity of a respiratory-active cell. The investigated systems include HeLa cells [100] and cells from the cell line SW-480 [101]. The respiratory activity was monitored in response to the injection of cell poisons and the kinetics of the uptake of these compounds has been mapped at the single-cell level. Since oxygen is converted at the UME and the living cell, it is crucial to distinguish between a depletion of dissolved O_2 by the respiratory activity of the cell and changes in the working distance d. In a series of papers, Shiku et al. [102, 103] used such measurements to investigate the development of in-vitro-fertilized bovine embryos. In this case the UME was used to perform differential measurements close to the embryo and in the bulk of the culture medium. The respiratory activity was correlated with the survival chances of the embryos. More recently, Matsue and coworkers [85] used specially designed microstructures to cultivate cells from a human erythroleukemia cell line and to measure the O_2 consumption without the problem of protruding cells (Fig. 27.10).

27.2.2.3
Monitoring the Local Release of Signaling Substances

There is a long history of using carbon microelectrodes in detecting the release of neurotransmitters from nerve cells with high temporal resolution [7]. In combination with SECM such experiments could be refined. Hengstenberg et al. [104] used

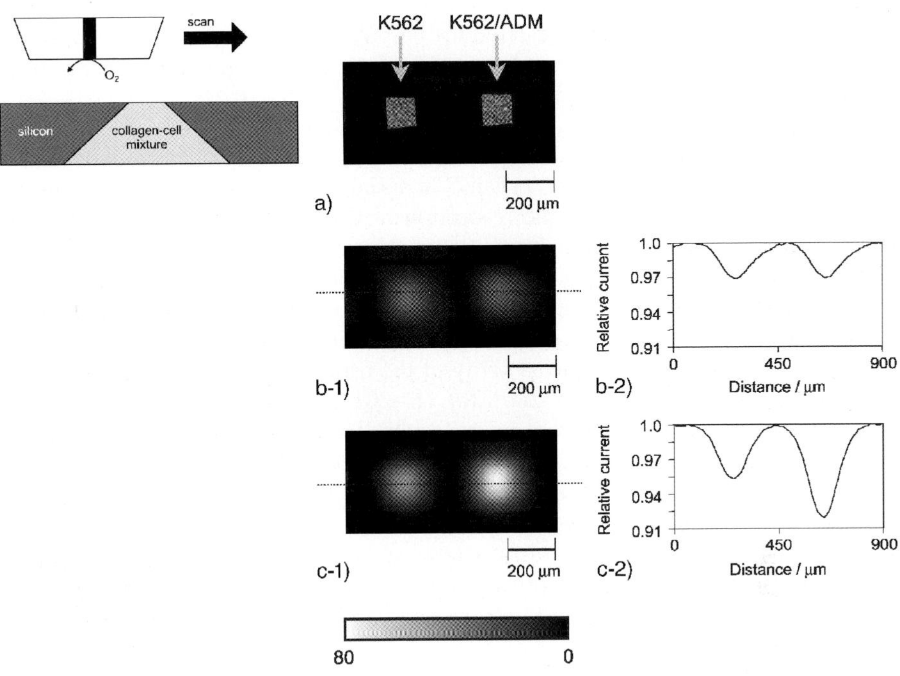

Fig. 27.10. Schematic of the silicon chip for cell assays (*left*) and example of measurements (*right*). (**a**) Optical micrograph of a collagen cell matrix with different human erythroleukemia cell lines incubated in the holes. (**b**) SECM images of oxygen consumption by the cells. Light shades mean high oxygen consumption. (**c**) Same as (**b**) but after 3 days of incubation with $1\,\mu\mathrm{mol\,l^{-1}}$ adriamycin. One cell line is insensitive to this drug and continues to grow, whereas the respiratory activity of the cell line in the left hole is inhibited. From [85] Copyright 2003, American Chemical Society

carbon-fiber microelectrodes in combination with a shear-force-based SECM system to obtain a low-resolution topographic image of PC12 cells and to position the microelectrode above a specific part of the cell. Upon chemical stimulation with K^+, the release of catechol amines could be monitored without scanning the UME. As the main advantage for using SECM compared to earlier experiments the authors state that the cells are not mechanically stimulated during the microelectrode positioning. In a similar way the release of adrenaline and noradrenaline from single secretory vesicles of chromaffin cells was measured after stimulation [105]. Cells were selected with the help of optical microscopy and the miniaturized carbon electrodes were positioned using a shear-force distance control (see Sect. 27.4).

In a related approach, nitric oxide release from endothelial cells was studied [82]. Among other effects, NO guides the growth of blood capillaries. In order to detect NO, an electrocatalytic sensor with a diameter of $50\,\mu\mathrm{m}$ based on substituted nickel porphyrins was employed. This sensor was combined with a conventional Pt microelectrode (diameter $10\,\mu\mathrm{m}$) into a dual electrode probe. The Pt electrode was used for positioning the NO sensor close to the surface based on hindered diffusion of

O_2 (27.2). The NO sensor then detected the NO release from the cells. However, to date further miniaturization of the presented NO sensor is limited since the reported currents for the NO detection are in the pA range.

SECM has been used to study the resorption of bone by specialized cells known as osteoclasts [106–108]. Bone is a complex material consisting of calcium-phosphate-based minerals (hydroxyapatite), organic material (mostly collagen) and various cells. The osteoclasts carry out the resorption of bone by the secretion of protons and hydrolytic enzymes, which break down the minerals and proteins releasing, among other species, Ca^{2+} (Fig. 27.11a). Bone is a dynamic system with a balance between the resorption and formation of bone. An imbalance in these processes is involved in many diseases, such as osteoporosis characterized by a loss of bone mineral. Assays for the resorptive activity of the osteoclasts are usually carried out by incubating the cells on bone slices for 18–20 h and then determining the area and number of resorption pits from micrographs. This assay is slow and imprecise. The rate of resorption has been more directly probed using SECM with liquid membrane calcium-selective microelectrodes to determine the Ca^{2+} released from the hydroxyapatite into the culture medium during resorption (Fig. 27.11a). The increase in Ca^{2+} due to the resorptive activity of the cells was detected within approximately 10 min of incubation. The retardation of resorptive activity after fluoride treatment was studied and conclusions about its mechanism were obtained.

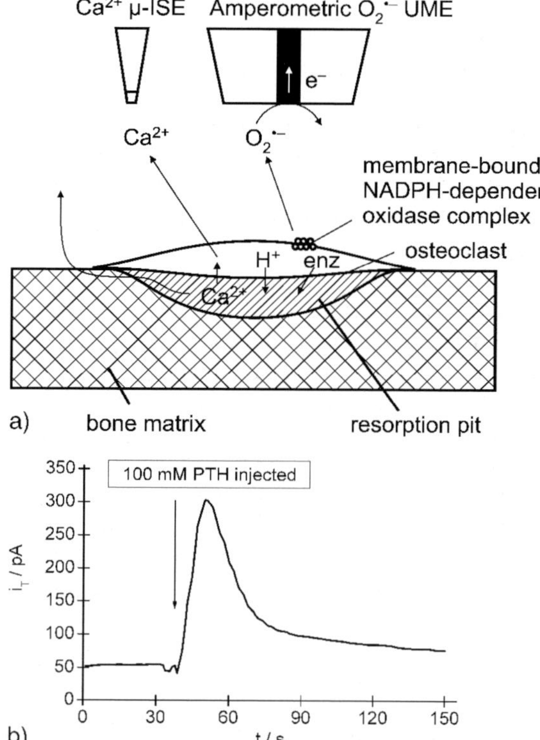

Fig. 27.11. (**a**) Schematic of the processes involved in osteoclastic resorption of bone. Ca^{2+} release can be monitored by a potentiometric Ca^{2+} ion-selective electrode, the superoxide release after stimulation can be recorded by an amperometric electrode. (**b**) Current–time trace for the superoxide release by osteoclasts after stimulation with parathyroid hormone (Reproduced by permission from the Society for Endocrinology from [50])

Free radicals, especially the superoxide anion, are known to be important in the regulation of osteoclast activity and can be detected using an amperometric sensor consisting of cytochrome c covalently attached to a gold electrode [50]. This method was used to detect the acute response of the cells to stimulation by parathyroid hormone (PTH) as a burst of superoxide anion generation. Other work has concentrated on the regulation of the membrane-bound NADPH-dependent oxidase responsible for superoxide generation [50]. The effects of inhibitors of protein kinases, membrane-permeable analogs of cAMP and cholera toxin on the stimulation of superoxide anion production by pertussis toxin (PTH) and ionomycin were studied. cAMP-dependent inhibition was found to be dominant in controlling the superoxide anion production. For this type of collection-mode experiment the SECM apparatus can provide real-time information and can be used in combination with an optical microscope to allow a fast positioning of the UME close to the cells. With such a setup it was demonstrated that the timescale of the superoxide anion burst following the stimulation by PTH was consistent with superoxide diffusion from the osteoclast to the tip (Fig. 27.11b). In contrast to the generally accepted notion of an indirect mediaton of superoxide release by other cells of the osteoblastic lineage, the short timescale of the SECM response together with the observed effects of cholera and pertussis toxins on the superoxide release suggest that the superoxide anion burst results from a direct action of PTH on the osteoclast via a G protein coupled receptor.

27.2.2.4
Probing the Redox Capacity of Individual Cells

Mirkin et al. [109] demonstrated that the redox recycling of a mediator directly couples to the internal metabolism of metastatic and nonmetastatic human breast cells. Several imaging modes could be demonstrated [109]. Hydrophilic mediators cannot penetrate the lipid cell membrane. An image using these mediators will provide an image of the cell topography based on (27.2). If a redox-active species is distributed inside and outside the cell, electrolysis at the UME may deplete the solution in the vicinity of the biological cell and change the concentration gradient. The cell may respond with a release of this compound and the time trace i_T may be indicative of the transport kinetics of the compound across the cell membrane. The mediator can be regenerated at the cell wall by an incorporated redox system, which shuttles the charge across the cell wall. Finally neutral mediators such as menadione and napthophinone may cross the lipid cell membrane and directly titrate the redox equivalents inside the cell [110]. Feng et al. [111] observed differences between normal and metastatic tumor cells that they related to the different protein expression profiles of the cells. The approach was also expanded to investigate bacteria [112]. The SECM experiments were treated with a theory usually applied for charge transfer occurring at liquid/liquid interfaces. It remains to be demonstrated to which limit such simplifying models can provide a deeper insight into complex internal regulatory processes of cells. Nagamine et al. [113, 114] monitored the metabolic regulation of bacteria that were immobilized within collagen microspots and exposed to osmotic stress. Changing the osmotic conditions changed the permeability of the cell membrane toward the hydrophilic electron mediator $[Fe(CN)_6]^{4-}$. It was found that the SECM response changed when cells were growing at a medium with D-glucose as the only carbon source. This suggests that the electron flow in the respiratory chain, with which the mediator interferes, depends on the metabolic pathway upstream of the respiratory chain [113].

27.2.3
Investigation of Mass Transport Through Biological Tissue

The SECM response is strongly dependent on the mass transport in the working solution. Therefore, it is conceivable that this technique can be used to image localized mass transport. Localized mass transport occurs in biological tissue, especially increased transport rates may be observed above follicles or pores. Typically, the biological tissue is mounted as a barrier between a donor and an acceptor compartment (Fig. 27.12). The donor compartment contains a redox active species that is driven through the tissue by diffusion (Fig. 27.12a), electroosmosis (Fig. 27.12b), or hydrostatic pressure (Fig. 27.12c) and is detected by the SECM probe in the acceptor compartment in a GC-mode-based configuration. The transport across skin, dentine and the oxygen transport in cartilage were intensively studied. An excellent review on this topic has been published [115]. White and coworkers [116, 117] studied the transport of various species across skin. These are important studies in respect to novel concepts for drug delivery.

The flux models were experimentally verified by studies on synthetic membranes (mica and track-etched membranes). Quantification through an individual pore can be obtained by (27.9)–(27.11). For these investigations the pores are localized in a survey scan, then the UME is directly positioned over a pore center and an approach curve is record. This experimental approach can be fitted to

$$i_\mathrm{T}(d) = 4nFDr_\mathrm{T}\frac{2c_\mathrm{S}}{\pi}\arctan\left(\frac{r_\mathrm{T}}{d}\right) . \qquad (27.12)$$

Scott et al. [53] identified regions of increased mass transport in hairless mouse skin. It became clear that iontophoretic transport is highly localized and bound to hair follicles. Studies on bulk skin tissues revealed that the transport pathways carry a net negative charge and lead to a preferential transport of cations across skin [118]. The transport of cations induces an electroosmotic flow of the solvent. Therefore the diffusional transport of neutral molecules is modulated by the solvent flow. The relative contributions could be quantified. Figure 27.13 shows the transport of hydroquinone through a single hair follicle of hairless mouse skin by combined diffusion and electroosmotic flow. The skin separates a donor compartment with neutral hydroquinone and a receptor compartment initially not containing hydroquinone. The pure diffusional transport is observed if no iontophoresis current is applied (Fig. 27.13b). If the applied iontophoresis current induces a solvent flow against the diffusional flux

Fig. 27.12. Typical setup for monitoring the transport to biological tissue. The driving force for transport can be provided by a concentration gradient between donor or acceptor compartment (**a**), a current between the donor or acceptor compartment (**b**) or a pressure gradient between the compartments (**c**)

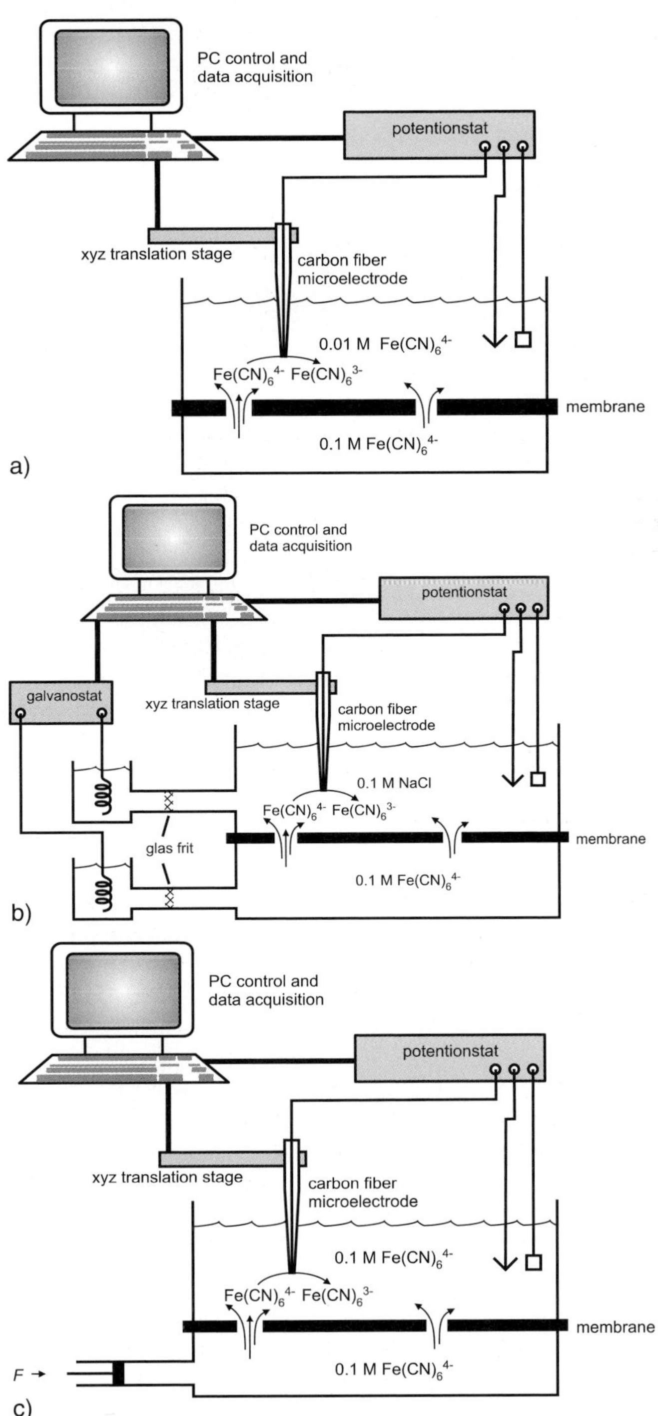

PC control and
data acquisition
potentionstat
xyz translation stage
carbon fiber
microelectrode
0.01 M Fe(CN)$_6$$^{4-}$
Fe(CN)$_6$$^{4-}$ Fe(CN)$_6$$^{3-}$
membrane
0.1 M Fe(CN)$_6$$^{4-}$
a)

PC control and
data acquisition
potentionstat
galvanostat
xyz translation stage
carbon fiber
microelectrode
0.1 M NaCl
Fe(CN)$_6$$^{4-}$ Fe(CN)$_6$$^{3-}$
glas frit
membrane
0.1 M Fe(CN)$_6$$^{4-}$
b)

PC control and
data acquisition
potentionstat
xyz translation stage
carbon fiber
microelectrode
0.1 M Fe(CN)$_6$$^{4-}$
Fe(CN)$_6$$^{4-}$ Fe(CN)$_6$$^{3-}$
membrane
F
0.1 M Fe(CN)$_6$$^{4-}$
c)

Fig. 27.13. SECM GC images of electroosmotic and diffusional transport of hydroquinone across a single hair follicle in hairless mouse skin. The skins separates a donor compartment with neutral hydroquinone and a receptor compartment that initially does not contain hydroquinone. The *middle image* (**b**) corresponds to pure diffusional transport in the absence of an iontophoresis current. In the *top image* (**a**) the iontophesis current induces a solvent flow from the receptor compartment into the donor compartment, in the *bottom image* (**c**) the current direction is reversed leading to a solvent flow into the receptor compartment. Neutral hydroquinone is transported with the solvent flow and by diffusion. Reprinted with permission from [117], Copyright 2000 Springer

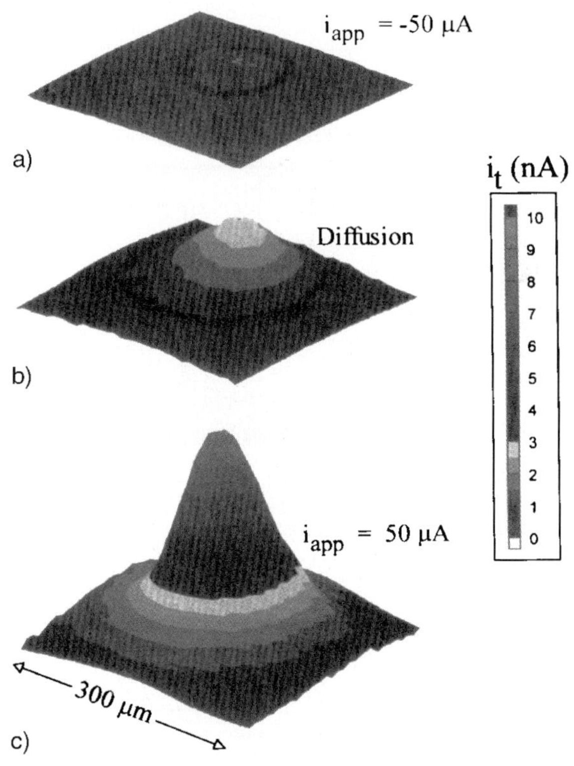

Fig. 27.14. SECM GC images of fluid flow through tubules in dentine. (**a**) Schematic of the experiment. (**b**) SEM image of a dentine slice; (**c**) SECM image of an untreated dentine slice without applied pressure. (**d**) SECM image of an untreated dentine slice with a hydrostatic pressure of 2 kPa to the donor compartment. (**e**) SECM image of a dentine slice after treatment with calcium oxalate without applied pressure. (**f**) SECM image of a dentine slice after treatment with calcium oxalate with a hydrostatic pressure of 2 kPa to the donor compartment. Parts (**c**)–(**f**) reprinted with permission from [120], Copyright 1995 American Chemical Society, Part (**b**) reprinted with permission from [119], Copyright 1996 The Royal Society of Chemistry

(out of the receptor compartment), the transport rate of hydroquinone is decreased (Fig. 27.13a). If the current polarity is reversed so that the solvent with the neutral compounds flows into the receptor compartment, the transport rate is markedly increased (Fig. 27.13c). Changes of transport patterns with differentiation of skin from rats could also be demonstrated.

Dentine is a calcerous material located between the enamel and the pulp of a tooth. It contains pores of 1–2 µm (Fig. 27.14b). Fluid flow inside these tubules is associated with dentinal hypersensitivity. Blocking of this flow is one strategy to treat this condition. Macpherson et al. [119, 120] investigated the convective transport within the tubes. A dentine slice was placed between two electrolyte solutions of identical composition. The UME was used to oxidize $[Fe(CN)_6]^{4-}$ (= Red in Fig. 27.14a). Without any external driving force, the features in the image represent

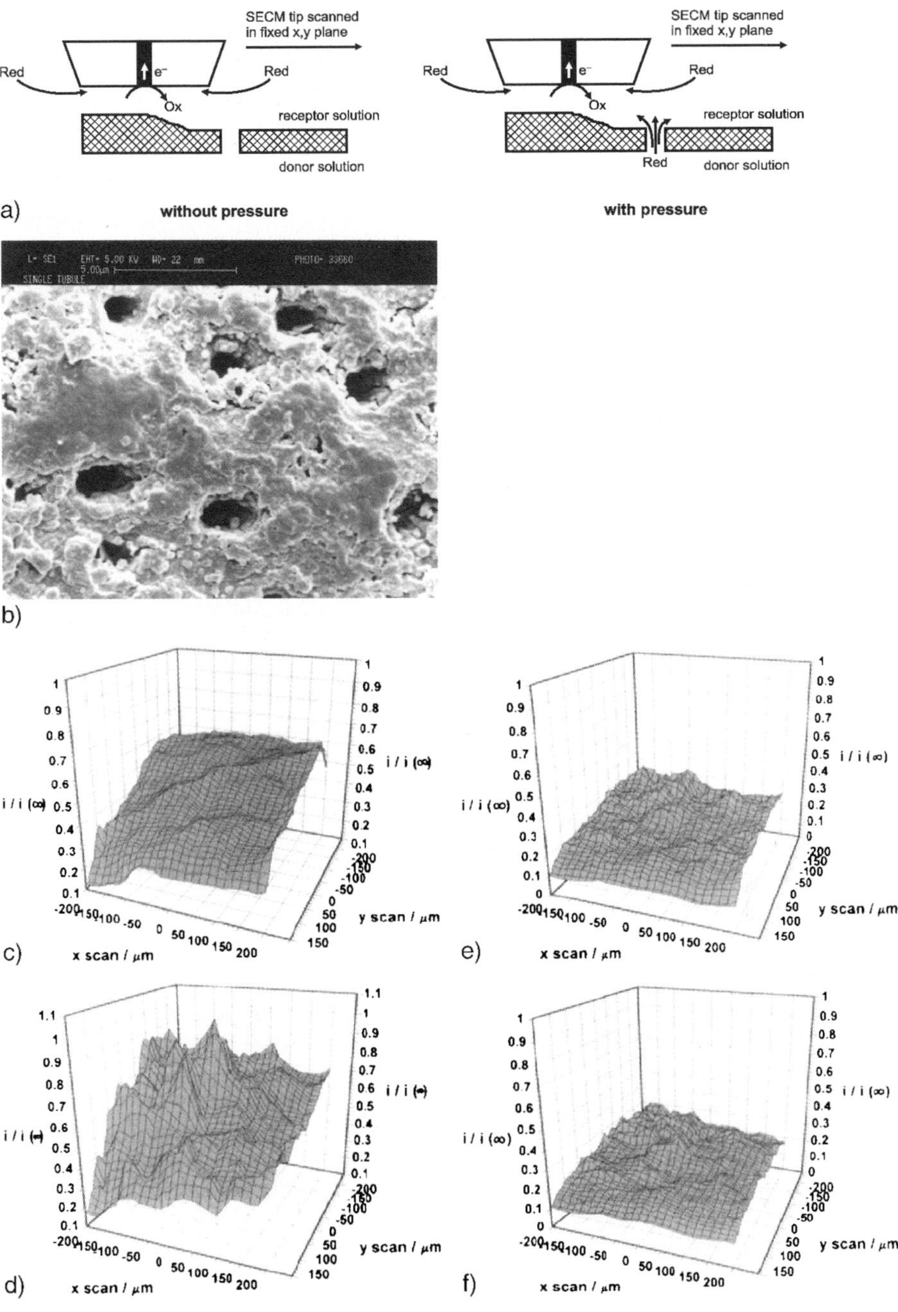
SECM tip scanned
in fixed x,y plane
Red
e⁻
Ox
Red
receptor solution
donor solution
a)
without pressure
SECM tip scanned
in fixed x,y plane
Red
e⁻
Ox
Red
Red
receptor solution
donor solution
with pressure
L = SE1   EHT = 5.00 KV   WD = 22 mm   PHOTO = 33660
5.00µm
SINGLE TUBULE
b)
c)
x scan / µm
y scan / µm
i / i (∞)
d)
x scan / µm
y scan / µm
i / i (∞)
e)
x scan / µm
y scan / µm
i / i (∞)
f)
x scan / µm
y scan / µm
i / i (∞)

essentially the topographic features of the dentine slice (Fig. 27.14c). By applying a pressure gradient across the dentine slice, a convective flow of the solution can be induced. This leads to increased currents above those dentine pores that are open and to an overall increased level of current (Fig. 27.14d). Treatment of the dentine with calcium oxalate can block the pore by deposition of insoluble calcium oxalate. After such a treatment, there is almost no difference between the images recorded with and without applied hydrostatic pressure (Fig. 27.14e,f). Pure diffusional transport inside dentine was studied by Nugues and Denuault [121]. They could identify occlusions of pores and correlate the current to numerical simulations of pore ensembles.

Experiments similar to the investigation of skin and dentine are also applicable to technical membrane systems.

Mass transport in cartilage was studied by Macpherson and coworkers [122]. Understanding of the redistribution rate of water is central to the load-bearing function of cartilage in the body. The physiological and pathophysiological behavior of cartilage is also related to the transport of oxygen and other nutrients and metabolites by diffusion and convection that can be studied by SECM.

A further increase in the resolution of such measurements and a significantly improved chemical selectivity was demonstrated by Kueng et al. [51,52,123], who integrated an amperometric enzyme sensor into an SFM tip and probed the release of glucose and ATP from artificial membranes under physiologically relevant conditions.

27.3
Application to Technologically Important Electrodes

27.3.1
Investigation of Passive Layers and Local Corrosion Phenomena

27.3.1.1
Detecting Precursor Sites for Pitting Corrosion

SECM can be seen as an ideal tool for corrosion research. It can distinguish between active and passive regions at the sample and detect the concentration of various relevant species such as Fe^{3+}, Fe^{2+}, O_2 and H_2. At most metals a thin oxide layer prevents the continuous dissolution of the metal in aqueous solutions. Local damage of this layer can lead to a rapid local dissolution of the metal while maintaining extreme concentrations of reagents and pH values (pitting corrosion). The initiation of the local breakdown of the passive layers is not yet completely understood. SECM was able to detect precursor sites for pitting corrosion on steel [124–128], Ti [129–133], Ta [134, 135] Ni [136], and Al [137, 138] by their enhanced heterogeneous electron-transfer kinetics to dissolved compounds such as Br^-, I^-, $[Ru(NH_3)_6]^{3+}$, nitrobenzene, etc. The formed reaction products can be detected in the GC mode. On Ti/TiO_2 such regions of enhanced rates of heterogeneous electron transfer turned out to be preferred sites for pitting corrosion under more positive potentials [129, 130]. A significant advantage of the SECM method is its destruction-free character, which allows the identification of precursor sites before the onset of otherwise visible pitting corrosion. By quantification of the fluxes using

(27.10)–(27.12), it was found that 69% of the current at a macroscopic electrode ($0.79\,cm^2$) were passing through 11 precursor regions that totaled only 0.1% of the total area [131]. The precursor site can be associated with inclusions of other elements [133]. Wiliams et al. [139] developed a method to follow the ongoing pit in the vicinity of sulfur-rich inclusions on steel by a combination of complemen-

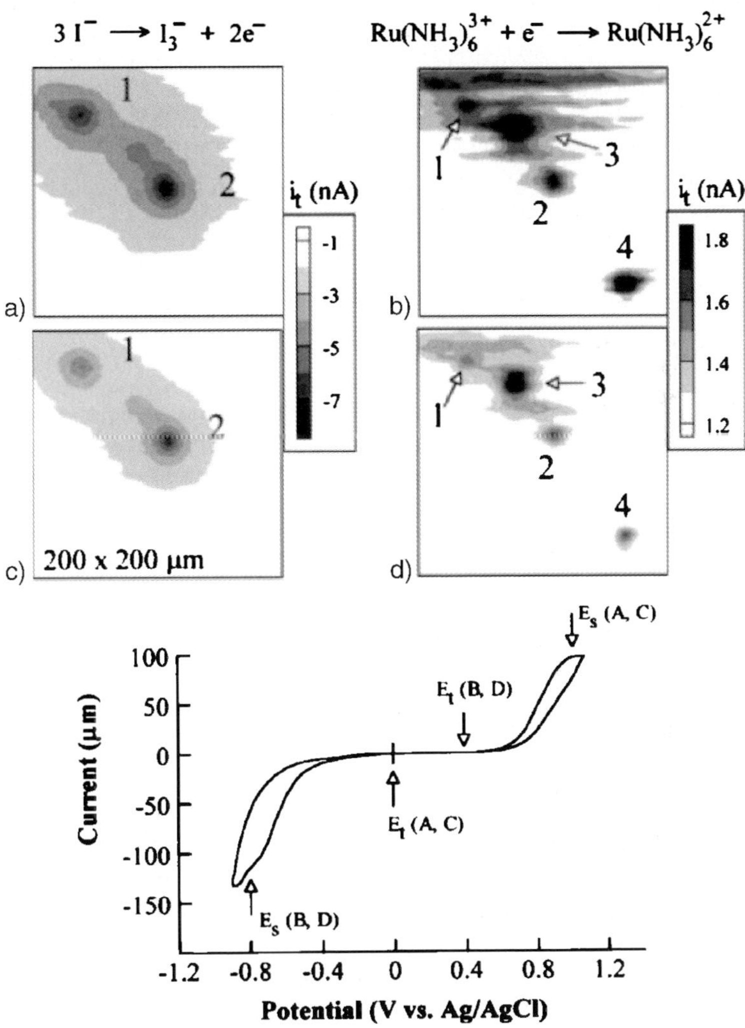

Fig. 27.15. Precursor sites for pitting corrosion on $200 \times 200\,\mu m^2$ Ta/Ta_2O_5 imaged in the GC mode in a solution containing *both*, 2.5 mM $[Ru(NH_3)_6]^{3+}$ and 10 mM I^-. Oxidation of I^- (**a**) proceeds at sites 1 and 2 only, $E_S = 1.0\,V$, $E_T = 0.0\,V$. Reduction of $[Ru(NH_3)_6]^{3+}$ (**b**) is found on sites 1–4 ($E_S = -0.8\,V$, $E_T = +0.4\,V$). By changing E_S and E_T it was switched repeatedly between images such as that shown in (**a**) or in (**b**). The cyclic voltammogram of the macroscopic electrode (**c**) shows the sum of the reaction at all sites and does not reveal the chemical selectivity of the sites; $v = 20\,mV\,s^{-1}$, ordinate label should be "Current (µA)" [G.W.]. Reprinted with permission from [135], Copyright 1999 American Chemical Society

tary microscopic techniques. Sulfur species were locally detected by reaction of microelectrode-generated I_3^- [125, 140]. SECM could demonstrate that SiC particles are preferred sites for oxygen reduction at Al-based composite material [141]. For inclusion-free materials, the structural nature of the precursor sites could not yet be determined. Passive layers of different metals show interesting differences. While Ti/TiO_2 has no chemical sensitivity with respect to dissolved compounds, precursor sites on Ta/Ta_2O_5 showed a remarkably different sensitivity towards I^- and $[Ru(NH_3)_6]^{3+}$ (Fig. 27.15) [135] When the sample was held at $E_S = 1.0\,V$ (oxidation of I^-) and the UME at $E_T = 0.0\,V$ (reduction of I_3^-), two sample regions that could oxidize I^- became visible (Fig. 27.15a,c). When the sample was held at $E_S = -0.8\,V$ in order to reduce $[Ru(NH_3)_6]^{3+}$ and the UME was held at $E_T = +0.4$ (oxidation of $[Ru(NH_3)_6]^{2+}$), four regions were detected at which electron transfer occurred at the sample (Fig. 27.15b,d). By changing E_S and E_T it was switched repeatedly between images.

27.3.1.2
Triggering and Monitoring of Active Local Corrosion

Pitting corrosion can be triggered locally by high chloride concentrations. SECM has been used to produce locally high chloride concentrations by reduction of trichloroacetic and dichloroactetic acids and to follow the subsequent development of the pit over time [142, 143].

SECM has also been used to monitor active corrosion processes either by identification of corrosion products by cyclic voltammetry at the UME, or by the use of potentiometric ion-selective electrodes. Besides steel [144], examples include dental fillings from amalgams and metallic implants [145, 146], titanium [147], alloys [148–154], potentiometric sensor materials of AgI [155], silicon in etching solutions [156], organic coatings [157, 158], enzyme–mediator–graphite composite materials for amperometric biosensors [159], and diamond-like carbon-coated optical waveguides from zinc selenide [160].

Different passivating properties of crystallites on Ti/TO_2 have been identified with the feedback mode of the SECM [161]. The crystallites could be clearly identified by their different heterogeneous kinetics for the $[Fe(CN)_6]^{3-}$ reduction. The pattern correlated with the crystallites detected in optical microscopy [147].

A new approach that currently emerges is the use of an AC signal at the microelectrode. Although the nature of such signals is currently still under discussion [162], it turned out to be useful for detecting small defects in technical materials and coatings [93, 151, 163].

27.3.2
Investigation of Electrocatalytically Important Electrodes

SECM has recently attracted a lot of attention as a tool to investigate electrocatalytically important electrodes connected to the development of fuel cells. The electrochemical hydrogen-oxidation reaction (HOR) and the oxygen-reduction reaction (ORR) are both electrocatalytical reactions. Technically, many problems are associated with catalytic materials. They are sensitive to poisoning by other compounds, and Pt-based catalysts are of limited availability. Therefore, alloys need to be tested that could replace or complement Pt-based catalysts.

27.3.2.1
Investigation of Hydrogen-Oxidation Reaction

The HOR reaction has been studied in the GC and feedback mode [164–168]. The H^+/H_2 redox couple was used as mediator. Unlike outer-sphere redox couples that have been discussed previously, the kinetics of the H^+/H_2 strongly depend on the catalytic properties of the electrode. In a typical experiment H^+ is reduced at the UME and the formed H_2 is then oxidized at the sample that is held at a specific potential. By evaluating approach curves according to (27.2)–(27.6) the potential-dependent kinetics of the HOR at the sample could be reconstructed [164]. The measurements are carried out in a steady-state regime. This is advantageous because the response is comparatively free of interferences common in alternative experiments and caused by ohmic drop, double-layer charging currents or mechanical difficulties associated with cyclic voltammetry at high scan rates or rotating-disk experiments, respectively.

27.3.2.2
Investigation of Oxygen-Reduction Reaction

The ORR is a complicated multistep reaction. In a still simplified version at least three reactions have to be considered in acidic solution

$$O_2 + 2e^- + 2H^+ \rightarrow H_2O_2 \tag{27.13}$$
$$H_2O_2 + 2e^- + 2H^+ \rightarrow 2H_2O \, , \tag{27.14}$$
$$O_2 + 4e^- + 4H^+ \rightarrow 2H_2O \, . \tag{27.15}$$

The rates of these reactions are also very dependent on the electrode material. The reaction is investigated in the tip-generation/substrate collection mode (TG/SC, Fig. 27.2b) [169–171]. Specifically, the UME is placed in proximity to the sample and molecular oxygen is produced by applying a constant current at the UME in acidic medium (Fig. 27.16). The oxygen diffuses to the sample and is reduced there. By plotting the sample current vs. the UME position, a mapping of the electrocatalytic efficiency is obtained. This has been used to scan chips at which electrocatalysts of systematically varied composition were prepared by combinatorial methods [169] (Fig. 27.17). The different spots represent different total compositions of the system Pd-Au-Co. The relative composition at each catalyst spot can be obtained by projecting three lines from an individual spot as indicated by the arrows in the lower-right part of each panel to the three axes for the composition (in atomic ratio). The bright spots are obtained for those material combinations that are effective oxygen-reduction catalysts. Potential-dependent properties can be obtained by performing the experiment with different substrate potentials. In contrast to the feedback mode, the TG/SC mode does not require that the product of the reaction at the sample

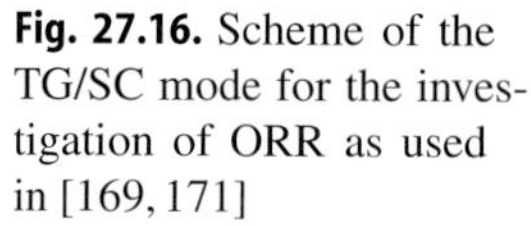

Fig. 27.16. Scheme of the TG/SC mode for the investigation of ORR as used in [169, 171]

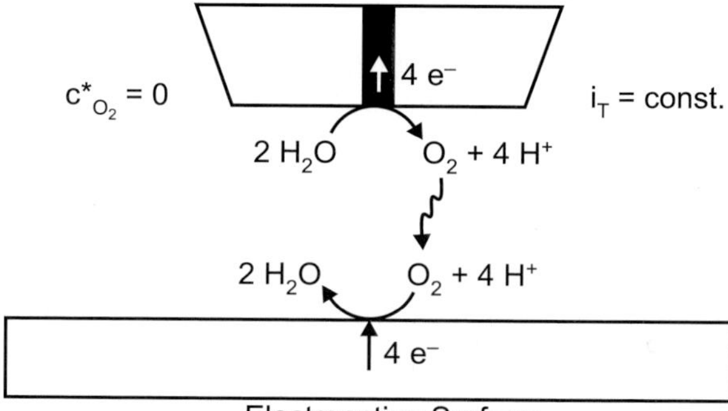

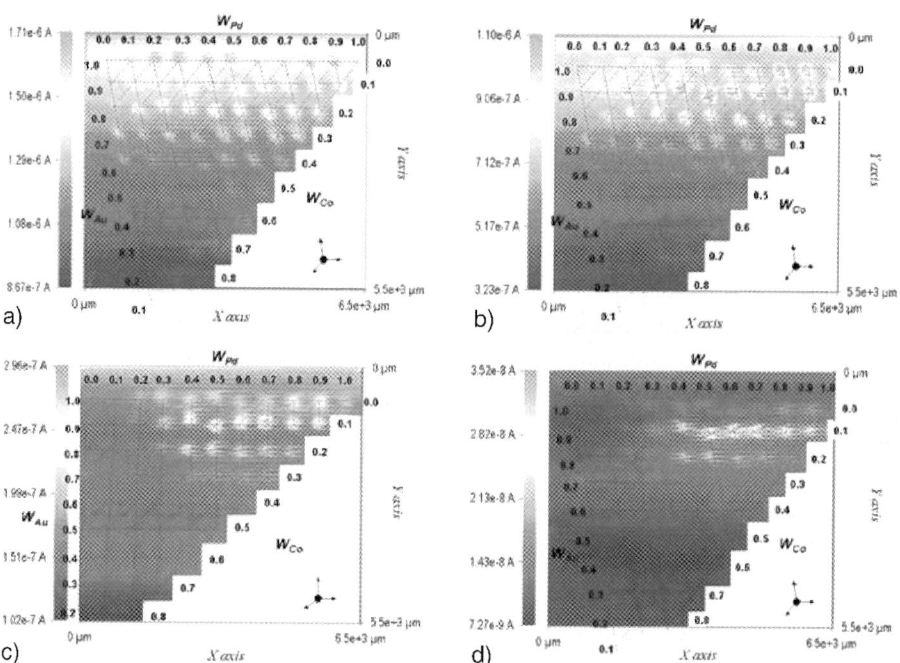

Fig. 27.17. SECM TG/SC images of oxygen reduction activity measured on Pd-Au-Co arrays in 0.5 M H$_2$SO$_4$. $d = 30\,\mu$m, constant UME current $= +160$ nA, $E_S = 0.2$ (**a**), 0.4 (**b**), 0.6 (**c**), 0.7 (**d**) V vs. HRE. W_M is the atomic ratio of the metal M in the spot. Reprinted with permission from [169], Copyright 2005 American Chemical Society

diffuses back to the UME to close the chemical feedback loop. This allows the investigation of irreversible reactions such as ORR. ORR in alkaline medium around pH 12 can also be investigated in the feedback mode [172]. Hydroxide ions serve as mediator. They are oxidized to O$_2$ at the UME and O$_2$ is reduced to OH$^-$ at a platinum sample. The applicability of this mode is not only limited by the narrow pH range but also by the fact that many metal samples will give H$_2$O$_2$ as one of the reduction products at the sample.

27.4
Conclusion and Outlook: New Instrumental Developments and Implication for Future Applications

SECM has been applied to study a large variety of heterogeneous reactions at solid/liquid interfaces. The examples of high practical relevance include studies of corrosion mechanisms of passivated metals, heterogeneous catalysis for fuel-cell materials, catalysis by immobilized enzymes used in biosensors and metabolic activity of single cells and intact organs. The SECM image provides a direct representation of interfacial reactivity even in those cases when the topography of the interface does not changes during the reaction, e.g. during an electron transfer from an electrode to a dissolved compound without accompanying deposition or dissolution processes. Furthermore, detailed kinetic analysis can be obtained by recording approach curves and by fitting the data to models of the interfacial reaction and mass-transfer processes. This ability has largely been exploited in fundamental electrochemistry, however, it was not the focus of this review. Finally, the possibility to modify surfaces by a large variety of chemically *well-defined* surface modifications holds a great potential for prototyping of advanced functional surfaces.

Currently, there are perhaps two main obstacles for a broader application of the technique: First, the lateral resolution and secondly, the level of fundamental knowledge in electrochemistry required to perform successful SECM experiments. Most studies have been performed in the micrometer range for a number of reasons: i) Preparation and characterization of electrodes with defined geometry is challenging for electrodes smaller than 1 μm. Such electrodes are required for quantitative work. ii) The working distance is proportional to r_T in the feedback mode. For submicrometer probes, a constant-distance working mode is required (vide infra). iii) The relevant effects are bound to the micrometer dimension (biological cells, biosensor arrays). Of course, there is a need for higher lateral resolution, in particular for a broader application in material science, where the investigation of reactions at individual grain boundaries would be an important objective.

Therefore, many attempts are currently being made to improve the lateral resolution and to ease the operation of SECM instruments and data interpretation. Highly desirable are mechanisms for guiding an UME over protruding samples. This requires a signal that is independent of the measured current (because the current is strongly influenced by the chemical properties of the sample). Most promising are methods based on shear forces (vide supra). Another development is to integrate UME or other chemical sensors into SFM tips or to combine other complementing scanning probe techniques. This direction is covered in the Chapt. 8 by Kranz of this volume. When discussing future possibilities to increase the lateral resolution, one should keep in mind some limitations for optimization. In the GC mode the resolution is controlled by the diffusion field of the active *sample* regions. A large active region will make it impossible to resolve small features in its vicinity even with a very small probe. In the feedback mode, the resolution is primarily controlled by the size of the active UME area. Besides this, the working distance and the size of the insulating shielding have a considerable influence. The smaller the working distance the better the achievable lateral resolution. Since the normalized distance is important, the working distance has to be reduced proportionally to the UME radius.

This makes it clear why current-independent distance-control schemes are the main prerequisite for enhancing resolution. The insulating shielding should be as thin as possible for high lateral resolution [173]. However, the smaller the insulating shielding the lower the sensitivity, which is also important for detecting small features on the sample. In many cases compromises have to be found for specific problems. The sensitivity is also significantly influenced by the kinetics at the sample and that adds another fundamental limitation to lateral resolution. Consider Fig. 27.3, curve 3. This curve represents a case where the reactivity of the sample surface can just be distinguished from an inert, insulating matrix. It corresponds to a normalized rate constant $\kappa = k_{\mathrm{eff}} r_{\mathrm{T}}/D = 0.3$. Since diffusion coefficients in aqueous solution do not vary too much, one can make a simple estimate, what is the minimum heterogeneous rate constant that must be present at the sample so that a contrast in an image is obtained using a UME of a specific r_{T} (Table 27.3). At an $r_{\mathrm{T}} = 5$ nm, the reaction rate constant at the sample must be as high as $3\,\mathrm{cm\,s}^{-1}$ that it corresponds to an almost diffusion-controlled reaction at conventional electrodes! SECM will become "blind" to all slower reactions at this UME size. While considerable progress has been reported to fabricate such UME, e.g. in Refs. [174, 175], the time for successful imaging experiments with them is still to come. However, imaging with electrodes of $r_{\mathrm{T}} \approx 50$ nm seems realistic for selected samples [174, 175]. Considering the interdependence of image features and heterogeneous chemical kinetics, it becomes clear that the strength of SECM will be imaging of chemical reactions. Therefore, the size of the electrode needs to be selected according to the investigated problem and despite the exciting progress, which can be achieved with nanometer-sized electrodes, there will still be many applications using micrometer-sized electrodes. Issues of resolution cannot be discussed independently of the sample and the reaction to be studied.

Table 27.3. Required heterogeneous rate constant k_{eff} to result in an approach curve such as curve 3 ($\kappa = 0.3$) in Fig. 27.3. $D = 5 \times 10^{-6}\,\mathrm{cm\,s}^{-1}$

$r_{\mathrm{T}}/\mu\mathrm{m}$	12.5	5	1	0.5	0.1	0.05	0.025	0.005
$k_{\mathrm{eff}}/\mathrm{cm\,s}^{-1}$	0.0012	0.003	0.015	0.03	0.15	0.3	0.6	3

Acknowledgements. The work on SECM in the authors' laboratory has been supported over the years by Deutsche Forschungsgemeinschaft, the State of Lower Saxony, the VW Foundation, the Alexander von Humboldt Foundation, the Fonds of the Chemical Industries and the Hanse Institute of Advanced Studies. G.W. would like to express sincere thanks to the University of Oldenburg for granting a sabbatical semester and in particular to Carl H. Hamann for taking up the teaching load during that time! We are grateful for the critical comments of our colleagues C. Kranz (Georgia Institute of Technology), I. Zawisza and A. Kittel (University of Oldenburg) that helped to improve the manuscript. The authors are thankful to A.J. Bard, T. Matsue, H. Shiku, H.S. White, J. Macpherson, and P. Unwin for supplying electronic versions of the original figures for this review.

References

1. Bard AJ, Mirkin MV (eds) (2001) Scanning Electrochemical Microscopy. Marcel Dekker, Inc., New York, Basel

2. Barker AL, Gonsalves M, Macpherson JV, Slevin CJ, Unwin PR (1999) Anal Chim Acta 385:223
3. Wittstock G (2003) Imaging Localized Reactivities of Surfaces by Scanning Electrochemical Microscopy. In: Wandelt K, Thurgate S (eds) Solid-Liquid Interfaces, Macroscopic Phenomena - Microscopic Understanding. Springer-Verlag, Berlin, Heidelberg, p 335
4. Wittstock G (2001) Fresenius J Anal Chem 370:303
5. Horrocks BR (2003) Scanning Electrochemical Microscopy. In: Unwin PR (ed) Instrumentation and Electroanalytical Chemistry. Wiley-VCH, Weinheim, p 444
6. Fleischmann M, Stanley P, Rolison DR, Schmidt PP (1987) Ultramicroelectrodes. Datatech Systems, Inc., Morganton, NC
7. Adams RN (1990) Prog Neurobiol 35:297
8. Engstrom RC, Weber M, Wunder DJ, Burges R, Winquist S (1986) Anal Chem 58:844
9. Liu H-Y, Fan F-RF, Lin CW, Bard AJ (1986) J Am Chem Soc 108:3838
10. Kwak J, Bard AJ (1989) Anal Chem 61:1794
11. Bard AJ, Fan F-RF, Kwak J, Lev O (1989) Anal Chem 61:132
12. Oesterschulze E, Abelmann L, van den Bos A, Kassing R, Schwendler N, Wittstock G, Ziegler C (2006) Sensor Technology for Scanning Probe Microscopy and New Applications. In: Bhushan B, Fuchs H, Kawata S (eds) Applied Scanning Probe Methods II. Springer, Berlin, Heidelberg, New York,
13. Saito Y (1968) Rev Polarogr (Jpn) 15:177
14. Amphlett JL, Denuault G (1998) J Phys Chem B 102:9946
15. Bard AJ, Mirkin MV, Unwin PR, Wipf DO (1992) J Phys Chem 96:1861
16. Bard AJ, Fan F-RF, Pierce DT, Unwin PR, Wipf DO, Zhou F (1991) Science 254:68
17. Wang J, Wu L-H, Li R (1989) J Electroanal Chem 272:285
18. Horrocks BR, Wittstock G (2001) Biological Systems. In: Bard AJ, Mirkin MV (eds) Scanning Electrochemical Microscopy. Marcel Dekker, New York, Basel, p 445
19. Shiku H, Ohya H, Matsue T (2002) Encyclop Electrochem 9:257
20. Zhao C, Wittstock G (2004) Anal Chem 76:3145
21. Anthony C (1996) Biochem J 320:697
22. Ye L, Haemmerle M, Olsthoorn AJJ, Schuhmann W, Schmidt HL, Duine JA, Heller A (1993) Anal Chem 65:238
23. Nistor C, Rose A, Wollenberger U, Pfeiffer D, Emneus J (2002) Analyst 127:1076
24. Wijayawardhana CA, Wittstock G, Halsall HB, Heineman WR (2000) Anal Chem 72:333
25. Pierce DT, Unwin PR, Bard AJ (1992) Anal Chem 64:1795
26. Kranz C, Wittstock G, Wohlschläger H, Schuhmann W (1997) Electrochim Acta 42:3105
27. Pierce DT, Bard AJ (1993) Anal Chem 65:3598
28. Shiku H, Takeda T, Yamada H, Matsue T, Uchida I (1995) Anal Chem 67:312
29. Shiku H, Matsue T, Uchida I (1996) Anal Chem 68:1276
30. Wittstock G, Wilhelm T, Bahrs S, Steinrücke P (2001) Electroanalysis 13:669
31. Zaumseil J, Wittstock G, Bahrs S, Steinrücke P (2000) Fresenius J Anal Chem 367:352
32. Kwak J, Bard AJ (1989) Anal Chem 61:1221
33. Wei C, Bard AJ, Mirkin MV (1995) J Phys Chem 99:16033
34. Mirkin MV, Arca M, Bard AJ (1993) J Phys Chem 97:10790
35. Mirkin MV, Fan F-RF, Bard AJ (1992) J Electroanal Chem 328:47
36. Zhao C, Sinha JK, Wijayawardhana CA, Wittstock G (2004) J Electroanal Chem 561:83
37. Wilhelm T, Wittstock G (2000) Mikrochim Acta 133:1
38. Wittstock G, Schuhmann W (1997) Anal Chem 69:5059
39. Horrocks BR, Schmidtke D, Heller A, Bard AJ (1993) Anal Chem 65:3605
40. Horrocks BR, Mirkin MV, Pierce DT, Bard AJ, Nagy G, Toth K (1993) Anal Chem 65:1213
41. Horrocks BR, Mirkin MV (1998) J Chem Soc 94:1115
42. Shiku H, Hara Y, Matsue T, Uchida I, Yamauchi T (1997) J Electroanal Chem 438:187

43. Wilhelm T, Wittstock G (2001) Electrochim Acta 47:275
44. Zhao C, Wittstock G (2005) Biosens Bioelectron 20:1277
45. Wijayawardhana CA, Wittstock G, Halsall HB, Heineman WR (2000) Electroanalysis 12:640
46. Wittstock G, Jenkins SH, Halsall HB, Heineman WR (1998) Nanobiology 4:153
47. Wittstock G, Yu K-j, Halsall HB, Ridgway TH, Heineman WR (1995) Anal Chem 67:3578
48. Zhao C, Wittstock G (2004) Angew Chem Int Ed Engl 43:4170
49. Antonenko YN, Pohl P, Rosenfeld E (1996) Arch Biochem Biophys 333:225
50. Berger CEM, Horrocks BR, Datta HK (1998) J Endocrinology 158:311
51. Kueng A, Kranz C, Mizaikoff B (2005) Biosens Bioelectron 21:346
52. Kueng A, Kranz C, Lugstein A, Bertagnolli E, Mizaikoff B (2005) Angew Chem, Int Ed 44:3419
53. Scott ER, White HS, Phipps JB (1993) Anal Chem 65:1537
54. Scott ER, White HS, Phipps JB (1991) J Membr Sci 58:71
55. Nowall WB, Dontha N, Kuhr WG (1998) Biosens Bioelectron 13:1237
56. Rosenwald SE, Dontha N, Kuhr WG (1998) Anal Chem 70:1133
57. Tiefenauer LX, Padeste C (1999) Chimia 53:62
58. Mandler D, Meltzer S, Shohat I (1996) Israel J Chem 36:73
59. Mandler D (2001) Micro- and Nanopatterning using the Scanning Electrochemical Microscope. In: Bard AJ, Mirkin MV (eds) Scanning Electrochemical Microscopy. Marcel Dekker, New York, Basel, p 593
60. Tenent RC, Wipf DO (2003) J Electrochem Soc 150:E131
61. Nowall WB, Wipf DO, Kuhr WG (1998) Anal Chem 70:2601
62. Kranz C, Gaub HE, Schuhmann W (1996) Adv Mater 8:634
63. Wilhelm T, Wittstock G (2003) Angew Chem Int Ed 42:2247
64. Sauter S, Wittstock G (2001) J Solid State Electrochem 5:205
65. Torisawa YSY-S, Shiku H, Yasukawa T, Nishizawa M, Matsue T (2005) Biomaterials 26:2165
66. Shiku H, Uchida I, Matsue T (1997) Langmuir 13:7239
67. Oyamatsu D, Kanaya N, Shiku H, Nishizawa M, Matsue T (2003) Sensors And Actuators B B91:199
68. Kaya T, Nagamine K, Matsui N, Yasukawa T, Shiku H, Matsue T (2004) Chem Commun, p 248
69. Ogasawara D, Hirano Y, Yasukawa T, Shiku H, Matsue T, Kobori K, Ushizawa K, Kawabata S (2004) Chem Sens 20:139
70. Motochi N, Hirano Y, Abiko Y, Oyamatsu D, Nishizawa M, Matsue T, Ushizawa K, Kawabata S (2002) Chem Sens 18:172
71. Kasai S, Yokota A, Zhou H, Nishizawa M, Onouchi T, Niwa K, Matsue T (2000) Anal Chem 72:5761
72. Wijayawardhana CA, Ronkainen-Matsuno NJ, Farrel SM, Wittstock G, Halsall HB, Heineman WR (2001) Anal Sci 17:535
73. Wang J, Song F, Zhou F (2002) Langmuir 18:6653
74. Carano M, Lion N, Abid J-P, Girault HH (2004) Electrochem Commun 6:1217
75. Wang J, Zhou F (2002) J Electroanal Chem 537:95
76. Turcu F, Schulte A, Hartwich G, Schuhmann W (2004) Angew Chem Int Ed Engl 43:3482
77. Turcu F, Schulte A, Hartwich G, Schuhmann W (2004) Biosens Bioelectron 20:925
78. Fortin E, Mailley P, Lacroix L, Szunerits S (2006) Analyst
79. Kranz C, Ludwig M, Gaub HE, Schuhmann W (1995) Adv Mater 7:38
80. Szunerits S, Knorr N, Calemczuk R, Livache T (2004) Langmuir 20:9236
81. Tsionsky M, Cardon ZG, Bard AJ, Jackson RB (1997) Plant Physiol 113:895
82. Isik S, Etienne M, Oni J, Bloechl A, Reiter S, Schuhmann W (2004) Anal Chem 76:6389

83. Torisawa Y-S, Shiku H, Kasai S, Nishizawa M, Matsue T (2004) Int J Cancer 109:302
84. Torisawa Y-S, Shiku H, Kasai S, Nishizawa M, Matsue T (2003) Chem Sens 19:28
85. Torisawa Y-S, Kaya T, Takii Y, Oyamatsu D, Nishizawa M, Matsue T (2003) Anal Chem 75:2154
86. Yasukawa T, Kaya T, Matsue T (1999) Anal Chem 71:4631
87. James P, Garfias-Mesias LF, Moyer PJ, Smyrl WH (1998) J Electrochem Soc 145:64
88. Hirano Y, Oyamatsu D, Yasukawa T, Shiku H, Matsue T (2004) Electrochemistry (Tokyo) 72:137
89. Lee Y, Bard AJ (2002) Anal Chem 74:3626
90. Yamada H, Fukumoto H, Yokoyama T, Koike T (2005) Anal Chem 77:1785
91. Ludwig M, Kranz C, Schuhmann W, Gaub HE (1995) Rev Sci Instrum 66:2857
92. Brunner R, Bietsch A, Hollenricher O, Marti O (1997) Rev Sci Instrum 68:1769
93. Etienne M, Schulte A, Schuhmann W (2004) Electrochem Commun 6:288
94. Etienne M, Schulte A, Mann S, Jordan G, Dietzel ID, Schuhmann W (2004) Anal Chem 76:3682
95. Bauermann LP, Schuhmann W, Schulte A (2004) Phys Chem Chem Phys 6:4003
96. Ballesteros Katemann B, Schulte A, Schuhmann W (2004) Electroanalysis 16:60
97. Ballesteros Katemann B, Schulte A, Schuhmann W (2003) Chem Eur J 9:2025
98. Hengstenberg A, Blöchl A, Dietzel ID, Schuhmann W (2001) Angew Chem Int Ed Engl 40:905
99. Yasukawa T, Kaya T, Matsue T (1999) Chem Lett 9:975
100. Nishizawa M, Takoh K, Matsue T (2002) Langmuir 18:3645
101. Yasukawa T, Kondo Y, Uchida I, Matsue T (1998) Chem Lett 767
102. Shiku H, Shiraishi T, Aoyagi S, Utsumi Y, Matsudaira M, Abe H, Hoshi H, Kasai S, Ohya H, Matsue T (2004) Anal Chim Acta 522:51
103. Shiku H, Shiraishi T, Ohya H, Matsue T, Abe H, Hoshi H, Kobayashi M (2001) Anal Chem 73:3751
104. Hengstenberg A, Blöchl A, Dietzel ID, Schuhmann W (2001) Angew Chem Int Ed 40:905
105. Pitta Bauermann L, Schuhmann W, Schulte A (2004) Phys Chem Chem Phys 6:4003
106. Berger CEM, Rathod H, Gillespie JI, Horrocks BR, Datta HK (2001) J Bone Miner Res 16:2092
107. Berger CEM, Horrocks BR, Datta HK (1999) Molec Cellular Endocrin 149:53
108. Berger CEM, Horrocks BR, Datta HK (1999) Electrochim Acta 44:2677
109. Liu B, Rotenberg SA, Mirkin MV (2000) Proc Natl Acad Sci USA 97:9855
110. Liu B, Rotenberg SA, Mirkin MV (2002) Anal Chem 74:6340
111. Feng W, Rotenberg SA, Mirkin MV (2003) Anal Chem 75:4148
112. Liu B, Cheng W, Rotenberg SA, Mirkin MV (2001) J Electroanal Chem 500:590
113. Nagamine K, Matsui N, Kaya T, Yasukawa T, Shiku H, Nakayama T, Nishino T, Matsue T (2005) Biosens Bioelectron 21:145
114. Nagamine K, Kaya T, Yasukawa T, Shiku H, Matsue T (2005) Sens Actuators B B108:676
115. Bath BD, White HS, Scott ER (2001) Imaging Molecular Transport Across Membranes. In: Bard AJ, Mirkin MV (eds) Scanning Electrochemical Microscopy. Marcel Dekker, New York, Basel, p 343
116. Bath BD, Scott ER, Phipps JB, White HS (2000) J Pharm Sci 89:1537
117. Bath BD, White HS, Scott ER (2000) Pharmaceut Res 17:471
118. Burnette RR, Marrero D (1986) J Pharm Sci 85:1186
119. Macpherson JV, Beeston MA, Unwin PR, Hughes NP, Littlewood D (1995) J Chem Soc Faraday Trans I 91:1407
120. Macpherson JV, Beeston MA, Unwin PR, Hughes NP, Littlewood D (1995) Langmuir 11:3959
121. Nugues S, Denuault G (1996) J Electroanal Chem 408:125

122. Gonsalves M, Barker AL, Macpherson JV, Unwin PR, O'hare D, Winlove PC (2000) Biophysical Journal 78:1578
123. Kueng A, Kranz C, Mizaikoff B (2004) Biosens Bioelectron 19:1301
124. Zhu Y, Williams DE (1997) J Electrochem Soc 144:43
125. Lister TE, Pinhero PJ (2003) Electrochim Acta 48:2371
126. Lister TE, Pinhero PJ (2005) Anal Chem 77:2601
127. Tanabe H, Yamamura Y, Misawa T (1995) Mater Sci Forum 185-188:991
128. Luong BT, Nishikata A, Tsuru T (2003) Electrochemistry 71:555
129. Casillas N, Charlebois SJ, Smyrl WH, White HS (1994) J Electrochem Soc 141:636
130. Casillas N, Charlebois SJ, Smyrl WH, White HS (1993) J. Electrochem. Soc. 140:142
131. Basame SB, White HS (1998) J Phys Chem B 102:9812
132. Basame SB, White HS (1995) J Phys Chem 99:16430
133. Garfias-Mesias LF, Alodan M, James P, Smyrl WH (1998) J Electrochem Soc 145:2005
134. Basame SB, White HS (1999) Anal Chem 71:3166
135. Basame SB, White HS (1999) Langmuir 15:819
136. Paik C–H, Alkire RC (2001) J Electrochem Soc 148:B276
137. Serebrennikova I, Lee S, White HS (2002) Faraday Discuss 121:199
138. Serebrennikova I, White HS (2001) Electrochem Solid State Lett 4:4
139. Williams DE, Mohiuddin TF, Zhu YY (1998) J Electrochem Soc 145:2664
140. Paik C–H, White HS, Alkire RC (2000) J Electrochem Soc 147:4120
141. Pech-Canul MA, Pech-Canul MI, Wipf DO (2004) J Electrochem Soc 151:B299
142. Still JW, Wipf DO (1997) J Electrochem Soc 144:2657
143. Wipf DO (1994) Colloid Surf A 93:251
144. Tanabe H, Togashi K, Misawa T, Mudali UK (1998) J Mater Sci Lett 17:551
145. Gilbert JL, Zarka L, Chang E, Thomas CH (1998) J Biomed Mater Res 42:321
146. Gilbert JL, Smith SM, Lautenschlager EP (1993) J Biomed Mater Res 27:1357
147. Fushimi K, Okawa T, Azumi K, Seo M (2000) J Electrochem Soc 147:524
148. Park JO, Paik C–H, Alkire RC (1996) J Electrochem Soc 143:174
149. Seegmiller JC, Buttry AD (2003) J Electrochem Soc 150:B413
150. Schulte A, Belger S, Etienne M, Schuhmann W (2004) Mater Sci Eng A A378:523
151. Belger S, Schulte A, Hessing C, Pohl M, Schuhmann W (2004) Materialwiss Werkst 35:276
152. Schulte A, Belger S, Schuhmann W (2002) Mater Sci Forum 394-395:145
153. Davoodi A, Pan J, Leygraf C, Norgren S (2005) Electrochem Solid State Lett 8:B21
154. Lister TE, Pinhero PJ, Towbridge TL, Mizia RE (2005) J Electrochem Soc 579:291
155. Toth K, Nagy G, Horrocks BR, Bard AJ (1993) Anal Chim Acta 282:239
156. Shreve GA, Karp CD, Pomykal KE, Lewis NS (1995) J Phys Chem 99:5575
157. Ballesteros Katemann B, Inchauspe CG, Castro PA, Schulte A, Calvo EJ, Schuhmann W (2003) Electrochim Acta 48:1115
158. Bastos AC, Simoes AM, Gonzalez S, Gonzalez-Garcia Y, Souto RM (2005) Prog Org Coat 53:177
159. Gründig B, Wittstock G, Rüdel U, Strehlitz B (1995) J Electroanal Chem 395:143
160. Janotta M, Rudolph D, Kueng A, Kranz C, Voraberger H-S, Waldhauser W, Mizaikoff B (2004) Langmuir 20:8634
161. Fushimi K, Okawa T, Seo M (2000) Electrochemistry 68:950
162. Baranski AS, Diakowski PM (2004) J Solid State Electrochem 8:683
163. Ballesteros Katemann B, Schulte A, Calvo EJ, Koudelka-Hep M, Schuhmann W (2002) Electrochem Commun 4:134
164. Jambunathan K, Shah BC, Hudson JL, Hillier AC (2001) J Electroanal Chem 500:279
165. Shah BC, Hillier AC (2000) J Electrochem Soc 147:3043
166. Zhou J, Zu Y, Bard AJ (2000) J Electroanal Chem 491:22

167. Ahmed S, Ji S, Petrik L, Linkov VM (2004) Anal Sci 20:1283
168. Zoski CG (2003) J Phys Chem B 107:6401
169. Fernandez JL, Walsh DA, Bard AJ (2005) J Am Chem Soc 127:357
170. Fernandez JL, Mano N, Heller A, Bard AJ (2004) Angew Chem Int Ed Engl 43:6355
171. Fernandez JL, Bard AJ (2003) Anal Chem 75:2967
172. Liu B, Bard AJ (2002) J Phys Chem B 106:12801
173. Sklyar O, Wittstock G (2002) J Phys Chem B 106:7499
174. Sklyar O, Treutler TH, Vlachopoulos N, Wittstock G (2005) Surf Sci 597:181
175. Treutler TH, Wittstock G (2003) Electrochim Acta 48:2923

28 Nanomechanical Characterization of Structural and Pressure-Sensitive Adhesives

Martin Munz · Heinz Sturm

List of Abbreviations and Symbols

DCB	Double cantilever beam test
DMT	Derjaguin–Muller–Toporov model
DGEBA	Diglycidyl ether of bisphenol A
DDS	Diaminodiphenylsulfone
DMTA	Dynamic-mechanical thermal analysis
EBL	Electron beam lithography
EDX	Energy-dispersive analysis of X-rays
FMM	Force modulation microscopy
γ-APS	γ-aminopropyl trimethoxysilane
IAZ	Indent-affected zone
IP	Interphase
JKR	Johnson–Kendall–Roberts model
LFM	Lateral force microscopy
MBE	Mechanical bias effect
M–D	Maugis–Dugdale model
MEMS	Microelectromechanical system
M-LFM	Modulated LFM
nBEAA	n-butylester of abietic acid
PC	Polycarbonate
PEP	Poly(ethylenepropylene)
PEP/60	PEP with 60 wt% nBEAA
PEP/80	PEP with 80 wt% nBEAA
PMC	Polymer–matrix composite
PPS	Poly(phenylenesulfide)
PS	Polystyrene
PVP	Poly(vinylpyrrolidone)
PSD	Power spectral density
PSA	Pressure-sensitive adhesive
RH	Relative humidity
RT	Room temperature
SEM	Scanning electron microscopy
SFM	Scanning force microscopy
SKPFM	Scanning Kelvin probe force microscopy
SMFM	Shear modulation force microscopy
SVM	Scanning viscoelasticity microscopy

TEM	Transmission electron microscopy
A	Oscillator strength
A_0	Amplitude of free oscillation (intermittent contact mode)
A_{sp}	Amplitude of controlled oscillation (intermittent contact mode)
a	Contact radius
$\tilde{a}$	Effective contact radius (shear deformation)
α	Tip opening half-angle
b	Cantilever bending
c	Radius of the annular plastic zone (cavity model)
d	Diameter of the residual imprint
d_{sp}	Average tip–surface distance (intermittent contact mode)
δ	Vertical deformation of the tip–sample contact
Δz	Vertical displacement
E	Young's modulus
E'	Storage component of the Young's modulus
E_s^b	Sample Young's modulus value in the bulk
f_a	Chemical functionality of the amine molecule
f_e	Chemical functionality of the epoxy molecule
F	Applied normal load
F_f	Friction force (sliding friction)
$F_{pull\text{-}off}$	Pull-off force
G	Shear modulus
G^*	Reduced shear modulus
G_{Ic}	Strain energy release rate
γ	Damping constant
h	Step height
H^b	Bulk hardness value
K	Reduced modulus of the tip–sample contact
κ	Effective spring constant
k_B	Boltzmann constant
k_c	Cantilever stiffness (spring constant)
k_{ts}	Tip–sample contact stiffness
k_{ts}^{lat}	Tip–sample contact stiffness for shear deformation
m	Effective mass of the simple harmonic oscillator
M_c	Average molecular weight between crosslinks
ν	Poisson's ratio
N_a	Amine molar number
N_e	Epoxy molar number
ω	Angular frequency
ω_0	Fundamental resonance frequency
r	Amine–epoxy mixing ratio
r_{sp}	Setpoint ratio (intermittent contact mode)
$\tilde{r}_{sp}$	Alternative definition of r_{sp} for very soft samples
R	Radius of curvature of the tip apex
$R(t)$	Random force (Langevin equation)
T	Absolute temperature [K]
τ	Adhesion strength

T_g	Glass transition temperature
T_{gu}	T_g of the uncrosslinked system
u	Displacement of the simple harmonic oscillator
w	Thermodynamic adhesion energy per unit area
w_{DDS}	Width of the DDS concentration decay
w_{DZ}	Width of the dead zone
w_{IP}	Width of the interphase
w_{MBE}	Width of the MBE zone
w_{tip}	Penetrating width of the tip
x_i	Position of the interfacial borderline
ξ	Drag coefficient
z_{ind}	Indentation depth

28.1
Introduction

Adhesives find applications in a large variety of products. In a strict sense, adhesives are used for the formation of a joint between two adherends. In a wider sense, also the polymers used in polymer–matrix composites (PMCs) are adhesives. In this case, the adherends are the stiffening filler particles or fibers, and the adhesive is used as a matrix material. Furthermore, in the case of coatings, a polymer system similar to an adhesive is applied in order to achieve a surface finishing for decorative, protective, or other functional purposes. For instance, smart surfaces can be generated by using coatings with the ability to sense changes of environmental parameters, such as humidity or temperature. In the case of coatings, the adhesive forms a free surface and an interface with the substrate, i.e. with the adherend.

Essentially, two different sorts of adhesives can be distinguished. Structural adhesives are used for building up stiff and strong structures. Typically, for such adhesives the process of adhering involves chemical reactions, resulting in strong covalent bonds. For instance, epoxies are used as adhesives, high-performance coatings, potting and encapsulation compounds, and as a matrix material for composites. On the contrary, pressure-sensitive adhesives (PSAs) act by virtue of physical interactions only and are mainly used for covering surfaces with a removable layer, the tape, for purposes such as protection or decoration.

The large variety of different applications requires a well-aimed optimization of the adhesive properties. Aside from the final mechanical properties, other essential requirements are wettability of the adherend, compatibility with the processing requirements, or resistance against chemical aging. Beyond more conventional characterization techniques, molecular-scale analysis of adhesives is a promising approach for further optimization. Furthermore, with the advent of microelectromechanical systems (MEMS), the thickness of adhesive layers needs to be scaled down to the submicrometer range. These developments require the availability of suitable measurement techniques, providing information on morphology, chemistry as well as local mechanical properties.

Issues to be addressed are the deformation behavior upon tensile stress, heterogeneities due to adhesive composition and processing, or heterogeneities due to interactions with the surface of the adherend. In particular, the latter can result in the formation of a third phase, the interphase, with properties different from those of the adherend and the bulk adhesive. The term interphase (in short, IP) was introduced by Sharpe [1] to describe the transition region between the adhesive and the adherend in adhesive joints and between fiber and matrix in composites.

Independent of the particular function of the adhesive layer, a basic requirement for its long-term reliability is mechanical integrity. This includes the integrity of the outer surface and the interface with the substrate. For instance, microcracks or other flaws present at the interface may be the starting point for a damage process leading to catastrophic failure, such as delamination. Thus, assessment of the mechanical performance of a polymer coating requires knowledge of its delamination behavior, and dedicated testing methods have to be provided. For instance, from the height profile across the residual imprint resulting from a nanoindentation experiment, the area of delamination and contact dimensions can be measured accurately [2]. Moreover, in combination with suitable mechanical models, by calculation of the strain energy release rate such studies can provide information on the toughness of the interface between polymer coating and substrate. For the indentation experiments on a polystyrene (PS) film covering a glass substrate, the topography images revealed crack arrest marks. These were associated with polymer crazing at the crack tip. Both crazing and plastic deformation were shown to add to the total dissipated energy [2].

Also, progress in the identification of epoxy microstructures is due to the availability of topography imaging based on scanning force microscopy (SFM). The question of the existence of such microstructures is matter of a long debate. In the 1970s it was reported that cured epoxy can show nodular structures [3–5]. These observations were made by means of scanning electron microscopy (SEM). However, it was argued that the structures may be an artefact induced by interaction with the electron beam [6]. Also, effects from the sample-preparation techniques necessary for the SEM investigations, such as coating with electrically conductive layers or etching, can be a possible source of artefacts. Furthermore, no established theory exists of the conditions under which these structures occur. Although to the best of the authors' knowledge the latter issue is still under discussion, investigations utilizing SFM-based topography imaging clearly demonstrated the existence of nodules [7–10]. Nodules of various average sizes were found, ranging from several tens of nanometers [7, 10, 11], over several hundreds of nanometers [7, 9], up to several micrometers [9]. Furthermore, VanLandingham et al. [7] studied epoxy samples of various amine–epoxy concentration ratios and found a decrease in the average nodule size with increasing amine concentration. According to a theory by Lüttgert and Bonart [5] taking into account the liberation of exothermic reaction energy, the formation of nodules can be described in terms of a nucleation reaction mechanism. In the vicinity of a spontaneous crosslinking reaction the temperature rises and, in turn, so does the reaction rate. Similar to the growth of spherulites in the melt of a semicrystalline polymer, the growth of crosslinked clusters, later to be identified as the nodules, is stopped when they encounter the front of other spreading clusters.

Thus, topography data provided by SFM can deliver valuable information also for the characterization of adhesives and coatings. Beyond the mere imaging of topography and morphology, SFM is increasingly used for purposes of localized mechanical analysis. A number of operation modes for local mechanical analysis are available, enabling the measurement of various quantities, such as adhesion, stiffness, shear stiffness, or friction. Thus, along with the nanometer-scale spatial resolution, SFM allows for crucial issues to be addressed, such as the characterization of IPs.

It should be emphasized that this chapter does not provide a comprehensive review of the field. Rather, basic approaches are presented with a few selected references. In this way, the potential of several SFM-based techniques is demonstrated. Essentially, the chapter is focused on examples of nanomechanical analysis of adhesives.

Applications not covered by this chapter are SFM-based studies of topography and morphology of adhesives, or the interesting field of delamination effects induced by electrochemical reactions. Here, scanning Kelvin probe force microscopy (SKPFM) provides valuable insight by probing the local work function. Since the height and the electrical potential are measured simultaneously, the technique allows for a direct comparison between the reaction sites (potential) and subsequent delamination (topography) [12]. Furthermore, when ions diffuse into the polymer/metal interface, the interfacial electrical resistance is reduced in the respective area and the corresponding potential is shifted to more negative values [12].

The chapter is organized as follows: In Sect. 28.2, a short introduction is given to the SFM-based techniques discussed in the following, including the essential theoretical foundations. In Sect. 28.3, some general aspects of nanoindentation experiments in the vicinity of interfaces are discussed. As a typical structural adhesive, epoxy and its IPs are outlined in Sect. 28.4. In Sect. 28.5, various SFM-based techniques and their application to PSAs are presented, including modulated lateral force microscopy (M-LFM) and the analysis of thermomechanical noise. Finally, in Sect. 28.6 some concluding remarks are given.

28.2
A Brief Introduction to Scanning Force Microscopy (SFM)

28.2.1
Various SFM Operation Modes

The core element of a scanning force microscope is a micromanufactured cantilever with a very sharp tip at its free end. Forces acting between tip and sample result in a cantilever deformation, which typically is detected by means of a laser beam reflected off its rear side. Owing to the distance dependence of the force interactions between tip and sample, distance changes between tip and sample surface can be measured via the cantilever deflection. In particular, when moving the cantilever across a surface, its height profile is transferred into changes of the cantilever deflection. However, the deflection can be kept constant by compensating for the height differences of the surface topography. This can be achieved by a feedback loop, including piezoelectric actuators for adjusting the vertical position of the cantilever. These height adjustments reflect the sample topography. In particular, a topography image can be stored, thus delivering the topographic height differences in length units. Depending on the distance between tip and sample surface, three different basic operation modes are distinguishable, namely contact mode, intermittent contact mode (tapping mode), and noncontact mode. The principles of these basic SFM modes are outlined in a number of monographs [13,14] and review articles [9,15–17] (just to name a few).

Mechanical surface characterization while operating in contact mode is possible by suitable extensions. In the simplest and most widespread approach, while scanning the lateral forces acting in the tip–sample contact are measured via the resulting torsion of the cantilever. This operation mode is referred to as lateral force microscopy (LFM). By applying a lateral displacement modulation of either cantilever or sample position, a sinusoidal LFM signal can be generated. Correspondingly, this mode can be called modulated lateral force microscopy (M-LFM) [18,19]. The essential elements of the experimental setup are depicted in Fig. 28.1. As compared to conventional LFM, M-LFM has several advantages:

(i)   at excitation frequencies well above $\sim 5\,\mathrm{kHz}$, a further increase in the signal-to-noise ratio is achieved because of the decay of the $1/f$ noise
(ii) the usage of a sinusoidal excitation enables the use of respective signal amplification techniques, such as lock-in amplification, resulting in valuable enhancements of the signal-to-noise ratio
(iii) with the employment of phase-sensitive amplifiers, two signals are available, namely an amplitude and a phase signal or, equivalently, signals related to the real part and the imaginary part of the complex response
(iv) as input parameters, both the amplitude and the frequency of displacement modulation can be varied, thus providing the potential for a detailed analysis of materials properties, such as viscoelasticity
(v)   depending on the amplitude of displacement modulation, information on both the shear stiffness and the frictional properties of the tip–sample contact can be gained.

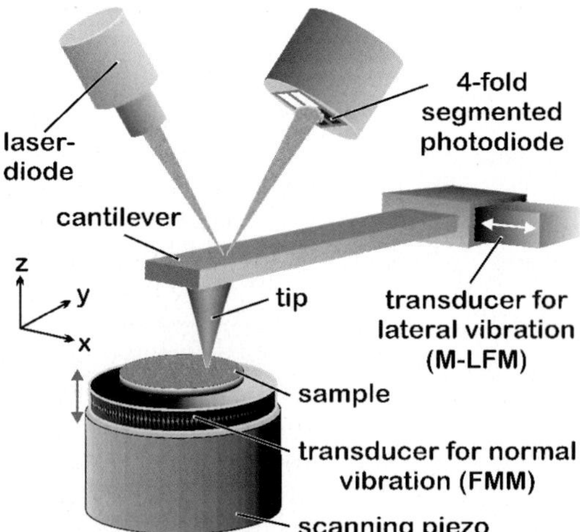

Fig. 28.1. Schematic representation of the SFM setup for modulated lateral force microscopy (M-LFM) and force modulation microscopy (FMM). The displacement modulation is applied in the direction perpendicular to the cantilever length axis or in the vertical direction, respectively

The items (i) and (ii) are of a technical nature, primarily, whereas the items (iii)–(v) are of a methodological nature. In particular, item (v) is discussed in greater detail further below (Sect. 28.5.2).

Similarly to M-LFM, a periodical displacement modulation in the direction perpendicular to the sample surface can also be applied (Fig. 28.1). Via the resulting dynamic cantilever bending, the displacement modulation is converted into a modulation in normal force (force modulation microscopy, FMM). The amplitude of the respective cantilever deformation is a measure of the contact stiffness. For frequencies well below the fundamental cantilever resonance frequency, a point-mass model holds and the cantilever can be represented by a spring of stiffness k_c. Then, the arrangement of cantilever and tip–sample contact is described as a series of two springs, namely the cantilever spring, k_c, and the spring representing the stiffness of the tip–sample contact against deformation in normal direction, k_{ts}. Given a certain displacement Δz, both springs will deform, with the deformation of the cantilever being higher the stiffer the tip–sample contact.

Frequently, FMM is performed while scanning the tip across the surface, thus generating images of lateral variations in the surface stiffness [9, 10, 19–33]. Moreover, a modified version of FMM, where the force modulation is achieved magnetically, was also demonstrated to enable a quantitative determination of the local Young's modulus [34]. Takahara and coworkers [25] referred to the FMM as scanning viscoelasticity microscopy (SVM), thus emphasizing the ability to deliver signals related to elastic as well as viscous properties.

However, a stationary mode can also be applied, where the tip is kept over a certain spot but moved in the vertical direction. In this case, the tip–sample interaction potential is driven through and a force–distance curve is recorded. Several properties can be sensed with these experiments, including elastic, plastic, and adhesive properties [35, 36]. With the vertical displacement modulation being activated simultaneously, amplitude–distance curves as well as phase–distance curves can be recorded also. Such a sort of point analysis is frequently chosen in order to analyze the effect of different loads on the stiffness measurements. For instance, this approach may prove useful for optimizing the stiffness contrasts by adjusting the load used when imaging stiffness variations. In a similar manner, amplitude–distance curves were used for adjusting the operation parameters of the intermittent contact mode [37–41] (Sect. 28.5.3).

Finally, it should be noted that a precondition for M-LFM measurements in the sticking regime is that the tip is not scanned simultaneously across the sample. Thus, in contradiction to FMM, often M-LFM measurements are performed stationary, i.e. without a simultaneous scanning motion. If frictional properties are of major interest, however, M-LFM imaging can deliver valuable information as well [19, 42–44]. In particular, as compared to LFM measurements, the M-LFM signals are much less affected by topography and calculation of the depth of the friction loop is not required [18, 43]. More recently, torsional cantilever resonances were also demonstrated as a valuable approach for measurement of friction and contact stiffness. In the so-called torsional resonance mode (TR mode), a rotation around the length axis of the cantilever is generated by means of two piezoelectric elements mounted at the cantilever holder and vibrating out of phase. A review of the technique was given by Reinstädtler et al. [45].

28.2.2
Contact Mechanics

Contact mechanics theories are frequently used to describe the elastic deformations occurring in the tip–sample contact. Depending on the applied load, the adhesion forces, and the elastic properties of the materials involved, the indentation depth and the contact radius can be calculated.

In general, the tip apex is thought of as a sphere of radius R. A combined modulus of the tip–sample contact, consisting of the sphere of the tip material (with Young's modulus, E_t, and Poisson's ratio, ν_t) and a plane surface of the sample material (with Young's modulus, E_s, and Poisson's ratio, ν_s) can be calculated. This modulus, usually referred to as the reduced modulus, is given by:

$$K = \frac{4}{3}\left(\frac{1-\nu_t^2}{E_t} + \frac{1-\nu_s^2}{E_s}\right)^{-1} . \tag{28.1}$$

Denoting the area density of the thermodynamic adhesion energy with w, and the normal load with F, then the contact radius, a, can be calculated from the following equation, which results from the contact theory by Johnson, Kendall and Roberts (JKR model) [46]:

$$a^3 = \frac{R}{K}\left\{F + 3\pi w R + \sqrt{6\pi w R F + (3\pi w R)^2}\right\} . \tag{28.2}$$

As a result of the adhesive forces, for vanishing load the contact radius is nonzero. A tensile force is required to detach the tip from the sample surface. This force, the so-called pull-off force, $F_{\text{pull-off}}$, can be considered as a measure of the adhesive forces acting in the tip–sample contact [35]. However, it should be kept in mind that $F_{\text{pull-off}}$ can also be affected by the viscoelastic behavior of the sample [47].

Other theories of contact mechanics exist. In the limit of negligible adhesive forces, $w = 0$, the JKR model is identical with the Hertz model [48]. Furthermore, the presumption of short-range forces underlying the JKR model is violated when dealing with comparatively stiff materials where the extent of elastic surface deformations is much smaller than the reach of surface forces. In this limit, the model by Derjaguin, Muller and Toporov (DMT) holds [49]. As shown by Muller et al. [50] and Maugis [51], the transition between the JKR and the DMT model is governed by a single parameter, which is referred to as the elasticity parameter. It is given by the ratio between the extent of elastic surface deformations and the reach of the surface forces. The Maugis–Dugdale model (M–D) encompasses both the JKR and the DMT model. For experiments with very soft materials, such as elastomers, the deformations are relatively large and the JKR model is suitable [52]. Further details of the various contact mechanics theories can be found in review articles [35,53].

For materials with strongly rate-dependent behavior, significant energy dissipation can occur in the course of a loading–unloading cycle. As demonstrated by Giri et al. [54], these viscoelastic losses increase with the cycle frequency. Essentially, the subsequent increase and decrease in contact area can be thought of as a receding and an advancing crack, respectively. Johnson developed a combination of linear viscoelastic fracture mechanics with the JKR contact mechanics model [47]. A major result of this model is that an energetic imbalance exists between the closing and the opening crack, thus resulting in cyclic energy dissipation.

In SFM, measurement of the contact stiffness is achieved indirectly. Essentially, the cantilever is utilized as a reference spring. Assuming for the sake of simplicity a purely elastic material, the setup can be described by two springs arranged in series, with one spring representing the cantilever of stiffness k_c and the other spring representing the tip–sample contact of stiffness k_{ts}. The latter increases with the product of contact radius, a, and reduced modulus, K:

$$k_{ts} = \frac{3}{2}aK \ . \tag{28.3}$$

For a given vertical displacement, Δz, a deformation in the tip–sample contact, δ, and a bending of the cantilever, b, occur. The latter is measured. If k_c is known, then k_{ts} can be calculated:

$$k_{ts} = k_c \frac{b}{\Delta z - b} \ . \tag{28.4}$$

Notably, b is large for high values of k_{ts}, since a stiff sample corresponds to low values of $\delta = \Delta z - b$.

Rather than a vertical displacement increasing linearly with time, within an FMM experiment a sinusoidal vertical displacement is applied, and the tip–sample contact stiffness can be calculated from the measured bending amplitude of the cantilever. In the quasistatic regime, i.e. below the resonance frequency of the system consisting of cantilever and tip–sample contact, inertial effects are negligible and (28.4) applies.

Similarly to FMM, in a shear modulation experiment, a lateral rather than a vertical displacement modulation is applied (M-LFM) [55]. Usually, this is done along the direction perpendicular to the cantilever length axis. In this case, the resulting lateral force results in a cantilever torsion that can be transferred into a second signal, separate from the signal related to cantilever bending.

Depending on the amplitude of the lateral displacement modulation and the surface characteristics, M-LFM can be operated in two different regimes, namely the sticking regime and the sliding regime. In the sticking regime, the sinusoidal lateral displacement results in a shear modulation, and the shear stiffness, k_{ts}^{lat}, of the tip–sample contact is given by [53, 55–57]:

$$k_{ts}^{lat} = 8\tilde{a}G^* \ . \tag{28.5}$$

Here, $\tilde{a}$ denotes the effective tip–sample contact radius, which is closely related to the contact radius a [58]. G^* is the reduced shear modulus. Similar to the reduced Young's modulus, G^* can be calculated from the shear modulus of the tip, G_t, and sample, G_s, and the respective values of the Poisson's ratios, ν_t and ν_s:

$$\frac{1}{G^*} = \frac{2 - \nu_t}{G_t} + \frac{2 - \nu_s}{G_s} \ . \tag{28.6}$$

In the sliding regime, a frictional force, F_f, acts that according to the adhesion model by Bowden and Tabor [15, 59] is proportional to the tip–sample contact area, A:

$$F_f = \tau \cdot \pi a^2 \ , \tag{28.7}$$

with the adhesion strength τ. Usually, the contact area is approximated by πa^2. The description of the adhesive interactions in terms of a strength expresses the idea of adhesive bonds existing between tip and sample to be broken before the tip can be moved to an adjacent position. Indeed, τ was found to increase with the surface adhesion energy, w [53]. Interestingly, the microscopic friction law of (28.7) is opposed to Amonton's law known from the macroscopic world, where the friction force increases with the load rather than with the contact area. Rough surfaces can be thought of as ensembles of asperities, and the two laws are connected via the transition from a single asperity to a rough surface [60].

It can be inferred that when operating in the sticking regime, the measured M-LFM amplitude increases with the shear stiffness of the tip–sample contact, k_{ts}^{lat}, and when operating in the sliding regime, the measured M-LFM amplitude increases with the frictional force, F_F. Correspondingly, M-LFM measurements in the sticking regime were also referred to as shear modulation force microscopy (SMFM) [61].

At the stick–slip transition, the coupling strength between tip and sample is reduced and a pronounced phase shift is observed (Sect. 28.5.2). Frequently, an onset of the stick–slip transition is observable from the phase shift, due to microslip occurring in the outer regions of the tip–sample contact area [57, 58]. In a plot of the torsional cantilever signal versus time, the stick–slip transition is recognizable from a plateau (Fig. 28.12b,d). The respective constant value of the frictional force is referred to as the dynamic friction force [43, 62]. With increasing adhesion strength, τ, and contact radius, a, the stick–slip transition shifts to higher values of the displacement amplitude. As a consequence, on a soft and sticky surface the stick–slip transition occurs at a larger lateral displacement than on a stiff surface where attractive interactions are restricted to weak van der Waals forces.

28.2.3
Extracting Information from Thermomechanical Noise

Due to the localized input of mechanical energy into the tip–sample contact, sensing local mechanical properties by exerting external forces may alter the system under investigation by driving it out of thermal equilibrium. Furthermore, as compared to the molecular length scale, displacement as low as 1 nm is still large and the elastic regime can be easily exceeded [63]. However, nonlinear deformations lead to considerable complications in the data analysis procedure.

As an alternative, the thermomechanical noise can be taken benefit of. Owing to the thermal energy, $k_B T$, and the limited cantilever stiffness, k_c, cantilever vibrations exist even without any deliberate activation. According to the equipartition theorem, the respective mean square elongation of the free cantilever is $\langle z^2 \rangle = k_B T / k_c$ [36, 63–70]. It is worthwhile noting that with the optical lever technique the inclination at the point of the laser focus is measured rather than the deflection itself [69].

In normal SFM experiments, these thermally activated oscillations add to the noise floor, but measuring the frequency distribution of the noise can provide valuable information also. Indeed, measurement of the noise power spectral density (PSD) of a free cantilever is a proven method for calibrating the cantilever stiffness, k_c [36, 68–71]. Basically, the method relies on fitting Lorentzian resonance curves to the noise spectra. Similarly, with the tip being in interaction with a sample surface, information related to the sample properties is extractable [36, 63, 72–76]. When operating in a fluid cell, the interaction of the cantilever with the liquid medium has to be taken into account as well as the tip–sample interaction. As is readily visible from the broadening of the resonance peak, a strong viscous damping of the cantilever vibration occurs. Due to the amount of fluid dragged along with the cantilever, the effective mass is the sum of the sphere mass and the extra mass of fluid [70, 77]. The latter is frequency dependent.

In a simplified approach, the cantilever can be modeled as an elastically suspended sphere exposed to random forces exerted by the surrounding medium. Then the Langevin equation of motion can be applied, which assumes random forces, $R(t)$, acting on a harmonic oscillator:

$$m\frac{d^2u(t)}{dt^2} + \xi\frac{du(t)}{dt} + \kappa u(t) = R(t) \, . \tag{28.8}$$

Here, $u(t)$ denotes the displacement, m the effective mass, κ the effective spring constant, and ξ the drag coefficient. Irritatingly, the latter is frequently also referred to as the friction coefficient, although the underlying physics is completely different from that of sliding friction.

The thermally activated motion of the system reflects its mechanical properties. This relation is covered by the fluctuation-dissipation theorem, which states that the noise power spectral density (PSD) is equal to the imaginary part of the system's mechanical susceptibility [63–65]. As a result, the noise PSD of the harmonic oscillator is given by resonance curves of the Lorentzian form:

$$\frac{d\left|u^2(\omega)\right|}{d\omega} = \frac{k_B T}{\pi}\frac{\xi}{(\kappa - m\omega^2)^2 + \xi^2\omega^2} = \frac{A\gamma}{\left(\omega_0^2 - \omega^2\right)^2 + \gamma^2\omega^2} \, , \tag{28.9}$$

where $\omega_0 = \sqrt{\kappa/m}$ is the fundamental resonance frequency, $\gamma = \xi/m$ the damping constant, $k_B T$ the thermal energy, and $A = k_B T/\pi m$ the oscillator strength. The parameters m, κ and ξ can be deduced by fitting (28.9) to the measured noise spectra. Owing to the coupling of the cantilever to its environment, it acts as a probe for viscoelastic interactions.

Notably, the position of the resonance peak depends on both ω_0 and γ. For strongly damped systems the resonance peak shifts to lower frequencies and is less distinct due to broadening. In particular, this affects the accuracy of the effective mass, m, since the relative weight of the inertial term decreases with increasing damping [73]. Aside from the operation in a well-defined state of thermal equilibrium, another advantage of the technique over soliciting at a certain frequency is the multiplex approach, since a wide range of frequencies is sampled at the same time.

28.3
Fundamental Issues of Nanomechanical Studies in the Vicinity of an Interface

Nanoscale measurements of tip–sample interactions offer great possibilities for characterizing local physical properties, such as stiffness or frictional behavior. For instance, this capability is of great importance for the characterization of polymer blends and copolymers, where the different components can be distinguished from each other by means of their mechanical properties. Thus, there is no need for additional contrast-enhancing techniques, such as staining used in transmission electron microscopy (TEM), including the corresponding time for sample preparation and the artefacts potentially coming along with such preparation techniques.

On the other hand, the directness of the SFM-based approach also implies a number of potential pitfalls, i.e. tip–sample interactions that may lead to variations in measurement signal not related to true variations in materials properties. In particular, such effects may occur for measurements in the vicinity of a vertical boundary, which is the typical geometry of IP characterization experiments. Essentially, the discussion of such artefacts can be separated into three different cases, namely the case of a step at the interface, the case of a flat surface in combination with large indentation depths, and the case of a flat surface in combination with small indentation depths.

First, however, a quantitative evaluation of property changes with decreasing distance from the interface requires a precise identification of the interface. More precisely, the interfacial borderline needs to be defined, i.e. the cross section between the surface under investigation and the interface, both of which are two-dimensional in nature.

28.3.1
Identification of the Interface

In many cases, the topography image of a cross section reveals the edge of the filler particles of a composite due to a distinct step. Depending on the particular procedure for preparing the surface, the step frequently reflects the different mechanical properties of the composite components. For instance, upon mechanical polishing a higher wear rate is likely to be observed on the softer matrix, resulting in a sort of washing-out effect. Thus, with increasing time of polishing, the surface roughness can be reduced, but at the same time the interfacial steps are likely to be increased. As a result, long periods of polishing do not necessarily deliver the most suitable topography, and the polishing time needs to be optimized as well as the other polishing parameters such as force, velocity, or fineness of the polishing slurry. Also, the results of dedicated cutting techniques, such as ultramicrotomy, may reflect the different mechanical properties of soft polymer and stiff filler particles. In particular, the different coefficients of thermal expansion come into play if the temperatures of cutting and SFM measurement differ. Cutting at a temperature well below room temperature (RT) can be a necessity, e.g. if a rubbery component is contained. As such, its glass transition temperature is lower than RT, and upon the action of shear forces applied in the course of the cutting procedure, smearing of the soft rubbery components is likely to occur.

If the interfacial borderline can be associated with a clear topographic feature, then the topography image can serve as a reference. As mentioned above, however, from the point of view of mechanical SFM measurements in the vicinity of the interfacial borderline, a height profile as flat as possible is desirable. Although being a difficult task to achieve, in the case of an almost flat interfacial borderline, the conditions for nanoindentation are improved but identification of the interfacial borderline from topographic information alone may prove very difficult. In such a case, a contrast other than topography is required. As a reference, the contrast used for interface identification should be different from the one to be used for IP evaluation. A further requirement for a suitable reference signal is the occurrence of a pronounced change at the interfacial borderline.

For instance, these requirements are fulfilled when dealing with electrically conductive filler particles embedded in an insulating polymer matrix. Provided an electrically conductive SFM tip and appropriate wiring exist, the electrical current across the tip–sample contact can be measured, in addition to the mechanical contrasts. For instance, such a type of experiment was performed for carbon fibers embedded in poly(phenylenesulfide) (PPS) [26]. When scanning across the interfacial borderline, a distinct jump in electrical current was observed. The positional accuracy of the deduced interfacial borderline is dependent on the particular tip shape, the interfacial topography, and the occurrence of tunneling currents. For a typical tip apex with a radius of curvature of $R \sim 10$ nm, any deviations from this idealized description in terms of a semisphere, such as protrusions, can be expected to be a few nanometers at maximum. Similarly, the reach of tunneling currents between the tip and the filler particle is of the order of several nanometers. The different situations of current measurement at a perfectly flat interface and a step-like interface are depicted in Fig. 28.2b,c, respectively. In addition to tunneling currents, in the case of a step, a mechanical contact can occur between the edge of the filler particle and the tip shank. Thus, depending on the height of the step, h, the opening half-angle of the tip, α, the position of the current drop can be shifted with respect to the fiber edge by a distance, w_{DZ}, in a way that the electrically conductive phase appears to be extended over this additional distance. Since the zone of width w_{DZ} is difficult to attribute to either phase unambiguously, it is referred to as the dead zone [26]. In

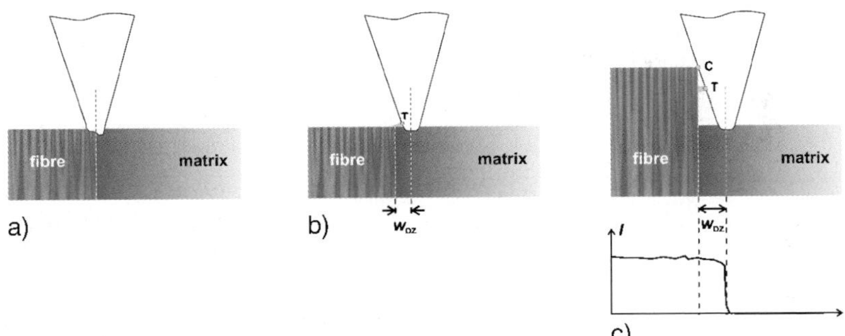

Fig. 28.2. Schematic representation of different interfacial tip–sample contact situations. Without any restriction of generality, the case of a fiber/matrix interface is shown. In (**a**) and (**b**), the fiber/matrix interface is perfectly flat, whereas in **c** a step of height $h > R$ exists at the interface. (**a**) The tip center axis is in agreement with the interfacial borderline, i.e. the tip apex is in contact with both fiber and matrix material. (**b**) The tip apex is in contact with matrix material only, but the tip center axis is very close to the interfacial borderline. Parts of the tip are within reach of tunneling currents between tip and fiber edge. (**c**) Due to an interfacial step of height h, mechanical contacts with the fiber edge can occur, although the tip apex is in contact with matrix material only. The width w_{DZ} depends on the height h as well as the opening half-angle α of the tip. Adapted from [26]

the case of a nonzero step, a further contribution to the dead zone exists due to the region not accessible by the tip because of geometrical restrictions.

For $h = 0$, w_{DZ} is in the order of several nanometers. This uncertainty in the position of the interfacial borderline, however, is negligible if the IP is extended over several hundreds of nanometers, or even more. For instance, in the case of the system carbon fiber/PPS, IP stiffness gradients were found with a typical width of ~ 150 nm.

For the purpose of completeness, in Fig. 28.2a, the situation of the tip apex being in direct contact with both the fiber and the matrix is shown. Williams et al. [78] referred to this case as composite interaction, as opposed to the situations with the tip in mechanical contact with matrix material but sensing the fiber (contact interaction), and the situation with the tip being in contact with IP material without being affected by the stiff filler (interphase interaction).

In more general terms, the extent of the tip–sample interaction of interest may produce an uncertainty in the detected lateral position. This lateral uncertainty may increase with the bluntness of the tip as well as with the interfacial step. These are the contributions by the tip–sample contact geometry. The more fundamental contribution is given by the nature of the particular tip–sample interaction. For instance, in the case of stiffness measurements, the lateral extent of the tip–sample interaction results from the spatial distribution of the mechanical stresses beneath the tip–sample contact area.

28.3.2
Implications of the Interface for Indentation Measurements

28.3.2.1
The Case of an Interfacial Step

As mentioned before, cross-sectional surfaces of polymer–matrix composites (PMCs) frequently exhibit steps between the different components. As a consequence of a zone not being accessible by the tip, the width of the dead zone, w_{DZ}, gives the minimum detectable width of the IP, w_{IP}. With the opening half-angle α of the tip and the height h of the step, the width of the dead zone is given by $w_{DZ} = h \tan(\alpha)$ (Fig. 28.3a).

Notably, the situation depicted in Fig. 28.3a is for the special case of a tip not being affected by lateral forces acting in the tip–sample contact, i.e. under the presumption of either very low friction forces or of a cantilever that is stiff against torsional deformation. For a nonzero tilt of the tip as induced by lateral forces, however, w_{DZ} depends on the direction of scanning (Fig. 28.3b,c). For stiffness measurements, additional complications can arise from changes of the size of the tip–sample contact area. For instance, a pronounced increase in contact area can occur when the tip touches the side-wall of the filler particle (Fig. 28.3b), thus resulting in a contact stiffness apparently increased.

28.3.2.2
The Case of a Flat Surface and Large Indentation Depths

Even in the absence of an interfacial step, the change in indentation depth resulting from a strong difference in modulus may lead to misinterpretation of the data if the

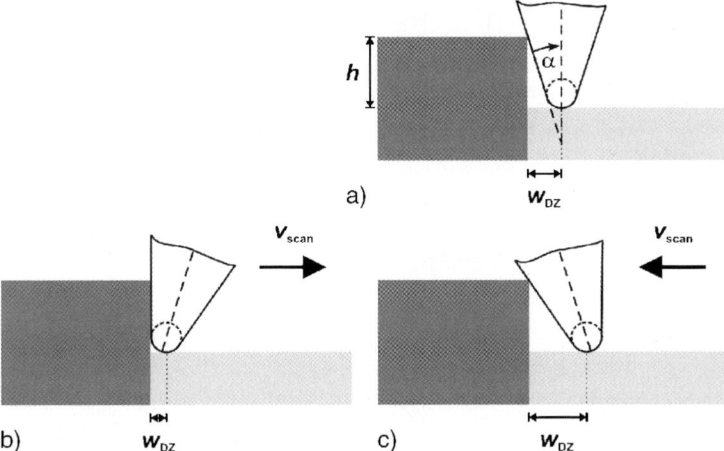

Fig. 28.3. Schematic representation of the possible variations in the tip–sample contact area when scanning across a step. (**a**) Case of an untilted tip. (**b**) Tip tilt for the case of scanning from *left to right*. (**c**) Tip tilt for the case of scanning from *right to left*. Beside the height change, h, the width of the dead zone, w_{DZ}, can be strongly affected by both the opening half-angle of the tip, α, and the tilt angle of the tip

effect of the tip shape is not taken into account properly. Hodzic et al. [79] analyzed the effect of the finite width of the indenter and its geometry on the measurement of the IP by means of nanoscratching. The combined analysis of the profile of indentation depth and the coefficient of friction showed that the signals recorded at the position of the center of the tip are also affected by the contact forces acting on its periphery. Thus, given the relatively large full opening angle of the Berkovich indenter of $2\alpha \sim 130.6°$ [80–82], at large indentation depths the measured position of the matrix-sided edge of the IP is shifted with respect to the true one. A Berkovich indenter is a three-sided pyramid made up of diamond.

The depth of indentation, z_{ind}, should be compared with the radius of curvature of the apex of the tip, R. If $z_{ind} > R$, then the global tip shape has to be considered for describing the tip–sample contact, i.e. in the case of a conical tip the opening half-angle α. Thus, an indentation is classified as deep if $z_{ind} > R$. In the opposite case of $z_{ind} < R$, only the foremost end of the tip gives significant contributions to the tip–sample interactions, and R is the key parameter [16].

When scanning from a compliant polymer matrix toward a stiff filler (such as a fiber), during a certain period of time the tip slides uphill with its edge at the forefront being in contact with the edge of the fiber. As a consequence, during this period the slope of the depth profile is governed by α.

Depending on the ratio between the penetrating width of the tip, w_{tip}, and the width of the IP, w_{IP}, essentially three different cases can be distinguished (Fig. 28.4). In the case of $w_{IP} \sim 0$, i.e. in the absence of any IP, the depth profile includes a zone of nonzero width where the penetration depth changes from the large depth on soft matrix material to the low depth on the stiff filler material (Fig. 28.4a). This zone of depth change is due to the interaction between the fiber edge and the tip shank, i.e. by the forces acting outside the contact area with the tip apex.

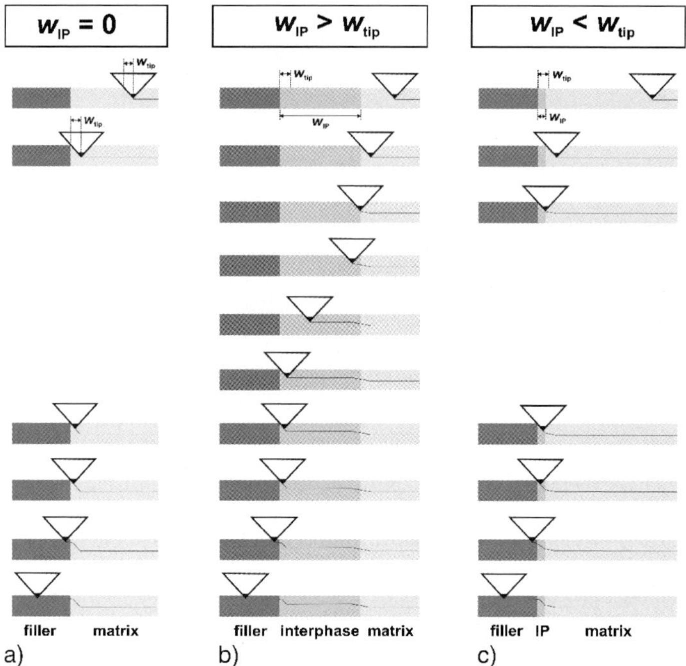

Fig. 28.4. Schematic representation of the effect of indentation depth on the recorded height profile when scanning across a filler/matrix interface. The apex of the tip is indicated by the *black filled triangle*. (**a**) The case of a sharp interface, i.e. the width of the IP, w_{IP}, is zero. (**b**) The case of an IP much wider than the zone affected by the change in indentation depth, $w_{IP} > w_{tip}$. (**c**) The case of a comparatively narrow IP, $w_{IP} < w_{tip}$. Adapted from [79]

In the limit $w_{IP} > w_{tip}$, two separate zones of depth change can be observed, namely at the interface between matrix and IP and at the interface between IP and fiber (Fig. 28.4b). In the intermediate case of $w_{IP} < w_{tip}$, the two depth changes at the fiber/IP interface and at the IP/matrix interface are overlapping. In this case, separation of the IP properties from the measured depth profile requires a suitable correction procedure to be applied to the measured data. Since any such correction procedure depends to some degree on assumptions on the deformations in the tip–fiber contact, the preferable case is the one with $w_{IP} > w_{tip}$. That is, the experiment should be conducted in a way that the effects due to finite tip size are neglectable.

In these terms, it can be concluded that the penetrating width of the tip, w_{tip}, should be as low as possible, which may be achieved by using a slender tip and by applying a normal load as low as possible. Typically, this is the case for SFM-based indentation experiments, where the condition $z_{ind} < R$ is met by a suitable choice of the cantilever spring constant, k_c, i.e. of the normal load. For composites containing phases ranging in their stiffness values over many decades (from several MPa as typical for elastomers up to hundreds of GPa as typical for many metals or ceramics), however, the condition $z_{ind} < R$ is unlikely to be met for the most compliant phases. For instance, normal automotive brake pads consist of a thermosetting matrix with dispersed particles of various types, ranging from stiff and solid to soft and rubbery [19].

28.3.2.3
The Case of a Flat Surface and Small Indentation Depths

Beyond any effects induced by a height difference between fiber and matrix, in the vicinity of the interface, the indentation experiment on soft polymer can be affected by the presence of the stiff component, e.g. the fiber. Since a significant part of the imposed load is transferred to the stiff component, the restriction imposed by the fiber on the adjacent matrix causes a stiffening effect. Apparently, within a certain zone adjacent to the interface an increase in the local modulus is observed. As induced by the redistribution of mechanical stresses, these apparent variations have to be differentiated from true property variations occurring within the IP.

In their investigation of the interface between aluminum and epoxy, Li et al. [83] detected a monotonic increase in the apparent modulus with decreasing distance from the interfacial borderline. In a similar manner, by studying an epoxy reinforced with glass fibers, Kumar et al. [84] observed a linear increase in the apparent modulus with decreasing distance from the fiber edge. Although the cutting angle with respect to the fiber axis showed some influence as well, they identified the magnitude of the normal load as the dominating parameter. Using lower loads helped reducing the effect of biasing the measurement by the close fiber. As shown by Li et al. [83] using nanoindentation experiments, the width of the zone affected by the mechanical interaction with the filler is larger than the contact impression. Thus, it was inferred that plastic flow is occurring during the indentation process. Li et al. [83] referred to the zone affected by the load transfer across the interface as the indent-affected zone (IAZ). In their study of the influence of the fiber on the indentation measurement next to it, Hodzic et al. [79] used the expression fiber bias effect. For the sake of generality, here the expressions mechanical bias effect (MBE) and MBE zone are used. Accordingly, the width of the latter is denoted as w_{MBE}.

As demonstrated by Li et al. [83], w_{MBE} can be determined from a plot of the load at a selected indentation depth versus the nominal distance from the interface (Fig. 28.5a). As the interface is approached starting from the bulk of the compliant component, at some critical distance the load at the selected depth starts to increase. This critical distance can be identified with w_{MBE}.

The critical distance was deduced from the intersection of the two linear curves describing the MBE zone and the bulk polymer, respectively. Obviously, the value of w_{MBE} is not a fixed number but it depends on various parameters such as the radius of curvature of the tip, the total load (including the adhesion forces), and the surface stiffness. Indeed, w_{MBE} was found to increase with the depth of indentation [83]. Supposedly due to the surface roughness, the data with indentation depth as low as 50 nm showed a strong scatter, and no obvious gradient could be observed. As mentioned before, the MBE is caused by transfer of stress from the tip–sample contact to the adjacent phase. Indeed, calculations of the subsurface stress distribution show that it is extended well beyond the contact radius [56].

The simplest model for describing the stresses and strains occurring in the contact zone of two purely elastic materials is the Hertz model, see [56] and the references

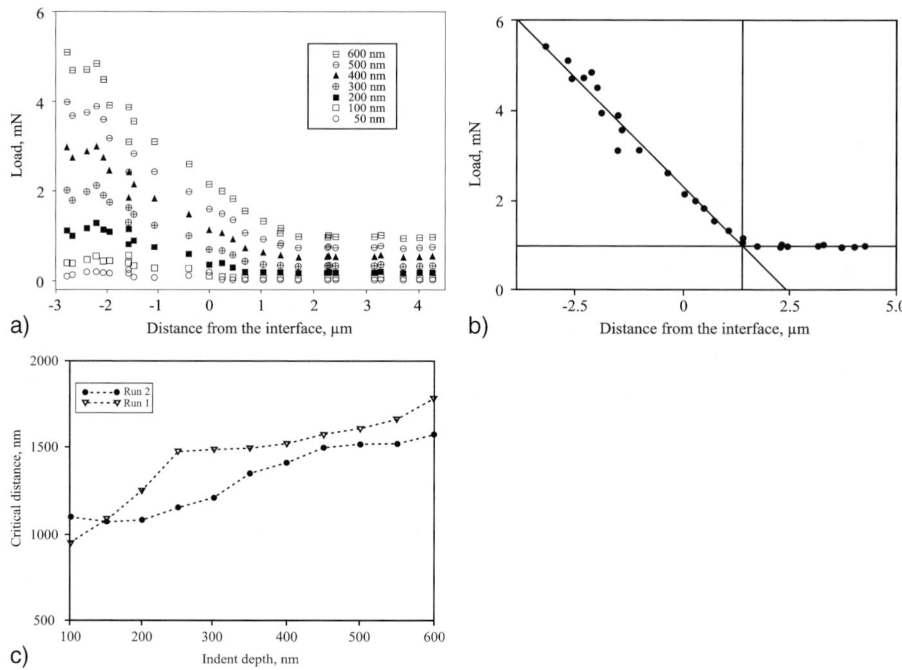

Fig. 28.5. Experiments on the determination of the zone affected by the mechanical bias effect (MBE), as measured at the interface between aluminum and epoxy. (**a**) The load at a selected depth as a function of the nominal distance from the interface. (**b**) Deduction of the width of the MBE zone, w_{MBE}. The data are shown for an indentation depth of 400 nm. (**c**) The critical distance (i.e. width of the MBE zone, w_{MBE}) versus the depth of indentation. Reprinted with kind permission from [83]. Copyright 2002, Brill Academic Publishers

given therein. The shape of the subsurface stress distribution is similar to that of a club, with nonzero stresses also for radial distances $r > a$, as measured from the center of the contact. The relevance of the subsurface stresses becomes obvious from the fact that the region of maximal values of the principal shear stress is not adjacent to the tip apex, but below the surface at a depth of $0.48a$ (for $\nu = 0.30$) [85]. If the local pressure exceeds a critical value proportional to the yield stress of the material, onset of plastic flow will occur within this subsurface region. The extent of the plastic deformation zone increases with further loading.

Starting from the observation that subsurface deformations under a blunt indenter can be described in terms of hemispherical contours of equal strain, a cavity model was developed where hydrostatic pressure acts within a hemispherical core just underneath the contact surface [56, 85, 86]. The hydrostatic core is surrounded by a zone of plastic deformations, with the plastic strains gradually diminishing until they match the elastic strains. The boundary between the outer elastic zone (so-called elastic hinterland) and the annular plastic zone is given by some radius c, with $c \sim 3.2a$. As a result, in the case of elastic–plastic indentation an MBE may occur if the close edge is within a distance of $\sim 3a$ from the symmetry axis of the indenter. Of course, the precise value of c depends on the particular plastic deformation behavior of the material under investigation. In general, its mechanical behavior cannot be expected to follow the simplified elastic-perfectly plastic pattern.

Experimentally, the width of the MBE zone was evaluated for polycarbonate (PC) using a Berkovich indenter [32]. A cross section of the PC/steel interface was prepared by cutting and polishing. A number of indents were made along a line slightly tilted with respect to the interfacial boundary line. Afterwards, the indentation marks were imaged in a scanning mode (Fig. 28.6a), and the diameter, d, of each imprint was measured. For $x/d < 1$, the profiles of the apparent modulus versus the distance, x, from the interfacial borderline showed a significant increase (Fig. 28.6b). The distance, x, was measured from the center of the imprint. This observation of increasing apparent modulus was done for several maximum load values. Concluding, for indentations with a Berkovich indenter on PC it is recommended that the distance of the imprint center from the interfacial borderline should be no smaller than the imprint diameter, d. Although the plastic deformation behavior of PC may differ from that of other amorphous polymers and the critical ratio of $x/d \sim 1$ may

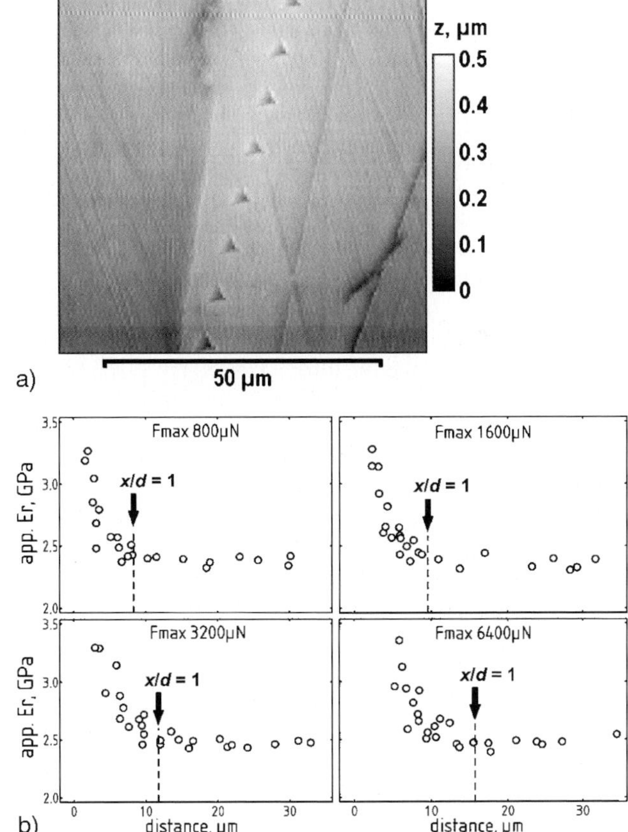

Fig. 28.6. Series of indentation experiments on the cross-sectional surface of an interface between steel and polycarbonate (PC). (**a**) Topography image, showing the line of residual imprints, slightly tilted with respect to the interfacial borderline. In the image, the steel region is located on the *left*. (**b**) Plots of the apparent reduced modulus, E_r, versus the distance, x, from the interfacial borderline. The values of the maximum load, F_{max}, were 800, 1600, 3200, and 6400 µN, respectively. Reproduced from [32] with kind permission from Wiley-VCH, Weinheim. Copyright 2005

not hold exactly in general, in the sense of a rule of thumb it can be considered as a guideline. In particular, the distance x should be significantly larger than $d/2$.

28.4
Property Variations Within Amine-Cured Epoxies

28.4.1
A Brief Introduction to Epoxy Mechanical Properties

Since their invention in the 1940s, epoxies have found a wide range of applications, in particular as structural adhesives, as matrix material of PMCs, and in coatings [87]. In addition to their practical relevance, recent experimental results have shown that epoxy systems can form IPs as wide as several micrometers or even hundreds of micrometers [9, 88, 89], if applied as a liquid formulation. Depending on a number of parameters such as the initial viscosity of the formulation, the curing temperature, and the interfacial driving forces for preferential adsorption of epoxy resin or curing agent, pronounced IPs may develop. In the following, a short introduction to some epoxy properties is given, in order to provide the basic background for the interpretation of the IP stiffness gradients discussed in Sect. 28.5.2.

A widely used epoxy resin is diglycidyl ether of bisphenol A (DGEBA). At the two terminal positions, DGEBA contains the epoxy groups (frequently also referred to as oxirane groups) that provide the ability for chemical crosslinking. With an average molecular weight of $\sim 170\,\mathrm{g/mol}$ per epoxy group, the so-called epoxy equivalent weight (EEW), DGEBA molecules are oligomers. The addition reaction with crosslinking agents such as amine molecules results in crosslinking as well as in polymerization, i.e. in a significant increase in molecular weight. Higher homologs of DGEBA are available and their higher molecular weight can be used for tuning the viscosity of the epoxy resin as well as the properties of the cured epoxy, depending on the requirements of the particular application. The structural formulae of the lowest homolog of the DGEBA molecule and of the aromatic molecule diaminodiphenylsulfone (DDS) are given in Fig. 28.7. Some properties of the DGEBA–DDS system will be discussed further below. DDS is a diamine, with its two amine groups ($-NH_2$) situated at the terminal positions.

Although not further elucidated in the following, it should be noted that the choice of the particular amine can show a strong influence on the properties of the amine–epoxy system [90–93]. Due to the bulky benzene rings, typically in aromatic amines the segmental mobility is more restricted than in aliphatic amines. In particular, this may affect the glass transition temperature, T_g. For DGEBA-type epoxy resin cured to full degree with DDS, the T_g value is $\sim 220\,°\mathrm{C}$ [94, 95]. Furthermore, the rate of curing reaction tends to be lower for aromatic amines.

The mechanical properties of amine-cured epoxy systems were studied for various amine–epoxy systems and cure schedules [93, 94, 96–99]. In most studies, both the small-strain and the large-strain properties were investigated. Concerning the glass transition temperature and small-strain properties, such as modulus, yield stress and yield strain, the major results of the various studies can be stated to be consistent with each other.

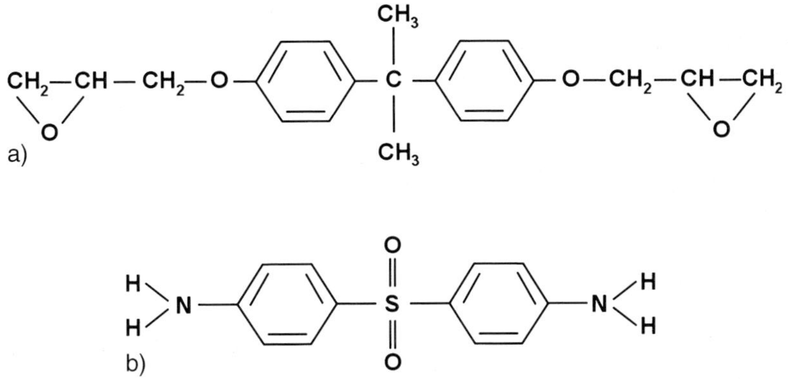

Fig. 28.7. Chemical structures of (**a**) the DGEBA molecules, and (**b**) the aromatic amine DDS

A simple and widespread approach for varying the epoxy network structure is by variation of the amine–epoxy concentration ratio. Each of the amine hydrogen atoms can react with one epoxy group. With the DGEBA molecule and the diamine molecule containing two of the respective groups, the chemical functionalities are $f_a = 4$ and $f_e = 2$, respectively. Thus, an equimolar mixture requires $N_e = 2\,\text{mol}$ of DGEBA for $N_a = 1\,\text{mol}$ of the diamine. Accordingly, the amine–epoxy mixing ratio, r, can be defined as follows [95]:

$$r = \frac{N_a}{N_e} \cdot \frac{f_a}{f_e} \,. \tag{28.10}$$

The equimolar (stoichiometric) case is given by $r = 1$. In the sub-stoichiometric regime, $r < 1$, epoxy is in excess, whereas in the over-stoichiometric regime, $r > 1$, amine is in excess. It should be noted, however, that side-reactions may exist that can consume additional amounts of either epoxy or amine [100].

For epoxy, T_g is widely found to reflect the crosslink density. The increase in T_g with decreasing values of the average molecular weight between crosslinks, M_c, is described by Nielsen's empirical equation [94, 98, 101]:

$$T_g - T_{gu} = \frac{C}{M_c} \,, \tag{28.11}$$

where T_{gu} denotes the glass transition temperature of the uncrosslinked system. C is a constant factor, the typical value of which is 3.9×10^4 K g/mol.

For temperatures above T_g, epoxy behaves like a rubber and rubber elasticity theory can be applied [101]. In consistency with this theory, above T_g the epoxy modulus was found to increase with the crosslink density. In the glassy state, i.e. below T_g, additional effects come into play. When investigated as a function of the amine–epoxy concentration ratio, the epoxy modulus shows a minimum around the stoichiometric concentration ratio. In the excess-epoxy regime, $r < 1$, a pronounced increase in the epoxy modulus is observed [88, 97, 102]. This finding was attributed to the antiplasticization effect. Notably, this effect is characterized by both an increased modulus at RT and a decreased T_g value. Since T_g increases with the crosslink density (28.11), this shows that in the glassy regime the modulus is governed by further parameters.

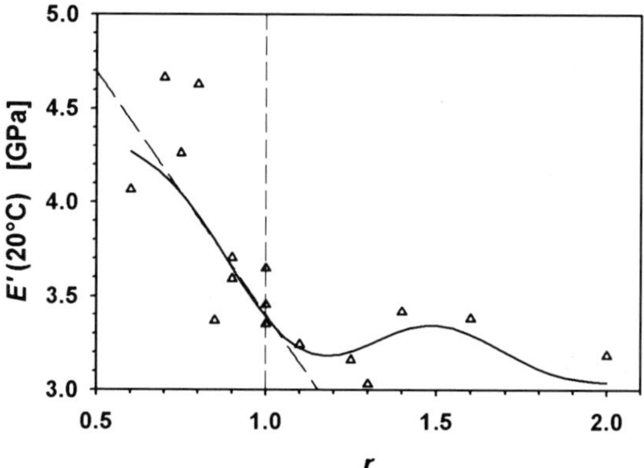

Fig. 28.8. Modulus–concentration relationship, as extracted from dynamic-mechanical thermal analysis (DMTA). The DMTA measurements were performed in torsional pendulum configuration. For the conversion from shear modulus to Young's modulus, a Poisson's ratio of $\nu = 0.42$ was used [88]

In Fig. 28.8, the change of the RT modulus with amine–epoxy concentration ratio, r, is shown for DGEBA-type epoxy resin cured with DDS [88]. The system was cured isothermally for 2 h at 170 °C, resulting in a final degree of conversion of $\sim 80\%$. Due to the fact that the system was not fully cured, the minimum of the curve is located slightly above the stoichiometric ratio, $r = 1$. Approximately, a linear function can be used for describing the epoxy modulus changes within the substoichiometric regime, $r < 1$. Within the range $0.6 < r < 1.0$ the total modulus change is $\Delta E' \sim 1.0\,\mathrm{GPa}$.

Furthermore, the antiplasticization effect is linked to a suppression of the β-transition, thus indicating the connection with segmental motions and local free volume [88, 92]. The β-transition was attributed to local motions of the hydroxypropylether sequence ($-CH_2-CHOH-CH_2-O-$), resulting from the addition reaction between epoxy groups and amines [91]. As such, the β-transition is affected by the crosslink density and the degree of off-stoichiometry, since nonreacted epoxy or amine groups remain as dangling ends (due to their terminal position). Indeed, the high-temperature wing of the β-transition was associated with a sort of local-scale cooperative motion [92]. Dangling ends, however, are in general associated with a higher fraction of free volume and degree of mobility. Thus, it can be inferred that their motions are activated at a lower temperature and they can be attributed to the low-temperature wing of the β-transition. Indeed, in the case of samples highly deficient in amine, a small separate peak at a temperature slightly below the main β-transition peak was observed for the loss modulus curve [88].

As a result of the different behavior in the glassy and the rubbery regime, the relationship between amine–epoxy concentration ratio and the epoxy modulus depends on temperature. The transition from glassy to rubbery behavior was described by Skourlis and McCullough for the case of DGEBA-type epoxy cured with a cycloaliphatic diamine [98]. When approaching T_g, the two local maxima below and above the stoichiometric formulation diminished and a very broad maximum around the stoichiometric concentration ratio was observed.

In contrast to the rubbery state, in the glassy state the epoxy modulus appears to be governed by the intermolecular packing rather than by the crosslink density. This is because the resistance to initial deformation arises from the intermolecular forces.

Below T_g, molecular mobility is strongly reduced and the conformation is likely to correspond to the one at T_g. Thus, for highly crosslinked samples, the difference between T_g and RT is large, resulting in a relatively higher fractional free volume at RT [96]. This in turn is equivalent to a lower intermolecular packing density.

According to the Eyring theory, yielding occurs as viscous flow in which the activation barrier for local shear displacements is decreased by the applied stress [103]. Mayr et al. [93] analyzed the shear yield stress as a function of the strain rate and found a reasonable agreement with the Eyring equation. As a result of their analysis, the average activation volume is $\sim 2.2\,\mathrm{nm}^3$, which corresponds approximately to the volume occupied by two DGEBA molecules linked to two amino units (two monoamines or one diamine) [93].

28.4.2
Epoxy Interphases

28.4.2.1
The Epoxy Interphase with Copper

An example of IP characterization using FMM is given in Fig. 28.9. The epoxy IP in the vicinity of the Cu microelectrodes was studied. The latter were created by employing electron beam lithography (EBL) in combination with the so-called lift-off technique [10, 32, 104]. After epoxy curing, the Si substrate was detached and the released surface was investigated. In Fig. 28.9, the topography image, the force image, and the FMM amplitude image of a region containing the edge of the microelectrode are shown [104]. The edge of the Cu structure is close to the upper edge of the images. Within the imaged surface area, the topography is very flat and only minor force variations occurred (Fig. 28.9b). Along the cross-sectional line indicated in Fig. 28.9c, the stiffness profile was evaluated. As is visible from Fig. 28.9c,d, with decreasing distance from the interfacial borderline the stiffness decreases until a minimum is reached.

Owing to the flatness of the analyzed area and the considerable distance of the stiffness minimum from the interfacial borderline, the zone of increasing stiffness can also be clearly attributed to the epoxy. Within the region of analysis, the absolute value of the local topography slope angles was found to be less than $3°$. Considering the typical performance of the topography feedback, this was identified as a threshold value below which contrasts in the FMM amplitude induced by the topography are unlikely to occur [9]. Variations in the FMM-amplitude measured across a wave-like epoxy surface were found to be in reasonable agreement with this rule of thumb [32]. By restricting the analysis of FMM data to regions with absolute slope angles $< 3°$, the rule can be applied as a measure of quality assurance.

In Fig. 28.9d, the profile of Fig. 28.9c is given in greater detail, including the results from the fitting procedure (solid line). The region where measurement artefacts induced by the nearby Cu edge could not be ruled out, was not taken into account for the fitting procedure. Each of the two decays on the left and on the right side of the minimum was fitted with a semi-Gaussian profile. The minimum is located at a distance of $\sim 0.43\,\mu\mathrm{m}$ away from the as-defined position of distance zero. The latter is marked with the arrow given in the graph of Fig. 28.9c. The outer

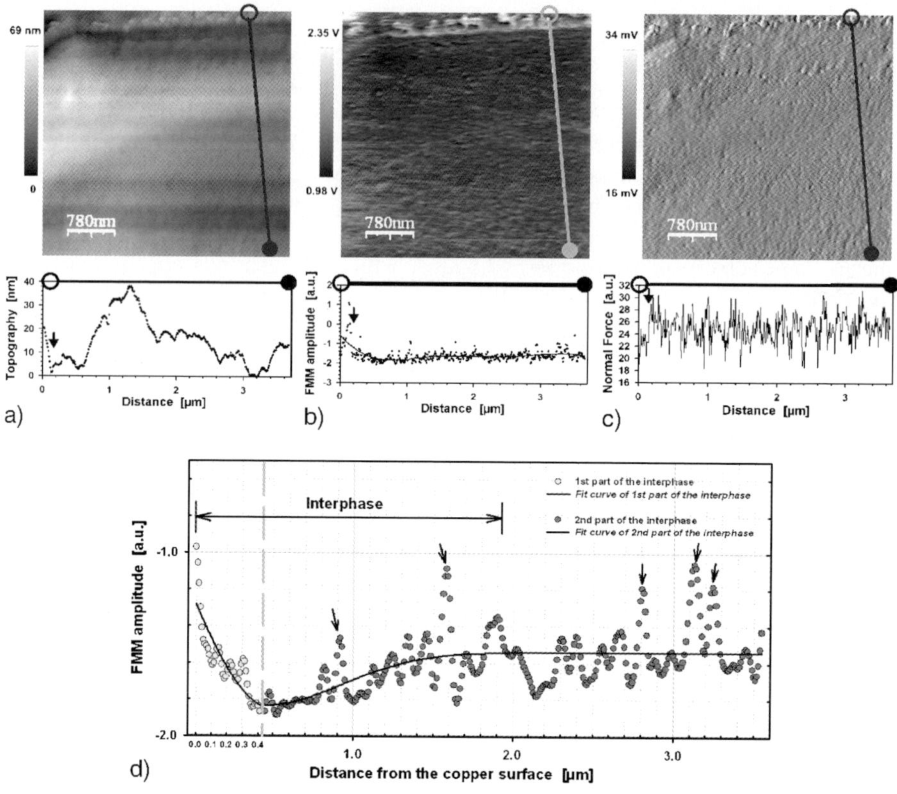

Fig. 28.9. FMM-based study of a copper–epoxy composite. The images and the cross-sectional profiles along the *solid line* are given, of (**a**) topography, (**b**) normal force, and (**c**) amplitude signal. The copper edge is near the upper edge of the images, as visible from the pronounced amplitude (stiffness). In (**d**), a close view on the stiffness profile is given, including the curves resulting from the fitting procedure. The distance was set to zero at the position indicated with an *arrow* in the plot of (**c**). The *arrows* given in (**d**) mark the position of spikes that resulted from the small troughs in the surface topography. The epoxy was prepared from the resin Epikote 828 and the amine curing agent Epikure F205. Adapted from [104]

edge of the IP was defined by the point where the relative deviation from the bulk stiffness is less than 1%.

The spikes marked with arrows (Fig. 28.9d), resulted from small troughs present in the epoxy surface. At these sites the tip–sample contact area was strongly increased, thus leading to a sudden rise in the contact stiffness and in the related amplitude signal.

Concluding, it can be stated that a two-zone IP was detected, with a zone of decreasing stiffness and a more distant zone of slowly increasing stiffness until the epoxy bulk stiffness is reached. The respective widths of the two IP zones are ~ 0.4 and $\sim 1.5\,\mu$m. In particular, these results show that epoxy IPs can be extended over several micrometers and that nonmonotonic property variations can occur.

In calculations of interfacial stress distributions, however, usually only monotonic modulus profiles are taken into account [105–107], if at all.

The epoxy IP may result from several interfacial interactions. Aside from preferential segregation of either amine or epoxy resin and the resulting heterogeneities in the network of crosslinks (Sect. 28.5.1), particular chemical interactions with the surface of the copper structure can occur. In the vicinity of copper substrates, Hong and Wang [108] observed an oxidative degradation of polymers. The formation of carboxylic acids can lead to soluble metal salts, which may diffuse into the polymer matrix and contribute to its degradation. Furthermore, as reported by Aufray and Roche [109], amine chemisorption onto oxidic or hydroxidic metallic surfaces leads to the formation of organometallic complexes having a lower functionality than the pure amines. As compared with the network resulting from pure amine and epoxy, the network resulting from the reaction of such complexes with epoxy was found to have a lower T_g. On the other hand, the polymerization reaction of the amine–epoxy system can be catalyzed by the hydroxyl groups present on the metal surface. As a result, in its vicinity the polymerization and the stiffness of the epoxy may be enhanced [110].

Concerning compositional variations in the amine–epoxy system due to preferential adsorption, in a first approximation a respective concentration–property relationship can be applied, such as shown in Fig. 28.8. In this case, departure from the stoichiometric amine epoxy concentration ratio should result in an increase in the epoxy modulus, in particular within the excess-epoxy regime (substoichiometric). Indeed, a similar relationship was found for the system Epikote 828 – Epikure F205 used in this case [104]. Epikote 828 is an epoxy resin of type DGEBA.

Information on spatial variations in chemical composition is required for resolving the mechanisms involved in IP formation. A useful technique providing a lateral resolution in the submicrometer range is energy-dispersive analysis of X-rays (EDX). In EDX, the X-rays generated by the interaction of the electron beam with the sample are used for mapping spatial variations in the chemical composition of the surface under investigation. For instance, EDX was applied to a DGEBA-type epoxy resin cured with DDS [88, 111, 112] (Sect. 28.5.1). In this case, the composite was an epoxy/thermoplastic/epoxy sandwich system. Owing to its large atomic weight as compared to C and N, the S atom contained in DDS has a much better EDX sensitivity and allows for tracking spatial variations in the amine concentration (Fig. 28.7). Also the DGEBA–DDS system was cured in the presence of Cu structures and a corresponding EDX analysis was performed [104]. A rather complicated amine concentration profile was found with a depletion zone next to the interface and an adjacent enrichment zone. The total width of the concentration profile was $\sim 4\,\mu m$, i.e. it is comparable to the one of the above stiffness analysis. However, it should be noted that the aromatic amine DDS (Fig. 28.7) is of a different chemical nature from that of Epikure F205 (from Resolution Europe), which contains modified cycloaliphatic amines. In particular, the different reaction kinetics and curing temperatures of the two systems can strongly affect the final IP width, and transferring the results from the investigation of one system to the other one may be erroneous. As shown by Yang and Pitchumani [113], a key parameter for describing the diffusion-reaction mechanism is the Damköhler number that gives the ratio of the chemical reaction rate to the adsorption rate. Both these rates are temperature dependent.

28.4.2.2
The Epoxy Interphase with Poly(vinylpyrrolidone)

A major drawback of nanoindentation experiments based on SFM is the ill-defined tip shape. In particular, calculation of the sample modulus requires knowledge of the tip–sample contact area (28.3). At this point, the smallness of typical SFM tips turns out to be a major drawback. Instead, if a reduction in lateral resolution is acceptable, it can prove useful to employ a less sharp but more well-defined tip. For instance, a Berkovich-type indenter was utilized for mapping modulus variations across the IP of an epoxy/thermoplastic composite [88].

The DGEBA–DDS system was cured in the presence of a poly(vinylpyrrolidone) (PVP) film. After curing, a cross section of the sandwich-like epoxy/PVP/epoxy sample was prepared employing an ultramicrotome. The mapping was done using a nanoindentation system attached to a 2D-scanning unit [32]. According to the evaluation procedure of Oliver and Pharr [114], the Young's modulus was deduced from a fit to the initial part of the unloading curve. The technique was also outlined in several recent reviews [8, 80, 81].

A plot of the sample modulus, E_s, as a function of the distance from the epoxy/PVP–IP is given in Fig. 28.10a. Two major observations can be made. First, the total IP width is $\sim 230\,\mu m$ [115]. Thus, the IP is much wider than the lateral extent of the residual imprints, and the usage of the Berkovich indenter is justified. Secondly, with decreasing distance from the interface, the epoxy modulus increases and the total modulus change is $\sim 1.1\,GPa$, which corresponds to $\sim 40\%$ of the bulk modulus value, E_s^b. Within the range $|x - x_i| < 175\,\mu m$, the Young's modulus increases monotonically with decreasing distance $|x - x_i|$. Hereby, x_i denotes the position of the interfacial borderline, which was identified using an EDX map of N showing a clear signal change [88]. Within the range $175\,\mu m < |x - x_i| < 230\,\mu m$, the modulus varies only slightly and a deviation from the monotonic trend seems to exist.

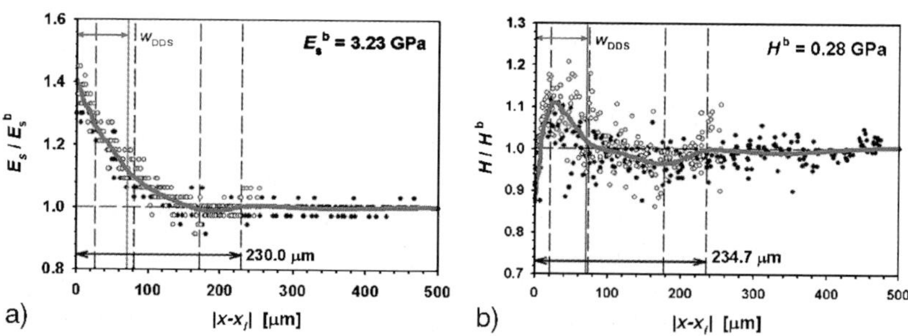

Fig. 28.10. Results of the nanoindentation study performed on the cross section of an epoxy/PVP/epoxy sandwich sample. A Berkovich indenter was used. The total size of the indentation field was $170 \times 810\,\mu m^2$. (a) Plot of the Young's modulus versus the distance from the interfacial borderline. (b) Respective plot of the hardness. Data measured both on the left-sided IP (*open circles*) and the right-sided IP (*full circle*) are given. The *solid lines* resulted from a smoothing procedure. The data are normalized with respect to the bulk values, E_s^b and H^b, respectively. The quantity w_{DDS} denotes the width of the amine depletion zone, as deduced from EDX measurements. Adapted from [115]

Considering the concentration-modulus characteristic given in Fig. 28.8, it can be inferred that within the IP amine is depleted (regime $r < 1$). This correlation is consistent with the amine concentration analysis by means of EDX [88]. This revealed a widely extended amine concentration decay, although its width of $w_{DDS} \sim 71\,\mu m$ is much lower than that of the modulus profile. Thus, additional long-range effects have to be taken into account [88]. Furthermore, it should be noted that across the PVP layer a nonzero and constant amine concentration was detected. Hence, it can be concluded that amine was absorbed by the thermoplastic layer.

By contrast to a metal filler surface that basically represents an impenetrable barrier (although a porous surface layer may exist), the thermoplastic enables interdiffusion. The degree of interdiffusion can be expected to increase with the mobility of the amine molecules or related species as well as with the mobility of the host molecules, i.e. the PVP molecules. The interdiffusion was enhanced by the fact that the curing temperature of $\sim 170\,°C$ was close to the glass transition temperature, T_g, of the PVP. The number average molecular weight of the PVP used was $\sim 360,000$ [116]. Furthermore, the interdiffusion process is likely to be driven by strong interaction forces with the carbonyl group contained in PVP, which is characterized by a strong polarity. In their EDX study, Oyama et al. [111, 112] also observed amine-concentration variations extended over tens of micrometers.

Similar to the profile of the Young's modulus, the hardness profile shows variations extended over $\sim 235\,\mu m$ (Fig. 28.10b). However, within this distance range, the hardness profile exhibits pronounced slope changes. This finding indicates the existence of different characteristic IP zones, including a comparatively narrow zone next to the interface (width $\sim 21\,\mu m$), and three further zones that are $\sim 53, 103$ and $58\,\mu m$ wide, respectively. Interestingly, the total width of the two nearest zones is similar to the width of the amine depletion zone, w_{DDS}. Furthermore, inspection of the residual imprints reveals that the imprints located within the zone next to the interfacial borderline are ~ 1.3 times wider than those located in the other zones [115]. The presence of various IP zones indicates the existence of several effects. For instance, in addition to the local depletion in the amine curing agent, other effects are likely to be superimposed, such as PVP molecules that diffused into the epoxy network and increased its plasticity.

Although the details of these complex interactions are not yet fully understood, the existing data demonstrate that within the IP the epoxy mechanical properties can vary in a complex manner. Application of concentration–property relationships (as deduced from macroscopic testing) in combination with results from chemical mapping (provided a sufficient spatial resolution) may indicate in what way the local mechanical properties can be affected. However, insight into the complex interplay of several effects requires localized mechanical analysis. Thus, nanomechanical studies can deliver not only input data for calculations of interfacial stress distributions, but may also provide valuable information for developing a deeper insight into interfacial interactions, which in turn is a precondition for well-aimed tuning of IP properties.

28.4.2.3
Moisture-Induced Aging of the Epoxy Interphase with a Glass Fiber

A key aspect of the long-term reliability of composite materials is the degree to which their performance is affected by moisture. Depending on the application, the moisture may result either from the humidity of the surrounding gas or from immersion in water. Essentially, the resistance of the interface to water penetration is governed by the interface bond strength. As a result of weak bonds, a larger fraction of free volume or the presence of polar groups, in the vicinity of interfaces the diffusion of water tends to be enhanced. Possibly, the amount of water absorbed by a composite material is larger than the amount of water absorbed by the same amount of matrix material. The uptake of water by the composite can result in swelling, plasticization, or chain scission by hydrolysis. Thus, sorption of water should be avoided, for instance by application of coatings acting as penetration barriers.

Hodzic et al. [117–119] applied nanoindentation and nanoscratching in order to study the changes of the IP properties upon water-induced aging. Three different systems of PMCs were investigated, where in each case the filler component was glass fibers sized with a silane coupling agent. The different polymers used for the matrix were a polyester and two different sorts of phenolic resins. The samples were stored in water at 20 °C, with durations ranging from 3 to 10 weeks. The samples were investigated before and after the exposure to water.

In short, the nanoindentation test proved particularly useful for the detection of significant changes in the material properties, whereas the nanoscratch test also revealed those parts of the IP that were slightly different from the matrix [118]. The force–penetration curves measured on polyester matrix, the transition region, and glass can be readily distinguished from each other (Fig. 28.11). Furthermore, after 10 weeks of aging in water, on glass as well as on the transition region the load required for ∼ 33 nm of penetration was strongly reduced.

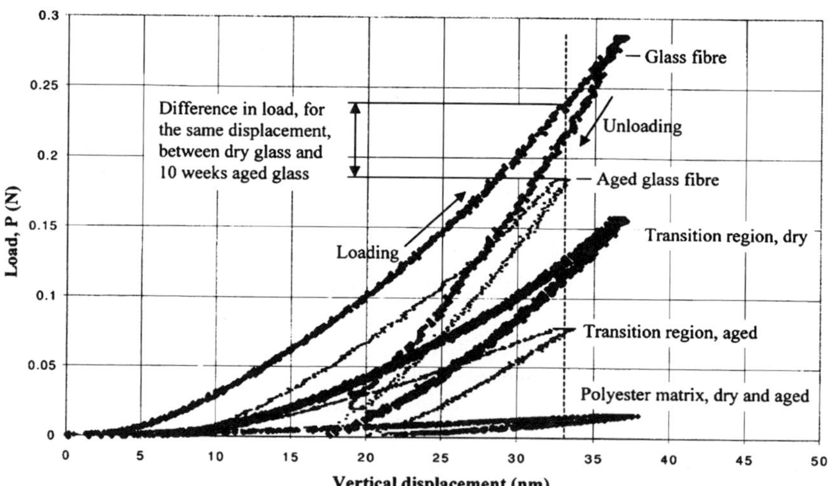

Fig. 28.11. Force-displacement curves as measured on various characteristic regions of a polyester/glass system, before and after 10 weeks of aging in water. Both the glass fiber and the IP are greatly affected by water degradation. Reprinted from Micron 32; Hodzic A, Kalyanasundaram S, Kim JK, Lowe AE, Stachurski ZH: Application of nanoindentation, nanoscratch and single-fiber tests in investigation of interphases in composite materials. 765–775, Copyright 2001, with kind permission from Elsevier

In particular, for the phenolic resin Resinox CL1880 a significant increase in the IP width was found after 10 weeks of water storage. Resinox CL1880 was developed as a fire-resistant material with good mechanical properties for marine and other structures [119]. According to the nanohardness tests, after 10 weeks of water-induced aging, the IP width increased from $\sim 0.8\,\mu m$ to $\sim 5.7\,\mu m$, i.e. by a factor of ~ 7.1 [117, 119]. Both an expansion of the IP region and a decrease in hardness were observed. It was concluded that water dissolved the physisorbed portion of the silane multilayer, and that the unreacted silane chains diffused deeper into the matrix polymer [117].

The observations from the local mechanical analysis of the IP were correlated with results from experiments on the fracture behavior. In particular, the fracture toughness was measured using the double cantilever beam (DCB) testing in mode I configuration [119]. With increasing duration of water storage, these tests showed decreases in the initial value of the strain energy release rate, G_{Ic}, as well as in the so-called R-curve, a plot of G_{Ic} versus crack length. Furthermore, for the system Resinox CL1880/glass-fibers, fractography revealed multiple debonding, with crack propagation predominantly driven through the IP region.

In addition to nanoindentation, Griswold et al. [120] also used phase imaging in intermittent contact mode in order to detect IP properties variations. They studied an epoxy/glass-fiber model composite as a function of the sizing thickness of the γ-aminopropyl trimethoxysilane (γ-APS) coupling agent. For adsorption from an aqueous solution of 5 wt %, an IP thickness of ~ 888 nm was found from the phase images. Within the tested concentration range between 0.1 and 5.0 wt %, the IP thickness was found to increase. From the observed phase shift, it was concluded that within the IP the stiffness is lower than on the matrix. Indeed, in the literature it was reported that reactions between the organo-functional portion of the silane molecule and the matrix can occur, see [120] and the references therein. Thus, in the vicinity of the fibers the epoxy curing may be inhibited by the silane sizing.

In contradiction to phase imaging, however, for some samples nanoindentation experiments indicated an apparent IP stiffness higher than that of the matrix. This was the case for sizings prepared from solutions of γ-APS concentrations lower than 3 wt % with corresponding IPs less wide than ~ 375 nm. The difference between the results from phase imaging and nanoindentation was attributed to the mechanical bias effect (MBE).

28.5
Pressure-Sensitive Adhesives (PSAs)

28.5.1
A Brief Introduction to PSAs

A pressure-sensitive adhesive (PSA) is a viscoelastic material that in solvent-free form remains permanently sticky [121]. Provided a slight pressure is exerted, without any chemical reaction such material will adhere instantaneously to many solid surfaces. No activation by heat or solvent is required, which makes PSAs particularly easy to use [41, 122]. Ideally, PSAs will resist detachment with clean removal [123].

Owing to these properties, PSAs represent a widespread class of adhesives, with applications such as self-adhesive tapes, labels, or foils [122].

The property of an adhesive that enables it to form a bond of measurable strength immediately after being brought into contact with the adherend is referred to as tack [121]. Thus, as a sort of "instantaneous" adhesion or quick stick, it differs from ultimate strength, the development of which requires a longer time [124]. For quantitative analysis, tack is measured as the impulse per unit area necessary to separate two planes connected through a viscoelastic layer. Macroscopic tests of adhesive performance showed that the efficiency of the bonding process increases with the flowability of the adhesive and with the contact time [125]. During the debonding process, however, a major factor is the debonding velocity. For instance, peel strength increases with the peeling velocity, since the viscoelastic losses occurring within the adhesive layer are controlled by the deformation rate. Essentially, PSAs obtain their special properties from the adhesion hysteresis, i.e. the difference between the energy gained in forming the interactions and the energy dissipated during the fracture of these bonds [122]. In addition to a strong viscous component in their viscoelastic deformation behavior, a basic requirement for PSAs is softness, in order to allow establishment of molecular contacts even on rough surfaces.

The majority of commercially available PSAs can be classified into three different groups, depending on the types of polymers used basically for the PSA formulation. These groups are natural rubbers, acrylates, and styrenic block copolymers [122].

PSAs based on natural rubbers are blends of unvulcanized elastomers and the proper concentration of low molecular weight resins [122, 126]. The latter are added to enhance the tackiness of the PSA, i.e. to act as tackifiers. Commonly, tackifiers have molecular weights ranging from ~ 200 to ~ 1500 and structures that are large and rigid [124]. From the mechanical point of view, the effect of adding a tackifier is a decrease in the average entanglement density, leading to a decrease in the modulus in the plateau region, i.e. the region separating the region of glassy behavior from the terminal flow region [122]. In particular, as a result of the reduced average entanglement density, the flowability of the adhesive is enhanced and it can more easily deform to wet the surface.

Similar to elastomeric PSAs, the different components of acrylic PSAs are utilized to fine tune the mechanical properties. Typically, acrylic PSAs are random copolymers of a long side-chain acrylic, a short side-chain acrylic, and an acrylic acid [122]. The latter two components are used to adjust T_g and the elongational properties, respectively. If, resulting from emulsion polymerization, acrylic PSAs are applied in the form of latex spheres with diameters ranging from $\sim 50\,\mathrm{nm}$ to several micrometers. In order to comply with environmental requirements, nowadays water is preferred as the liquid medium carrying the latex particles.

Formation of a clear and homogeneous phase requires particle coalescence, i.e. disappearance of the particle interfaces. Such interparticle fusion is achieved by particle deformation and chain interdiffusion [41]. A necessity for such processes is a sufficient segment mobility that is provided only in the rubbery phase, i.e. at temperatures well above T_g. However, formation of a completely homogeneous film depends on a number of factors of which the T_g is only one. Also, the evaporation rate of the aqueous solvent can be decisive [127]. In particular, the initial stages of film formation are associated with loss of solvent. Subsequent drying leads to collapse of the voided structure initially formed to create a densely packed structure [127]. Probably as a result of the reduced surface area caused by extensive particle deformation, evaporation rates were found to be retarded for temperatures well above the latex T_g. In addition to the interplay of the various film-formation processes, complete coalescence can be prevented by the surface energetics of the particle membranes. For instance, the particle membranes can be stabilized by enrichment of processing additives at the particle surface [128]. In particular, an increase in the T_g of the particle membrane was stated as a result of surfactant-related hydrogen bonds [127].

The tackiness of PSAs was shown to be strongly associated with the PSA's ability to form fibrils upon debonding [74, 122, 129]. In this way, the separation process involves significant plastic flow and a large degree of energy dissipation. The latter can be measured as the hysteresis area between approach and retraction of a probe such as a flat punch or the SFM tip. The long tails characteristic of tack curves measured on PSAs are related to the process of fibrillation [74, 122, 130].

28.5.2
Heterogeneities of an Elastomer–Tackifier PSA as Studied by Means of M-LFM

Tack adhesion is not well understood at the molecular level [126]. While the adhesive performance is dependent on the viscoelastic properties of the bulk, those of the surface layer appear to be very important also. Possible effects altering the adhesive-film microstructure are bulk-phase separation or surface segregation [131]. In particular, due to segregation effects the composition of the surface regions may differ from that of the bulk. For instance, tackifier enrichment was found on the PSA surface [131, 132]. Furthermore, the adhesive properties may be altered by aging or effects related to humidity.

A detailed surface mechanical characterization of a typical PSA system based on a natural rubber was performed by Foster et al. [62, 123, 125, 131–133] using SFM. In addition to nanoindentation, LFM and M-LFM were shown to provide useful information. The study was focused on poly(ethylenepropylene) (PEP) filled with various amounts of the tackifier n-butyl ester of abietic acid (nBEAA). The latter belongs to the family of wood resin derivatives widely used in commercial tackifiers [123]. The 50–70 μm thick adhesive films were prepared by solution casting on microscope slides. For the system containing 80 wt % nBEAA (in short, PEP/80), a two-phase morphology was found. From the LFM images revealing a higher friction on the domains, it was inferred that the domains are enriched in tackifier [123]. On the contrary, for the system containing 60 wt % tackifier (in short, PEP/60) a homogeneous surface was found, thus indicating the presence of a surface layer [123].

The effects of the stick–slip transition on the M-LFM amplitude were demonstrated for a SiO_2 surface as well as for a PEP–nBEAA sample [62]. The former was used as a reference sample exhibiting purely elastic tip–sample interactions. In the sticking regime, the force amplitude is sinusoidal, similar to the excitation signal (Fig. 28.12a). In the slip regime, however, truncation of the sinusoidal waveform is observed (Fig. 28.12b). Furthermore, the plateau value decreases with increasing excitation amplitude. Thus, the stick–slip transition can be readily observed from the corresponding change in the waveform of the response signal. In Fig. 28.12c,d, the lateral force response curves are given for the matrix and the domain, respectively,

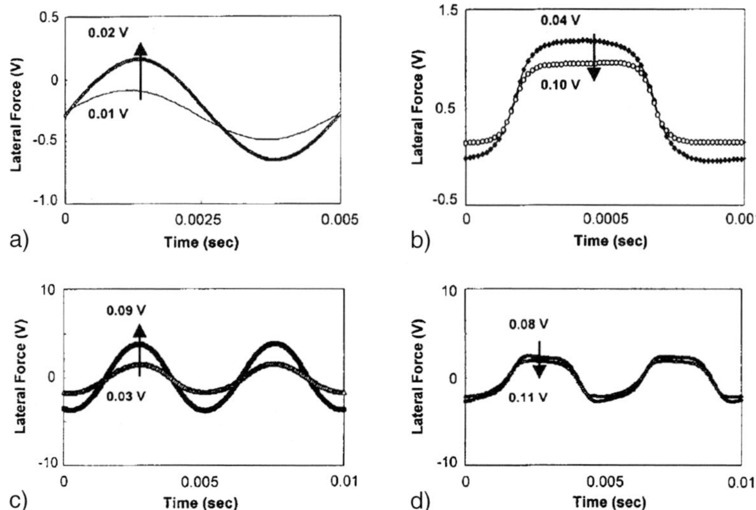

Fig. 28.12. Characteristics of stick–slip motion. The M-LFM signal as a function of time for various cases: (**a**) Silicon oxide: sticking regime (200 Hz, 1 nN), (**b**) Silicon oxide: slip regime (1 kHz, 3 nN), (**c**) the matrix of PEP with 80 wt % nBEAA (PEP/80; 200 Hz, 1 nN), (**d**) domain of PEP with 80 wt % nBEAA (PEP/60; 200 Hz, 1 nN). Notably, in the slip regime the M-LFM amplitude decreases with increasing excitation amplitude. Reprinted with permission from Langmuir 2002, 18, 1865–1871. Copyright 2002 American Chemical Society

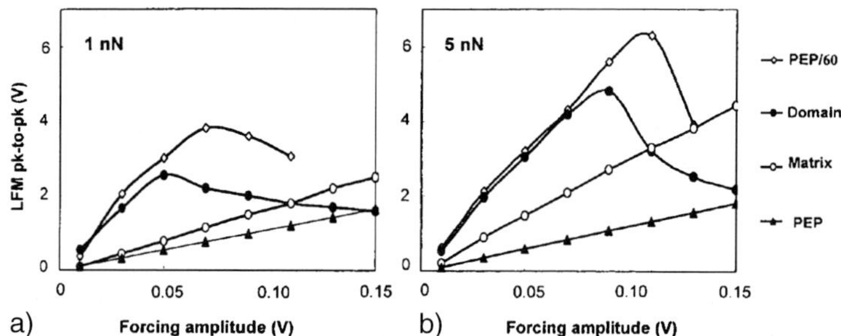

Fig. 28.13. Variation of the M-LFM amplitude with excitation amplitude, as measured on the surface of PEP with 60 wt % of nBEAA (PEP/60), PEP with 80 wt % of nBEAA (PEP/80), and PEP. Modulation frequency 200 Hz. The measurements were performed for two different normal loads, namely 1 nN (**a**) and 5 nN (**b**). Reprinted with permission from Langmuir 2002, 18, 1865–1871. Copyright 2002 American Chemical Society

for a PEP/80 sample. Although the excitation conditions were the same on the matrix and the domain, a stick–slip transition was observed on the domain only. Since the domain rich in tackifier is highly adhesive, the occurrence of the stick–slip transition indicates a comparatively small contact area, i.e. a high stiffness.

The variation of the M-LFM force amplitude with the amplitude of the exciting signal is given in greater detail in Fig. 28.13. From force–distance curve experiments, regions rich in tackifier are known to be much stiffer than pure PEP [123, 131].

Accordingly, the initial slope of the respective curve is significantly larger than that of the matrix rich in PEP. For the curves measured on the latter as well as on pure PEP, no stick–slip transition is observed, and the response amplitude increases with the excitation amplitude in a monotonic manner. Over the complete sticking regime the response amplitude is significantly lower than on the sample PEP/60 and on the domains of the PEP/80 sample. Furthermore, as is visible from the respective curves, above the stick–slip transition the response amplitude decreases with increasing excitation amplitude.

Similar results were found for the higher load of 5 nN (Fig. 28.13b). The response amplitudes, however, are generally increased and the stick–slip transitions occur at larger driving amplitudes.

Notably, the M-LFM amplitude is higher on the matrix of PEP/80 than on pure PEP, thus indicating that the matrix contains a non-negligible amount of tackifier [62]. As compared to the bulk, on the surface the tackifier concentration of the matrix is expected to be increased. Similarly, the surface of the PEP sample containing 60 wt % tackifier seems to be uniformly enriched in tackifier. As a result, it shows a stick–slip transition similar to that of the tackifier-rich domains on the PEP sample containing 80 wt % tackifier.

Apparently, a contradiction exists between the lateral forces measured in conventional LFM mode and in M-LFM mode: In LFM, the lateral forces are higher in the matrix regions, whereas in the M-LFM mode, the lateral force amplitude is greater on the domains rich in tackifier. Moon et al. [62] argued that due to the higher stickiness and the higher stiffness of the domains, in LFM mode the lateral forces are dominated by the deformation behavior, whereas in the M-LFM mode they are dominated by the adhesive forces. However, the paradoxical results may express just the different nature of LFM and M-LFM. In LFM, the tip is scanned across the surface. Thus, basically the origin of the lateral forces recorded is a sliding motion. This can be described by an average velocity, namely the velocity of scanning, although a closer view of the lateral force changes over time may reveal a series of minute stick–slip transitions [134]. As a sliding friction experiment, LFM delivers in the first instance information on the frictional properties of the tip–sample contact, similar to the sliding regime of the M-LFM experiment. Generally speaking, friction forces encompass contributions from elastic or plastic deformations as well as adhesive interactions. In the sticking regime of an M-LFM experiment, however, the shear stiffness of the tip–sample contact is measured rather than a frictional quantity [55, 135].

Since a stick–slip transition was observed only for the PEP/60 surface and the domains of the PEP/80 surface, within the given range of forcing amplitudes the primary nature of the M-LFM experiments was of the sticking one, i.e. the outcome is mainly given by the shear stiffness, and the different force amplitudes reflect the different stiffness values. Taking the much higher stickiness of the tackifier-rich regions for granted, even the occurrence of the stick–slip transitions can be explained in the light of stiffness. Since the total adhesion force is given by the product of contact area specific adhesion force, an increased specific adhesion force can be overcompensated by a decreased contact area (which is due to a larger stiffness). Thus, on the regions rich in tackifier the stick–slip transition is observed, although the specific adhesion force is significantly larger than on pure PEP.

As a result, measurements of the M-LFM amplitude as a function of the excitation amplitude, such as shown in Fig. 28.13, deliver multiple information, including the shear stiffness of the tip–sample contact. Furthermore, as noted also by Moon and Foster [62], a variation in the excitation frequency provides the opportunity to study the viscoelastic behavior of the adhesive. With increasing frequency, the surface mechanical behavior increasingly is of elastic rather than of viscous character, in particular, such changes are reflected by the M-LFM phase signal.

Beyond the investigation of the adhesive morphology and the corresponding local variations in its mechanical properties, M-LFM was also employed for studying the effects of water on the mechanics of the tip–sample contact. Under ambient conditions, water has to be taken into account for the largest part of adherends, in particular when dealing with micro- and nanoscale contacts, where water is attracted due to capillary interactions. Moreover, frictional forces were observed to change with humidity in a complex manner that does not follow a simple monotonic trend.

As reported by Piner and Mirkin [136], the effect of the water meniscus on LFM measurements can be considerable. Transport of water between tip and sample affects the thickness of the water film. Beyond the humidity and the chemical nature of the sample surface, also the time available for water migration was recognized as a relevant source of variations in lateral force. On hydrophilic substrates, the surface water film can act as a boundary lubricant, resulting in a decrease in the friction forces [134]. In the limit of low humidities, however, the capillary force can be the governing parameter leading to an increase in the adhesion force [137–139] and in the friction force [138, 140] with rising humidity.

As is visible from Fig. 28.14, also for M-LFM measurements on the PSA system PEP/60 a strong dependence on relative humidity is observed. For RH = 45%, the M-LFM amplitude increases over the full range of excitation amplitudes that shows that the stick–slip transition is not reached. For RH = 21%, however, the stick–slip transition occurs for a forcing amplitude of $\sim$ 1.13 V, as is obvious from

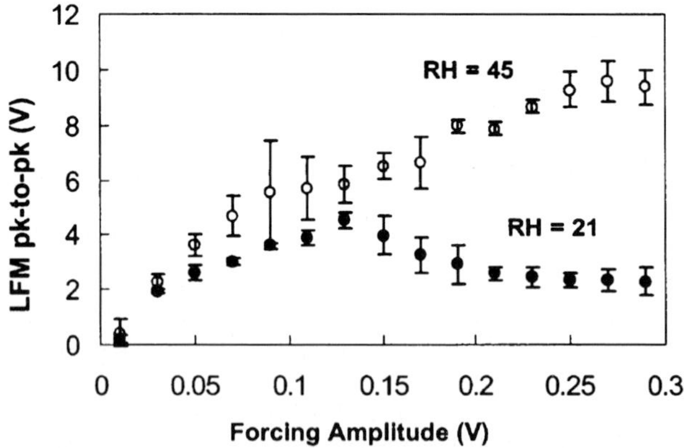

Fig. 28.14. Variation of the M-LFM amplitude with excitation amplitude, as measured on the surface of PEP with 60 wt % of nBEAA (PEP/60). Modulation frequency 200 Hz. The measurements were done for two different relative humidities, RH, namely 21% and 45%. Reprinted with permission from Langmuir 2002, 18, 8108–8115. Copyright 2002 American Chemical Society

the decrease in the M-LFM-amplitude for higher forcing amplitudes. Thus, for the higher relative humidity, RH = 45%, the tip–sample interaction is stronger and the stick–slip transition is prevented.

These findings are indicative of hydrophilicity that can result from a surface enrichment of tackifier molecules [132]. Indeed, in the case of pure PEP a completely different behavior was observed. Within the given range of excitation amplitudes no stick–slip transition was detected, but the M-LFM amplitude decreased with increasing humidity [132]. As a result of the hydrophobic character of PEP, the water is more likely to act as a lubricant than to enhance the tip–sample interaction.

To develop a deeper understanding of the effect of normal load on the M-LFM measurement, the M-LFM amplitude and phase were measured in parallel with force curves [125]. The corresponding curves are given in Fig. 28.15. From the force–distance curves a pronounced difference in the adhesive behavior is visible between the measurements on pure PEP and PEP/60, see Fig. 28.15a,b. On PEP/60, the

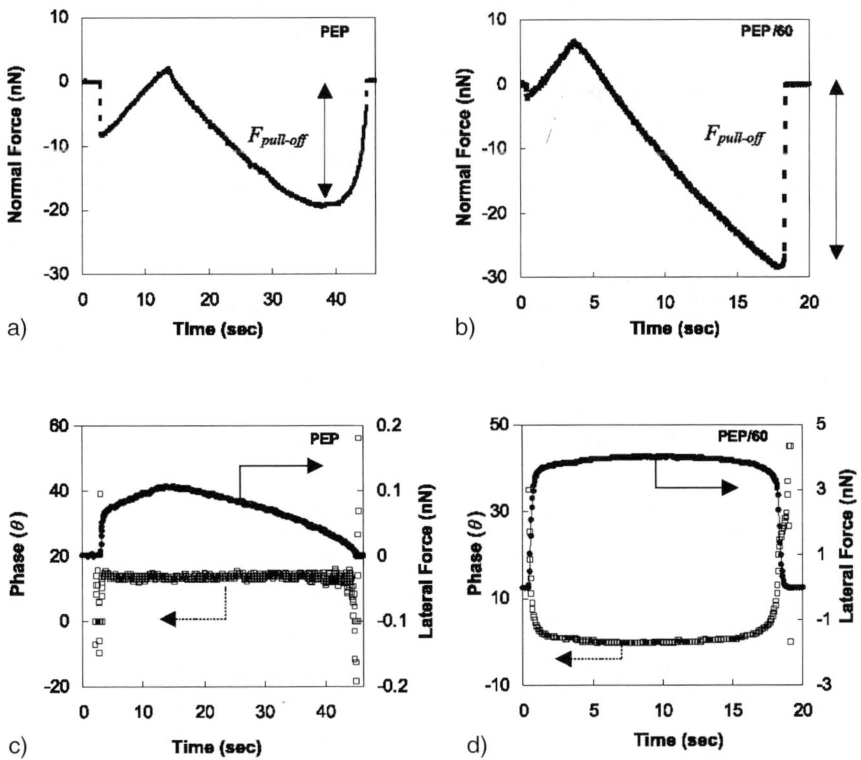

Fig. 28.15. Indentation experiments with superimposed M-LFM: Variations of the normal load (**a**),(**b**), the M-LFM amplitude (**c**),(**d**), and of the M-LFM phase (**c**),(**d**). The measurements were made on PEP (**a**),(**c**) and on PEP/60 (**b**),(**d**), respectively. Note the different force scales in (**c**) and (**d**). Reprinted from Polymer 45, S. Moon, S. Swearingen, M.D. Foster, Scanning probe microscopy study of dynamic adhesion behavior of polymer adhesive blends, 5951–5959, Copyright 2004, with permission from Elsevier

pull-off force is 1.5–2.0 times larger than on PEP. The M-LFM amplitude measured on PEP/60 is even ~ 30 times larger than on PEP (Fig. 28.15c,d, note the different force scales). Although the M-LFM amplitude depends on both the shear stiffness and the adhesion force, changes of the latter appear to show a particularly strong impact.

A deeper understanding of the relationship between adhesion force and M-LFM response can be achieved by a separate consideration of the bonding and the debonding processes [125]. Both processes are governed by the deformation rates. Since firm bonding requires wetting of the adherend surface first, the bonding process benefits from increasing contact time and flowability of the adhesive. Concerning the debonding process, the peel strength is affected by viscoelastic losses, i.e. it is rate dependent.

Since the energy loss upon debonding was shown to increase with increasing velocity of unloading (Sect. 28.2.2), the time behavior of the bonding process follows a trend opposite to the time behavior of the debonding process. The results of a respective analysis [125] are given in Fig. 28.16.

For both PEP and PEP/60, the M-LFM amplitude decreases with increasing velocity. Concerning the pull-off force, however, the two materials differ in their behavior. For PEP and the sampled range of velocities, $F_{\text{pull-off}}$ increases with velocity, whereas for PEP/60 a minimum is observed. It can be inferred that PEP/60 is affected by two competing mechanisms, namely contact area formation and viscoelastic losses. Below $\sim 50\,\text{nm/s}$ its behavior is dominated by the contact formation, and above this velocity it is dominated by the energy dissipation related

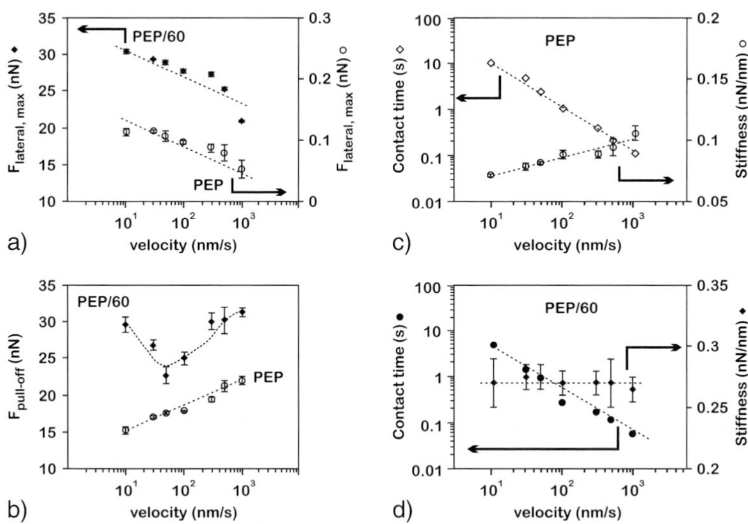

Fig. 28.16. The maximum lateral force (**a**), the pull-off force (**b**), the contact time (**c**),(**d**), and the stiffness (**c**),(**d**) as a function of the loading velocity. The measurements were performed on PEP (*open circles*) and on PEP/60 (*solid diamonds*). Reprinted from Polymer 45, S. Moon, S. Swearingen, M.D. Foster, Scanning probe microscopy study of dynamic adhesion behavior of polymer adhesive blends, 5951–5959, Copyright 2004, with permission from Elsevier

to the deformation processes occurring upon debonding. The fact that the pull-off behavior of PEP is not markedly affected by the contact time, may result from both the lower tackiness as well as the lower stiffness. The contact area achievable with the limited tackiness is lower and it is established faster owing to the stronger deformability. As given in Fig. 28.16c,d, for PEP/60 the stiffness has a constant value of $\sim 0.27\,\mathrm{nN/nm}$, whereas the PEP stiffness increases slightly with velocity, but is lower than $\sim 0.1\,\mathrm{nN/nm}$. The higher stiffness of PEP/60 was attributed to crosslinking of tackifier with aging [125].

28.5.3
The Particle Coalescence Behavior of an Acrylic PSA as Studied by Means of Intermittent Contact Mode

The above SFM-based studies on PSAs were performed in contact mode, thus allowing for the characterization of shear and frictional properties. Due to the high tackiness of freshly prepared films, however, stable imaging was impossible [131]. With the aim to extend the investigation of their PSA system to nonaged samples, Mallégol et al. [41] employed the intermittent contact mode. Owing to the limited tip–sample contact time, lateral forces can be distinctively reduced; a benefit that is frequently taken advantage of when investigating delicate samples, such as highly compliant surfaces or nanoscale objects only gently adsorbed onto the substrate.

However, stable imaging and reliable contrasts require a careful choice of the imaging parameters, namely the amplitude of the freely oscillating cantilever, A_0, the amplitude ratio between the interacting and the free cantilever, $r_{\mathrm{sp}} = A_{\mathrm{sp}}/A_0$, and the cantilever stiffness, k_{c}. In the particular case of a soft and tacky PSA surface, contradictory requirements must be considered [41]. Because of the softness, only light tapping is acceptable, i.e. slight damping of the oscillation amplitude, A_{sp}. Because of the tackiness, however, sufficient mechanical energy needs to be provided in order to avoid catastrophic sticking of the tip to the surface. Since the potential energy of the cantilever increases with its stiffness, k_{c}, as well as with the square of its oscillation amplitude, a stiff cantilever should be excited in a relatively strong manner.

Indeed, it was shown that imaging of highly tacky PSA surfaces is possible with this approach [41]. In order to take into account the comparatively large indentation depths, z_{ind}, due to the softness of the PSA, an alternative definition of r_{sp} was used. With the average tip–surface distance, $d_{\mathrm{sp}} = A_{\mathrm{sp}} - z_{\mathrm{ind}}$, the setpoint ratio was redefined as $\widetilde{r}_{\mathrm{sp}} = d_{\mathrm{sp}}/A_0$. Thus, $z_{\mathrm{ind}} = A_0(1 - \widetilde{r}_{\mathrm{sp}})$, which shows that the indentation depth increases with decreasing values of $\widetilde{r}_{\mathrm{sp}}$. Imaging artefacts related to indentation were diminished by setting $\widetilde{r}_{\mathrm{sp}}$ as close as possible to the value one, i.e. by operating distinctively in the soft tapping regime. The hardness of tapping was demonstrated to significantly affect the image quality by altering the measured shape of the latex particles. As a consequence, it is recommended to perform scanning under various tapping conditions, in order to ensure the accuracy of the measured topography. Reducing indentation-related imaging artefacts by restricting measurements to soft cantilevers was shown to be of limited value, since the latex film surface can be partly covered with a layer containing species of low molecular weight, such as exudated surfactants. Phase images recorded in intermittent contact mode exhibited a strong contrast between the particles and a surrounding medium filling out the interstices [41, 128]. From the distinctly different mechanical properties a liquid-like behavior was attributed to the interstitial medium, smoothing out the particle contours.

"Looking through" such top layers in order to achieve a sort of subsurface imaging requires stiff cantilevers. With increasing indentation depth, however, a distortion of the particle shape was observed. Thus, an increase in the free amplitude, A_0, needs to be counterbalanced by an increase in $\tilde{r}_{sp}$. For values of $\tilde{r}_{sp}$ very close to unity, however, the feedback loop was reported to be unstable and the surface contact was easily lost [41]. After all, in a similar manner as to the surface properties the tapping conditions have to be adjusted to the information aimed for.

A well-aimed optimization of the tapping parameters can be achieved by inspection of respective amplitude–distance curves. An example is given in Fig. 28.17 [41]. The cantilever stiffness, k_c, and the free amplitude, A_0, were 48 N/m and 163 nm, respectively. The measured cantilever oscillation amplitude decreases with decreasing distance from the sample surface. For the purpose of reference, the straight solid line gives the amplitude as measured on the stiff surface of a Si wafer.

Assuming a negligible deformation in the contact between tip and Si wafer, the vertical distance between both curves gives the depth of indentation, z_{ind}, into the soft PSA film. At the point I, z_{ind} has its maximum value. It can be considered as the demarcation between the regime of larger distances where the amplitude decreases

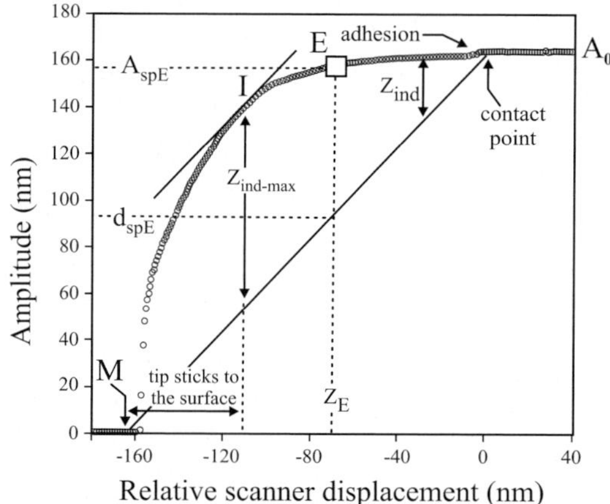

Fig. 28.17. Amplitude–distance curve recorded for optimizing the intermittent contact mode operation on a PSA. The investigated PSA was a waterborne acrylic latex. The free amplitude of cantilever oscillation was $A_0 = 163$ nm. The cantilever spring constant was $k_c \sim 48$ N/m. When approaching the sample and tip, the cantilever oscillation amplitude decreases. The corresponding gradient depends on the sample stiffness and stickiness. In the case of negligible deformations of the tip–sample contact, a linear curve is observed (*solid line*), i.e. the scanner displacement is fully translated in a decrease in the amplitude. Point E represents the situation of a feedback with $d_{sp} = 93$ nm. Point I is the point of maximum indentation depth, $z_{ind-max}$. At point M, zero amplitude is observed on a stiff sample. On the PSA, however, the amplitude dropped to zero before point M was reached, due to the stickiness of the surface. Reprinted with permission from Langmuir 2001, 17, 7022–7031. Copyright 2001 American Chemical Society

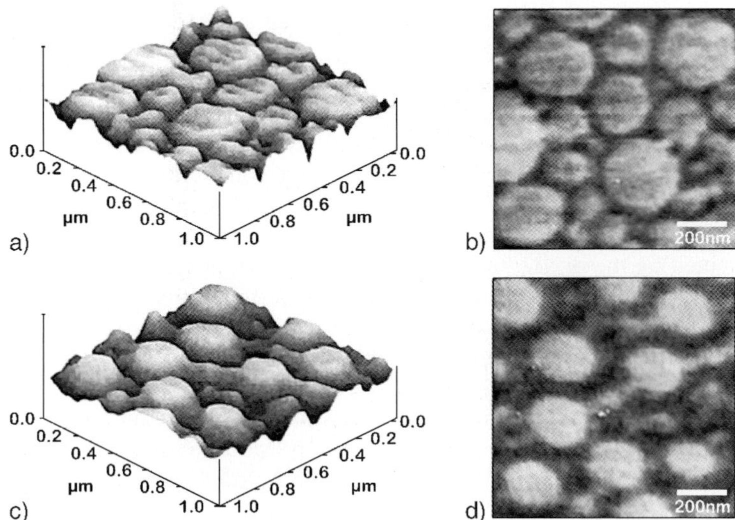

Fig. 28.18. Topography images of a PSA as measured in intermittent contact mode. In addition to the *gray scale* in (**a**) and (**c**), the images are given in pseudo-3D representation (slice view) also. As compared to the images given in (**a**) and (**b**), the images in (**c**) and (**d**) show slightly smaller particles that are more circular and less flat. These differences show the effect of variations in the setpoint ratio, $\tilde{r}_{\mathrm{sp}}$, which was 0.85 and 0.64, respectively. In both cases, the free amplitude was $A_0 = 135$ nm. Correspondingly, the indentation depths were $z_{\mathrm{ind}} = 18$ nm and 44 nm, respectively. The cantilever spring constant was $k_{\mathrm{c}} = 48$ N/m. Reprinted with permission from Langmuir 2001, 17, 7022–7031. Copyright 2001 American Chemical Society

only slowly because of the softness of the PSA, and the regime of smaller distances where the amplitude decreases rapidly because of the stickiness of the PSA. Notably, due to sticking of the tip to the sample, the amplitude drops to zero at a distance slightly larger than the one expected from the reference curve (point M).

An example of the changes in the measured topography with tapping parameters is given in Fig. 28.18.

A striking difference can be stated for the topography images given in Fig. 28.18a,b and 28.18c,d. In the former images, the appearance of the latex particles is flatter and more irregular. In the latter images, the particle shape has a spherical character and the particle size seems to be smaller. The indentation depths of the two measurements were 18 and 44 nm, respectively, as calculated from the free amplitude and the respective setpoint ratios. It can be inferred that in the case of the higher indentation depth, the particles and in particular their edges were deformed in a way that the particle curvature appears to be increased. Indeed, from the onset of particle coalescence, some degree of flattening of the initially spherical particles can be expected. Indeed, for the height images measured with tapping parameters corresponding to a larger indentation depth, a higher roughness was found [41]. Essentially, under the conditions of gentle tapping a particle shape is detected that resembles a cylinder more than a sphere. With $T_{\mathrm{g}} \sim -30\,^{\circ}\mathrm{C}$, at RT the mechanical behavior of the latex spheres should exhibit a strongly viscous character, thus allowing for particle deformation and interdiffusion [41].

Essential prerequisites for film formation are particle coalescence and interdiffusion, i.e. chain migration between adjacent particles. However, these processes can be inhibited by the presence of exudated surfactants and possibly other water-soluble species. By trapping between particle/particle boundaries via bond formation, their enrichment at the particle surface can stabilize the particles against coalescence. Segregation of the surfactant is driven by a poor compatibility with the polymer and by lowering the interfacial energy of the polymer/air and polymer/substrate interfaces [141]. The distribution of surfactants in waterborne films was also shown to be affected by the drying process, since the surfactants can be carried with the water flux to the film surface and build up around the particles [128]. As a result, a top layer of $\sim 60\,\mathrm{nm}$ thickness was formed with surfactant in excess and a reduced degree of coalescence. In a combined approach using SFM, Rutherford backscattering spectroscopy (RBS), and magnetic resonance profiling, Mallégol et al. [128] characterized the water and surfactant concentration profiles. Interestingly, rinsing with water led to a significantly reduced surfactant concentration in the top layer and a pronounced morphology change where individual particle contours were no longer observed. Furthermore, capillary pressure was identified as the essential driving force for transporting the surfactants into particle boundaries.

28.5.4
Evidence for the Fibrillation Ability of an Acrylic PSA from the Analysis of the Noise PSD

Benmouna et al. [74,75] performed a local analysis of the viscoelastic properties of acrylic PSA films by analyzing the noise PSD. The PSAs investigated are of acrylic nature, with the granular structure maintained even after an extended period of aging. The mean particle diameter of the initial particles was $\sim 300\,\mathrm{nm}$ [75]. The noise PSD was measured for various distances between tip and sample, both for approach and retraction (Fig. 28.19). On approach, some amount of hydrodynamic damping is observed (Fig. 28.19a). As is visible from the shift of resonance frequency, on retraction the effective spring constant is increased (Fig. 28.19b). This is a result of the filament bridging the gap between the tip and the PSA film.

Benmouna et al. [75] generated 2D maps of the PSA viscoelastic properties in the (x,z)-plane. At a given lateral position, the cantilever is approached to the sample surface until a firm contact is established and retracted subsequently. For each step of the force–distance curve, the cantilever noise spectrum is recorded and analyzed online. After each force–distance curve, the lateral position is incremented along the x-axis and another approach–retraction cycle is initiated. The increments were $10\,\mathrm{nm}$ in the z-direction and $40\,\mathrm{nm}$ in the x-direction. The data were measured on a PSA film aged for 6 weeks. As is visible from Fig. 28.20a, the majority of filaments extended over a distance of $\sim 1\,\mu\mathrm{m}$. Interestingly, the resonant frequency of each filament undergoes several stepwise transitions, thus indicating rupture events. These were attributed to rupture of the junctions between individual particles in the film and in the filament [74].

Obviously, the resonance frequency shows less scatter than the bandwidth. Essentially, the bandwidth follows the patterns of the resonance frequency. Thus, the resonance frequency is the preferable quantity. Furthermore, it can be stated that

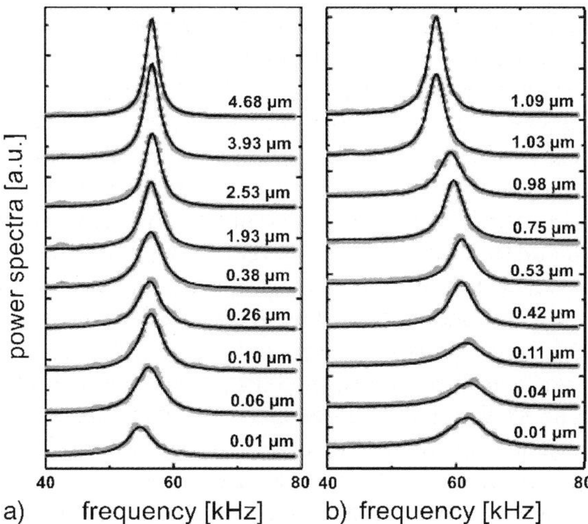

Fig. 28.19. Noise power spectral density (PSD) data and corresponding fits of (28.9). (**a**) As measured on approach and (**b**) on retraction. As calculated from the equipartition theorem (Sect. 28.2.3), the cantilever spring constant was $k_{\mathrm{c}} \sim 0.06\,\mathrm{N/m}$. Resonance frequency and bandwidth of the free cantilever were 56.7 and 2 kHz, respectively. Reprinted with permission from Langmuir 2003, 19, 10247–10253. Copyright 2003 American Chemical Society

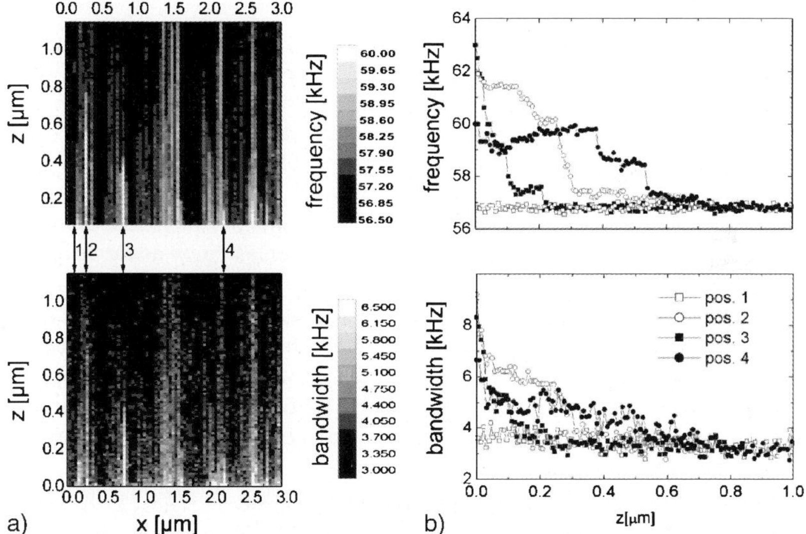

Fig. 28.20. (**a**) *Gray-scale* representation of the filament density along the *x*-axis. The *brighter stripes* correspond to filaments pulled by the SFM tip. The distance curves were recorded upon retraction. *Top*: resonance frequency, *bottom*: resonance bandwidth. (**b**) Resonance frequency (*top*) and bandwidth (*bottom*) for various *x*-positions: 0.08 μm (position 1), 0.28 μm (position 2), 0.80 μm (position 3), and 2.20 μm (position 4). Position 1 exemplifies the case of an unsuccessful pull (no filament). The elongation rate was 1.5 nm/s. Reprinted with permission from Langmuir 2003, 19, 10247–10253. Copyright 2003 American Chemical Society

along the x-axis the filaments are grouped with a period of $\sim 0.4\,\mu m$, which fits to the size of the individual particles constituting the film. Hence, the granular structure of the film is reflected by the spatial distribution of the filaments, indicating the existence of preferential zones for filament formation. However, the question of whether the filaments are generated more likely at the particle centers or at the interstices can be decided only on the basis of 3D maps, i.e. maps providing lateral information in both x- and y-directions. From first experiments with a simplified procedure allowing for acceptable measurement times, it was concluded that the filaments are more likely to occur on the particle centers [75].

Furthermore, the fibrillation ability was demonstrated to decrease with increasing time of aging [75]. Both the number of filaments and their average length were observed to decrease. Since the degree of coalescence increases with time of aging, the fibrillation ability appears to be correlated with the coalescence. Obviously, in the limit of a completely coalescent film with a high degree of interdiffusion, the cohesive forces are strong and fibrils are not very likely to be pulled. Thus, it can be inferred that a high level of cohesion inhibits tackiness. As stated by Dimitrova et al. [74], tack requires a delicate balance between adhesive tip–sample forces and cohesive forces inside the sample. The ability to draw filaments requires flowability of the chains, a property that is prevented by a large number of entanglements or chemical crosslinks. Hence, weak interparticle junctions appear to be beneficial for tack [74].

28.6
Conclusions

The potential of nanoscale characterization for the analysis of issues in adhesion science was demonstrated. Concerning pressure-sensitive adhesives (PSAs), data from modulated LFM (M-LFM) measurements were discussed in combination with results from conventional LFM and force–distance curves. The intermittent contact mode was shown to allow for imaging of surfaces even as tacky as freshly prepared PSAs. Finally, a combination of force–distance curves with analysis of thermomechanical noise was used as a testing method for the fibrillation ability, a feature closely related to tackiness. These examples prove that SFM and related techniques can be a valuable tool for the engineer or scientist who is in charge of optimizing PSAs for particular applications. In combination with well-established characterization techniques, such nanomechanical studies may contribute to the development of structure–property relationships.

Nanoindentation was used for detecting and analyzing stiffness variations across epoxy interphases (IPs). The relevance of IPs in such thermosetting systems becomes obvious from the large IP width and total modulus change across the IP. In the case of the epoxy/thermoplastic system a monotonic modulus decay extended over $\sim 175\,\mu m$ was found, and the total modulus change across the IP was ~ 1.1 GPa. The characteristics of the modulus gradient were shown to comply with a concentration-modulus relationship deduced from a series of epoxy reference samples. For an epoxy/copper composite a nonmonotonic stiffness profile was found, extended over $\sim 2.3\,\mu m$. Based on concentration profiling, these IPs were attributed to imbalances in the amine–epoxy concentration ratio, as induced by preferential adsorption at the interface. However, such chemical effects may be superimposed to some degree with thermal ones, since the reinforcing phase can act as a sink for the heat liberated in the course of the exothermic curing reaction. The relative amount of the thermal effect depends on a number of factors, such as the thermal conductivity, the heat capacity of the filler particles, and the rate of heat liberation.

Concerning the nanoindentation experiment, the mechanics of the tip–sample contact is possibly affected by the interface, either as a result of variations in contact area or of the stress transfer to the adjacent stiff phase (mechanical bias effect). Variations in contact area can be caused either by topography or by variations in normal force (due to the finite time constant of the topography feedback). These effects should be borne in mind when analyzing nanoindentation data measured in close vicinity of an interface. The considered case was the one of the surface under investigation being oriented perpendicularly to the interface. In this configuration the axis of indentation is perpendicular to the main gradient in interphasial property variations, which typically is oriented in the direction normal to the interface. The major benefit of this approach lies in the direct access to the tip-interface distance as a key parameter for IP characterization. In the opposite case of indenting a film, however, additional complications arise due to the axis of indentation being parallel to the main axis of interphasial gradients.

References

1. Sharpe LH (1972) J Adhes 4:51
2. Li M, Carter CB, Hillmyer MA, Gerberich WW (2001) J Mater Res 16:3378
3. Racich JL, Koutsky JA (1977) Boundary layers in thermosets. In: Labana S (ed.) Chemistry and properties of crosslinked polymers. Academic Press, New York, p 303
4. Mijovic J, Koutsky JA (1979) Polymer 20:1095
5. Lüttgert KE, Bonart R (1978) Prog Colloid Polym Sci 64:38
6. Dušek K (1996) Angew Makromol Chem 240:1
7. VanLandingham MR, Eduljee RF, Gillespie JW Jr (1999) J Appl Polym Sci 71:699
8. VanLandingham MR, Villarrubia JS, Guthrie WF, Meyers GF (2001) Macromol Symp 167:15
9. Munz M, Cappella B, Sturm H, Geuss M, Schulz E (2003) Adv Polym Sci 164:87
10. Chung J, Munz M, Sturm H (2005) J Adh Sci Techn 19:1263
11. Giraud M, Nguyen T, Gu X, Van L,ingham MR (2001) Effects of stoichiometry and epoxy molecular mass on wettability and interfacial microstructures of amine-cured epoxies. In: Emerson JA (ed.) Proc 24th Ann Meeting of the Adhesion Society, Blacksburg VA, p 260
12. Wapner K, Grundmeier G (2005) Scanning Kelvin probe studies of ion transport and de-adhesion processes at polymer/metal interfaces. In: Possart W (ed.) Adhesion – Current Research and Application. Wiley-V CH, Weinheim, p. 507
13. Magonov SN, Whangbo M-H (1996) Surface analysis with STM and AFM, 1st edn. VCH Verlagsgesellschaft mbH, Weinheim
14. Meyer E, Hug HF, Bennewitz R (2004) Scanning probe microscopy – the lab on a tip, 1st edn. Springer, Berlin Heidelberg New York
15. Jandt KD (1998) Mater Sci Eng R21:221
16. Sheiko SS (2000) Adv Pol Sci 151:61
17. Alessandrini A, Facci P (2005) Meas Sci Technol 16:R65
18. Sturm H (1999) Macromol Symp 147:249

19. Munz M, Schulz E, Sturm H (2002) Surf Interface Anal 33:100
20. Radmacher M, Tillmann RW, Fritz M, Gaub HE (1992) Science 257:1900
21. Overney RM, Meyer E, Frommer J, Güntherodt H-J, Fujihira M, Takano H, Gotoh Y (1994) Langmuir 10:1281
22. Kajiyama T, Tanaka K, Ohki I, Ge S-R, Yoon J-S, Takahara A (1994) Macromolecules 27:7932
23. Nysten B, Legras R, Costa J-L (1995) J Appl Phys 78:5953
24. Labardi M, Allegrini M, Marchetti E, Sgarzi P (1996) J Vac Sci Technol B 14:1509
25. Kajiyama T, Tanaka K, Ge S-R, Takahara A (1996) Prog Surf Sci 52:1
26. Munz M, Sturm H, Schulz E, Hinrichsen G (1998) Composites Part A 29:1251
27. Jourdan JS, Cruchon-Dupeyrat SJ, Huan Y, Kuo PK, Liu GY (1999) Langmuir 15:6495
28. Munz M, Sturm H, Schulz E (2000) Surf Interface Anal 30:410
29. Tomasetti E, Legras R, Henri-Mazeaud B, Nysten B (2000) Polymer 41:6597
30. Ge S, Guo L, Rafailovich MH, Sokolov J, Overney RM, Buenviaje C, Peiffer DG, Schwarz SA (2001) Langmuir 17:1687
31. Bonse J, Munz M, Sturm H (2004) IEEE Trans Nanotechn 3:358
32. Munz M, Chung J, Kalinka G (2005) Mapping epoxy interphases. In: Possart W (ed.) Adhesion – current research and applications. Wiley-V CH, Weinheim, p 103
33. Price WJ, Kuo PK, Lee TR, Colorado R Jr, Ying ZC, Liu GY (2005) Langmuir 21:8422
34. Piétrement O, Troyon M (2000) Trib Lett 9:77
35. Cappella B, Dietler G (1999) Surf Sci Rep 34:1
36. Butt H-J, Cappella B, Kappl M (2005) Surf Sci Rep 59:1
37. Chen X, Davies MC, Roberts CJ, Tendler SJB, Williams PM, Davies J, Dawkes AC, Edwards JC (1998) Ultramicroscopy 75:171
38. Stapff I, Weidemann G, Schellenberg C, Regenbrecht M, Akari S, Antonietti M (1999) Surf Interface Anal 27:392
39. Bar G, Ganter M, Brandsch R, Delineau L, Whangbo M-H (2000) Langmuir 16:5702
40. Chen X, Davies MC, Roberts CJ, Tendler SJB, Williams PM, Burnham NA (2000) Surf Sci 460:292
41. Mallégol J, Dupont O, Keddie JL (2001) Langmuir 17:7022
42. Colchero J, Luna M, Baró AM (1996) Appl Phys Lett 68:2896
43. Krotil H-U, Weilandt E, Stifter Th, Marti O, Hild S (1999) Surf Interface Anal 27:341
44. Sturm H, Schulz E, Munz M (1999) Macromol Symp 147:259
45. Reinstädtler M, Kasai T, Rabe U, Bhushan B, Arnold W (2005) J Phys D: Appl Phys 38:R269
46. Johnson KL, Kendall K, Roberts AD (1971) Proc R Soc London A 324:301
47. Johnson KL (2000) ACS Symp Ser 741:24
48. Hertz H (1881) J Reine Angew Math 92:156 (Reproduced in: Miscellaneous papers by Heinrich Hertz (1896) Macmillan, London, p 146)
49. Derjaguin BV, Muller VM, Toporov YP (1975) J Colloid Interface Sci 53:314
50. Muller VM, Yushenko VS, Derjaguin BV (1980) J Colloid Interface Sci 77:91
51. Maugis D (1992) J Colloid Interface Sci 150:243
52. Johnson KL, Greenwood JA (1997) J Colloid Interface Sci 192:326
53. Dedkov GV (2000) phys stat sol (a) 179:3
54. Giri M, Bousfield DB, Unertl WN (2001) Langmuir 17:2973
55. Carpick RW, Ogletree DF, Salmeron M (1997) Appl Phys Lett 70:1548
56. Johnson KL (1985) Contact mechanics, Cambridge University Press, Cambridge
57. Lantz MA, O'Shea SJ, Welland ME, Johnson KL (1997) Phys Rev B 55:10776
58. Johnson KL (1997) Proc R Soc London A 453:163
59. Bowden FP, Tabor D (1964) The friction and lubrication of solids – part 2. Clarendon Press, Oxford

60. Greenwood JA (1992) Contact of rough surfaces. In: Singer IL, Pollock HM (eds) Fundamentals of friction: macroscopic and microscopic processes. Kluwer Academic Publishers, London, p 37

61. Ge S, Pu Y, Zhang W, Rafailovich M, Sokolov J, Buenviaje C, Buckmaster R, Overney RM (2000) Phys Rev Lett 85:2340

62. Moon S, Foster MD (2002) Langmuir 18:1865

63. Gelbert M, Roters A, Schimmel M, Rühe J, Johannsmann D (1999) Surf Interface Anal 27:572

64. Reif F (1965) Fundamentals of statistical and thermal physics, 1st edn. McGraw-Hill, New York

65. Kubo R, Toda M, Hashitsume H (1985) Statistical physics, vol. 2. Springer, Heidelberg

66. Saulson PR (1990) Phys Rev D 42:2437

67. Sarid D (1991) Scanning force microscopy, 1st edn. Oxford University Press, Oxford

68. Hutter JL, Bechhoefer J (1993) Rev Sci Instrum 64:1868

69. Butt H-J, Jaschke M (1995) Nanotechnology 6:1

70. Walters DA, Cleveland JP, Thomson NH, Hansma PK, Wendman MA, Gurley G, Elings V (1996) Rev Sci Instrum 67:3583

71. Stark RW, Drobek T, Heckl WM (2001) Ultramicroscopy 86:207

72. Roters A, Gelbert M, Schimmel M, Rühe J, Johannsmann D (1997) Phys Rev E 56:3256

73. Roters A, Schimmel M, Rühe J, Johannsmann D (1998) Langmuir 14:3999

74. Dimitrova TD, Johannsmann D, Willenbacher N, Pfau A (2003) Langmuir 19:5748

75. Benmouna F, Dimitrova TD, Johannsmann D (2003) Langmuir 19:10247

76. Benmouna F, Johannsmann D (2004) Langmuir 20:188

77. Chen GY, Warmack RJ, Thundat T, Allison DP, Huang A (1994) Rev Sci Instrum 65:2532

78. Williams JG, Li F, Miskioglu I (2005) J Adhes Sci Technol 19:257

79. Hodzic A, Stachurski ZH, Kim JK (2000) Polymer 41:6895

80. Briscoe BJ, Fiori L, Pelillo E (1998) J Phys D: Appl Phys 31:2395

81. Fischer-Cripps AC (2000) Vacuum 58:569

82. VanLandingham MR, Juliano TF, Hagon MJ (2005) Meas Sci Techn 16:2173

83. Li F, Williams JG, Altan BS, Miskioglu I, Whipple RL (2002) J Adhes Sci Technol 16:935

84. Kumar R, Cross WM, Kjerengtroen L, Kellar JJ (2004) Compos Interfaces 11:431

85. Bhushan B (1999) Principles and applications of tribology, Wiley-Interscience, New York

86. Baltá Calleja FJ, Fakirov S (2000) Microhardness of polymers, 1st edn. Cambridge University Press, Cambridge

87. Lee H, Neville K (1967) Handbook of epoxy resins, 3rd edn. McGraw-Hill, New York

88. Munz M, Sturm H, Stark W (2005) Polymer 46:9097

89. Krüger JK, Müller U, Bactavatchalou R, Liebschner D, Sander M, Possart W, Wehlack C, Baller J, Rouxel D (2005) Mechanical interphases in epoxies as seen by nondestructive high-performance Brillouin microscopy. In: Possart W (ed.) Adhesion – current research and applications. Wiley-V CH, Weinheim, p 125

90. Cuddihy EF, Moacanin J (1970) J Polym Sci A 8:1627

91. Heux L, Halary JL, Lauprêtre F, Monnerie L (1997) Polymer 38:1767

92. Heux L, Lauprêtre F, Halary JL, Monnerie L (1998) Polymer 39:1269

93. Mayr AE, Cook WD, Edward GH (1998) Polymer 39:3719

94. Min BG, Hodgkin JH, Stachurski ZH (1993) J Appl Polym Sci 48:1303

95. White SR, Mather PT, Smith MJ (2002) Polym Eng Sci 42:51

96. Gupta VB, Drzal LT, Lee CYC, Rich MJ (1985) Polym Eng Sci 25:812

97. Meyer F, Sanz G, Eceiza A, Mondragon I, Mijovič J (1995) Polymer 36:1407

98. Skourlis TP, McCullough RL (1996) J Appl Polym Sci 62:481

99. Plangsangmas L, Mecholsky JJ Jr, Brennan AB (1999) J Appl Polym Sci 72:257

100. Finzel MC, Delong J, Hawley MC (1995) J Polym Sci: Part A: Polym Chem 33:673

101. Nielsen LE (1969) J Macromol Sci – Rev Macromol Chem C 3:69

102. Mijovič J, Koutsky JA (1979) Polymer 20:1095
103. Crist B (1998) Plastic deformation in polymers. In: Cahn RW, Haasen P, Kramer EJ (eds) Materials science and technology – A comprehensive treatment. Vol. 12 Structure and properties of polymers. Wiley-V CH, Weinheim, p 446
104. Chung J (2006) Nanoscale characterization of epoxy interphase on copper microstructures, PhD thesis, TU Berlin, available from URL: http://nbnresolving.de/urn:de:kobv:83-opus-12637
105. Theocaris PS, Sideridis EP, Papanicolaou GC (1983) J Reinf Plast Compos 4:396
106. Anifantis NK (2000) Compos Sci Techn 60:1241
107. Vörös G, Pukánszky B (2002) Macromol Mater Eng 287:139
108. Hong SG, Wang TC (1994) J Appl Polym Sci 52:1339
109. Aufray M, Roche AA (2005) Properties of the interphase epoxy-amine/metal: influences from the nature of the amine and the metal. In: Possart W (ed) Adhesion – Current Research and Application. Wiley-V CH, Weinheim, p 89
110. Dillingham RG, Boerio FJ (1987) J Adhes 24:315
111. Oyama HT, Lesko JJ, Wightman JP (1997) J Polym Sci Part B: Polym Phys 35:331
112. Oyama HT, Solberg TN, Wightman JP (1999) Polymer 40:3001
113. Yang F, Pitchumani R (2003) J Appl Polym Sci 89:3220
114. Oliver WC, Pharr GM (1992) J Mater Res 7:1564
115. Munz M (2006) Evidence for a three-zone interphase with complex elastic-plastic behavior – nanoindentation study of an epoxy/thermoplastic composite, accepted for publication in Journal of Physics D: Applied Physics
116. Bühler V (2001) Kollidon – polyvinylpyrrolidone for the pharmaceutical industry, 6th edn. BASF-Pharma Ingredients, Ludwigshafen
117. Hodzic A, Kim JK, Stachurski ZH (2001) Polymer 42:5701
118. Hodzic A, Kalyanasundaram S, Kim JK, Lowe AE, Stachurski ZH (2001) Micron 32:765
119. Hodzic A, Kim JK, Lowe AE, Stachurski ZH (2004) Compos Sci Techn 64:2185
120. Griswold C, Cross WM, Kjerengtroen L, Kellar JJ (2005) J Adhes Sci Technol 19:279
121. ASTM (1977) Standard Definition, Designation D 907-74. In: Skeist I (ed.) Handbook of adhesives, 2nd edn. Van Nostrand Reinhold Company, New York, p xi
122. Creton C (2003) MRS Bull 28:434
123. Paiva A, Sheller N, Foster MD, Crosby AJ, Shull KR (2000) Macromolecules 33:1878
124. Autenrieth JS (1977) Resins for rubber-based adhesives. In: Skeist I (ed.) Handbook of adhesives. 2nd edn. Van Nostrand Reinhold Company, New York, p 229
125. Moon S, Swearingen S, Foster MD (2004) Polymer 45:5951
126. Paiva A, Foster MD, von Meerwall ED (1998) J Polym Sci: Part B: Polym Phys 36:373
127. Cannon LA, Pethrick RA (1999) Macromolecules 32:7617
128. Mallégol J, Gorce J-P, Dupont O, Jeynes C, McDonald PJ, Keddie JL (2002) Langmuir 18:4478
129. Creton C, Lakrout H (2000) J Polym Sci: Part B: Polym Phys 38:965
130. Creton C, Roos A, Chiche A (2005) Effect of the diblock content on the adhesive and deformation properties of PSAs based on styrenic block copolymers. In: Possart W (ed.) Adhesion – current research and applications. Wiley-V CH, Weinheim, p 337
131. Paiva A, Sheller N, Foster MD, Crosby AJ, Shull KR (2001) Macromolecules 34:2269
132. Moon S, Foster MD (2002) Langmuir 18:8108
133. Jia S, Foster MD (2002) J Appl Polym Sci 84:400
134. Persson BNJ (1998) Sliding friction – Physical principles and applications, 1st edn. Springer, Berlin Heidelberg
135. Wahl KJ, Stepnowski SV, Unertl WN (1998) Tribology Lett 5:103
136. Piner RD, Mirkin CA (1997) Langmuir 13:6864
137. Xu L, Lio A, Hu J, Ogletree DF, Salmeron M (1998) J Phys Chem B 102:540

138. Bouhacina T, Desbat B, Aimé JP (2000) Tribology Lett 9:111
139. Xiao X, Qian L (2000) Langmuir 16:8153
140. Opitz A, Ahmed SI, Schaefer JA, Scherge M (2002) Surf Sci 504:199
141. Belaroui F, Grohens Y, Boyer H, Holl Y (2000) Polymer 41:7641

29 Development of MOEMS Devices and Their Reliability Issues

Bharat Bhushan · Huiwen Liu

29.1
Introduction to Microoptoelectromechanical Systems

A microoptoelectromechanical system (MOEMS) is a miniaturized system combining optics, electrical, and mechanical elements, fabricated with collective techniques derived from the microelectromechanical systems (MEMS) fabrication processes. The demand for optical devices together with advances in MEMS technology have spurred rapid development of MOEMS [1–3]. Figure 29.1 shows the relationship between MOEMS and microoptics, microelectronics, and micromechanics systems. Microelectronic integrated circuits can be thought of as the "brains" of a system and MOEMS augments this decision-making capability with "eyes" and "arms," to allow microsystems to sense and control the environment. Electronics (sensors) gather information from the environment through measuring optical, mechanical, thermal, biological, chemical, and/or magnetic properties. The electronics then process the information derived from the sensors and through some decision-making capability direct the optical elements and actuators to respond by moving, positioning, regulating, pumping, and filtering, thereby controlling the environment for some

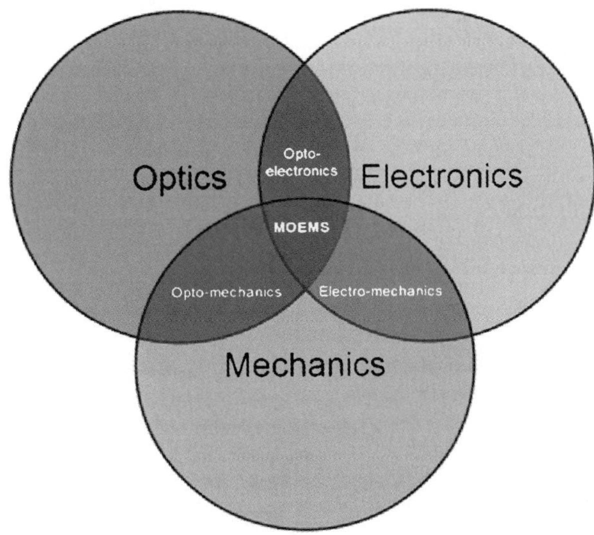

Fig. 29.1. The relationship between microoptoelectromechanical systems (MOEMS) and microoptics, microelectronics, and micromechanics [4]

desired purpose [4, 5]. Movement of a microoptical element permits the dynamic manipulation of a light beam. This dynamic manipulation can involve amplitude or wavelength modulation, temporal delay, interference, diffraction, reflection, refraction, or simple spatial realignment. Any two or three of these operations can be combined to form a complex operation on the light beam. The ability to carry out these operations, using miniaturized optical, mechanical, and electrical elements, is one of the key attributes that distinguishes MOEMS from classic physical optics [6–8].

There are three primary characteristics that make MOEMS an important technological development: (1) the batch process by which the systems are fabricated, (2) the size of the elements in the systems is on the microscale/nanoscale, and (3) perhaps the most distinctive, is the possibility to endow the optical elements in the system with the ability for precise and controlled motion [6, 8]. Since MOEMS devices are manufactured using batch fabrication techniques similar to those used for integrated circuits, unprecedented levels of functionality, reliability, and sophistication can be placed on a small chip at a relatively low cost.

Today, an increasing number of academic and industry laboratories are working together in research and development of MOMES devices. Many different kinds of MOEMS have emerged in the research laboratory and some have made it to the market. The typical MOEMS devices include [4]:

– Arrays of micromirrors for digital image processing
– Deformable membranes for adaptive optics
– Optical switches, attenuators, and shutters used for optical communications
– Microscanners for image processing and obstacle detection
– Tunable optical sources and filters, reflection modulators, spectral equalizers
– Guided optical devices
– Optical elements assembles: fiber-chip connection, optoelectronic hybrid integration.

It is believed that MOEMS along with the development of MEMS are expected to have a major impact on our lives, much like the way that the integrated circuit has affected information technology. The application of MOEMS will include but is not limited to the following areas [1–3, 6, 8]:

– Military, such as inertial navigation units on a chip for navigation
– Telecommunications and information technology
– Biomedical
– Automotive
– Industrial maintenance and control
– Space and astronomy
– Environmental monitoring
– Mass data storage devices and systems for storage densities of terabytes per square centimeter.

In this review, first the structure and mechanisms of some typical MOEMS devices will be summarized, then the reliability issues related to these devices will be discussed.

29.2
Typical MOEMS Devices: Structure and Mechanisms

29.2.1
Digital Micromirror Device and Other Micromirror Devices

The digital micromirror device (DMD) was invented by Texas Instruments in 1987. The invention of the DMD led to the development of the digital light processing (DLP) projection displays launched on the market in 1996 [9–12]. DLP projection displays present bright and seamless images that have high fidelity and stability. The DMD is one of the few successfully commercialized devices in the emerging field of MEMS. Today, many products based on the DLP technology are sold in various segments of the market, including computer projectors, high-definition television (HDTV) sets, and movie projectors (DLP cinema) [12].

The DMD chip is a MEMS array of aluminum alloy micromirrors, monolithically integrated onto and controlled by an underlying silicon complementary metal oxide superconductor (CMOS) static random access memory (SRAM) array [13, 14]. The micromirror superstructure is fabricated through a series of aluminum metal depositions, oxide masks, metal etches, and organic spacers. The organic spacers are subsequently removed to leave the micromirror structure free to move using an oxygen and fluorine plasma etching [13, 15, 16]. The DMD has a layered structure, consisting of a micromirror layer, a yoke and hinge layer, and a metal layer on a CMOS memory array. An exploded view of the DMD layer and the corresponding atomic force microscopy (AFM) surface height images are presented in Fig. 29.2 [17].

One DMD chip consists of half a million to more than two million of these independently controlled reflective micromirrors (micromirror size is on the order of 14-μm square and 15-μm pitch) which switch forward and backward at a frequency on the order of 5–7 kHz. The micromirror is rotated as a result of electrostatic attraction between the micromirror structure and the underlying electrodes. Figure 29.3 shows a schematic of two pixels of a DMD [13]. The micromirror and yoke are connected to a bias/reset voltage. The address electrodes are connected to the underlying CMOS memory through contacts. Movement of the micromirror is accomplished by storing a 1 or a 0 in the memory cell applying a bias voltage to the micromirror/yoke structure. When this occurs, the micromirror is attracted to the side with the largest electrostatic field differential, as shown in Fig. 29.3. To release the micromirror, a short reset pulse is applied that excites the resonant mode of the structure and the bias voltage is removed. This combination results in the micromirror leaving the landing site. The micromirror lands again when the bias voltage is reapplied. Each SRAM cell corresponds to a micromirror and allows each micromirror to be individually addressed so as to rotate $\pm12°$ with respect to the horizontal plane, limited by a mechanical stop [12, 13]. Contact between cantilever spring tips at the end of the yoke (four are present on each yoke) with underlying stationary landing sites is required for the two digital (binary) operations.

Other companies, such as Samsung and Reflectivity, are also working on some similar aluminum-array-based MOEMS to perform a function like the DMD. Another type of micromirror array device based on polysilicon developed by Sandia National Laboratory is shown in Fig. 29.4. The size of the micromirror is about

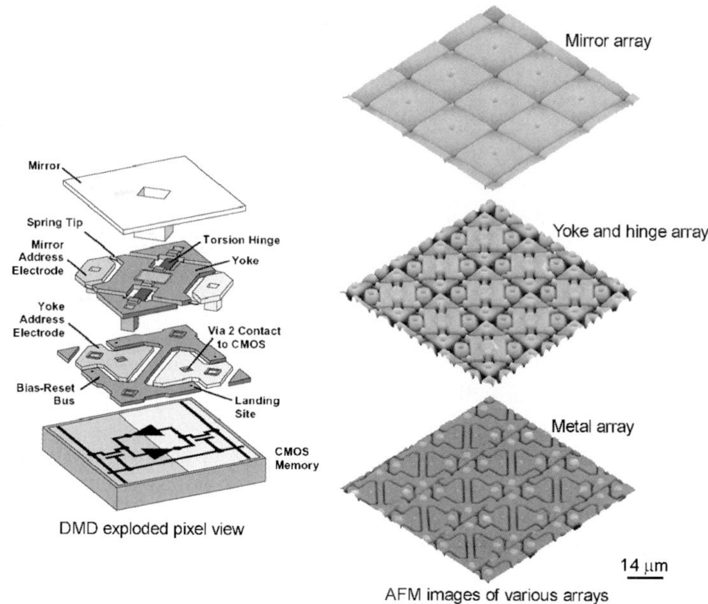

Fig. 29.2. *Left*: An exploded view of the digital micromirror device (DMD) layered structure. *Right*: Atomic force microscope (AFM) images for the micromirror, yoke and hinge, and metal layer arrays [17]

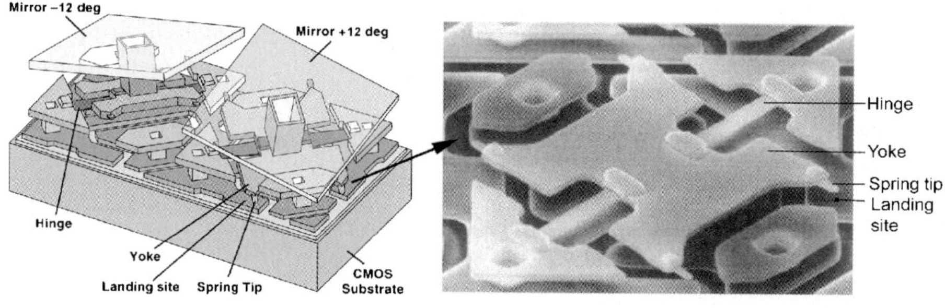

Fig. 29.3. *Left*: Two pixels of a DMD. The micromirror is attracted to the side with the largest electrostatic field differential. *Right*: Scanning electron microscope pictures of the key elements of a hinge layer [12, 13, 17]

$100\,\mu m \times 100\,\mu m$ with 1-µm gaps between adjacent mirrors. The mirrors can be tilted 10°. These mirrors are sensitive to IR radiation and, as a result, will be able to detect faint signals from the first billion years after the Big Bang [18, 19]. The mirrors are built by depositing thin films of polycrystalline silicon on a silicon wafer. The first layer contains connection wires, and the others are mechanical layers that allow the MEMS device to move. A start-up company called Miradia in San Jose, CA, USA, is working on similar silicon micromirror array device for display applications. Their mirror size is close to the size of the DMD.

Fig. 29.4. Polysilicon-based micromirror array developed by Sandia National Laboratory [18]

29.2.2
MEMS Optical Switch

Internet users face a similar traffic jam as we encounter while driving on the freeway. At hubs, the system must convert the light pulses into electronic packages to permit them to travel through the electronic switch. Once rerouted, the packages are converted back into light. Obviously, any sort of electronic system at the end of an optical fiber will inevitably cause a traffic jam because it lacks the capacity to deal with the optical fiber's bandwidth. The conversion and reconversion slows down the entire process. One solution is to avoid the transitions between optical and electronic transmitters by using optical switches. A strong demand will be created on optical switching for networking, telecommunications, and wireless technologies [20].

Lucent's Bell Laboratories developed the world's first practical MEMS optical switch [21]. The device is based on a tiny pivoting bar with a gold-plated mirror at one end that fits in a tiny space between two hair-thin optical fibers lined up end to end (Fig. 29.5, left). The micromirror fits in a space, about one tenth as wide as a human hair, between two hair-thin optical fibers lined up end to end. When the switch is off, the mirror rests below the cores of the two fibers, allowing lightwave signals to travel from the core of one to the other. To turn the switch on, a voltage is

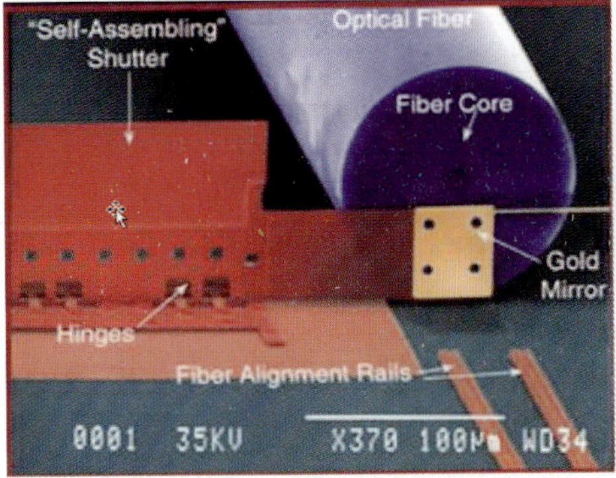

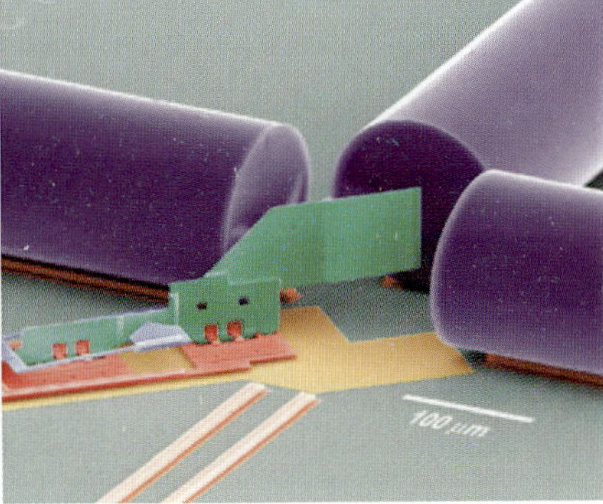

Fig. 29.5. Microoptical switch devices developed by Lucent's Bell Laboratories. The micromirrors are driven by electrostatic force [21]

applied at the other end of the bar, beneath an attached plate; the electrostatic forces pull the plate down, lifting the bar so the mirror reflects the light instead of letting it move from one fiber to the other. Another similar MOEMS switch developed by the same laboratory is shown in Fig. 29.5 (right). A micromirror fits in a tiny space between three hair-thin optical fibers. Two of them are lined up end to end, another is in the perpendicular direction. The micromirror is also driven by electrostatic force. When it is in the off state, it deflects the signal to its left toward the rear fiber. When it is in the on position, the signal continues in a straight line to the fiber on the right.

Wave Star Lambda Router is another example of a MEMS optical switch used for network communication. It uses 256 miniature mirrors, each no larger than the head of a pin, to steer light signals from one optical fiber to another (Fig. 29.6) [21, 22]. This promises the ability to direct traffic through optical networks 16 times faster.

Figure 29.7 shows a MOEMS micromirror optical switch developed by Sandia National Laboratory. It is driven by a combination of a microengine with a microtransmission. With the torque increase resulting from this combination, the micromirror can be lifted by the mechanism alone [23]. This device was designated for switching laser or IR signals. In addition to the examples discussed here, other

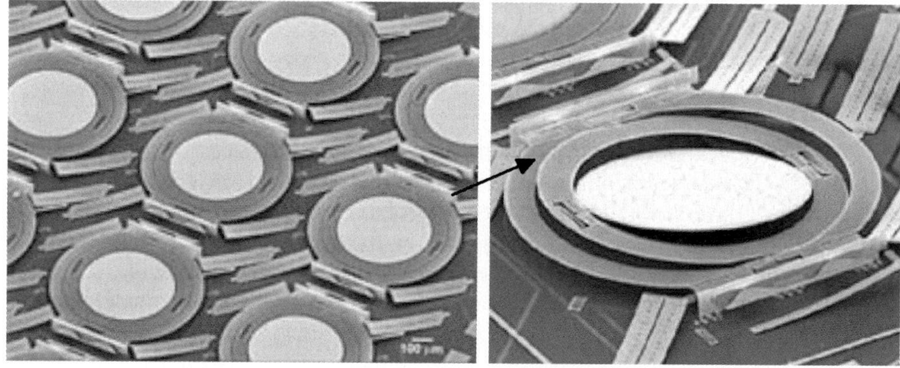

Fig. 29.6. Wave Star Lambda Router, which is a MOEMS device integrated from a lot of micromirrors and designated for network communication [22]

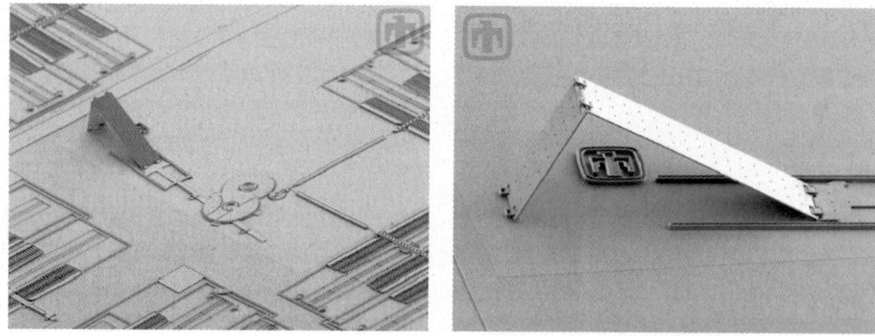

Fig. 29.7. A micromirror optical switch designed by Sandia National Laboratory. The device is driven by a combination of a microengine with microtransimission [23]

companies, including Xerox, Optical Micromachines, Optical MOEMS, and Agilent Technologies, are also developing MOEMS-based optical switches. It is expected that MOEMS switches will provide one of the critical techniques for optical network communication.

29.2.3
MEMS-Based Interferometric Modulator Devices

The interferometric modulator (iMoD) technique was developed by Iridigm (acquired by Qualcom in 2004) in 1996 and is projected to reach the market in 2006. This technique can be used for development of novel displays [24–29]. The iMoD element uses interference to create color in the same way that butterfly wings create shimmering iridescent colors (Fig. 29.8). Since the iMoD contains no pigments, the color is inherently stable and cannot fade like inks or dyes when exposed to light.

The iMoD element is a simple MOEMS device that is composed of two conductive plates (Fig. 29.9). One is a thin-film stack on a glass substrate; the other is a metallic membrane suspended over it. There is a gap between the two that is filled with air. The iMoD element has two stable states. When no voltage is applied, the plates are separated, and light hitting the substrate is reflected as shown in Fig. 29.9. When a small voltage is applied, the plates are pulled together by electrostatic at-

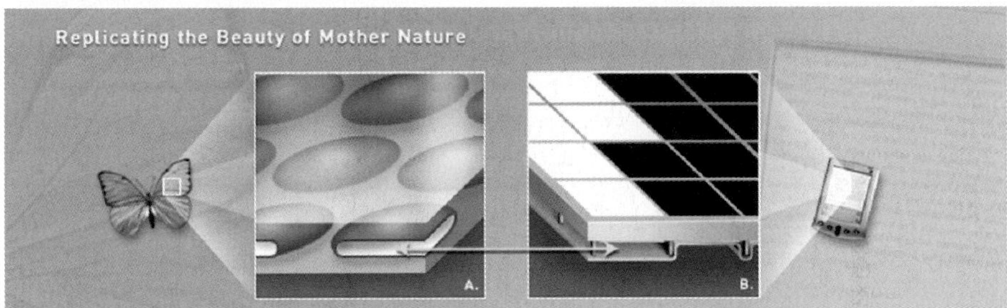

Fig. 29.8. Comparison between an interferometric modulator (iMoD) device and a butterfly wing. They create color using the same principle by interference [24]

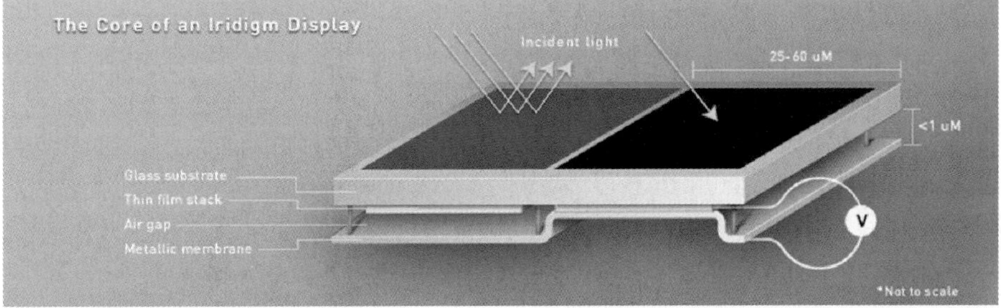

Fig. 29.9. Two pixels of the iMoD device. Applying a voltage on the two conductive plates can adjust their distance, which in turn changes the appearance of the elements from bright to dark [24]

traction and the light is absorbed, turning the element black. This is the fundamental building block from which Iridigm displays are made.

iMoD elements are typically 25–60 µm on a side (400–1000 dots per inch); therefore, many iMoD elements are ganged and driven together as a pixel, or subpixel, in a color display (Fig. 29.10). The color of the iMoD element is determined by the size of the gap between the plates. As shown, the blue iMoD has the smallest gap and the red one has the largest. To create a flat-panel display, a large array of iMoD elements are fabricated in the desired format and packaged. Finally, driver chips are attached at the edge to complete the display. The iMoD display technique increases brightness and contrast, and enables saturated colors to be displayed, high-resolution formats, smooth video playback, and low power consumption.

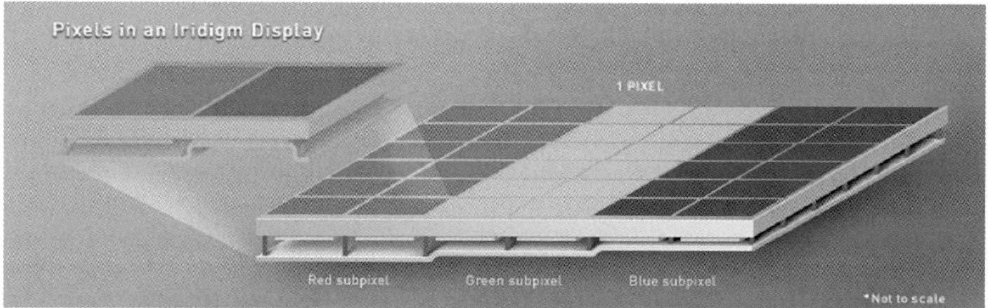

Fig. 29.10. The color pixel formation mechanisms of the iMoD device [24]

29.2.4
Grating Light Valve Technique

The grating light valve (GLV) device was developed by Silicon Light Machines [30]. It is a unique MOEMS that acts as a dynamic, tunable grating to precisely vary the amount of laser light that is diffracted or reflected. The GLV technology was originally developed for use in extremely high resolution digital display products. The GLV device creates bright and flawless images with high contrast and supreme color reproduction. Sony Corporation is developing a product based on Silicon Light Machines' technology to accelerate the development of its next-generation display products that utilize lasers as a light source. Except for the display application, GLV technology can also be used in high-end printer and optical communication. MEMS techniques are used to fabricate tiny reflective elements on the surface of a silicon chip. Each of these elements is made up of multiple ribbon-like structures, which can actually be moved up or down over a very small distance (only a fraction of the wavelength of light) by controlling electrostatic forces (Fig. 29.11). The ribbons are arranged such that each element is capable of either reflecting or diffracting light. This allows an array of elements, when appropriately addressed by control signals, to vary the level of light reflected off the surface of the chip. This control of light can be analog (variable control of light level) or digital (switching of light on or off).

Fig. 29.11. Grating light valve device developed by Silicon Light Machines. Each of these elements is made up of multiple ribbon-like structures, which can actually be moved up or down over a very small distance [30]

29.2.5
Continuous Membrane Deformable Mirrors

Whenever an optical signal travels through air, it runs the risk of being distorted when it moves through turbulence in the atmosphere. Adaptive correction is vital if a high-fidelity signal is required. Otherwise, this turbulence limits the useful range of the optical signal as the strength of the signal degrades with distance. Continuous membrane deformable mirrors (CMDM), proposed by MEMS Optical, can perform such kind of correction in real time, which will greatly extend the useful range of the signal (Fig. 29.12). This is especially important in the free-space communication area, where a longer useful signal range means fewer transmitting stations [31]. The major applications of CMDM are free-space communications, astronomical telescopes, and focal length modification. Another similar MOEMS device developed by the same company is designated to be applied in tunable lasers (Fig. 29.13). When the device is installed in a laser cavity, the micromirror position can be moved by up to $10,100\,\mu$m, which allows a simple and effective way of changing the path length inside the laser cavity.

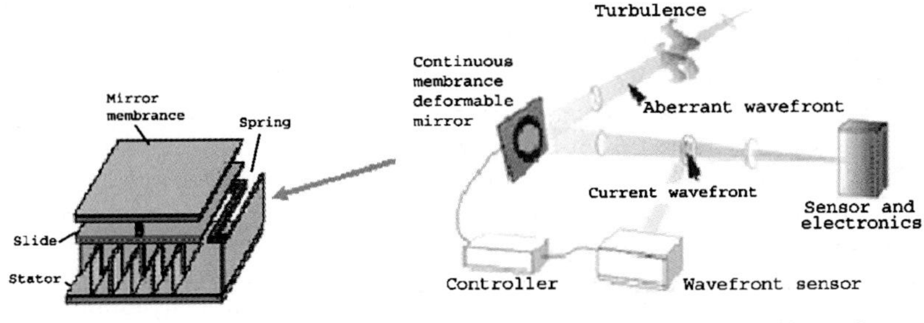

Fig. 29.12. Continuous membrane deformable micromirrors, which is a MOEMS device designated for adaptive light correction [31]

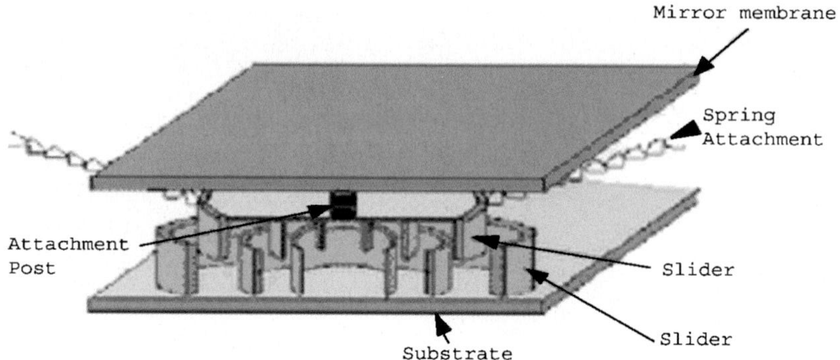

Fig. 29.13. A continuous membrane deformable micromirror designated for the tunable laser application [31]

29.3
Reliability Issues of MOEMS

Bringing MOEMS technology to the market has its difficulties. An important challenge in achieving successful commercial MOEMS products is associated with their reliability. For example, the lifetime requirements of the DMD micromirrors for commercial application have grown substantially. A lifetime estimate of over 100,000 h with no degradation in image quality is the norm [13, 32]. At a mirror modulation frequency of 7 kHz, each micromirror element needs to switch about 2.5 trillion cycles. The combination of micromechanical, electrical, and optical elements of the MOEMS has led to some unique challenges in the area of reliability as compared with other semiconductor products. Today, in many cases we really lack knowledge of the MOEMS failure mechanisms. This limitation presents a challenge in developing practical MOEMS products. The studies of MEMS have revealed the profound influence of adhesion, stiction/friction, wear, fracture ,and fatigue on the reliability [33–42]. MOEMS is a kind of unique MEMS device with moving optical devices; thus all of the failure modes discovered in MEMS can appear in MOEMS. In this section, some of the typical failure modes and their technical solution related to the MOEMS devices presented in Sect. 29.2 will be discussed.

29.3.1
Stiction-Induced Failure of DMD

For a DMD, in isolated cases, the meniscus and van der Waals forces between the contacting elements can be so large that they cannot be overcome by the reset pulse induced force. This can cause the problem of a micromirror not switching properly, resulting in a dark or bright blemish on the projection screen. Figure 29.14a shows the AFM images of a normal and a stuck micromirror. To minimize stiction between contacting surfaces, a spring-tip structure is used in order to use the spring-stored elastic energy to pop up the tip during pull-off (Fig. 29.14b) [13]. A self-assembled

Fig. 29.14. (a) AFM images of a stuck micromirror of a DMD compared with a normal micromirror array [17]. (**b**) The spring tip used in the DMD to reduce stiction [13]

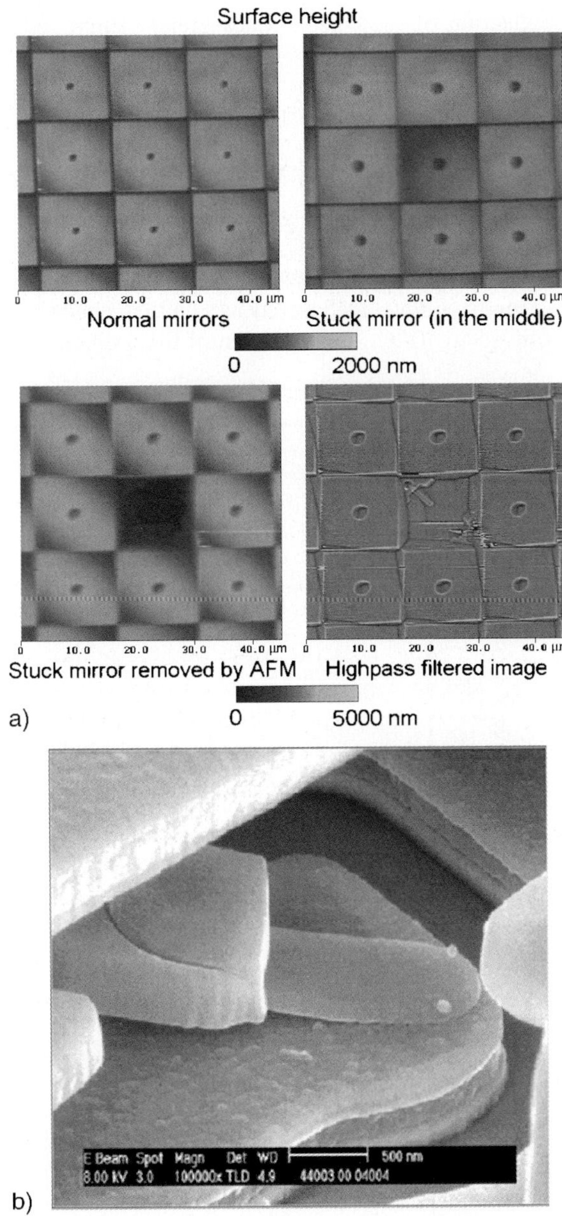

monolayer (SAM) of perfluorinated n-alkanoic acid ($C_nF_{2n-1}O_2H$), e.g., perfluorodecanoic acid, $CF_3(CF_2)_8COOH$, applied by a vapor-phase deposition process is used on the spring tips and landing sites to make the surfaces hydrophobic in order to reduce the formation of a meniscus [43–49]. Furthermore, the entire DMD chip set is hermetically sealed in order to prevent particulate contamination and excessive condensation of water at the spring tips and landing sites [13, 42]. The application of these techniques greatly improved the reliability of the DMD devices. The study by Liu and Bhushan [17] using AFM shows that the defects on the SAM could be one of the potential reasons that lead to the hard stiction of the micromirror.

29.3.2
Thermomechanical Issues with Micromirrors

MOEMS devices can be used in very different temperature ranges. For example, the micromirror developed by MEMS Optical needs be applied in the temperature range from −5 to 70 °C. The variation of the temperature can cause the mirror curvature

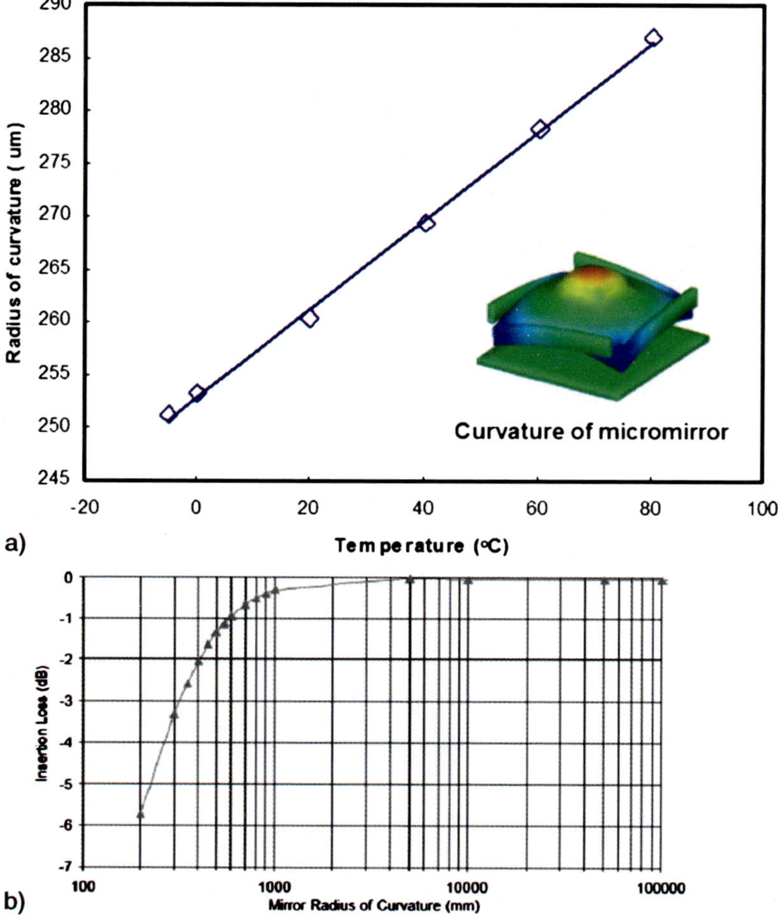

Fig. 29.15. Simulation results: (**a**) variation of micromirror curvature with temperature; (**b**) variation of insertion loss and the micromirror radius of curvature for a two micromirror MOEMS switch [50]

Table 29.1. Effect of temperature on micromirror optical properties [49]

Mirror size (μm^2)	Temperature (°C)	Focus (waves)	Astigmatism	Coma (waves)
250 × 250	23	0.82	0.18	0.006
	36	0.69	0.14	0.005
	50	0.52	0.07	0.000
200 × 250	23	0.80	0.36	0.014
	35	0.66	0.27	0.014
	50	0.50	0.17	0.018

change owing to stress (Fig. 29.15a) [50]. The curvature change will affect coupling efficiency or insertion loss (Fig. 29.15b). In addition to that, the curvature change can also significantly influence other optical properties of the mirror, such as focus, astigmatism, and coma (Table 29.1).

Another example for thermal-induced failure is the micromirrors developed by Sandia National Laboratory [18] which are designed for space application. The operating temperatures in space can be −243 °C or lower. That means these mirrors have to be built in a special way so that they will not break at such an extreme condition. As we discussed earlier, the mirrors are built by depositing thin films of polycrystalline silicon on a silicon wafer, and a gold layer is deposited on the top to enhance the reflection of IR radiation. As the temperature is reduced, the gold layer will shrink faster than the polysilicon layer. This could cause very high stress that breaks or deforms the micromirror, or the gold coatings could peel away. One of the solutions is to reduce the gold layer thickness so that it does not cause excessive stress, but yet is thick enough to reflect theIR radiation.

29.3.3
Friction- and Wear-Related Failure

As discussed in Sect. 29.2.2, the MOEMS switch developed by Sandia National Laboratory is driven by the microengine. In laboratory tests, strong friction- and wear-induced failure has been found in the microengine device operated in air [51–53] (Fig. 29.16). And it is also found that the friction and wear performance is related to the humidity and frequency. It is suggested that application of SAMs and W coating can improve the wear resistance and reduce the friction [54, 55].

29.3.4
Contamination-Related Failure

As the elements of an MOEMS device are so small at the micrometer scale, one particle at 100-μm scale could be larger than several of the elements; therefore, one such contaminated particles can cause catastrophic failure. Compared with such kinds of huge particle, some of the microparticles or contamination that was introduced in the fabrication process is very hard to find and more effort needs to be spent to get rid of it. The AFM images in Fig. 29.17 show some of the tiny

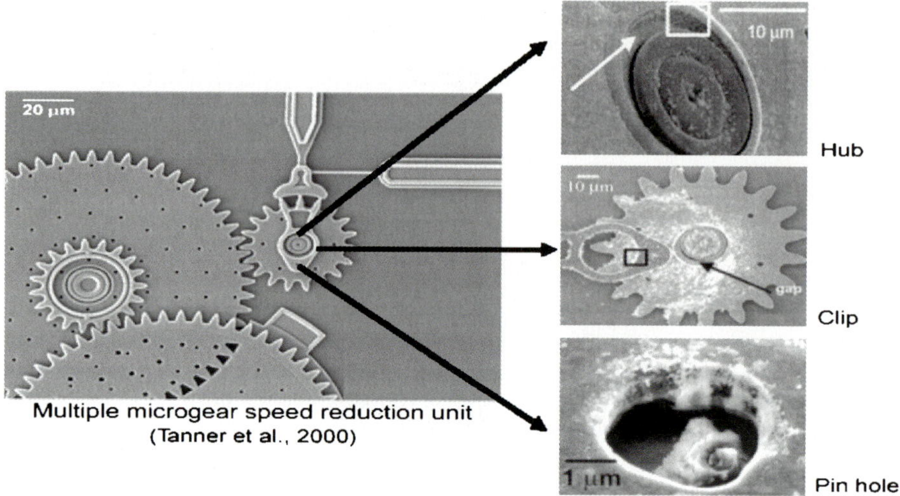

Fig. 29.16. Wear evidence on hub, clip, and pin hole on a microgear used in the microengine to drive the MOEMS switch designed by Sandia National Laboratory [51–53]

Fig. 29.17. AFM images show the contamination particles at the edge of the micromirror of a DMD. Such kinds of particles can lead to the nonsmooth micromirror tilting [56]

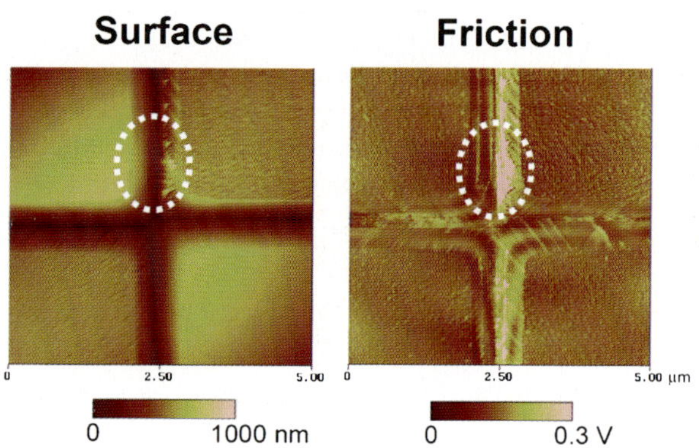

contamination particles at the edge of the micromirror. The studies by Liu and Bhushan [56, 57] in 2004 using AFM found that such contamination particles can cause so-called soft stiction. This can be reflected from the nonsmooth AFM load–displacement curve of the micromirror which is shown in Fig. 29.18. Reduction of the particle counts in a clean-room environment and improvement of the etching and release process are the key for reducing such kinds of failure mode. Besides stiction and the contamination effect on the reliability of the DMD, the fatigue of the DMD hinge was also studied. It is believed that fatigue has less impact on the lifetime of the DMD [13, 14, 58].

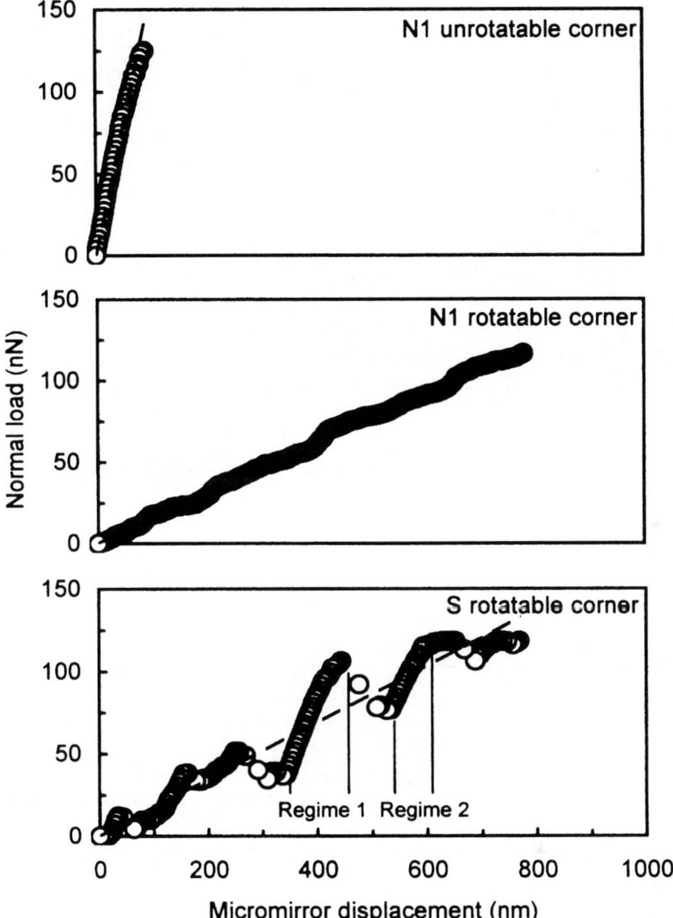

Fig. 29.18. AFM force–distance curves which are collected on the unrotatable corner (*top*) and rotatable corner (*middle*) of a normal mirror. As a comparison, the force–distance curve collected on a soft stick mirror (*bottom*) is also given [56]

29.4
Summary

It is believed that MOEMS along with the development of MEMS are expected to have a major impact on our lives, much like the way that the integrated circuit has affected information technology. Some typical MOEMS devices were introduced in this chapter. Their structure and mechanisms were briefly discussed. Some of the failure mechanisms of the MOEMS were also discussed. It is believed that MOEMS failure mechanisms studies and the development of novel stiction, friction, wear, and residual stress reduction techniques will become critical for commercialization of MOEMS.

References

1. Hunsperger RG, (2002) Integrated Optics: Theory and Technology, 5th edn, Springer, Berlin, Heidelberg, New York
2. Bhushan B (ed) (2004) Handbook of Nanotechnology, Springer, Berlin, Heidelberg, New York
3. Cho H (2003) Opto-Mechatronic Systems Handbook, CRC, Boca Rota
4. http://www.mems-exchange.org
5. http://www-leti.cea.fr/commun/europe/MOEMS/moems.htm
6. http://www.darpa.mil/MTO/MOEMS/index.html
7. http://www.researchandmarkets.com/reports/28420/28420.htm
8. http://www.memsoptical.com/techinfo/memstut1.htm
9. Hornbeck LJ, Nelson WE (1988) Bistable deformable mirror device, OSA Technical Digest Series Vol. 8: Spatial Light Modulators and Applications, p 107–110
10. Hornbeck LJ (1999) A digital light processing^{TM} update-status and future applications. Proc Soc Photo-Opt Eng 3634:158–170
11. Hornbeck LJ (1990) Deformable mirror spatial light modulation. Proc SPIE 1150:86–102
12. Hornbeck LJ (2001) The DMD^{TM} projection display chip: a MEMS-based technology. MRS Bull 26:325–328
13. Douglass MR (1998) Lifetime estimates and unique failure mechanisms of the digital micromirror devices (DMD). In: 1998 International Reliability Physics Proceedings, IEEE Catalog No. 98CH36173. Presented at the 36th Annual International Reliability Physics Symposium, Reno, p 9–16
14. Bailey WE, Baker JC (1996) Fabrication method for digital micromirror devices using low temperature CVD, US Patent 5,526,951, June 18
15. Choi JK (2001) Method for manufacturing digital micro-mirror devices (DMD) packages, US Patent Application 2001/0041381 A1, November 15
16. Hornbeck LJ (1994) Low reset voltage process for DMD, US Patent 5,331,454, July 19
17. Liu H, Bhushan B (2004) Nanotribological characterization of digital micromirror devices using an atomic force microscope, Ultramicroscopy, 100:391–412
18. http://www.sandia.gov/media/NewsRel/NR1999/space.htm
19. http://ngst.gsfc.nasa.gov
20. http://www.lucent.com/pressroom/lambda.html
21. http://www.bell-labs.com/news/1999/february/23/mems.jpeg
22. http://www.spie.org/web/ oer/august/aug00/home.html
23. http://www.techonline.com/community/ed_resource/feature_article/20655
24. http://www.iridigm.com/tech_overview.htm; http://www.qualcomm.com/qmt
25. Miles MW (2004) Large area manufacture of MEMS based displays, DTIP of MEMS & MOEMS, Montreux, May 12–14
26. Miles MW, Larson E, Chui C, Kothari M, Gally B, Batey J (2003) Digital paper^{TM} for reflective displays. J Soc Inf Display 11:209–215
27. Gally BJ (2004) Wide-Gamut color reflective displays using iMoD interference technology, SID Symp Digest, 35:654
28. Miles MW (2000) Digital paper: reflective display using interferometric modulation. SID Symp Digest 31:32
29. Miles MW (1997) A new reflective FPD technology using interferometric modulation. J Soc Inf Display 5:379–382
30. http://www.siliconlight.com/htmlpgs/homeset/homeframeset.html
31. http://www.memsoptical.com/prodserv/products/microlensar.htm

32. Douglass MR (2003) DMD reliability: a MEMS success story. In: Ramesham R, Tanner D (eds) Proceedings of Reliability, Testing, and Characterization of MEMS/MOMES II, SPIE, Bellingham, p 1–11

33. Arney S (2001) Designing for MEMS Reliability. MRS Bull 26:296–299

34. Bhushan B (1998) Tribology Issues and Opportunities in MEMS, Kluwer, Dordrecht

35. Bhushan B (1999) Handbook of Micro/Nanotribology, 2nd edn, CRC, Boca Raton

36. Bhushan B (2001) Macro and microtribology of MEMS materials. In: Bhushan B (ed) Modern Tribology Handbook, Vol 2, p 1515–1548, CRC, Boca Raton

37. Boer MPD, Mayer TM (2001) Tribology of MEMS. MRS Bull 26:302–304

38. Fujimasa I (1996) Micromachine, (Oxford Science Press, Oxford)

39. Kayali S, Lawton R, Stark BH (1999) MEMS reliability assurance activities at JPL, EEE Links 5:10–13

40. Man KF (2001) MEMS reliability for space applications by elimination of potential failure modes through testing and analysis. http://www-rel.jpl.nasa.gov/Org/5053/atpo/products/Prod-map.html

41. Man KF, Stark BH, Ramesham R (1998) A Resource Handbook for MEMS Reliability. Rev A JPL Press, Pasadena

42. Bhushan B (ed) (2005) Nanotribology and Nanomechanics: An Introduction, Springer, Berlin, Heidelberg, New York

43. Wallace RM, Henck SA, Webb DA (1996) PFPE coating for micro-mechanical devices, US Patent 5,512,374, April 30

44. Hornbeck LJ (1997), Low surface energy passivation layer for micromachined devices, US Patent 5,602,671, February 11

45. Henck SA (1997) Lubrication of digital micromirror devices. Tribol Lett 3:239–247

46. Lee SH, Kwon MJ, Park JG (1999) Preparation and characterization of perfluoro-organic thin films on aluminium. Surf Coat Tech 112:48–51

47. Lee KK, Cha NG, Park JS et al. (2000) Chemical, optical and tribological characterization of perfluoropolymer films as an antiadhesion layer in micromirror arrays. Thin Solid Films 377–378:727–732

48. Gudeman CS (2001) Vapor phase low molecular weight lubricants, US Patent 6,251,842, B1, June 26

49. Robbins RA, Jacobs SJ (2001) Lubricant delivery for micromechanical devices, US Patent 6,300,294 B1, Oct. 9

50. Talghader JJ (2000) MOEMS and miniaturized system II, Proc SPIE 4561:20–27

51. Tanner DM, Peterson KA, Irwin LW et al. (1998) Linkage design effect on the reliability of surface micormachined microengine driving a load. Proc SPIE 3512:215–226

52. Tanner DM, Walraven JA, Irwin LW et al. (1999) The effect of humidity on the reliability of a surface micromachined microengine, In: IEEE International Reliability Physics Symposium, San Diego, p 189–197

53. Tanner DM, Miller WM, Eaton WP et al. (1998) The effect of frequency on the lifetime of a surface micromachined microengine driving a load, In: IEEE International Reliability Physics Symposium, Reno, p 26–35

54. Hankins MG, Resnick PJ, Clews PJ, Mayer TM, Wheeler DR, Tanner DM (2003) Vapor deposition of amino-functionalized self-assembled monolayer on MEMS, In: Ramesham R, Tanner DM (eds.), Proceedings of SPIE, Reliability, Testing, and Characterization of MEMS/MOEMS II, SPIE, Bellingham, p 238–247

55. Mani SS, Fleming JG, Walraven JA, Sniegowski JJ et al. (2000) Effect of W coating on microengine performance, In: Proceedings of the 38th Annual International Reliability Physics Symposium, IEEE, New York, p 146–151

56. Liu H, Bhushan B (2004) Investigation of nanotribological and nanomechanical properties of the digital micromirror device by atomic force microscope. Vac J Sci Technol A 22:1388–1396
57. Bhushan B, Liu H (2004) Characterization of nanomechanical and nanotribological properties of digital micromirror devices, Nanotechnology, 15:1785–1791
58. Liu H, Bhushan B (2004) Bending and fatigue study on a nanoscale hinge by an atomic force microscope. Nanotechnology 15:1246–1251

Subject Index

Aarhus STM   *VII* 216
AC imaging techniques   *VI* 135
acoustic microscopy   *V* 278
acrylate   *VII* 330
action spectroscopy   *VI* 56
active
    cantilever   *V* 1, *V* 2
    corrosion   *VII* 290
    phase   *VII* 214
    regions   *VII* 288
    sites   *VII* 197, *VII* 198, *VII* 207, *VII* 212
activity   *VII* 207
actuation   *VII* 99–101, *VII* 108
    phase   *V* 15
    piezoelectric   *VII* 99, *VII* 100, *VII* 109
    thermal   *VII* 99
adhesion   *VI* 288, *VI* 292, *VI* 294, *VI* 296,
        VI 297, *VI* 312, *VI* 313, *VI* 315–317,
        VI 320, *VII* 41, *VII* 46, *VII* 47,
        VII 52–56, *VII* 58, *VII* 60, *VII* 62,
        VII 65, *VII* 67, *VII* 68, *VII* 70, *V* 267
    hysteresis   *V* 282
    molecule   *VI* 170
    strength   *VII* 310
adhesive   *VII* 303–305, *VII* 307, *VII* 308,
        VII 310, *VII* 320, *VII* 329–331,
        VII 333–335, *VII* 342
    contact   *VI* 175
    layer   *VI* 67
adrenaline   *VII* 281
adsorbate structure   *VII* 216
adsorbates   *VII* 199
adsorption   *VI* 184, *VII* 233, *VII* 235,
        VII 236, *VII* 242, *VII* 248
adsorption of proteins   *VI* 185, *VII* 233,
        VII 234
adsorption structure   *VII* 214
advancing and receding contact angles
        (contact angle hysteresis)   *VII* 1,
        VII 37

advancing contact angle   *VII* 2
affinity imaging   *VI* 196
AFM-SECM   *V* 231
AFM-SECM probes   *V* 236
AFM-SECM tip-integrated biosensors
        V 252
Ag   *VII* 211
Ag surfaces   *VI* 93
aging   *VII* 303, *VII* 327–329, *VII* 331,
        VII 337, *VII* 340, *VII* 342
AgNi   *VII* 213
Al   *VII* 288
Al$_2$O$_3$ *VI* 39
alkaline phosphatase   *VII* 275
alkanethiols   *VI* 70
alkanethiols on a Au(100)   *VI* 78
alloys   *VII* 290
amine   *VII* 304, *VII* 320–323, *VII* 325,
        VII 327, *VII* 343
aminosilanes   *VI* 140
amperometric
    biosensors   *V* 249
    electrodes   *VII* 261
    enzyme sensor   *VII* 288
    glucose sensor   *V* 250
amphiphile/protein   *VII* 228, *VII* 232–235
amplitude   *V* 99
amplitude density   *V* 59
amplitude modulation (AM)   *VI* 135, *V* 75
amplitude vs. distance curve   *V* 84
analog actuation   *V* 13
analog amplification   *V* 18
analog signal processing   *V* 10, *V* 11
analog-to-digital converter (ADC)   *V* 12
analytical functions   *VII* 269
angle   *VII* 11
anharmonic coupling   *VI* 52
anharmonicity   *VI* 53
anisotropy   *V* 128, *V* 138
antibody   *VII* 275

antiplasticization   *VII* 321, *VII* 322
applied normal load   *V* 187, *V* 217
approach curve   *VI* 186, *VII* 269
array   *VII* 83, *VII* 84, *VII* 91, *VII* 99,
     VII 104, *VII* 107, *VII* 108
array detector   *V* 56, *V* 57, *V* 61
atom species   *VI* 249
atomic force microscopy (AFM)   *VI* 101–
     106, *VI* 108, *VI* 110, *VI* 112, *VI* 113,
     VI 117–119, *VI* 121–123, *VI* 125,
     VI 131, *VI* 182, *VI* 247, *V* 149,
     V 154–156, *V* 158–164, *V* 166–168,
     V 175, *V* 185–187, *V* 193, *V* 194,
     V 196, *V* 215, *V* 216
   tip-integrated biosensors   *V* 249
ATP   *V* 252
ATP synthase   *V* 257
Au(111)   *VII* 201
   surface   *VI* 68
Au/Ni alloy   *VII* 217
Au/Ni surface alloy   *VII* 217
Au/Ni(111)   *VII* 217, *VII* 219
   surface alloy   *VII* 216, *VII* 218
azimuthal polarization   *VI* 268–272

batch microfabrication   *V* 239, *V* 241
bi-potentiostat   *VI* 66
bias voltage   *VII* 199
bifunctional AFM-SECM tip   *V* 244
bimorph effect   *V* 9
Binnig   *VII* 77
biochemical activity   *VII* 266
biochips   *VII* 275
biological interaction   *VI* 175
biomolecule   *VI* 49
bionanotechnology   *VI* 159
biosensors   *VII* 225, *VII* 226, *VII* 236,
     VII 243, *VII* 245, *VII* 247–250
bond-breaking selectivity   *VII* 211
bonding configuration   *VI* 43
bottom-up fabrication   *VI* 128, *VII* 137
boundary condition   *VII* 269, *V* 155, *V* 162,
     V 169, *V* 170, *V* 172, *V* 173, *V* 178,
     V 180, *V* 185
boundary element method (BEM)   *V* 234
bright brim   *VII* 203
Brillouin scattering   *V* 288
brim   *VII* 202
brim state   *VII* 203–206
Brownian motion   *V* 69
buffered aqueous solutions   *VII* 266
butanethiol SAMs   *VI* 79

C_2H_4 *VII* 207
C_4H_7S- *VII* 204
$C_4H_{44}S$ *VII* 203
C–H stretching mode   *VI* 41
C–S bond cleavage   *VII* 204
cadherins   *VI* 170
calibration procedure   *VI* 187
cantilever   *VI* 131, *VI* 183, *V* 3, *V* 62, *V* 99
cantilever shape   *V* 54
cantilever-shaped nanoelectrodes   *V* 236
capacitance sensors   *VII* 142
capillary neck   *VI* 136
capillary or gravitational waves   *VII* 12
capillary waves   *VII* 12, *VII* 19
carbon nanotube   *VI* 313, *VII* 135, *V* 238,
     V 307
   charge distribution   *VII* 180
   chirality   *VII* 136
   CNT-based NEMS   *VII* 146
   CNTs   *VII* 135, *VII* 136
   composite   *VI* 288, *VI* 290
   direct growth   *VII* 139
   external field alignment   *VII* 139
   fabrication   *VII* 137
   failure mode   *VII* 156
   feedback-controlled nanocantilevers
        VII 153
   helicity   *VII* 136
   manipulation   *VII* 138
   memory   *VII* 146
   multi-walled   *VII* 136, *VII* 147, *VII* 150,
        VII 178
   nanorelay   *VII* 152
   nanotweezer   *VII* 147
   oscillator   *VII* 158
   purification   *VII* 137
   random dispersion   *VII* 137
   rotational motor   *VII* 150
   self-assembly   *VII* 139
   single-walled   *VII* 136, *VII* 146, *VII* 177
   synthesis   *VII* 137
carbon-fiber composite   *VI* 312
carbon-fiber microelectrodes   *VII* 281
cartilage   *VII* 288
catalysis   *VII* 197, *VII* 198, *VII* 200, *VII* 207,
     VII 214
catalyst stability   *VII* 216
catalytic turnover number   *VII* 270
catalytically active edge   *VII* 202
catechol amines   *VII* 281

cell membrane *VI* 106, *VI* 112, *VI* 117,
VI 121, *VI* 122
cellular system *VII* 278
characteristic equation *V* 169, *V* 170,
V 172, *V* 173, *V* 213–215
charge-transfer *V* 313
chemical
identification *VI* 31
microsensor *VII* 261, *VII* 263
reactions on surfaces *VII* 199
sensitivity *VII* 290
chemical force microscopy (CFM) *VI* 158
chemisorbed organic molecular assembly
VI 69
chemisorption *VI* 69
chips *VII* 276
chloroplasts *VII* 280
chromaffin cells *VII* 281
chromatin *VI* 152
cis-but-2-ene-thiolates *VII* 204
CMOS-based *V* 3
CNT-AFM-SECM probe *V* 239
CO *VII* 218, *VII* 219
coalescence *VII* 330, *VII* 337, *VII* 339,
VII 340, *VII* 342
coating *VII* 303, *VII* 304, *VII* 320, *VII* 328
coherent anti-Stokes near-field Raman
imaging *V* 324
coherent anti-Stokes Raman scattering
(CARS) *VI* 279, *VI* 280, *V* 289
combinatorial methods *VII* 291
combined AFM-SECM measurement
V 244
combined scanning electrochemical/optical
microscopy *V* 246
combined SPM-SECM probe *V* 228
combined technique *V* 254
CoMoS *VII* 200, *VII* 205, *VII* 206
compartmentalized surface *VII* 275
complementary metal oxide semiconductor
(CMOS) *V* 1
composite *VI* 293
composite and homogeneous interfaces
VII 2
composite interface *VII* 8–12, *VII* 15–19,
VII 22, *VII* 23, *VII* 26, *VII* 37, *VII* 38
composite liquid–solid–air interface *VII* 26
composite material *VII* 290
composite solid–liquid–air interface *VII* 2,
VII 8, *VII* 11, *VII* 20, *VII* 28, *VII* 29,
VII 38

concentration cell *VII* 271
constant force imaging *VI* 134
constant height imaging *VI* 134
constant-current *VII* 199
constant-current imaging *VII* 278
constant-distance SECM *V* 252
constant-force mode *V* 14
contact angle *VI* 307, *VI* 308, *VI* 310,
VI 311, *VII* 1–8, *VII* 10, *VII* 11, *VII* 13,
VII 17–20, *VII* 23, *VII* 26–31, *VII* 34,
VII 35, *VII* 37, *VII* 38, *VII* 239, *VII* 248
contact angle hysteresis *VII* 1–3, *VII* 7,
VII 8, *VII* 11, *VII* 19, *VII* 29, *VII* 31,
VII 38
contact angle θ *VII* 20
contact area *VII* 308–310, *VII* 314, *VII* 315,
VII 324, *VII* 326, *VII* 332, *VII* 333,
VII 336, *VII* 337, *VII* 343
contact mechanics *V* 271, *V* 272
Burnham–Colton–Pollock theory *V* 271
Derjaguin–Muller–Toporov *V* 271
Hertz theory *V* 271
Johnson–Kendall–Roberts *V* 271
Maugis *V* 271
Sneddon *V* 271
contact mode *VII* 306, *V* 127, *V* 132
contact resonance *V* 127, *V* 128
contact resonance frequency *V* 153, *V* 159,
V 163, *V* 170, *V* 171, *V* 177, *V* 181,
V 185, *V* 186, *V* 213–215, *V* 217
contact stiffness *VII* 307, *VII* 309, *VII* 314,
VII 324, *V* 124, *V* 127, *V* 128, *V* 274,
V 277
lateral contact stiffness *V* 275, *V* 276
normal contact stiffness *V* 275–277
continuum mechanics *VII* 176
conventional IETS *VI* 36
corrosion *VII* 288, *V* 257
counterion correlation force *VI* 139
coupled torsional-bending analysis *V* 163,
V 177, *V* 214
covalent binding *VI* 184, *VI* 185
Cravilier method *VI* 67
crosslink density *VII* 321, *VII* 322
current-distance curve *V* 246
cyclic voltammetry *VII* 290
cyclic voltammogram *VII* 269

Damköhler number *VII* 325
dangling bond orbital *VI* 250
decanethiol *VI* 76

deep RIE   *VII* 96

defect structure   *VI* 71

deformation potential   *VI* 44

degrees of freedom (DOF)   *V* 151, *V* 154, *V* 162, *V* 188, *V* 190, *V* 193, *V* 197, *V* 200, *V* 207

dehydrogenation   *VI* 55, *VII* 207, *VII* 210, *VII* 213

density functional theory   *VII* 202, *VII* 204

dental fillings   *VII* 290

dentinal hypersensitivity   *VII* 286

dentine   *VII* 284, *VII* 286

depth of focus   *V* 53

desorption   *VI* 49

(FM) detection   *VI* 247

detection sensitivity   *V* 54, *V* 56, *V* 64, *V* 66

DFT   *VII* 202, *VII* 206, *VII* 210

diamine   *VII* 320, *VII* 321

diaminodiphenylsulfone (DDS)   *VII* 320, *VII* 322, *VII* 325, *VII* 326

diamond   *VII* 129

dielectrophoresis   *VII* 139

difference signal   *V* 60

diffusion   *VII* 81, *VII* 85, *VII* 86, *VII* 91, *VII* 284

diffusion layer   *VII* 271, *VII* 273

diffusion-limited current according   *VII* 269

diffusion-limited current at an UME   *VII* 264

diffusional flux   *VII* 286

digital signal processing   *V* 13

digital simulation   *VII* 262

diglycidyl ether of bisphenol A (DGEBA)   *VII* 320–322, *VII* 325, *VII* 326

dimensionless normalized units   *VII* 264

dip-pen nanolithography (DPN)   *VI* 158, *VII* 79–87, *VII* 89, *VII* 100, *VII* 102, *VII* 103, *VII* 105, *VII* 109, *VII* 111, *VII* 118, *VII* 127, *VII* 140

dipole scattering   *VI* 34

direct or indirect protein immobilization   *VI* 185

disc-shaped integrated nanoelectrodes   *V* 242

displacement conversion mechanism   *V* 24

dissociation   *VII* 207, *VII* 210

dissolution   *V* 246

DMD   *VII* 351, *VII* 352, *VII* 358–360, *VII* 362

DNA   *VI* 125, *VI* 147, *VII* 83, *VII* 98, *VII* 105–107, *VII* 276, *VII* 277

condensation   *VI* 150

dynamic   *VI* 153

gyrase   *VI* 151

structure, interactions and dynamics   *VI* 143

donor and acceptor compartment   *VII* 284

Doppler   *V* 108

DPPC   *V* 92

drag-out   *VI* 297, *VI* 298, *VI* 304–306, *VI* 312, *VI* 314

DRIE   *VII* 97

drift   *VI* 302, *VI* 303, *VI* 316, *VI* 317

driving forces   *VI* 1, *VI* 2, *VI* 5, *VI* 6, *VI* 28

driving frequency   *V* 154, *V* 167, *V* 176, *V* 177, *V* 179, *V* 181, *V* 184, *V* 185, *V* 187, *V* 199

drugs   *VI* 149

dsDNA   *VI* 147

dual electrode probe   *VII* 281

dual-electrodes   *VII* 278

dynamic contact angle   *VII* 8

dynamic force microscopy (DFM)   *VI* 101, *VI* 103, *VI* 106, *V* 75, *V* 97, *V* 99, *V* 108, *V* 110

dynamic force spectroscopy   *VI* 195

dynamic friction force   *VII* 310

dynamic measurement mode   *V* 62

dynamic mode   *V* 15, *V* 149, *V* 156, *V* 164, *V* 187, *V* 216

dynamic range   *V* 59

dynamic strength of the *LFA* 1/*ICAM* 1 complex   *VI* 200

EC-AFM-SECM   *V* 258

ECSTM-SECM   *V* 232

ECSTM-SECM probes   *V* 235

edge state   *VII* 202, *VII* 203

edge structure   *VII* 201

eigenvalue   *V* 149, *V* 153, *V* 154, *V* 169, *V* 173, *V* 174, *V* 190, *V* 192

eigenvector   *V* 154, *V* 192

elastic force   *VII* 153, *VII* 182

elastic modulus   *V* 276

   shear modulus   *V* 277

   Young's modulus   *V* 276, *V* 277

elasticity   *VII* 182, *V* 267, *V* 281

elastomer   *VII* 308, *VII* 316, *VII* 330, *VII* 331

electro-pen nanolithography   *VII* 80

electrocatalytical reactions   *VII* 291

electrochemical epitaxitial growth   *VI* 67

electrochemical etching	*V* 235

electrochemical scanning tunneling microscopy (ECSTM)	*VII* 261

electrochemical STM (EC-STM)	*VI* 66

electroless plating	*VI* 261

electrolyte solution	*VII* 261

electromotive force	*VII* 141

electron beam lithography (EBL)	*VII* 323, *V* 241

electron mediator	*VII* 266, *VII* 269

electron source	*VI* 32

electron tunneling	*VII* 142, *VII* 143

tunneling current	*VII* 154

electronic structure	*VI* 1, *VI* 4, *VI* 23, *VI* 25, *VI* 27, *VI* 28

electroosmosis	*VII* 284

electrophoresis	*VII* 152

electrophoretic paint	*V* 237

electrostatic (i.e. present in ionic bonds)

actuation	*VII* 140, *VII* 141, *VII* 188

attraction	*VI* 176

double-layer force (F_{dl})	*VI* 177

force	*VII* 138, *VII* 141, *VII* 147, *VII* 153, *VII* 179, *VII* 182

microscanner	*V* 39

empty orbital	*VI* 250

endothelial cells	*VII* 281

energy dissipation	*VI* 253, *V* 150, *V* 164, *V* 165, *V* 167, *V* 168, *V* 195, *V* 198, *V* 208, *V* 216

enzyme	*VII* 235, *VII* 236, *VII* 248, *VII* 272

enzyme label	*VII* 275

EPN	*VII* 81, *VII* 94

epoxy	*VII* 304, *VII* 305, *VII* 317, *VII* 320–327, *VII* 329, *VII* 342, *VII* 343

interphase	*VII* 323, *VII* 326

equipartition theorem	*VII* 310, *V* 67

ethiol	*VII* 237, *VII* 238, *VII* 251

ethylene	*VII* 207–209, *VII* 212

EXAFS	*VII* 206

experimental setup	*VII* 262

extraction of an excess amount of gold atoms	*VI* 72

Fabry–Perot interferometry	*V* 102, *V* 103

fano-shaped feature	*VI* 47

Faradaic current	*VII* 261

feedback	*V* 116, *V* 123, *V* 126

feedback mode	*VII* 264, *VII* 265, *VII* 272, *VII* 277, *VII* 291, *VII* 293

feedback-controlled	*VII* 153

ferrocenes	*VII* 268

fiber composite	*VI* 291

fiber–polymer composite	*VI* 295

fiber-based	*V* 237

field-enhancement	*V* 319

finite element (FE)	*V* 154, *V* 162, *V* 163, *V* 187–190, *V* 201, *V* 205, *V* 216

finite element method (FEM)	*V* 25, *V* 26

flame annealing	*VI* 67

flame-melting method	*VI* 67

flexural	*V* 116, *V* 118

fluctuation-dissipation theorem	*VII* 311

focal length	*V* 53

focused ion beam (FIB)	*VI* 262, *V* 28, *V* 34, *V* 36, *V* 41

focused ion beam (FIB) milling	*V* 241

focused ion beam (FIB)-assisted fabrication	*V* 241

focused ion beam (FIB)-assisted processing	*V* 241

focused ion beam (FIB)-milled bifunctional probe	*V* 260

focused spot	*V* 53, *V* 63, *V* 67

diameter	*V* 62, *V* 66

position	*V* 66

force between isolated proteins	*VI* 194

force curve	*VI* 182, *V* 60

force gradient	*V* 99

force histogram	*VI* 189

force modulation microscopy (FMM)	*VII* 307, *VII* 309, *VII* 323

force sensors	*V* 4

force spectroscopy	*VI* 102, *VI* 110, *VI* 113, *VI* 114, *VI* 116, *VI* 117, *VI* 119, *VI* 120, *V* 54

force–distance curve	*VII* 307, *VII* 332, *VII* 335, *VII* 340, *VII* 342, *V* 46, *V* 59, *V* 273, *V* 274

force-indentation curve	*V* 273

force-spectroscopy mode	*VI* 182

fountain probe	*VII* 84

Fourier series	*V* 80

fracture	*VI* 294, *VI* 295, *VI* 304, *VI* 305, *VI* 316, *VI* 317

frame electrode	*V* 242

frame electrode structures	*V* 242

free radicals	*VII* 283

frequency modulation (FM)	*VI* 135

frequency shift	*V* 106, *V* 107, *V* 149, *V* 158, *V* 168, *V* 170, *V* 171, *V* 216

frequency spectrum	*V* 62

friction   *VII* 305–307, *VII* 309–311, *VII* 315,
     VII 331, *VII* 333, *VII* 334, *VII* 337,
     VII 361, *VII* 363
friction force   *V* 216, *V* 282
friction force map   *V* 203
friction force microscopy (FFM)   *VII* 307,
     V 149, *V* 154, *V* 156, *V* 158, *V* 160,
     V 162, *V* 163, *V* 185, *V* 200, *V* 202–205,
     V 207, *V* 208, *V* 216
fuel cells   *VII* 290
functional group   *VI* 48

galactosidase   *VII* 271
gap mode   *VI* 265
2D gas phase   *VI* 73
gas purification   *VII* 215
Gaussian beam   *V* 53
Gaussian optic   *V* 52
GC   *VII* 291
GC mode   *VII* 272, *VII* 288, *VII* 293
gecko feet   *VII* 40–42, *VII* 48, *VII* 49,
     VII 52, *VII* 56, *VII* 70, *VII* 71
gene-therapy application   *VI* 150
generation-collection mode   *VII* 263
glass sheath   *VII* 264
glass slide   *VI* 67
glass transition temperature   *VII* 312,
     VII 320, *VII* 321, *VII* 327
glucose oxidase   *V* 247
glucose transport   *V* 252
GLV   *VII* 356
glycoprotein   *VI* 167
gold nanoparticles   *VII* 276
gold thin film   *VI* 67
governing equation   *V* 172, *V* 175, *V* 178
graphite   *VI* 1, *VI* 2, *VI* 4–9, *VI* 11–14,
     VI 16, *VI* 18–21, *VI* 23, *VI* 24, *VI* 26–28
grating   *V* 41
grating light valve   *VII* 356
guanine   *VII* 277
guard cells   *VII* 280

H$_2$S   *VII* 201
harmonic oscillator   *V* 63
harmonics   *V* 140
HDN   *VII* 200
HDS   *VII* 200
HDS catalysts   *VII* 201
head-to-head molecular arrangement   *VI* 73
heights of DNA   *VI* 136
HeLa cells   *VII* 280

heptadecane   *VI* 80
Hertz-plus-offset model   *V* 83
heterodyne laser Doppler interferometry
     V 103
heterodyne laser interferometry   *V* 103
heterogeneous   *VII* 197
     catalysis   *VII* 207, *VII* 220
     chemical reactions   *VII* 260
     kinetics   *VII* 290
     reaction rates   *VII* 265
     reactions   *VII* 293
hexadecane   *VI* 80
hierarchy   *VII* 42, *VII* 47, *VII* 67, *VII* 68,
     VII 70, *VII* 71
high aspect ratio silicon (HARS) tips   *V* 240
high-frequency dynamic force microscopy
     V 97
high-pressure   *VII* 215
high-pressure cell   *VII* 215
high-pressure STM   *VII* 214
higher frequency   *V* 99
higher-order vibration   *V* 62
highly oriented pyrolytic graphite (HOPG)
     VI 138, *VI* 144, *V* 154, *V* 165, *V* 195,
     V 203
hindered rotation mode   *VI* 42
hindered translational mode   *VI* 42
homogeneous (solid–liquid) and composite
     (solid–liquid–air)   *VII* 4
homogeneous and composite interfaces
     VII 2, *VII* 3
homogeneous interface   *VII* 5, *VII* 10,
     VII 11, *VII* 15–19
homogeneous solid–liquid interface   *VII* 2,
     VII 5, *VII* 9, *VII* 11–13, *VII* 15
Hooke's law   *VI* 131
hopping   *VI* 51
HOR   *VII* 291
horseradish peroxidase (HRP)   *VII* 275,
     V 247
HR-EELS   *VI* 38
human bladder cell line   *VI* 199
human breast cells   *VII* 283
humidity   *VII* 303, *VII* 328, *VII* 331,
     VII 334, *VII* 335
hybrid SICM-NSOM   *V* 253
hydration
     force   *VI* 175
     repulsion   *VI* 178
     STM   *VI* 145
hydrodenitrogenation   *VII* 200

hydrodesulfurization *VII* 200, *VII* 205
hydrogen *VII* 197
hydrogen-oxidation reaction *VII* 290
hydrogenation reaction *VII* 204
hydrophilic *VII* 2, *VII* 3, *VII* 11, *VII* 19,
    VII 35, *VII* 38
  surface *VII* 1
hydrophilicity *VII* 335
hydrophobic *VI* 176, *VII* 2, *VII* 3, *VII* 11,
    VII 31, *VII* 35, *VII* 37, *VII* 38, *VII* 228,
    VII 232, *VII* 237, *VII* 238, *VII* 240,
    VII 249, *VII* 251
hydrophobic force *VI* 180
hydrophobic/hydrophilic *VII* 1
hydrophobicity *VII* 1, *VII* 2, *VII* 35, *VII* 37,
    VII 248
hydrostatic pressure *VII* 284
hydrotreating *VII* 200, *VII* 207
hysteresis *V* 39, *V* 43, *V* 46, *V* 49, *V* 84

imaging enzyme activity *V* 246
immobilization *VII* 226, *VII* 227,
    VII 243–248, *VII* 252, *VII* 253
immobilization of DNA *VI* 139
immunoassay *VII* 272, *VII* 275
immunoglobulin *VI* 169
immunoglobulin superfamily *VI* 172
immunosensors *VII* 245, *VII* 248
iMoD *VII* 355, *VII* 356
impact scattering *VI* 34
in situ characterization *VII* 215
in situ STM observations of the SA process
    VI 75
in situ studies *VII* 216
in situ surface-characterization *VII* 215
inclined ICP-RIE *V* 34
inclusions *VII* 289
individual
    carbon fiber *VI* 304
    carbon nanotube *VI* 295, *VI* 296, *VI* 300,
        VI 311, *VI* 316
    MWCNT *VI* 289, *VI* 290, *VI* 300, *VI* 316
    nanotube *VI* 289, *VI* 300, *VI* 306, *VI* 314
    nanotube pull-out *VI* 302
    nanotube–polymer *VI* 305
    nanotube–polymer composite *VI* 300
    SWCNT *VI* 315
    tubule *VI* 306
inelastic process *VI* 38
insulating, semiconducting and conducting
    samples *VII* 260

insulation of the tip *VI* 66
integral membrane protein *VI* 167
integrate UME into SFM tips *VII* 293
integrated actuation *V* 8
integrated AFM *V* 2
integrated AFM-SECM probe *V* 244
integrated detection *V* 1
integrated microbiosensor *V* 250
integrated nanoelectrode *V* 260
integrins *VI* 171
intensity noise *V* 55
interaction regime *V* 86
interatomic force *V* 163, *V* 204
intercalators *VI* 149
intercellular adhesion molecule-1 (*ICAM* 1)
    VI 200
interdiffusion *VII* 327, *VII* 330, *VII* 339,
    VII 340, *VII* 342
interface *VI* 1, *VI* 5, *VI* 7, *VI* 9, *VI* 11,
    VI 13, *VI* 14, *VI* 18–21, *VI* 23, *VI* 26,
    VI 28
interfacial reactivity *VII* 293
interfacial strength *VI* 295–298, *VI* 304,
    VI 305, *VI* 311–314, *VI* 320
interferometry *VII* 140, *VII* 142
intermittent contact mode *VII* 306, *VII* 307,
    VII 329, *VII* 337, *VII* 342
internal metabolism *VII* 283
interphase *VII* 304, *VII* 314, *VII* 327,
    VII 342
ion-selective electrodes *VII* 272, *VII* 290
ion-selective microsensor *V* 252
iontophoresis current *VII* 284
iontophoretic transport *VII* 284
IP *VII* 305, *VII* 312, *VII* 314–317, *VII* 320,
    VII 323–327, *VII* 329, *VII* 342
irradiance distribution *V* 54, *V* 60
isolated enzymes *VII* 266

kinetics *VII* 272, *VII* 294
kink site *VII* 207

ladder *VI* 55
Langevin equation *VII* 311
Langmuir adsorption curve *VII* 209
Langmuir–Blodgett *VII* 227–229,
    VII 231–233, *VII* 252
laser diode *V* 52
lateral *V* 99
    bending *V* 149, *V* 150, *V* 153, *V* 154,
        V 156, *V* 159, *V* 160, *V* 162, *V* 177–181,

V 185–187, *V* 192, *V* 194, *V* 197–200,
    V 202, *V* 207, *V* 211, *V* 216
bending stiffness   *V* 184
contact stiffness   *V* 153, *V* 165, *V* 166,
    V 176, *V* 214
contact stiffness and viscosity   *V* 181
contact stiffness/viscosity   *V* 162, *V* 214,
    V 216
contact viscosity   *V* 153, *V* 176
excitation (LE) mode   *V* 149, *V* 152,
    V 154, *V* 156, *V* 159, *V* 160, *V* 162,
    V 163, *V* 177, *V* 179–182, *V* 184–189,
    V 199, *V* 200, *V* 214, *V* 216
force microscopy (LFM)   *VII* 77, *VII* 306,
    VII 307, *VII* 331, *VII* 333, *VII* 334,
    V 108
resolution   *VII* 262, *VII* 265, *VII* 293
latex   *VII* 330, *VII* 331, *VII* 337, *VII* 339
layer-by-layer growth   *VI* 87
leukocyte function associated antigen-1
    (*LFA* 1)   *VI* 200
LIGA microstructure   *V* 245
line scan   *V* 20
line tension   *VI* 307
linearity   *V* 59
liquid–gas interfaces   *VII* 260
liquid–liquid interfaces   *VII* 260, *VII* 283
living cell   *V* 253
loading rate dependence   *VI* 194
local density of state   *VII* 199, *VII* 202,
    VII 216
localized electron state   *VII* 202
lock-in-amplifier   *VI* 40
long-range interaction   *VI* 175
Lorenz force   *VII* 141
lubricant   *VII* 334, *VII* 335

magnetic force   *V* 138
magnetic microbeads   *VII* 267
magnetomotive   *VII* 142
magnetomotive actuation   *VII* 141
    Lorenz force   *VII* 141
magnetomotive detection   *VII* 142
manipulation   *VI* 49
mapping   *VI* 48
Mars–Van Krevelen mechanism   *VII* 216
mass transport   *VII* 284
materials gap   *VII* 198
mechanical bias effect (MBE)   *VII* 317–319,
    VII 329, *VII* 343
mechanical shear forces   *VII* 278

mediator   *VII* 264, *VII* 279
membrane proteins   *VI* 103–105, *VI* 117,
    VI 119
meniscus   *VI* 307, *VI* 308, *VI* 310, *VI* 311,
    VII 82, *VII* 83, *VII* 85–89, *VII* 98
16-mercaptohexanoic acid (MHA)   *VII* 82,
    VII 85, *VII* 102–104, *VII* 127, *VII* 128
mercaptotrimethyoxysilane   *VI* 67
metabolic regulation of bacteria   *VII* 283
metabolites   *VII* 266
metal oxide cluster   *VI* 84
metal–insulator–metal   *VI* 36
metal-coated optical fiber   *V* 257
metal-complex monolayer   *VI* 84
metallic implants   *VII* 290
metallized AFM tips   *V* 239
mica   *VI* 67, *VI* 138
Michaelis–Menten constants   *VII* 270
micro-Raman   *V* 299
microbeads   *VII* 275
microcavities   *VII* 278
microcomposite   *VI* 313
microcontact printing   *VII* 139
microelectromechanical systems (MEMS)
    VII 135, *VII* 303, *VII* 349, *VII* 351,
    VII 352, *VII* 355, *VII* 357, *VII* 360,
    VII 363, *V* 1
    optical switch   *VII* 353, *VII* 354
microfabrication   *V* 239
microfluidic   *VII* 84, *VII* 88
microfluidic probes   *VII* 86
micropatterned surfaces   *VII* 275
micropipette   *VII* 107
3D microstage   *V* 29
microstructured electrochemical cells
    VII 275
miniaturized biosensors   *VII* 273, *V* 250
Mo edge   *VII* 202, *VII* 205, *VII* 206
modal analysis   *V* 169, *V* 171, *V* 172
mode shape   *V* 63, *V* 64, *V* 173, *V* 174
model catalysts   *VII* 198
model systems   *VI* 1, *VI* 2, *VI* 4, *VI* 18,
    VI 19, *VI* 25, *VI* 27, *VI* 28, *VII* 214
modeling NEMS   *VII* 165
    analytical solutions   *VII* 184
    bridging scale method   *VII* 170
    concurrent multiscale modeling   *VII* 167
    continuum mechanics   *VII* 176
    coupling methods   *VII* 172
    elasticity   *VII* 182
    finite-kinematics regime   *VII* 187

governing equations   *VII* 182
MAAD   *VII* 168
molecular dynamics   *VII* 165
multiscale modeling   *VII* 166
quasi-continuum method   *VII* 170
small-deformation regime   *VII* 186
modify surfaces   *VII* 275
modulated lateral force microscopy (M-LFM)
   VII 305–307, *VII* 309, *VII* 310,
   VII 331–336
modulated nanoindentation   *V* 273, *V* 274
MOEMS   *VII* 349–351, *VII* 354–358,
   VII 360, *VII* 361, *VII* 363
molecular
   assemblies of inorganic molecules   *VI* 84
   assembly of alkanes   *VI* 80
   biology   *VI* 128
   combing   *VI* 147, *VI* 158
   crystals   *V* 309
   device   *VI* 1, *VI* 2, *VI* 4, *VI* 5, *VI* 19,
      VI 23, *VI* 27, *VI* 28
   force   *VI* 102, *VI* 110, *VI* 117
   c(2 × 8) molecular lattice with a 1 × 4 Au
      missing row   *VI* 79
   motor   *VI* 151
   recognition   *VI* 102, *VI* 103, *VI* 110,
      VI 117, *VI* 118, *VI* 120
   recognition force microscopy (MRFM)
      VI 158
molecule-to-molecule   *VI* 54
monolayer   *VI* 1, *VI* 2, *VI* 4–12, *VI* 14,
   VI 17–19, *VI* 21, *VI* 23–28
monomolecular layer   *V* 246
MoS_2 *VII* 200–202, *VII* 205, *VII* 206
motion equation   *V* 161, *V* 166, *V* 189–193,
   V 197, *V* 199, *V* 205
multilayer   *VI* 94
multiplier-accumulator architecture   *V* 14
multiwalled nanotube   *V* 239

n-butyl ester of abietic acid (nBEAA)
   VII 331
N-glycans   *VI* 168
nanocomposite   *VI* 288, *VI* 320
nanoelectromechanical systems   *VII* 135,
   VII 136, *VII* 146, *VII* 152, *VII* 153,
   VII 160, *VII* 163
nanofountain probe (NFP)   *VII* 80,
   VII 85–88, *VII* 90–95, *VII* 97, *VII* 98,
   VII 100, *VII* 102–109
   fabrication   *VII* 90, *VII* 92, *VII* 94–96

nanoindentation   *VII* 304, *VII* 305, *VII* 312,
   VII 317, *VII* 326, *VII* 328, *VII* 329,
   VII 331, *VII* 342, *VII* 343
nanomanipulation   *VII* 138, *VII* 155
nanopipette   *VII* 80, *VII* 84, *VII* 89, *VII* 105,
   V 237
nanorelay   *VII* 152, *VII* 154
nanoscratch   *VII* 315, *VII* 328
nanotechnology   *VI* 127
nanotube   *V* 131, *V* 238
nanotube composite   *VI* 291, *VI* 296, *VI* 313
nanotube–polymer   *VI* 314
nanotube–polymer composite   *VI* 288,
   VI 290, *VI* 291, *VI* 296
nanotweezer   *VII* 147
nanowire   *VII* 136, *VII* 160
   read-only memory   *VII* 161
   resonator   *VII* 160
near-field Raman spectroscopy   *V* 288,
   V 299
near-field ultrasonic methods   *V* 278
   acoustic force atomic microscopy   *V* 278,
      V 279, *V* 281
   heterodyne force microscopy   *V* 278,
      V 280
   scanning local acceleration microscopy
      V 278, *V* 279, *V* 281
   scanning microdeformation microscopy
      V 278
   ultrasonic force microscopy   *V* 278,
      V 280, *V* 282
negative differential resistance   *VI* 93
negative ion resonance   *VI* 35
Ni   *VII* 288
Ni carbonyl   *VII* 218, *VII* 220
Ni(111)   *VII* 207, *VII* 209, *VII* 211, *VII* 219
Ni(211)   *VII* 211
nitric oxide   *VII* 281
NO microsensor   *V* 253
nodules   *VII* 304
noise   *V* 7, *V* 55
noradrenaline   *VII* 281
normal mode   *VI* 52
normal vibration mode   *V* 63
NSOM   *VI* 257, *VI* 258
NSOM-SECM   *V* 237
NSOM-SICM   *V* 234
numerical differentiation   *VI* 40

O-glycans   *VI* 168
offset compensation   *V* 11, *V* 18

optical beam deflection　*V* 52
optical detection noise　*V* 55
optical detection sensitivity　*V* 54, *V* 65
optical lever　*VI* 132, *V* 52
optical near-field interaction　*V* 233
optical readout　*V* 1
optical switch　*VII* 350, *VII* 353–355
organic molecule　*VI* 49
(111)-oriented　*VI* 67
ORR　*VII* 291
oscillator　*VII* 158
osteoclasts　*VII* 282
Ostwald ripening　*VI* 72
oxide layer　*VII* 288
oxidoreductase　*VII* 269, *V* 247
oxygen　*VII* 279
　consumption　*VII* 280
　production　*VII* 280
　reduction　*VII* 290
　transport cartilage　*VII* 284
oxygen-reduction catalysts　*VII* 291
oxygen-reduction reaction　*VII* 290

parallel scanning　*V* 19
parameter analysis　*V* 176, *V* 180, *V* 181
partial differential equation　*VII* 269
parylene C　*V* 242
passive cantilever　*V* 1
passive layer　*VII* 290
passive regions　*VII* 288
Pd layer　*VI* 87
PECVD　*V* 242
PEP–nBEAA　*VII* 331
pH change　*V* 255
phase angle　*V* 149, *V* 153, *V* 155, *V* 158,
　　V 159, *V* 163, *V* 167, *V* 168, *V* 174,
　　V 176, *V* 214, *V* 216
phase separation　*VII* 220
phase shift　*V* 162, *V* 175–177, *V* 181–183,
　　V 196, *V* 214, *V* 217
photodiode　*V* 52
photoelectrochemical microscopy　*V* 257
photoemission spectroscopy　*VII* 215
photosynthetic oxygen production　*VII* 279
photothermal excitation　*V* 105
physisorbed　*VI* 1, *VI* 4–6, *VI* 8
physisorbed assemblies of organic moleucles
　　VI 70
physisorption　*VI* 137
piezoelectric detection　*VII* 142
piezoelectricity　*V* 2

piezoresistive detection　*VII* 142, *VII* 144
piezoresistive stress sensor　*V* 5
piezoresistive/piezoelectric　*V* 1
piezoresistor　*V* 2
pitting corrosion　*VII* 288, *VII* 290
plant lectin　*VI* 173
plasma membrane　*VI* 166
plasma membrane oligosaccharides　*VI* 199
plasmon　*VI* 261
point-mass model　*V* 161–163, *V* 166,
　　V 170, *V* 216
pointing noise　*V* 55
Poisson statistics　*V* 56
Poisson's ratio　*V* 128
polarization modulation infrared reflection
　　absorption spectroscopy　*VII* 215
poly(ethylene glycol) (PEG)　*VI* 101, *VI* 112
poly(ethylenepropylene) (PEP)　*VII* 331–
　　333, *VII* 335, *VII* 336
poly(phenylenesulfide) (PPS)　*VII* 313,
　　VII 314
poly(vinylpyrrolidone) (PVP)　*VII* 326,
　　VII 327
polyester　*VII* 328
polymer composite　*VI* 314
polymer–matrix composites (PMCs)
　　VII 303, *VII* 314, *VII* 320, *VII* 328
polymorphism　*VI* 71
polyoxometalates　*VI* 92
polystyrene composite　*VI* 313
position-sensitive photodetector　*V* 52
potentiometric electrodes　*VII* 261
potentiometric pH sensor　*V* 257
potentiometric sensor　*VII* 290
(bi)potentiostat　*VII* 263
power spectral density (PSD)　*VII* 310,
　　VII 311, *VII* 340
pressure gap　*VII* 198, *VII* 214, *VII* 215,
　　VII 218, *VII* 220, *VII* 221
pressure gradient　*VII* 288
pressure-sensitive adhesives (PSAs)
　　VII 303, *VII* 305, *VII* 329–331, *VII* 337,
　　VII 340
probe fabrication　*V* 234
proportional damping　*V* 190, *V* 194, *V* 198
prostate cell line　*VI* 198
protein　*VI* 166, *VII* 233–235, *VII* 243–247,
　　VII 249, *VII* 250, *VII* 252, *VII* 276
protein unfolding　*VI* 117–119, *VI* 122
protein/amphiphile　*VII* 245
protoblasts　*VII* 280

protruding cells *VII* 280
pseudomorphic Pd layer *VI* 89
PSMA antigen *VI* 198
pull-in voltage *VII* 153
pull-off force $F_{\text{pull-off}}$ *VI* 188, *VII* 308, *VII* 336
pull-out *VI* 294, *VI* 295, *VI* 298, *VI* 300, *VI* 304–306, *VI* 312, *VI* 314, *VI* 316, *VI* 317
pure torsional analysis *V* 153, *V* 163, *V* 171, *V* 180, *V* 181, *V* 184, *V* 185, *V* 198, *V* 214, *V* 216, *V* 217
pyrroloquinoline quinone (PQQ)-dependent glucose dehydrogenase *VII* 266

Q factor *V* 100
Q-control *V* 75
quadraplex DNA *VI* 147
quality factor (*Q*) *VI* 135, *VII* 135, *VII* 163, *V* 151, *V* 158, *V* 166, *V* 168, *V* 176, *V* 194
quantum limit *VII* 165
quasireversible redox couple *VII* 264

radial breathing mode *V* 287
radial polarization *VI* 268–273
Raman spectroscopy *V* 288, *V* 289
random access memory *VII* 146, *VII* 154
Rayleigh range *V* 53
reaction order *VI* 50
reaction pathway *VII* 210
reactivity measurements *VII* 213
read-only memory *VII* 161
real *z*-stage *V* 27
real-time imaging *VI* 153
real-time single-molecule enzymology *VI* 158
receding contact angle *VII* 2
recognition *VII* 225, *VII* 243, *VII* 248, *VII* 250
recognition image *VI* 120–122
reconstructed Au(111) surface *VI* 68
reconstruction *VI* 68
reference electrode *VI* 66
reinforcement *VI* 288, *VI* 290–292, *VI* 294, *VI* 295, *VI* 315, *VI* 317, *VI* 320
reliability *VII* 350, *VII* 358, *VII* 360, *VII* 362
repulsive electrostatic force *VI* 175
repulsive force *VI* 177
resistance *VII* 119–125

resonance amplitude *V* 15
resonance frequency *V* 15, *V* 32, *V* 49, *V* 63, *V* 149, *V* 153, *V* 155, *V* 156, *V* 158–160, *V* 167–170, *V* 172, *V* 176–180, *V* 185, *V* 186, *V* 215, *V* 216
resonant Raman scattering *V* 287
resonant tunneling *VI* 37
resonator *VII* 160
ultrahigh frequency resonators *VII* 135
resorption of bone *VII* 282
retract curve *VI* 186
Richard Feynman *VI* 127
RNA polymerase *VI* 151
roll-off angle *VII* 2
rotational motion *VI* 56
rotational motor *VII* 150
roughness *V* 272, *V* 283
rubber *VII* 321, *VII* 330, *VII* 331

S edge *VII* 206
S–H groups *VII* 204
SA of metal-complex molecules *VI* 85
salt bridge *VI* 139
SAM *VII* 85
sample-generation/tip-collection mode *VII* 271
scan direction *V* 156, *V* 203, *V* 208, *V* 212, *V* 213, *V* 217
scan velocity *V* 90
scanning electrochemical microscopy (SECM) *VII* 260
scanning electrochemical-scanning chemiluminescence microscopy *V* 249
scanning image *V* 19, *V* 20
scanning near-field optical microscopy *V* 287
scanning near-field optical microscopy technique *V* 288
scanning surface confocal microscopy (SSCM) *V* 254
scanning tunneling microscope tunneling current *VII* 77
scanning tunneling microscopy (STM) *VI* 1–14, *VI* 16–28, *VI* 65, *VI* 144, *VII* 77, *VII* 198, *VII* 199, *VII* 215, *VII* 220, *V* 287
movies *VII* 218
scanning tunneling spectroscopy (STS) *VI* 4, *VI* 14, *VI* 25, *VII* 200
SECM *V* 230, *V* 231
SECM-fluorescence imaging *V* 255

SECM-induced pH change   *V* 257
SECM/PEM   *V* 257
selectins   *VI* 195
selectivity   *VII* 207, *VII* 210, *VII* 211,
   VII 213
self-assembled monolayer (SAM)   *VI* 65,
   VII 81, *VII* 227
self-assembly   *VI* 1, *VI* 2, *VI* 4–6, *VI* 8,
   VI 10, *VI* 11, *VI* 13, *VI* 14, *VI* 21, *VI* 23,
   VI 25, *VI* 27, *VI* 28, *VI* 65, *VI* 128,
   VII 139
sensitivity   *VII* 271, *VII* 294, *V* 7
sensor-actuator crosstalk (SAC)   *V* 9
sensors   *VII* 78, *VII* 108
   biological sensors   *VII* 135
   chemical sensors   *VII* 135
   force sensors   *VII* 135
separation work   *VI* 193
SERS   *VI* 258, *VI* 260–264, *VI* 274, *VI* 275,
   VI 282
seta   *VII* 47, *VII* 48, *VII* 55–58, *VII* 60,
   VII 70
setae   *VII* 42, *VII* 44, *VII* 46, *VII* 47, *VII* 49,
   VII 54, *VII* 56, *VII* 57, *VII* 63, *VII* 70,
   VII 73
shape resonance   *VI* 35
shear force microscopy (ShFM)   *VI* 158
shear forces   *VII* 293
shear stiffness   *VII* 305, *VII* 306, *VII* 309,
   VII 310, *VII* 333, *VII* 336
shear strength   *VI* 294, *VI* 295, *VI* 304,
   VI 306, *VI* 312, *VI* 313, *VI* 317, *VI* 319
shear stress   *VI* 293–295
shear yield strength   *VI* 312
shear yield stress   *VI* 294
shear-force based SECM   *V* 231
shear-force distance control   *VII* 281
shear-force mode   *V* 252
shear-force-based system   *VII* 281
short-range force   *VI* 175
shot noise   *V* 55
SICM   *V* 232
SICM micropipette   *V* 253
SICM/patch-clamp study   *V* 261
signal-to-noise ratio   *V* 54, *V* 57
silicon   *VII* 290
silicotungstic acid (STA)   *VI* 93
simple *z*-stage   *V* 25
single
   carbon fiber   *VI* 312
   crystal   *VI* 67

engineering   *VI* 295
   molecule   *VI* 31
   molecules   *V* 314
   MWCNT   *VI* 289
   nanotube   *VI* 296, *VI* 298, *VI* 305, *VI* 309,
      VI 312, *VI* 320
   nanotube composite   *VI* 302
   tube   *VI* 296
single-chip AFM   *V* 16
single-chip CMOS AFM   *V* 16
single-crystal   *VII* 198
single-crystal surfaces   *VII* 207, *VII* 216,
   VII 220
single-fiber   *VI* 293–296
   composite   *VI* 295
   pull-out   *VI* 294, *VI* 295
single-molecule fluorescence spectroscopy
   (SECM-SMFS)   *V* 255
single-walled nanotube   *V* 239
site-selective chemistry   *VI* 54
skin   *VII* 284
small cantilevers   *V* 56
smart adhesion   *VII* 40, *VII* 41, *VII* 70
solid–liquid interface   *VI* 65, *VII* 293
spatial eigenvalue   *V* 63
spectroscopy   *VII* 242, *VII* 248, *VII* 249
spring constant   *VI* 132, *V* 3, *V* 100
spring constant calibration   *V* 69
ssDNA   *VI* 147
static AFM mode   *V* 156, *V* 200
static contact angle   *VII* 5, *VII* 6, *VII* 27,
   VII 35
statistical approach using a Poisson
   distribution   *VI* 192
steady-state diffusion-limited current
   VII 269
steam reforming   *VII* 217, *VII* 220
steel   *VII* 288, *VII* 289
step edge   *VII* 207–211, *VII* 218
steric force   *VI* 179
stick-slip   *VII* 310, *VII* 331–335, *V* 151,
   V 200, *V* 210–213
stiction   *VII* 358, *VII* 360, *VII* 363
stiction/friction   *VII* 358
stiffness   *VII* 89, *VII* 92, *VII* 94, *VII* 111,
   VII 116, *VII* 122
STM-pH measurement   *V* 257
stomatal complexes   *VII* 280
strained silicon   *VI* 258, *VI* 260, *VI* 272–274
streptavidin-coated magnetic microbeads
   VII 267

stress   *V* 307
stress transfer   *VI* 288, *VI* 293, *VI* 294,
       VI 315
structural molecular biology   *VI* 159
structure of SAMs   *VI* 78
structure sensitivity   *VII* 207
substituted alkanes   *VI* 6, *VI* 8, *VI* 9, *VI* 23
substrate-generation/tip-collection mode
       VII 263
sulfur vacancies   *VII* 205
sum frequency generation   *VII* 215
supercoiling   *VI* 147
superhydrophobic   *VII* 1, *VII* 2, *VII* 4,
       VII 19, *VII* 23, *VII* 31, *VII* 37, *VII* 38
superhydrophobicity   *VII* 2, *VII* 3, *VII* 19
superoxide anion   *VII* 283
surface   *VI* 1–5, *VI* 7–11, *VI* 16, *VI* 24,
       VI 25, *VI* 27
   catalysis   *VII* 221
   chemistry   *VII* 221
   modifications   *VII* 293
   stress   *VI* 253
surface-directed condensation   *VI* 151
surface-enhanced and tip-enhanced near-field
       Raman spectroscopy   *V* 314
surface-enhanced Raman scattering   *V* 287
surface-science approach   *VII* 198, *VII* 214,
       VII 216, *VII* 220
surfactant   *VII* 331, *VII* 337, *VII* 340
SW 480   *VII* 280
SWCNT composite   *VI* 315
SWCNT–polymer composite   *VI* 315
SWCNT-polyurethane acrylate (PUA)
       composite   *VI* 315
SWCNT-PUA composite   *VI* 315
SWNT   *VI* 267, *VI* 268
synthetic track-etched membrane   *V* 244

Ta   *VII* 288
tack   *VII* 330, *VII* 331, *VII* 342
tackifier   *VII* 330, *VII* 331, *VII* 333, *VII* 335
tapping   *VII* 306, *VII* 337–339, *V* 123,
       V 124, *V* 126
tapping mode   *VI* 135, *V* 62, *V* 75, *V* 149,
       V 154, *V* 156, *V* 192
temporary negative ion   *VI* 44
tensile
   strain   *VI* 315
   strength   *VI* 289, *VI* 290, *VI* 294
   stress   *VI* 293
TERS   *VI* 258–262, *VI* 268–274

Tersoff–Hamann   *VII* 199, *VII* 203
thermal actuation   *V* 9
thermal noise   *V* 66
thermodynamical equilibrium   *VII* 214
thermomechanical noise   *VII* 305, *VII* 310,
       VII 342
thiol   *VII* 238, *VII* 240–243, *VII* 247,
       VII 254
thiolate   *VII* 236
thiophene   *VII* 203–205
Ti   *VII* 288
time-lapse imaging   *VI* 153
tip and surface functionalization   *VI* 184
tip eccentricity   *V* 149, *V* 150, *V* 163, *V* 197,
       V 199, *V* 200, *V* 216
tip–sample force   *V* 159, *V* 162
tip–sample interaction   *V* 82, *V* 149–152,
       V 155, *V* 156, *V* 158–170, *V* 173–177,
       V 179, *V* 181, *V* 183, *V* 185, *V* 186,
       V 188, *V* 190, *V* 192, *V* 193, *V* 195,
       V 197–199, *V* 208, *V* 214, *V* 216, *V* 219
tip-enhanced near-field Raman spectroscopy
       V 305
tip-enhanced Raman scattering   *V* 287
tip-enhancement   *V* 305
tip-generation/substrate collection mode
       VII 263
tip-generation/substrate collection mode
       (TG/SC)   *VII* 291
tip-pressurized effect   *VI* 275, *VI* 283
tip-surface distance   *V* 162, *V* 205
titanium   *VII* 290
top-down fabrication   *VI* 128, *VII* 137
topographic map   *V* 149, *V* 163, *V* 201,
       V 203, *V* 205, *V* 207, *V* 208, *V* 210–213,
       V 216
topographical signal   *VI* 134
topography   *V* 155, *V* 156, *V* 158–160,
       V 202–205, *V* 210, *V* 212, *V* 213, *V* 216
torsion   *V* 149–151, *V* 153, *V* 154, *V* 156,
       V 159, *V* 162, *V* 175, *V* 177–181,
       V 183–185, *V* 192, *V* 194, *V* 197–200,
       V 203, *V* 207, *V* 211, *V* 216
torsional resonance (TR)   *V* 112, *V* 113,
       V 149
torsional resonance (TR) mode   *V* 113,
       V 116, *V* 121, *V* 150, *V* 152, *V* 154,
       V 156, *V* 159, *V* 162, *V* 163, *V* 171,
       V 176, *V* 178–182, *V* 184–187, *V* 189,
       V 196, *V* 197, *V* 199, *V* 200, *V* 214,
       V 216

total internal reflection fluorescence
  microsopy (TIRFM)　*VI* 159
transmission electron microscope　*VII* 215
transmission electron microscopy　*VII* 198
tranverse dynamic force microscopy (TDFM)
  VI 158
trial-and-error　*VII* 200
triple-stranded DNA　*VI* 147
tumor cells　*VII* 283
tuning fork　*VII* 278
tunneling　*VII* 199
tunneling current　*VI* 66
turnover frequency　*VII* 207
two-component catalysts　*VII* 216
two-segment detector　*V* 59

UHV　*VII* 214
ultrahigh vacuum　*VII* 198
ultramicroelectrodes (UME)　*VII* 261
ultramicrotomy　*VII* 312
ultrananocrystalline diamond (UNCD)
  VII 110–118, *VII* 120, *VII* 122–125,
  VII 127, *VII* 128
  fabrication　*VII* 112–114
unbinding force　*VI* 188, *VI* 197
unbinding probability　*VI* 197
universal filter　*V* 13
universal sensitivity function　*V* 66
upper detection limit　*V* 61

vacancy islands (VIs) of the gold surface
  VI 72
vacuum evaporation　*VI* 67

vacuum level　*VI* 32
van der Waals　*VI* 176, *VI* 307, *VI* 312,
  VI 313
  attraction　*VI* 175
  energy　*VII* 176
    Lennard-Jones potential　*VII* 176
  force　*VI* 179, *VI* 300, *VI* 304, *VI* 306,
    VI 307, *VI* 314, *VII* 136, *VII* 138,
    VII 162, *VII* 182, *VII* 185, *VII* 189, *V* 82
    Lennard-Jones potential　*VII* 178
vertical bending　*V* 149, *V* 150, *V* 153,
  V 156, *V* 158, *V* 159, *V* 163, *V* 168,
  V 192, *V* 194, *V* 195, *V* 197, *V* 202,
  V 213, *V* 216
vibration amplitude　*V* 62
vibration mode　*V* 62, *V* 64, *V* 66
vibrational mode　*VI* 33
viscoelasticity　*VII* 306, *VII* 307
vitro-fertilized bovine embryos　*VII* 280

wear　*VII* 80, *VII* 109, *VII* 110, *VII* 117,
  VII 358, *VII* 361, *VII* 363
wetting　*VI* 306, *VII* 1–3, *VII* 7, *VII* 8,
  VII 29, *VII* 31, *VII* 37, *VII* 38
wetting angle　*VI* 306, *VI* 310, *VI* 311
Wheatstone bridge　*V* 6, *V* 7
white noise　*V* 56

X-ray absorption spectroscopy　*VII* 215
X-ray diffraction spectroscopy　*VII* 215

Young's modulus　*VI* 289–291, *VI* 293

Printing: Krips bv, Meppel
Binding: Stürtz, Würzburg